BIODIESEL:
Basics and Beyond

BIODIESEL:

Basics and Beyond

A Comprehensive Guide to Production and Use for the Home and Farm

William H. Kemp

Aztext Press
2622 Mountain Road, Tamworth, Ontario Canada K0K 3G0
michelle@aztext.com • www.aztext.com

Library and Archives Canada Cataloguing in Publication

Kemp, William H., 1960-

Biodiesel: basics and beyond / William H. Kemp.

Includes index.

ISBN 0-9733233-3-7

1. Biodiesel fuels. I. Title.

TP359.B46K44 2006 662'.669 C2006-901915-0

Disclaimer:
The installation and operation of the biodiesel production system described herein, including the associated processing and testing chemicals required, involves a degree of risk. Ensure that all proper installation, regulatory, and safety rules are followed.

Electrical systems are subject to the rules of the National Electrical Code ™ in the United States and the Canadian Electrical Code ™ in Canada. Additional electrical and fire safety regulations regarding flammable or explosive substances may be in effect in your jurisdiction.

In addition, local utility, insurance, zoning, and many other issues must be dealt with prior to beginning any installation.

Waste stream products that are the result of the biodiesel production process may be subject to regulations under any environmental or waste management program in your jurisdiction. Ensure that all necessary environmental waste processing regulations are followed.

The author, contributors, and publishers assume no liability for personal injury, property damage, consequential damage or loss, from using the information in this book, however caused.

The views expressed in this book are those of the author personally and do not necessarily reflect the views of contributors or others who have provided information or material used herein.

Printed and bound in Canada.
Trade distributor: New Society Publishing www.newsociety.com

Acknowledgements

A book of the magnitude of ***Biodiesel: Basics and Beyond*** requires the technical knowledge and expertise of many individuals and companies, not to mention the patience of family during the writing and editing of the text.

I am indebted to and would like to thank: Steve Anderson Biodiesel; ASTM International for assistance with biodiesel quality standards; BioDiesel Technologies GmbH, Austria; Dr. Marc Dube, P.Eng., University of Ottawa Chemical Engineering Department; Zenneth Faye, Milligan Bio-Tech Inc., for supplying ASTM-certified biodiesel for engine testing and helping me understand the plight of oilseed farmers; Dan Friedman of Dr. Dan's Alternative Fuel Werks; Frontier Energy; Cherie and John Graham of Church Key Brewing Company; George Heinzle; Barry Hertz, P.Eng., University of Saskatchewan; Neil Hevenor, Sommers Motor Generator Sales Ltd., for his insight into the world of diesel engine sales, service, and testing; Ed Hogan of Natural Resources Canada, Energy Technology Branch; Steve Howell, MARC-IV Consulting and Chair of the ASTM D 6751 Biodiesel Quality Committee; Jeff Knapp for his technical review of the book; Sean McAdam, Veggie Gas.ca; Mercedes-Benz Canada; Dr. Martin Mittelbach, University of Graz, Austria, for his technical support and kind words of enthusiasm; Christina Moretto, Canadawide Scientific, for her assistance in researching and locating equipment and lab supplies required to make the biodiesel facility a reality; Olympia Green Fuels LLC; Christine Paquette, Past Executive Director, Biodiesel Association of Canada, for assistance, research, review, and access to her amazing Rolodex; Peacock Industries; Brent, Brandon, and Nick of the Postmaster Pub, Almonte, Ontario for their great food and WVO.... Thanks guys!; Dave Probert ; Dr. Martin Reaney, University of Saskatchewan; "The Boys" from Renewable Energy of Plum Hollow; "Bio" Lyle Rudensey; Daniel Sherwood, Arborus Consulting; Kenneth Sommers, Sommers Motor Generator Sales Ltd., for his kind assistance with engine testing and analysis; Dr. Andre Tremblay, P.Eng., Professor and Chair, Chemical Engineering Department, University of Ottawa; West Central; Paul Zammit and Michelle Murphy of the Fall River Pub and Grill and Country Gift Store for their deep fryer oil not to mention great food and company during our many work meetings.

If I have missed anyone, please accept my apologies and understand that your assistance is truly appreciated!

I am truly grateful for the assistance and mental horsepower of my fellow biodiesel researchers, including (from left) Dr. Marc Dube, P.Eng., University of Ottawa Chemical Engineering Department; Dr. Martin Mittelbach, University of Graz, Austria; Christine Paquette, Past Executive Director, Biodiesel Association of Canada; Daniel Sherwood, Arborus Consulting; and Dr. Andre Tremblay, P.Eng., Professor and Chair, Chemical Engineering Department, University of Ottawa. The fact that Dr. Albert Einstein is watching over everyone is also quite fitting; perhaps he was there in spirit.

A huge thank you must also be given to Joan McKibbin, my editor. Joan worked tirelessly under very demanding deadlines, as usual, and managed not to lose her cool. Thanks Joan!

I am also indebted to those who have had to work directly with me in the preparation of the text. A big thanks goes out to Cam Mather for his excellent line drawings, graphics, and layout skills, and Michelle Mather for her multiple creative inputs to the book production process, not to mention having to work with my demanding timelines.

A very special thanks to Lorraine Kemp for her great help behind the camera. (All images without photo credits are by her.)

I get my turn behind the camera, photographing the always good-spirited Lorraine, who besides being my lovely wife is also a great photographer. She is shown here behind the wheel of our Smart™ car, driving across the Canadian prairies as we researched this book. The diesel-powered Smart™ car went approximately 6,500 kilometers (4,000 miles) on just under US$200 worth of fuel. Efficiency first, biofuels second: that is the environmentally sustainable way.

I would be remiss without giving special thanks to Geoff Shewfelt and Jamie Wilson, who burned the candle at both ends helping to build the biodiesel facility and get the show up and running.

Contents

Preface

The motivation for writing ***Biodiesel: Basics and Beyond*** comes from my wife Lorraine's and my desire to lower our overall carbon fuel footprint on the planet while at the same time trying to maintain a relatively "normal" standard of living. The journey down this path culminated in our building an off-the-electrical-grid, energy-efficient, and sustainably built home and horse farm. You can learn more about renewable energy, energy efficiency, and my other book titles at www.aztext.com

As those of you who have read ***The Renewable Energy Handbook*** already know, one of the biggest problems with off-grid living is trying not to simply "fuel switch" from electricity to fossil fuel. The high cost and low availability of electrical energy from the renewable energy system makes most off-grid homes worse overall energy consumers than equivalent urban homes operating on grid. This is due to the effect of fuel switching from relatively high-efficiency electric appliances to lower efficiency ones powered by fossil fuels. For example, an electric water heater is often 10 to 20% more efficient than their fossil-fuel counterparts.

In addition, most off-grid homes are located in rural or low-density areas, necessitating the transportation of everything, including kids, garbage, food, and you.

Like most off-grid owners, we require a backup generator to provide battery charging when the weather doesn't cooperate in the low-sun and low-wind months of November and December. In our case a diesel generator was selected as the lowest-cost, cleanest, and lowest-greenhouse-gas-emitting option. (Surprised? Keep reading.)

We had already lowered our water consumption and reduced our garbage output shortly after moving in but we decided to do more, and the idea of producing or purchasing biodiesel came to mind. We were already sold on diesel cars having had a couple in recent years, and watching them become more powerful and cleaner burning only increased our resolve to stick with the technology. After some initial searching we became aware of Milligan Bio-Tech Inc. of Foam Lake, Saskatchewan and arranged to have a skid of their canola-based biodiesel fuel shipped to us. We splash blended the fuel at 20% concentration and were pleased with the results but quickly wanted to do more, deciding that it was time to start producing the fuel ourselves.

It didn't take long to realize that this stage of our journey was going to be rather rocky.

I read the only book available at the time on the subject and found that it was riddled with errors and omissions with the author making statements of "fact" that were completely baseless. Visits to home-based biodiesel producers were no better, revealing to me the disregard for issues of safety, fuel quality, waste disposal, and environmental sustainability that should have been obvious to people who were professing their concern for the environment as their reason for making the fuel in the first place. Most people were simply unaware that there was any problem at all.

I cannot fault people for not understanding the chemistry and environmental cycle related to the production of biodiesel. I can even see how people become immersed in the misinformation that is disseminated by various media and the Internet. But that does not mean I have to take the same direction.

Biodiesel: Basics and Beyond is based on the idea that current micro-scale technology has gone only so far and that with a bit of work and access to professional researchers and funding it is possible to expand the technology and prove or disprove many of the issues that have become legend.

A central aim of this book is to alter the status quo of making poor-quality biodiesel fuel while leaving an environmental mess caused by waste stream products. There is no reason why the micro-scale community can't make quality biodiesel, but without advancing the state of the art to solve these problems and developing a better strategy for testing processes people are doing themselves a disservice, one that may actually harm them.

I have made numerous batches of fuel that meets national quality standards using the equipment and processes described in this book without adversely affecting the environment or my safety. I hope that others will follow along in the same manner, demonstrating that micro-scale biodiesel production can be safe, satisfying, and responsible.

1
Introduction

The demand for energy in the United States and Canada is insatiable, and it appears no one has learned from the mistakes of the '70s. Look around: sport utility vehicles and minivans abound, with the result that fuel economy is nearing an all-time low. At the same time, the square footage of the average North American home has more than doubled and energy efficiency codes have been relaxed, requiring additional energy for heating and air conditioning.

Figure 1-1. The demand for energy in the United States and Canada is insatiable, and it appears no one has learned from the mistakes of the '70s. With gasoline and diesel prices expected to rise at least 10% per year, fuel will cost $75 per gallon ($20 per liter) 20 years from now. Why is society not moving towards higher-density housing and better public transit while we still have the capital and energy to make the transition?

From 1970 to the present, domestic oil supplies in the United States have continued to fall, with the result that vast quantities of imported oil are required in order to make up the deficit. Current world consumption is around eighty million barrels of oil per day and rising quickly because developing countries are demanding their share of the energy pie. Even though the majority of the world's oil is supplied by regions that are politically unstable and often unfriendly to the West, energy dollars are being exported as fast as the U.S. Treasury Department can print them. According to an October 25, 2003 report in *The Economist*, OPEC has drained the staggering sum of $7 trillion from American consumers over the past three decades. This massive amount does not even include industry subsidies, cheap access to government land for oil extraction, or military security required to get the sticky stuff safely into North America.

The uncertainty surrounding the current and future cost and supply of fossil fuels is causing governments, universities, and industry to search for alternate forms of transportation and heating energy.

Hybrids, hydrogen, and exotic technologies get great media coverage, but these approaches all require a massive investment in the existing automotive

Figure 1-2. Biodiesel is a proven alternative to petroleum-based diesel fuel. It reduces lifecycle greenhouse gas emissions as well as air toxins. Production capacity for this renewable, positive energy balance fuel is growing rapidly across North America, with dozens of large-scale producers supplying the expanding market.

and fuel system infrastructure. Delivering renewable ethanol and biodiesel fuels without having to retool the entire North American economy is an exercise well worth considering.

Biodiesel (a clean, renewable diesel-like fuel that minimizes exhaust emissions and air toxins) is one of the most thoroughly tested alternative fuels, with over 100 million kilometers of in-field use by organizations such as the U.S. military and school bus and private vehicular fleets. Biodiesel production capacity in North America now exceeds 1.2 billion liters and continues to grow annually. All of this production capacity is based on a growing number of large-scale plants, each with a minimum production rate of 4 million liters per year.

Figure 1-3. Home-based biodiesel brewers discovered that by collecting waste restaurant fryer oils and performing a chemical transesterification process they could manufacture low-cost, crude (poor quality) biodiesel. What has followed is a plethora of guide books and Web-based misinformation.

Simultaneously, there has been a surge in the production of "underground" biodiesel. Although it often occurs under the guise of "environmental sustainability," it is actually undertaken for economic reasons. Home brewers discovered that by collecting waste restaurant fryer oils and performing a chemical transesterification process they could manufacture low-cost, crude (poor quality) biodiesel and use it in automobiles at 100% (neat) concentrations. What has followed is a plethora of guide books and Web-based misinformation. Unfortunately, it is deceptively easy to make poor-quality biodiesel fuel while leaving behind a mess of toxic waste and coproducts.

Biodiesel: Basics and Beyond will provide the reader with the necessary knowledge and procedures to correct this dilemma.

Technical Background

The use of vegetable oils as a fuel for home heating and transportation is a hundred years old. It is well known that the original diesel engine developed by Rudolf Diesel operated on plant oils because of the lack of fossil fuel availability. In later years, chemical engineering methods made it possible to convert plant and animal oils into a petrodiesel-compatible fuel known as biodiesel.

In the early 1980s Europeans saw biodiesel production as one way of addressing the lack of natural fossil fuel supplies. Research quickly led to commercial availability of high-quality biodiesel fuel. Although slower to react, North America has also embraced large-scale production of this clean, low-carbon fuel (certified to American Standards and Testing Methods Standard D 6751). Both Europe and North America are beginning to advocate blending biodiesel and petrodiesel at rates not exceeding 20% biodiesel by volume. A recent press release from DaimlerChrysler's Jeep Division announced that diesel vehicles will be shipped from the factory fueled with a 5% (B5) blend of biodiesel fuel. Other vehicle manufacturers are following suit, with some already recommending heady levels of biodiesel of up to B20.

Figure 1-4. There is little factual data on the potential damage that crude or quality high-concentration biodiesel may cause in late-model automobiles or diesel-powered machinery. ***Biodiesel: Basics and Beyond*** *will investigate the facts on this issue.*

However, many North American automotive and engine manufacturers insist that late-model vehicles will not operate on or may be damaged by high concentrations of quality biodiesel. Indeed, many manufacturers will not honor engine warranties where any biodiesel fuel has been used.

North American vehicle manufacturers are a cautious bunch. Calls to the technical departments of VW of America and Mercedes-Benz yielded guarded replies when they were asked for their positions on biodiesel. VW indicated that they are "pleased to extend vehicle warranties for up to B5 on all US-market diesel vehicles." Mercedes stated that biodiesel fuels are not compatible with their vehicles and that the fuel should not be added even in low concentrations. These are interesting positions given that biodiesel has been used in Europe at concentrations of up to B100 for a number of years now.

This North American reluctance to advocate high-blend levels of biodiesel may appear to be at odds with the home brewers' use of neat (100%) biodiesel until it is realized that crude biodiesel can be manufactured for less than one-half the cost of petrodiesel fuel and that most people who make the fuel are using it in older, low-value automobiles. There is little factual data on the potential damage that crude or quality high-concentration biodiesel may cause in late-model automobiles or diesel-powered machinery.

Although the use of B5 blended biodiesel is considered a landmark decision by the major North American automotive manufacturers, hard-core B100 advocates look askance at this effort. Despite the conservative approach taken by the major automotive manufacturers, thousands of people claim to be operating their cars, trucks, and tractors on low- and high-quality B100 without any problem. Why the discrepancy?

According to Mr. Steve Howell, Chairman of the ASTM D 6751 Biodiesel Fuel Standard Committee: "You really need to have ASTM-certified fuel if you want your engine and fuel system components to operate correctly and over the long term. Sure, there are lots of people who use substandard fuel, but they are generally operating older vehicles that are not equipped with computerized engine management systems. Diesel engines are designed to last hundreds of thousands of miles, and homebrew and substandard fuels are going to reduce the life of these vehicles. Time after time after time, at every OEM (original equipment manufacturer) facility, we have found long-term problems resulting from the use of substandard fuels." Steve goes on to explain that the mono, di-, and triglycerides present in homemade biodiesel accelerate the instability of the fuel, causing more filter clogging and fuel injector coking problems and leading to expensive engine overhauls.

Given that low-quality biodiesel is known to cause engine and fuel system problems and that little test data is for high-quality, high-concentration biodiesel, I contracted a diesel engine supplier to perform real-life operational tests on a new diesel engine. During the testing phase, engine oil was sent to a laboratory for wear and engine degradation analysis. At the completion of the test phase, the engine was disassembled and examined piece by piece to determine if quality biodiesel fuel contributes to decreased engine life. The results offer a glimpse into the compromises and politics that surround the traditional fossil fuel industry.

Home Heating Option

According to the U.S. Department of Energy, heating fuel oil consumption in the United States is currently hovering around 7 billion gallons (26.5 billion liters) per annum, with the vast majority supplied by imported oil. Blending a mixture of 20% biodiesel with No. 2 would reduce fossil-fuel consumption and replace that amount of petrodiesel with domestic, renewable, clean-burning biodiesel. If this strategy were adopted across the United States, biodiesel consumption would increase by 1.4 billion gallons per year (5.3 billion liters per year), reducing carbon monoxide, hydrocarbon, and particulate emissions by approximately 20% while supporting domestic agricultural economies.

Figure 1-5. Biodiesel can be used in virtually any oil-burning furnace or boiler provided that fuel lines and gaskets are not made of natural rubber. Because biodiesel gels (like No. 2 heating oil), storage tanks should be installed indoors or underground. Alternatively, your fuel supplier can provide winter-blended heating oil containing both No. 1 and No. 2 heating oil with biodiesel blended to B20 to achieve the desired low-temperature storage performance. (Courtesy Frontier Energy)

Small-scale biodiesel producers can offset some or all of their home heating requirements depending on fuel storage location. In cold temperatures, biodiesel is subject to a thickening condition known as "gelling." Outdoor fuel tanks allow partial blending of biodiesel with heating oil, while indoor (warm) tanks allow high concentrations to be used.

The Farming Community

"Nothing runs like a Deere" is a familiar slogan heard across the rural and agricultural communities of North America. The John Deere Company recently announced that new equipment leaving their factories will contain blends of 2% biodiesel. While this effort is laudable, farmers and agricultural entrepreneurs are wondering why they can't produce quality biodiesel from both waste and virgin oilseed feedstock.

Crop farmers and dairy farmers require enormous amounts of diesel fuel throughout the growing season. ***Biodiesel: Basics and Beyond*** investigates how farmers can use excess or "off-spec" oilseed as a feedstock for the biodiesel production process. Careful monitoring of oilseed crop commodity prices may also allow small-scale producers to "trade" in traditional sales channels or "hedge" against the markets through biodiesel production. This is especially true given that oilseed crops such as soybeans can produce two value streams for the farmer.

Figure 1-6. Farmers can use a low-cost, cold-pressing technology to produce oil for biodiesel as well as for high-protein animal feed made from the coproduct seed hull. Using financial evaluation tools farmers can determine whether to sell into the commodity market or direct oilseed into biodiesel and animal feed production.

Using low-cost, cold-pressing technology, farmers can use soybeans to produce an oil stream output which can be used both for biodiesel or food production and for high-protein animal feed made from the coproduct seed hull. Using financial evaluation tools, farmers can evaluate whether to sell into the commodity market or direct oilseed into biodiesel and animal feed production.

Biodiesel: Basics and Beyond provides examples of cooperative production techniques as well as supplying prospective developers with essential financial reference material to determine if oilseed crops should be directed to commodity markets or biodiesel production.

The book draws on the collective experience of North American and European researchers, with the author developing and detailing the step-by-step processes and equipment required to produce biodiesel fuel that can meet North American standards. You will read about the construction and operation of a 40-gallon (150-liter) per-day scalable-size biodiesel production facility which includes a test and measurement laboratory for determining fuel quality. Each process or method is verified by independent government or university laboratories, ensuring the reader of the accuracy of both the data and the methodology.

Although most people are primarily interested in the economic benefits of biodiesel production, considerable information is given on waste product reduction and treatment including water recycling, methanol recovery, and the productive use of process technology byproducts to ensure that the production of biodiesel is not only economic but also ecofriendly.

Biodiesel: Basics and Beyond will not solve the world's fuel and energy supply problems, but it will provide homeowners, small-scale cooperatives and farmers, and agricultural communities with the data necessary to become energy independent and make their homes and businesses environmentally sustainable.

2 North American Energy Consumption

2.1 Energy *Inefficiency*

When the price of gasoline jumps for the tenth time in a month, most people tend to cut back on their driving—or at least to talk about it. Some people might even consider trading their vehicle in for something with better gas mileage than a Hummer. General Motors is teetering on the verge of bankruptcy while SUV sales plummet as a result of the relatively minor gasoline price shocks in the aftermath of hurricanes Katrina and Rita.

Family budgets are straining from the rising cost of living: taxes, mortgages, car payments, and energy bills for homes and vehicles. As a society we rarely consider the source of these costs. People hop on the bandwagon of the middle-class dream fueled by the two incomes that make it happen. Large homes in the suburbs require enormous amounts of heat, light, and air conditioning, and many people have two (full-size) cars to get to distant jobs. All of this eats away at our precious discretionary income—and our free time.

It is amazing to see how many people live in a McMansion home in the suburbs, drive two vehicles to work (adding to a work/travel day that is now 10 or more hours long), and come home exhausted. Saturday begins with a harried cup of coffee followed by a trip to the malls for a round of shopping

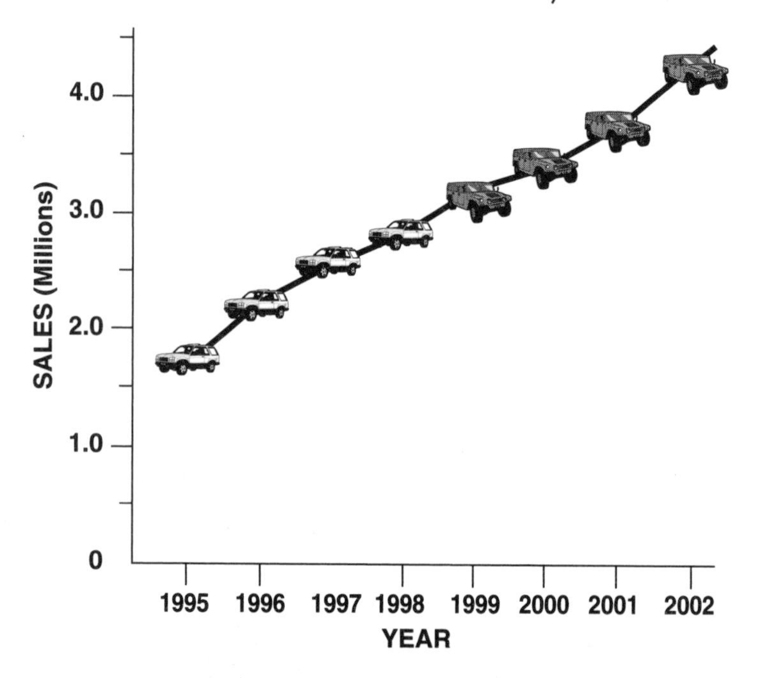

Figure 2.1-1. Cars, trucks, and SUVs are more about style, ego and one-upmanship than about transportation. Does anyone really need a 300-horsepower SUV for a city commute? (Data Source: Office of U.S. Transportation Technologies)

Figure 2.1-2. Large homes in the suburbs require enormous amounts of heat, light, and air conditioning, and many people have two (full-size) cars to get to distant jobs. All of this eats away at our precious discretionary income—and our free time.

for "necessities." Sunday is no longer a day of rest but rather a day of fatigue as we watch the tube and try to console ourselves about the impending start of another week.

The two large incomes are subject to equally large tax deductions and cover payments for the big home and late-model cars. (These are considered necessities as we feel obliged to equal the neighbors' visible consumption.) Upkeep, property taxes, and compound interest take the lion's share of the balance of discretionary income. The remainder is used to fund the joyless prosperity that reigns over middle-class society: harried packaged holidays (purchased on credit), hockey expenses, or whatever else is required by the middle-class treadmill.

Cheap energy is what makes all of this possible. Large homes, cars, exotic food (say, tomatoes or Italian parsley in January), distant travel, even the suburbs themselves are all made possible through the glorious gift of cheap energy from Mother Earth.

Figure 2.1-3 Cheap energy is what makes suburban middle-class life possible. Large homes, cars, exotic food (say, tomatoes or Italian parsley in January), and distant travel are all made possible through the glorious gift of cheap energy from Mother Earth.

How much longer will this gift continue? Without the benefit of hindsight, it is difficult to say exactly. What is clear is that Canada and the United States together have approximately 5% of the world's population but consume 35% of its energy as a result of massive inefficiency and excessive consumption.

Rapidly developing countries like China (with 1,285 million people or 4.4 times the population of the United States) and India (1,025 million people or 3.5 times the population of the United States) want their fair share of the energy and financial pie. And why shouldn't every one across the planet have the same access to homes, cars, education, medicine and luxury consumables that we in the west take for granted? It sounds fair and reasonable until we look at the energy, raw material, and environmental picture a little closer and realize it simply will not and cannot happen.

According to a report in the December 24, 2005 issue of *The Economist*, demand for luxury items is soaring in emerging economies such as Russia, India, Brazil, and China. Estimates indicate that in 2004, Chinese consumers already accounted for 11% of worldwide revenues for luxury-goods firms, and it is expected that by 2014 they will have overtaken American and Japanese consumers, becoming the world's leading luxury shoppers for such brands as Louis Vuitton, Audi, Prada and the like.

If the demand for energy to fuel industry, homes, and automobiles in emerging market economies continues to rise at current rates, it is unlikely that global fossil fuel supplies will be able to keep up with consumption. Of course, rising demand along with supply shortages will eventually cause economic slowdown, recession (with a resulting drop in energy demand), and possibly global conflict as energy-importing nations struggle to keep the lights on.

All of this means that rising energy costs and shortages of fuel, food, and consumables will be the order of day in the not-too-distant future. Although this is pretty hard to imagine given the square miles of "stuff" for sale at shopping malls, as the following short essays will discuss, the writing is on the wall.

How does biodiesel production fit into this picture? Sadly, it is not a magic bullet. Taken in isolation, biofuels in total (including ethanol, biomass, methane, and biodiesel) are but a blip on the world's energy map. Locally, they will provide some level of energy and financial diversification for agricultural communities and agribusiness, but globally their importance will be relatively insignificant given current energy consumption and growth trends.

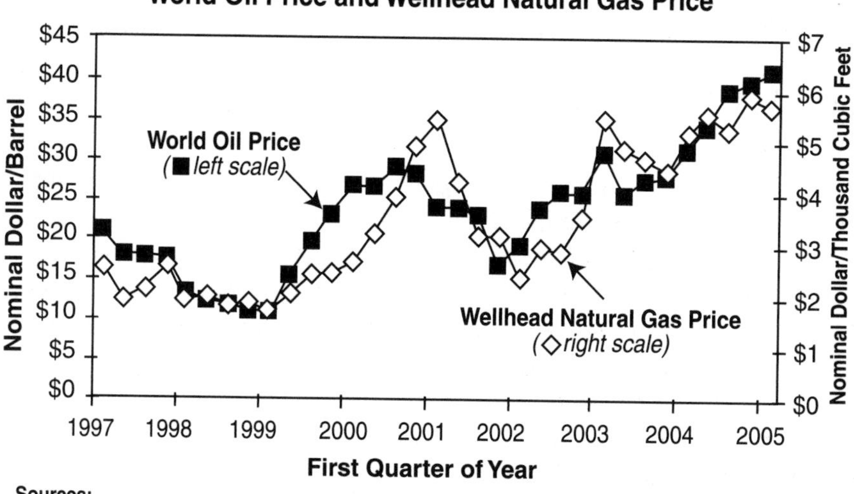

Figure 2.1-4. If demand for energy to fuel industry, homes, and automobiles in emerging market economies continues to rise at current rates, it is unlikely that global fossil fuel supplies will be able to keep up with demand. Nor will people be able to afford current levels of consumption.

However, given massive improvements in energy efficiency and policy, carbon caps, and higher energy prices (including fuel taxes), biofuels can form an important piece of the energy puzzle, providing clean-burning fuels for home heating and the transportation sector.

The Case for Energy Efficiency

North Americans need a new strategy, one that focuses on reducing energy use through efficiency and conservation rather than on increasing supply. Experience from across North America has proven that it is significantly cheaper to invest in energy efficiency than to build additional sources of energy supply. This theory holds true in any circumstance: it is more economical to implement an energy-efficiency program than to build a new power plant; and a family is better off retrofitting their home than installing additional or larger heating, cooling or energy production technologies.

Or as Amory Lovins, the energy guru from the Rocky Mountain Institute, is fond of saying: "It is far less expensive and environmentally more responsible to generate *negawatts* than megawatts."

Using less energy isn't about making drastic lifestyle changes or sacrifices. Conservation and efficiency measures can be as simple as improving insulation standards for new buildings, replacing incandescent light bulbs with compact fluorescent models, or replacing an old refrigerator with a more efficient one. In fact, energy efficiency often provides an improvement in lifestyle. A poorly insulated or drafty house may be impossible to keep comfortable no matter how much energy (and money) you use trying to keep warm.

California learned firsthand that saving energy means saving money and the environment. You may recall the rolling blackouts and severe power shortages that afflicted the state a few years ago. It was predicted that dozens of generating stations would be required on an urgent basis to solve the state's energy problems. At the time of writing, the total number of power stations built to solve the problem stands at zero. Faced with the realization that construction cycles for significant generating capacity would require several years, forward-thinking officials looked to energy efficiency instead. The state's energy–efficiency standards for appliances and buildings have helped Californians save more than $15.8 billion in electricity and natural gas costs. One–third of Californians cut their electricity use by 20% to qualify for a 20% rebate on their bills. The government also introduced a renewable-energy buydown and accompanying net-metering program that has seen thousands of clean, photovoltaic power systems installed on residential rooftops.[1]

In addition to saving electricity and reducing fossil fuel burning, California's conservation and efficiency efforts reduced greenhouse gas emissions by close to 8 million tonnes and nitrogen oxide emissions by 2,700 tonnes during 2001 and 2002.[2]

Not sure if you can make a significant difference at home? Consider the following points:

- Switching to compact fluorescent lamps will reduce lighting energy consumption by 80%.
- High efficiency appliances such as refrigerators, washing machines, and dishwashers can reduce energy consumption by a factor of five times.
- Using on-demand water heaters will reduce hot water energy costs by up to 50%.
- Low-flow showerheads, aerator faucets, and similar fittings will reduce water consumption and resulting heating costs by approximately one-half.
- A well-sealed and well-insulated house can reduce home heating and cooling costs by 50%.

- Adding a solar thermal water-heating system can further reduce hot water heating costs by 50%.
- Purchasing a fuel-efficient car will cut capital and financing costs, decrease operating costs, and reduce pollutants dramatically.

Figure 2.1-5. An energy-conserving lifestyle does not mean a spartan lifestyle. Choosing the most energy-efficient appliances and products that "do more with less" is how to make it happen. The Toyota Prius hybrid doubles gas mileage efficiency, reduces pollutants by 90%, and, most importantly, is driven by actress Cameron Diaz. (Courtesy Toyota Corporation)

Many of the items on this list are neither expensive nor difficult to implement. Best of all, using these products not only dramatically reduces smog and greenhouse gas emissions but also provides a rapid payback by inflation-proofing your home against rising energy costs and putting dollars in your wallet.

This is the *eco-nomic* approach: doing more with less.

It may sound like nickel-and-dime stuff, but in fact the opposite is true. The constant "leakage" of energy dollars here and there can become quite significant. The simple act of switching a houseful of common incandescent lamps to high-efficiency compact fluorescent models will put over 2,300 *after-tax* dollars in your pocket over the operating life of the lamps. If you are in a 35% tax bracket this translates into a savings of over $3,100, or roughly $300 per year, real money that can be used for more important things than simply paying utility bills.[3]

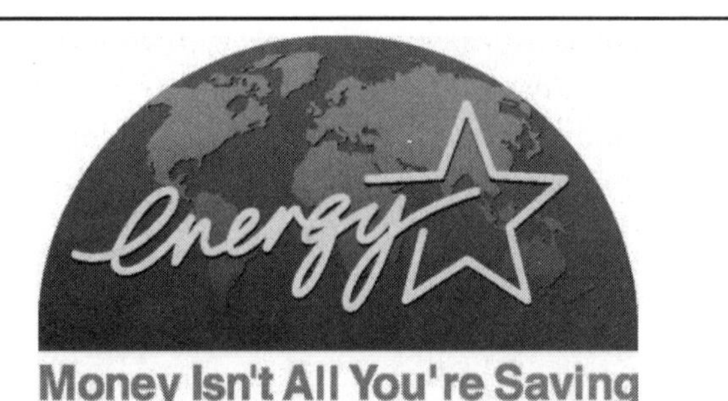

Figure 2.1-6. Purchasing computers, home-office equipment, major appliances or lighting that bears the Energy Star label ensure the lowest energy consumption and is always worth the investment.

Home heating and cooling consumes massive amounts of energy which can be greatly reduced by either purchasing or upgrading a home to meet improved energy-efficiency program requirements. Geo-exchange (also known as geothermal) systems are known to be the most energy-efficient and lowest-cost means of heating and cooling a home.

Figure 2.1-7. A professional energy-consumption assessment of your home will include a pressure door test. A computer-controlled fan will pressurize the house, calculate air leakage, and allow the assessor to identify problem areas in the home and suggest corrective action.

People often balk at these suggestions, saying that they are too expensive to consider or that the "payback time" is excessive. This is simply nonsense.

A natural gas or oil heating system in the average North American home can be replaced for approximately $15 to $20,000. However, the capital used to make the update or conversion should not come from your pocket. Let me explain....

One couple I met suggested that their 10-year-old oil furnace was ready to be replaced and that they wanted to add air-conditioning at the same time. Although they were excited about using a geo-exchange system, the $20,000 capital cost was, they estimated, beyond their financial ability.

What this couple did was typical of most families: they obtained the correct cost information but did not apply the financial data correctly. The proper methodology, I pointed out, was to go to the bank, have the home reassessed after including the value of the new geo-exchange system, and finance the upgrade with the bank's money.

To support the cash flow required to make the monthly payments, the couple could improve the energy performance of the home using an energy auditing program known as "Energuide for Houses" which is sponsored by the Canadian government (www.energuideforhouses.ca).[4] After improving the home's energy efficiency, the energy savings resulting from the switch from heating oil to geo-exchange could be used to pay for the home improvement loan.

After all was said and done, the couple ended up with a house that had improved its value by over $30,000 (geo-exchange system including air conditioning, improved energy efficiency, and lower operating costs) and they were able to service the loan and put a profit of about $40 per month in their pockets (due to energy savings between the old and new systems), all without spending one cent of their own money! And, to make matters even better, the "profit" realized due to the difference in cost between the old and new systems will increase with each increment in the cost of fossil fuel energy, thereby inflation-proofing their home.

Companies also complain that rising energy costs will damage the economy, destroy jobs, make North America less competitive, and push us back into the dark ages. The State of California and the Province of Ontario are two jurisdictions that have used this argument numerous times.

Ontario is particularly adept at this call. The regulated electricity sector has almost brought the province to the verge of bankruptcy because of the short-sighted actions of its politicians. The province's economy has relied on artificially low electricity rates through heavy subsidization of its poorly operating nuclear fleet to maintain its honor as the industrial crown jewel of the Canadian economy.[5]

By comparison, Germany continues to retain its position as the **world's** largest exporter of manufactured goods in 2005 and will also post a record trade surplus, even in light of the fact that energy prices are more than double the Canadian average.[6]

As if this isn't extraordinary enough, the majority of these German exports are advanced manufactured items that by North American standards would require massive amounts of energy input. However, Germans are able to produce these goods using less than half the energy required by North American firms performing the same task.[7]

While North Americans complain about high energy prices, Europeans innovate and use intelligence and technology to their advantage. President George W. Bush waxes poetic about not destroying *the American way of life* while increasing the nation's foreign trade deficit by sending trillions of dollars abroad to fund excessive oil consumption and environmental damage back home. He directs the world's largest and most expensive military fleet, which is becoming nothing more than a police force to ensure that oil shipping lanes remain open.

The American way of life is now in its sunset years. No amount of biodiesel or any other fuel source will allow our current gluttonous ways to continue. It is not Mr. Bush or any other world leader who will dictate energy policy, but rather Mother Earth, for She has the ultimate say.

2.2 The Issues of Pollution and Climate Change

The United States has less than 5% of the world's population (296 million out of 6.4 billion) and consumes a whopping 30% of *all* of the world's resources. This includes not only energy but also water, steel, aluminum, timber, and just about everything else you can imagine. When we enter a Wal-Mart or local grocery store, it is difficult to imagine that there could possibly be shortages of anything we desire. Based on the insane number of people who line up every Thanksgiving day before dawn to purchase the latest "must have" gizmo, shortages are the last thing on westerners' minds.

Unfortunately, present levels of consumption are not sustainable in the long term. Consider for a moment that if all of the world's population were to consume resources at the same level as the United States, we would require approximately 20 times current levels. In other words, we would require an additional five planets' worth of stuff and energy.

Indeed, with China, Russia, India, and many other emerging-economy nations with combined populations that are many times those of the United States and Canada, it will not be long before our resources dry up and eco-systems give up in frustration.

Many people simply do not believe this. I have personally heard educated people say that climate change, air pollution, and fossil fuel shortages are the fabrications of doomsayers and quacks. However, scientific evidence suggests that these environmental hazards are not a fabrication but pose a definite threat. Although frugality or erring on the side of caution is currently not in vogue in North America, I would suggest that this generation has a responsibility to our children and our children's children to not squander our resources.

Our current way of life includes the belief that cheap energy is our God-given right. Never mind that the cost of gasoline or imported heating oil does not include the huge subsidies lavished on the oil industry. The price of a gallon of gasoline neglects the ongoing American military presence in the Middle East, depletion subsidies, cheap access to government land, as well as monies to advance drilling and exploration technologies. And none of these "hidden" costs takes into account the environmental and health damage caused by the burning of fossil fuels. For example, air pollution, largely from the burning of fossil fuels, kills an estimated 5,600 Ontarians prematurely each year and results in hundreds of thousands of incidents of illness, absenteeism, and asthma attacks—costing the economy billions of dollars. [1]

Fossil fuels don't just power our cars. Home heating and coal- and natural gas-fired electrical power plants rely on fossil fuels as well. Contrary to popular belief, it is the electrical power-generation industry and not the transportation sector that contributes the largest amounts of smog, acid rain, and greenhouse gas emissions to the world's ecosystems.[2] With much of the developing world's population connecting to the electrical grid every day, it is obvious that world energy demand will mushroom as more (and larger) appliances are brought online, further exacerbating the problem. In the developed world, middle-class families are demanding more and more appliances that were considered luxury items only one generation ago. Central air conditioning, multiple refrigerators, computers, chest freezers, hot tubs, and swimming pools, all luxury or unimaginable items in our parents' day, are now necessities and consume enormous amounts of energy.

Energy efficiency and demand-side management of these electrical loads will go only so far. No matter how efficient our appliances are, energy is still required to power our homes. Reducing our reliance on fossil fuels and working towards cleaner energy supplies based on the efficient use of natural gas, biofuels, geothermal, and renewable sources will go a long way toward creating a carbon-neutral and low-smog energy supply.

Everyone has heard about them, but do you actually know what smog, climate change, and greenhouse gas emissions really are? Ask anyone who lived in Los Angeles in the early 1980s and they will tell you about dirty skies and endless haze. For someone with asthma or other respiratory sensitivities it meant staying indoors for countless hours or depending on medical inhalers to struggle with each breath. Steve Ovett wished the air had been cleaner in Los Angeles. Although he won medals at the Olympics in Moscow in 1980, the air in Los Angeles was his undoing. During the 1984 Olympics he collapsed during the 1500-meter race because of smog-induced asthma and spent two nights in hospital.

Smog pollution is created when the emissions from burning fossil fuels combine with atmospheric oxygen and ultraviolet light from the sun. Soot and particulate matter occur because poor engine and emission system designs spew unburned hydrocarbon fuels directly into the atmosphere. Estimates indicate that over half a million deaths worldwide each year are directly attributed to excessive levels of airborne pollution.

When smog and soot particles combine with raindrops, acid rain is formed. Clean rain water is neither acidic nor base, meaning that this life-giving element is non-corrosive and restorative to everything it touches. Acid rain, on the other hand, is precisely that—water which has become acidic. When it comes into contact with metal, rock, or plants the result is a corrosive action and the decay and destruction of a wide array of living and lifeless objects.

Smog has been greatly reduced in the developed world in recent years as a result of decreased sulfur concentrations in gasoline and diesel fuels. Improvements in automotive and coal-fired power station emissions have also gone a long way in reducing smog over the last 20 years. However, the problem is far from over. Increasing demand for coal-fired electrical energy and ever-increasing numbers of automobiles on the world's highways along with lack of emission controls in the developing world continue to exacerbate the problem.[3]

Carbon dioxide and methane are two greenhouse gases that are increasing in concentration in the atmosphere and are directly linked to global warming, although they are not currently regulated in the United States and

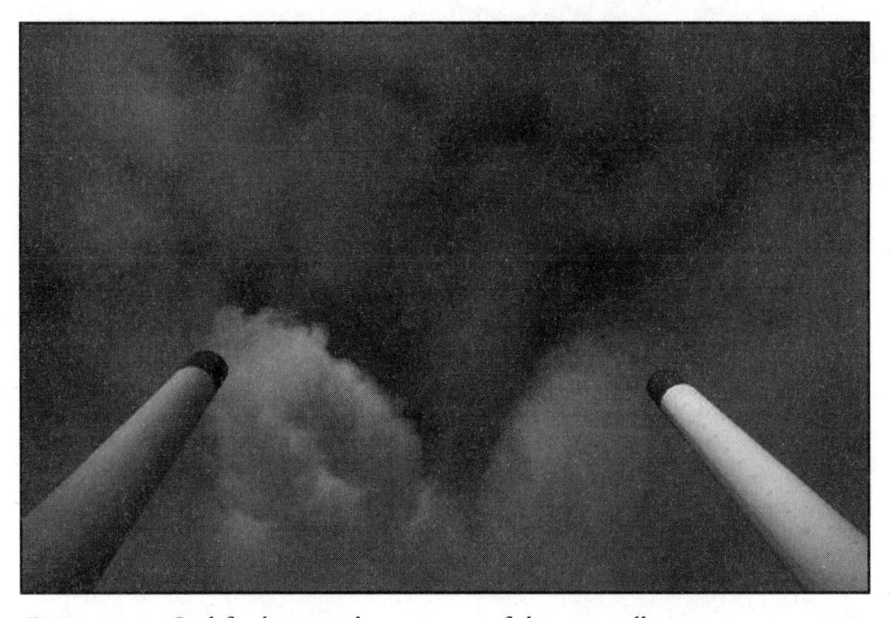

Figure 2.2-1. Coal-fired power plants are one of the most polluting energy sources in the world, notwithstanding the industry rhetoric about "clean coal technology." This is simply a marketing ploy to help gloss over the facts. Although smoke stack scrubbers and other advanced technological fixes can reduce smog-forming emissions, it is currently impossible to eliminate greenhouse gases from the exhaust plume. Carbon sequestration (a fancy word for pumping CO_2 into the ground) may work, but it is far from proven and not without technical and financial risks.

Canada. Carbon dioxide concentrations are higher than at any time in the last 420,000 years and have directly contributed to a rise of approximately 1 degree Celsius in the Earth's surface temperature over the last century. It should also be noted that the 1990s were the warmest decade on record, with 1998 recorded as the single warmest year of the last 1,000 years. In 2004, the average annual temperature rose by .45°C and the ten warmest years on global record have all occurred since 1990.[4]

Scientists predict that if the earth's average temperature were to rise by approximately 2°C, numerous devastating effects could occur:

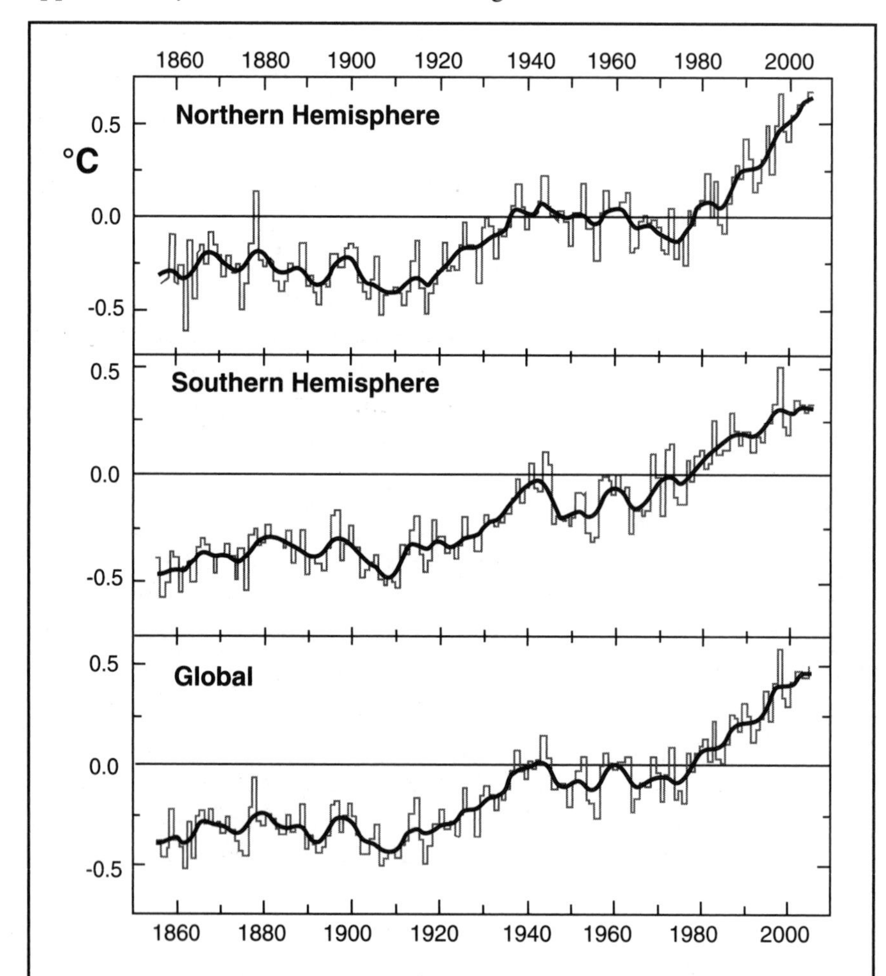

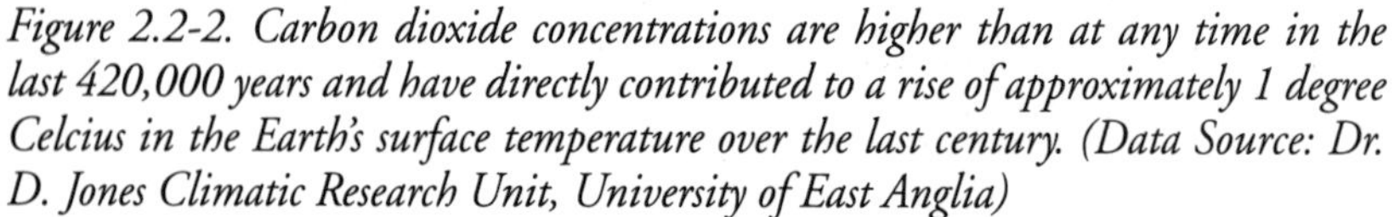

Figure 2.2-2. Carbon dioxide concentrations are higher than at any time in the last 420,000 years and have directly contributed to a rise of approximately 1 degree Celcius in the Earth's surface temperature over the last century. (Data Source: Dr. D. Jones Climatic Research Unit, University of East Anglia)

- increased magnitude and frequency of weather events
- rising ocean levels causing massive flooding and devastation of low-lying areas
- increased incidents of drought affecting food production
- rapid spread of non-native diseases

Perhaps we are beginning to see some of the effects now. Consider that the hurricane season of 2005 generated the most named storms ever, the most category 4 and 5 storms, and the record-breaking damage caused by Katrina and Rita.[5]

Some economists and politicians have gone on record as saying that climate change may actually be a good thing, for example by creating longer growing seasons in Canada or opening arctic sea lanes. Perhaps a longer growing season in Canada is a good thing, but not if it is accompanied by prolonged drought in the prairies. Arctic sea lanes are fine, provided the melting of sea ice doesn't flood the coastal areas where the million dollar penthouses are built.

No, the best change in the environment is no change at all. Society has adapted fairly well to the "normal" vagaries of the weather, and introducing the wild card of climate change into the game is hardly an ace up the sleeve.

So where do these gases come from? Carbon dioxide is a byproduct of the burning of any carbon-based fuel: gasoline, wood, oil, coal, or natural gas. With few exceptions, the world's current energy economy is fueled by carbon.

We have to go back half a million years and more to learn how this carbon fuel came to be. All carbon fuel sources began as living things in prehistoric times. Peat growing in bogs absorbed carbon dioxide from the air as part of the photosynthetic process of plant life. The rolling, heaving crust of the earth entombed the dead plant material. Sealed and deprived of oxygen, the plant matter could not rot. Over the millennia, shifting soils and ground heating compressed the organic material into soft coal that we retrieve today from shallow open-pit mines. Allowed to simmer and churn longer, under higher heat and pressure, oil and hard coal located in deep underground mines were created.

Provided these prehistoric fuel sources remain trapped underground there is no net increase in atmospheric carbon dioxide. However, the process of burning any of these fuels reduces the carbon stored in the coal or oil and drives off the trapped CO_2.

Conversely, the burning of renewable carbon-based fuels such as wood, biomass, and oilseed-based fuels such as biodiesel does not contribute to net greenhouse gas emissions. These plant materials "recycle" carbon dioxide from the atmosphere into carbon during the growing cycle. Allowed to die and rot or through burning, these plant materials release the same CO_2 that was originally (and relatively recently) absorbed. Although the burning of prehistoric fossil fuels such as coal and oil also gives back carbon dioxide the atmosphere lost millennia ago, these "new" emissions of "old" material increase concentration levels, causing global warming.

It *may* not be necessary for society to make a wholesale switch to a non-carbon-based energy economy using sources such as hydrogen in order to reduce the effects of global warming. By selecting on a voluntary basis a more fuel-efficient car, using biofuels, switching to compact fluorescent lamps, or improving the energy efficiency of our home we can drastically reduce personal greenhouse gas emissions and enjoy significant energy and cost savings along the way.

On a larger scale, it is necessary for governments around the world to cap carbon emissions, increase fossil fuel taxes, integrate a worldwide carbon trading program, and support the technological fixes necessary to achieve these goals.

The Kyoto Climate Change Protocol is a good start, but many of the world's developing and developed economies (including the United States and Australia) do not subscribe to the program. Unfortunately for climate change naysayers, Mother Nature and the carbon absorption capacity of the atmosphere could care less what politicians say. The destructive forces of nature will continue their relentless march until humans are either beaten into submission or banished to oblivion.

2.3
Geopolitics and Oil

Humans have very short memories. If you were of driving age in 1973 you will no doubt recall the long lineups and short supply of gasoline at your local filling station. Even if gasoline was available, OPEC raised the price of oil from $4.90 a barrel to $8.25 a barrel in that year. None of this had anything to do with diminishing world oil supplies, but came about as a result of Arab indignation over American support of Israel and the misguided energy policies of the Nixon administration. Given the geopolitical issues associated with having to rely on other countries to provide the energy to run the American domestic economy, one would assume that the need for energy self-sufficiency would be obvious, with a corresponding move towards mandatory conservation, efficiency, and regulatory programs. Unfortunately, this assumption is dead wrong.

Figure 2.3-1. It was not lack of supply but rather the misguided energy policies of the Nixon administration coupled with the Arab oil embargo over American support of Israel that led to severe oil supply shortages in the early 1970s. (Courtesy U.S. National Archives & Record Administration)

Energy consumption in the United States and Canada is insatiable, and it appears no one has learned from the mistakes of the 70s. Lulled into a false sense of security by falling oil prices and seemingly endless supplies, Sport Utility Vehicles (SUVs) and minivans abound, with the result that automotive fuel economy is nearing an all-time low. Need and want have become confusing signals to most of the middle class, who have grown up believing that "bigger is better" and that happiness can only be gained by ever-larger amounts of personal consumption.

Consider that the venerable 1925 Ford Model-T should be considered an economy car by today's standards, as it was able to achieve an astounding 25 miles-per-gallon fuel consumption. After 80 years of technological improvement, the 2005 Ford Expedition (a necessity according to more and more North Americans) gets a mere 12 mpg.

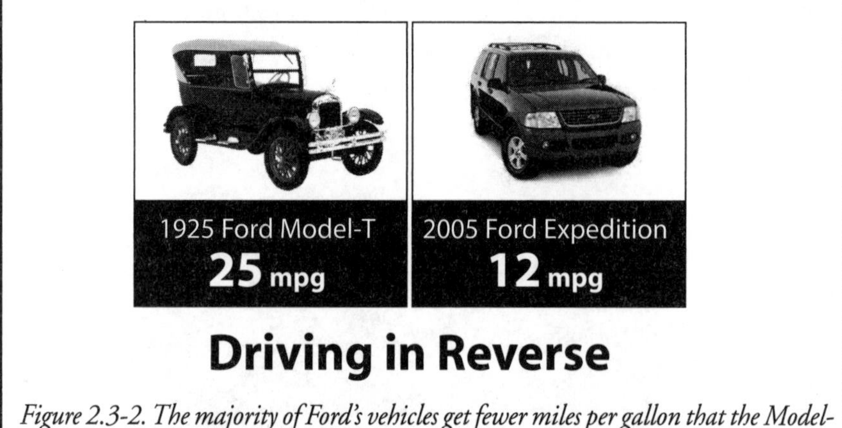

Figure 2.3-2. The majority of Ford's vehicles get fewer miles per gallon that the Model-T did 80 years ago. Given that the United States has little more than a three-year supply of domestic oil, energy efficiency programs are required on an urgent basis.

Consider too that the square footage of the average North American home has more than doubled during the past generation (while family size has halved), and energy efficiency standards for construction have stalled, with the result that additional energy is required for heating and air conditioning.

Over the same period, domestic oil production in the United States has dropped to less than half of total daily demand, requiring vast quantities of imported oil to make up the deficit. During 2005 the United States consumed 20,800,000 barrels of oil **per day**, of which only 7.5 million barrels (or 36% of total consumption) came from domestic sources. This consumption level equates to over 25 barrels per year for each of the 296 million inhabitants of the United States.[1]

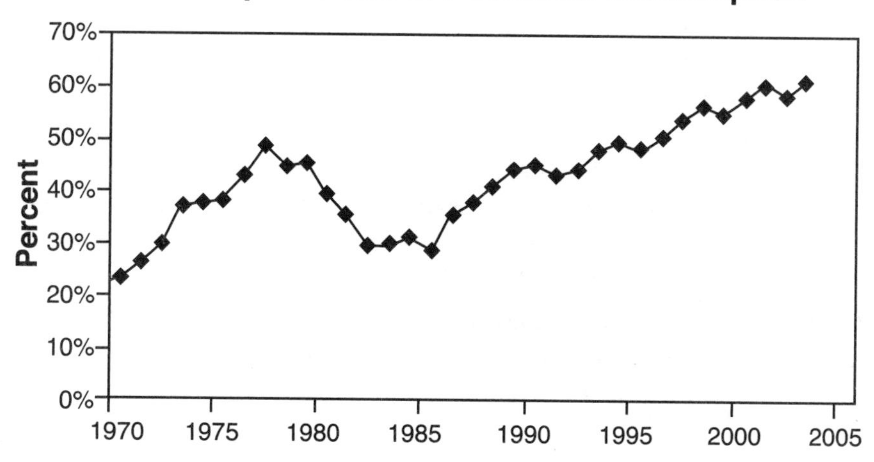

Figure 2.3-3. Over the past generation, domestic oil production in the United States has dropped to less than half of total daily demand, requiring vast quantities of imported oil to make up the deficit. (Source: U.S. Energy Information Administration, Annual Energy Review 2003)

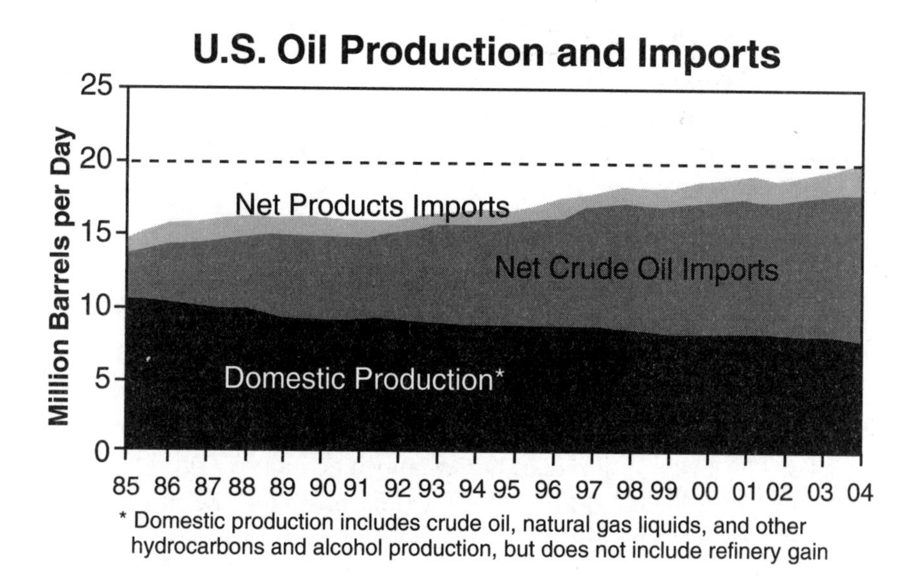

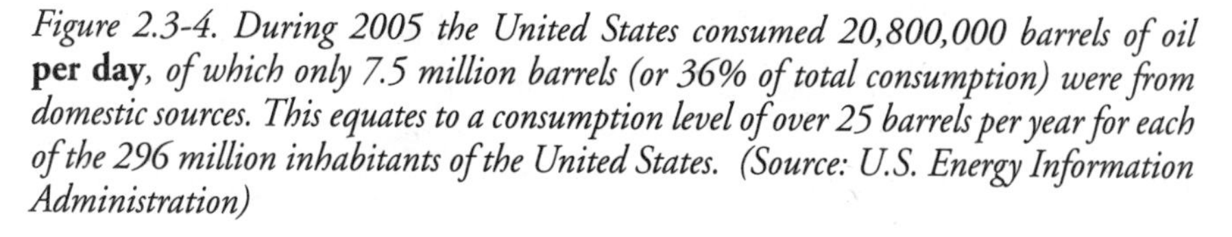

Figure 2.3-4. During 2005 the United States consumed 20,800,000 barrels of oil **per day**, *of which only 7.5 million barrels (or 36% of total consumption) were from domestic sources. This equates to a consumption level of over 25 barrels per year for each of the 296 million inhabitants of the United States. (Source: U.S. Energy Information Administration)*

Given that the United States reported oil reserves of 21.9 billion barrels of oil as of January 1, 2005, it has less than three years of domestic supply at current rates of consumption.[2] Imports **must** come to the rescue in order to maintain existing levels of use.

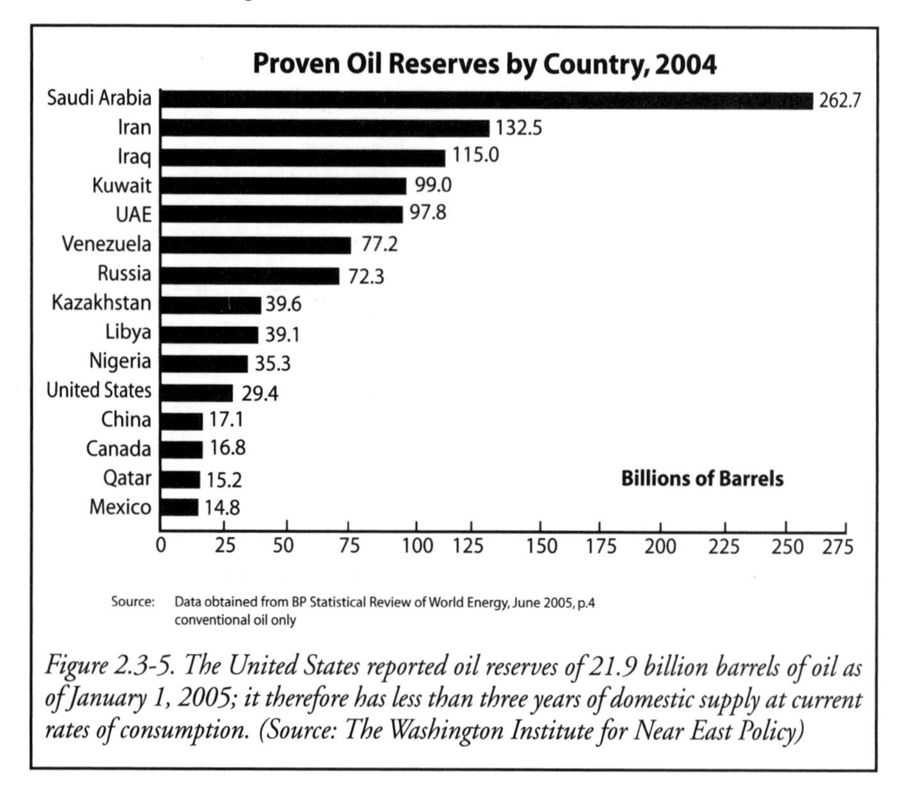

Figure 2.3-5. The United States reported oil reserves of 21.9 billion barrels of oil as of January 1, 2005; it therefore has less than three years of domestic supply at current rates of consumption. (Source: The Washington Institute for Near East Policy)

Even though the majority of the world's oil is supplied by regions that are politically unstable and often unfriendly to the West, energy dollars are being exported as fast as the U.S. Treasury Department can print them. According to an October 25, 2003 report in The Economist, OPEC has drained the staggering sum of $7 trillion from American consumers over the past three decades. This massive sum does not even include industry subsidies, cheap access to government land for oil extraction, or military security required to get the sticky stuff safely into North America. If this money had been pumped into energy efficiency, mass transit, and improved building standards, the U.S. trade deficit as well as conflicts in oil-rich regions of the developing world would be greatly reduced or even eliminated.

The list of current beneficiaries of U.S. energy demand may surprise most people. Based on information provided by the U.S. Energy Information

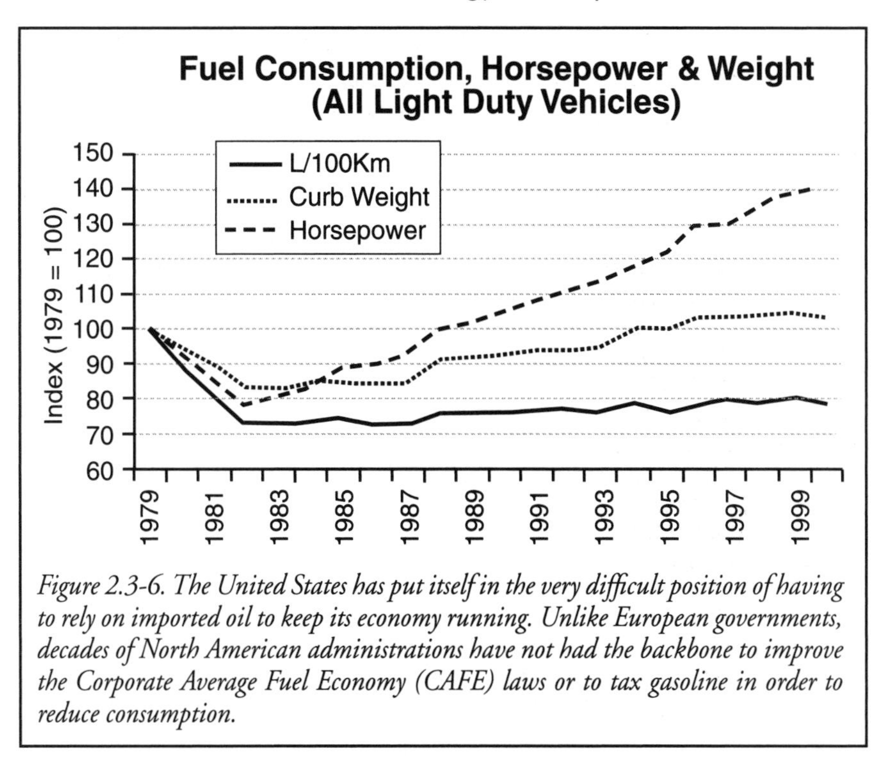

Figure 2.3-6. The United States has put itself in the very difficult position of having to rely on imported oil to keep its economy running. Unlike European governments, decades of North American administrations have not had the backbone to improve the Corporate Average Fuel Economy (CAFE) laws or to tax gasoline in order to reduce consumption.

Administration, countries providing crude oil imports to the United States for the period January through July 2005 were:[3]

Canada	1.61 million barrels per day
Mexico	1.59
Saudi Arabia	1.48
Venezuela	1.29
Nigeria	1.09

This equates to a staggering $620 million ***per day*** being drained from the American economy and significantly contributes to the U.S. trade deficit, currently hovering at $821 billion annually (fiscal 2005 figures). While this is good financial news for energy-exporting countries, Canada, as one of them, has cause to worry.

Canada to the Rescue?

The Western Canada Sedimentary Basin (WCSB) underlies, as the name implies, most of the western provinces and has been the main supply of oil production for the last two generations. Because of the effects of peak oil (see Chapter 2.4), production has been in steady decline and is expected to fall precipitously within the next decade. Industry is now in a race against the clock to replace conventional light oil supplies with "synthetic" oil from the Alberta tar sands. (The politically correct term is "oil sands," but given that the sticky bitumen is about as liquid as asphalt, I prefer the visually accurate moniker).

Extraction of the bitumen from the tar sands is accomplished using massive open pit mining or steam extraction from underground deposits. Upon extraction, the bitumen is mixed with other hydrocarbons to allow the slurry to flow through pipelines to refineries where it is upgraded into "synthetic crude oil."

Canada reported that proven oil supplies were 178.8 billion barrels in 2005, of which only 8.9 billion were from conventional sources. With current production hovering at around 3.1 million barrels per day, conventional low-cost, easy-access sources therefore represent approximately eight years of additional production as of January 2006.[4]

The remaining 169.9 billion barrels of oil are attributed to tar sands. Despite the potential financial windfall of developing the tar sands, there are a considerable number of economic and technical hurdles that could squelch project development.

Cost overruns, low world oil prices, lack of fresh water, carbon emissions and environmental issues, as well as availability of affordable energy to process the bitumen could easily bankrupt development. Because of the high capital and technology requirements for building the mining and upgrading plants, world oil prices must stay above approximately $30 per barrel and natural gas below $7 per 1,000 cubic feet in order to break even. The existing tar sand refineries gulp massive amounts of natural gas, another Canadian commodity that is on the endangered species list and is relied upon to provide heat for steam extraction and bitumen upgrading.

The WCSB provides more than liquid fossil fuels; it is also a main source of natural gas for both Canada and the United States. With the winter 2005/2006 natural gas futures price hovering around $15 per 1,000 cubic feet (up from $2 just a few years ago), it appears that natural gas costs and midterm supply shortages may become the devil-in-the-details that could stall the massive tar sands industry buildup. (There has been considerable talk about circumventing this problem by using nuclear power to provide the thermal energy required for bitumen extraction; however, that's a whole other essay.)

As of January 2005, proven Canadian natural gas reserves were ranked 19th in the world at 56.1 trillion cubic feet (tcf). (1 trillion is 1,000 billion or 1 million million). Although this is a staggeringly large number, domestic gas consumption is equally staggering. According to official figures, in 2002 Canada produced 6.6 tcf of natural gas, consuming 3 tcf and exporting the majority of the balance to the United States to help meet their demand of 22.4 tcf per year.[5]

Given that North American demand for natural gas is expected to rise to 34 tcf over the next six years, Canadian proven reserves will be exhausted over this same period. While there is no doubt that unconventional gas sources such as coal bed methane or liquefied natural gas may begin to replace conventional supplies, it is unlikely that this transition will be seamless and painless for the economies of both Canada and the United States.

What is perfectly clear is that North American fossil fuel liquids (gasoline, diesel, heating oil) as well as natural gas are going to be in short supply and increasingly expensive within a very short period of time. Blankets, coats, horse and buggy anyone?

Figure 2.3-7. It is perfectly clear that North American fossil fuel liquids (gasoline, diesel, heating oil) as well as natural gas are going to be in short supply and increasingly expensive within a very short period of time. Blankets, coats, horse and buggy anyone? Photo credit: University of Saskatchewan Archives.

Mexico Perhaps?

Mexican oil exports to the United States are a close second to those of Canada, so perhaps Mexico will be able to bridge the growing oil gap? Not so, according to Oil & Gas Journal. Proven Mexican oil reserves were 14.6

billion barrels as of January 2005, and depleting quickly. Production at the super-giant Cantarell oil field, which is one of the largest in the world, averaged 2.1 million barrels per day or 63% of Mexico's total oil production during 2005.[6] According to Matthew Simmons, CEO of Simmons & Company International, a Houston-based investment bank that specializes in the energy industry, "Cantarell, discovered in 1975, was the last oil field found anywhere whose daily production would exceed one million barrels."[7] According to technical data from the U.S. Energy Information Administration and Pemex (Mexico's state-owned, nationalized petroleum company), production will decline by 14% a year, regardless of any additional drilling or expansion undertaken to the oil field.[8]

Oil exploration in Mexico is hampered by costly exploration and difficulties between the company and its political bosses. Major economic reforms between the government and the company will be required before major new exploration, drilling, and development projects can begin. Given the fiscal and geotechnical challenges facing Mexico and Pemex, the midterm forecast is a continuing drop in crude oil exports, further aggravating the flow of oil into the North American energy market.

Regardless of Mexico's ability to produce additional crude supplies, it is still a net importer of refined petroleum products, importing over 311,000 barrels per day, of which gasoline comprises 50%. Demand for refined products will continue unless Pemex is able to invest at least $19 billion over the next eight years. Given the immediate requirements for exploration and drilling infrastructure, finding these substantial funds will be difficult at best.

The Persian Gulf Region

According to the U.S. Energy Information Administration, the Persian Gulf contains 715 billion barrels of proven oil reserves, representing over half (57%) of the world's oil supply, and maintains almost all of the world's excess oil production.[9] This excess capacity is required to buffer swings in supply and demand, acting as a rubber mallet to force oil markets in line with Saudi policy.

Forecasters hope that production in the Persian Gulf area will reach 26 million barrels per day by 2010 and 35 million by 2020. This proposed increase would place Persian Gulf oil capacity at 33% of world total by 2020, up from 28% in 2000.[10]

Provided the massive oil supply reserve and production data are correct, imports of crude oil and natural gas from the region should be able to satisfy the North American and world appetite for perhaps another 75 to 100 years.

"Not so fast," says Matthew Simmons. "No one has raised a murmur that these growth projections might be mere fantasy." Mr. Simmons' research concludes that these production levels are extremely unlikely and that Saudi Arabia, the world's current largest producer will not be able to maintain let alone increase its production capacity to meet rising world demand.[11]

Even if Saudi Arabia and the balance of the Persian Gulf states have sufficient reserves and production capacity, supply shocks offer another worry for oil consumers. It is no understatement that the United States economy is extremely vulnerable to Middle East energy shocks. Tight oil supplies from all global energy exporters mean that any sizable disruption in the supply of oil will cause prices to skyrocket. A recent report[12] estimates that oil prices will rise in excess of $7 per barrel for every one million barrels removed from supply lines, with sharp price hikes likely to trigger worldwide recession or at a minimum cause severe hardship to the financially disadvantaged at home and in the developing world.

Disruptions in the supply of energy can come from natural disasters such as the 2005 hurricanes Katrina and Rita, accidents such as refinery and infrastructure fires, or terrorism. A November 2005 report by the Washington Institute for Near East Policy has listed several key areas of concern related to attacks on Middle East oil infrastructure:[13]

- Al-Qaeda attacks on Persian Gulf and Iraqi oil facilities which Osama bin Laden has urged on the grounds that they are "the most powerful weapon against the United States"
- an exodus of oil workers occasioned by fear of terrorism and domestic unrest
- the spread of Iraqi instability into other oil-producing countries
- a confrontation with Iran over its nuclear program or other "problematic" behavior
- domestic instability or uncertain political transitions

Energy efficiency, conservation, renewable energy, and well-developed energy policies will go a long way toward reducing the risks surrounding geopolitics and oil; however, it would be fanciful to believe that North Americans and the world economy as a whole will ever completely wean themselves from oil. Given the technical and economic hurdles of transitioning away from fossil fuels, particularly from the Middle East, western nations will be well advised to develop policies that allow commerce with this politically unstable region, to improve domestic energy policy, and to be better prepared for oil supply disruptions when they do occur.

2.4
Peak Oil a.k.a. Production Limits

At a recent Calgary, Alberta oil patch meeting[1], Mr. Fatih Birol, chief economist of the International Energy Agency (IEA), praised Canada for its efforts in building the tar sands infrastructure, which helps place a ceiling on crude oil prices. "It plays a very important role in international oil diplomacy, but only a global conservation effort will keep energy costs from spiralling upward in coming years."

Although Canada has officially reported 178.8 billion barrels of proven oil reserves, second only to Saudi Arabia, over 95% of these reserves are in tar sands deposits and, as explained earlier, are subject to severe technical, energy, and financial constraints. Even if these resources can be developed, it is unlikely that Canada will ultimately be able to play the moderating role of counterbalancing the strength of the Organization of Petroleum Exporting Countries (OPEC). The reason: peak oil production limits.

When queried about the issue of peak oil, Mr. Birol chuckled, saying that "such concerns rise in step with the cost of oil. It's a fashion. It comes every 10 years, when we have high prices. Four times, we've reached a peak in the last 20 years!"

Mr. Birol's remarks may have been intended as a joke, a means of dealing with the issue, but sadly, they were anything but.

Defining Peak Oil

Peak oil is a frequently used term which is often taken out of context, so let's start this examination of the theory with an accurate definition:

"The Hubbert peak theory, also known as peak oil, is an influential theory concerning the long-term rate of conventional oil and other fossil fuels extraction and depletion. It predicts that future world oil production will soon reach a peak and then rapidly decline, notwithstanding improvements in oil extraction techniques or the technology used. The actual peak year will only be known after it has passed."

Peak oil theory is named after the late Dr. Marion King Hubbert who worked as a research geophysicist with the Shell Oil Company from 1943 to 1964. He was later a professor of geology and geophysics at Stanford University as well as a research geophysicist with the United States Geological Survey in Washington D.C.

During his long and distinguished career, Dr. Hubbert developed numerous theoretical concepts regarding the flow of fluids in the earth's crust, which led to the development of important techniques employed to locate oil and natural gas deposits, many of which are still in use today. He is, however, most famously remembered for his postulation that U.S. oil production would peak in the early 1970s and decline thereafter, no matter how well oil extraction technology improved.

Although initially scoffed at, Dr. Hubbert's theory proved to be amazingly accurate, predicting an oil shortage in the lower 48 states some 20 years before the event actually occurred.

Interestingly, despite the data confirming Dr. Hubbert's hypothesis, many politicians, economists, and energy planners do not subscribe to the theory of peak oil. But then again, maybe they do. Mr. Birol's comment about there being "four peaks in the last 20 years" was intended to refute the peak oil argument, while it unwittingly lends support to it.

In reality, there will not be one but hundreds if not thousands of peaks resulting from the development and extraction of individual oil and natural gas fields located around the world. Further, peak oil does not mean that

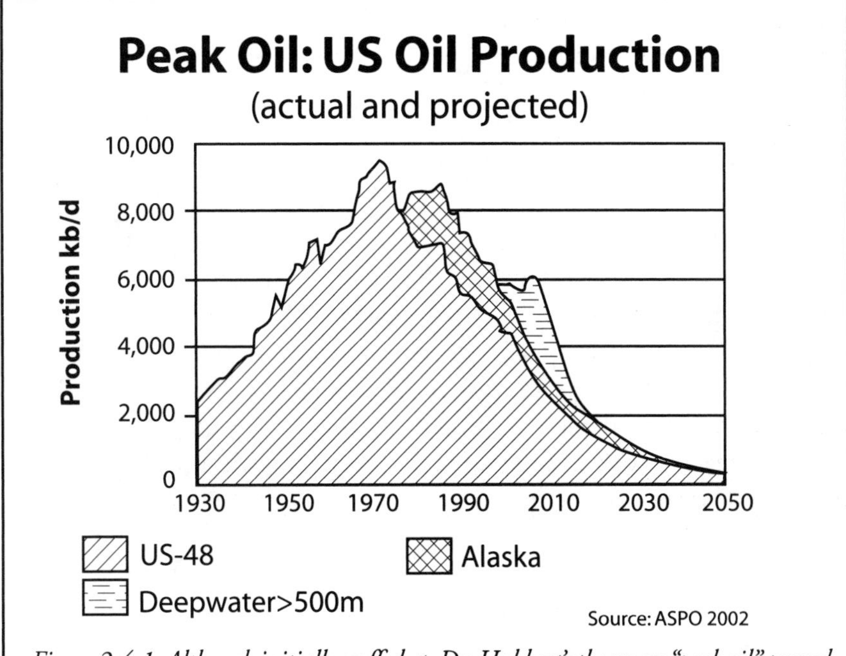

Figure 2.4-1. Although initially scoffed at, Dr. Hubbert's theory on "peak oil" proved to be amazingly accurate, predicting an oil shortage in the lower 48 states some 20 years before the event actually occurred. (Source: www.peakoil.org)

the world is running out of oil, but rather that there are intrinsic limits on how quickly oil and gas can be extracted from the earth. Indeed, there will be billions of barrels of oil, gas, and oil equivalents in the ground long after the world has reached "The Peak."

Understanding the Geophysics of Peak Oil

Many people assume that if an oil field is said to hold a specific number of barrels of oil (or natural gas) then those resources can be had by simply extracting them with the appropriate pumping technology. Although this is a reasonable assumption, it is completely incorrect.

Oil is not stored in the ground in neat and tidy underground drums, rivers, or lakes. Rather, it is normally contained within semi-impermeable oil-bearing rock and other strata of the earth's crust, often mixed with ground water, which complicates both the technological and the financial hurdles of well development and oil extraction.

Once oil is discovered at a given location, sensing and data acquisition technology is used to determine the gross estimated volume of the field. Upon completion of the gross volume estimation, oil extraction engineering

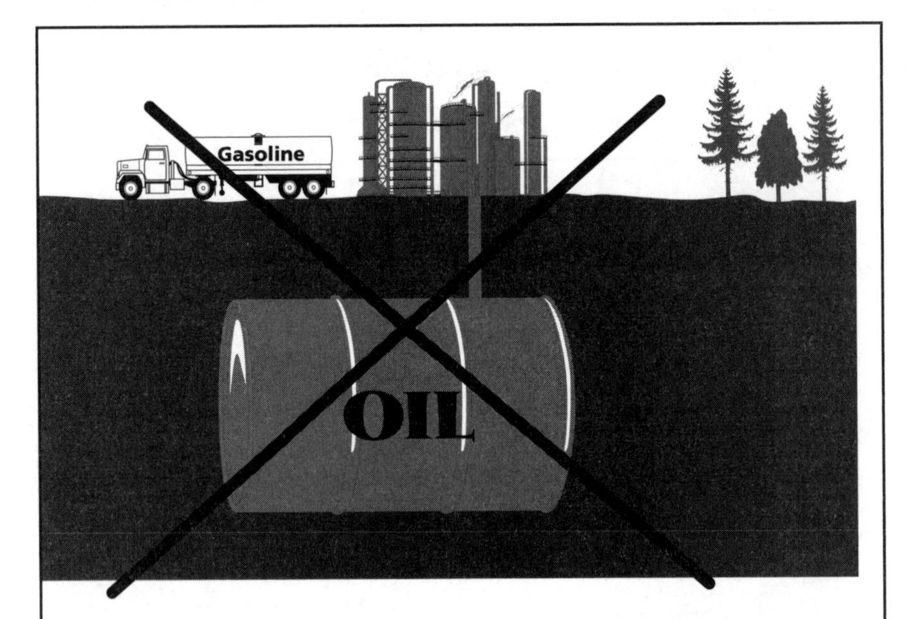

Figure 2.4-2. Oil is not stored in the ground in neat and tidy underground drums, rivers, or lakes. Rather, it is contained within oil-bearing rock and other strata of the earth's crust, often mixed with ground water, complicating both the technological and the financial hurdles of development and extraction.

studies and economic analysis must be applied to the reservoir. Oil fields that produce high-quality crude oil with little water under high natural pressure are considerably less expensive to develop and operate than fields that have high levels of water contamination and low pressure and produce low-grade oil. Application of the technological and economic criteria leads

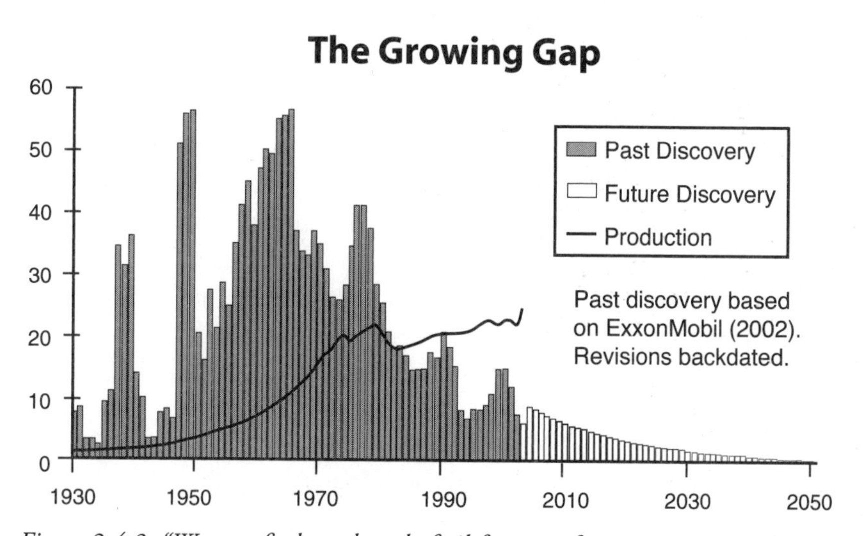

Figure 2.4-3. "We now find one barrel of oil for every four we consume. The general situation seems so obvious....How can governments be oblivious to the realities of discovery and their implications...given the critical importance of oil to our entire economy." (Source: Dr. Colin Campbell, President of ASPO (Association for the Study of Peak Oil and Gas) in his testimony to the British House of Commons, www.peakoil.org)

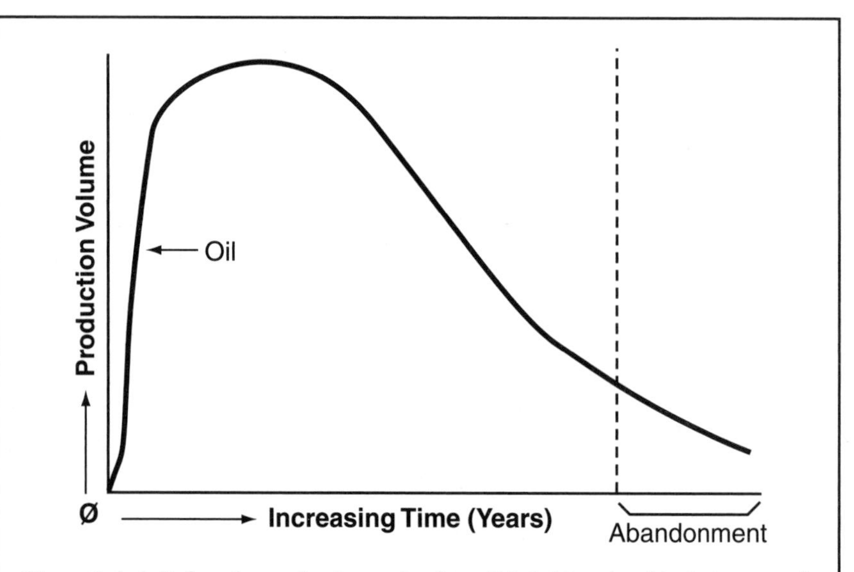

Figure 2.4-4. Before the production cycle of a well is initiated, a "depletion curve" such as the example shown is calculated, providing the oil company with the best balance between extraction of crude oil, co-products, operating costs, and return on investment. Unfortunately, the difficulty in determining the optimum rate of extraction is that the rate that maximizes ultimate (oil) recovery will probably not be the rate that maximizes return on investment.

to the "proven reserve" estimate, which will be considerably below the gross volume of the field.

Before the production cycle of a well is initiated, a "depletion curve" is calculated, providing the oil company with the best balance between extraction of crude oil, co-products, operating costs, and return on investment. According to Matthew Simmons, "The difficulty in determining the optimum rate of extraction is that the rate that maximizes ultimate (oil) recovery will probably not be the rate that maximizes return on investment." Figure 2.4-4 describes the life cycle of a typical water-driven oil well and helps to illustrate Mr. Simmons' point.

When a typical oil well first starts production, natural pressure within the reservoir assists oil in flowing to the well head. This pressure comes from the combined forces of the earth's crust surrounding the reservoir, ground water, and dissolved gases in the oil phase.

As pumping continues and the reservoir matures, internal pressure drops, resulting in a peak extraction rate. The void created by oil removal is rapidly filled with additional ground water and the regasification of the dissolved natural gas bubbles. Additional "pumping stations" only accelerate the drop in pressure, further reducing the extraction rate; the better humans become at extracting oil, the faster the reservoir reaches peak production and the faster we slide down the slippery slope to oil well abandonment.

As reservoir oil extraction and pressure decline continue, a point is reached where natural flow ceases and secondary measures, such as water, steam, or carbon dioxide injection are required to continue production, albeit at ever-declining rates. Applying obvious economic data of the day, including benchmark oil price and extraction cost, will determine where on the life cycle curve oil well abandonment will occur.

As one can imagine, the "abandonment point" on the reservoir's life cycle curve will vary considerably, and once reached it will leave behind huge amounts of "unrecoverable" oil. It is completely logical to abandon a well one day, only to return it to production should favourable economic conditions occur in the future.

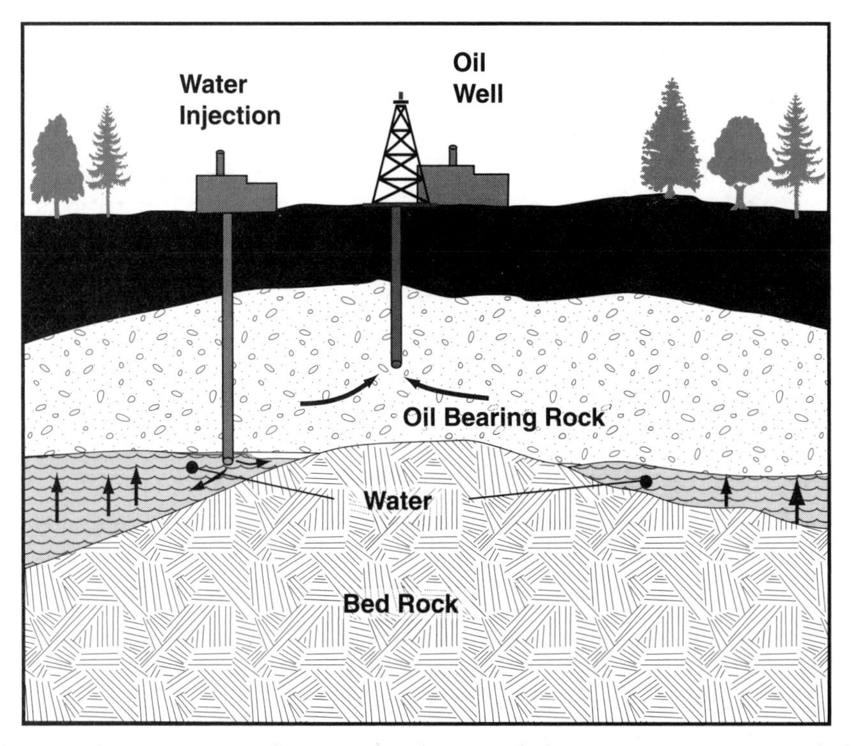

Figure 2.4-5. As reservoir oil extraction and pressure decline continue, a point is reached where natural flow ceases and secondary measures, such as water, steam, or carbon dioxide injection are required to continue production, albeit at ever-declining rates.

One example of abandonment and restart is in the once-prosperous North Sea area, where the major oil and gas companies have sold their positions to secondary oil producers who, with lower overhead costs, can continue to operate these wells during their twilight years.

A further example of end-of-lifecycle oil extraction comes from the U.S. lower 48 states where over half a million "marginal" or "stripper" wells produce thousands of barrels of water per day, but only a few barrels of oil, offering a perfectly sensible income to individual operators with minimal overhead costs.[2]

Even the massive U.S. Strategic Petroleum Reserve (SPR) is subject to the rules of peak oil.[3] In 1975, the Energy Policy and Conservation Act was passed, establishing the SPR to act as a buffer in case of major supply disruptions. To store this amount of oil, the government acquired several salt caverns near Lake Charles, Louisiana and Big Hill, Texas as well as other locations in the Gulf Coast area.

According to official data, the SPR, which currently has a target capacity of 700 million barrels of oil, is subject to the following maximum drawdown capacities:[4]

Days Since Beginning Drawdown	Daily Extraction Rate in Millions of Barrels per Day
0 to 90	4.3
91 to 120	3.2
121 to 150	2.2
151 to 180	1.3

Tinkering with the Data

Given that proven reserves represent "dollars in the ground," oil companies and governments have a vested interest in ensuring that the official recorded volumes for these assets are as large as possible. This is a particular problem where a major producing nation does not publish accurate, or any, proven reserve data. Saudi Arabia is a case in point. It is the largest and most important oil-producing nation, controlling nearly all of the world's spare production capacity. The secretive Saudi government does not publish proven reserve data, posing the threat of global economic chaos if their production peaks or if it cannot be raised to stay in step with worldwide demand.[5]

Further proof of corporate tinkering with proven reserve data was reported by Reuters in February 2005. Royal Dutch/Shell Group was struggling to rebuild investor confidence after five larger-than-expected restatements of reserves by 1.4 billion barrels of oil equivalent as well as the firing of several members of top management.[6]

Given the difficulty of accurately determining recoverable oil and gas reserves as well as peak production rates under optimum conditions, padding and misrepresenting or not reporting data only worsens the prospect of the world's ability to keep the supply running.

Without any doubt, peak oil production is a geotechnological speed limit placed on our ability to extract oil and gas from the earth. There may well be billions or even trillions of barrels of oil in the ground; however, rapidly rising consumption will eventually (some say presently) put the brakes on worldwide economic development.

It is a very interesting facet of human nature that we agree with the facts in hindsight but are too stubborn to accept that history will repeat itself as we face the same issues in the immediate future. Fuel shortages, sharply increased energy costs, and warring over remaining resources are the forecast for tomorrow's energy supply.

2.5 The Case for Biofuels

While the world waits breathlessly for hydrogen-powered fuel cells and advanced electric vehicles (which may or may not ever become mainstream), developments in biofuels are continuing at a surprising rate. Almost everyone has heard of ethanol and some enlightened souls may be familiar with biodiesel. While these technologies are not nearly as sexy as hydrogen and fuel cells, they are available now and offer numerous advantages over imported oil while also making it unnecessary to replace trillions of dollars worth of existing fuel infrastructure systems.

Modern internal combustion engines including those used in transportation vehicles do not have to run on gasoline. In fact, early automotive pioneers did not have access to refined gasoline and consequently used peanut oil and alcohol for fuel. Coincident with the development of the internal combustion engine was the discovery of large amounts of crude oil in the United States. With ready access to this low-cost energy source, the days of the gasoline-powered buggy were upon us.

Fossil fuels in the form of coal, oil, and natural gas originated eons ago as plant matter and marine plankton growing in the ancient world. All living plants use the sun's energy in a process known as photosynthesis to convert atmospheric carbon dioxide into carbon which is stored in the plant structure. When a plant dies and becomes trapped in mud or bogs, decomposition is stalled and under conditions of high heat and pressure fossilization takes place, forming coal. Marine plankton follows a similar fate, converting into various grades of crude oil and natural gas, trapping the stored carbon for eons. When these substances are burned as fuel, carbon dioxide is released, destroying the natural balance in the atmosphere. The wood-burning fireplace or stove also converts stored carbon back to heat and atmospheric carbon dioxide, but it is a renewable and carbon-neutral process. As new trees replace those used as fuel, they absorb carbon, ensuring that the level of emissions is not altered, offsetting the emissions from the wood fuel.

Variations in Crude Oil Grades

There are currently more than 30 grades of crude oil sold on the worldwide markets, with Brent and West Texas Intermediate (WTI) being the benchmarks for quality and price setting. Oil from the North Sea is traded on the London IPE (International Petroleum Exchange) spot market as Brent, while

WTI is traded on the NYMEX (New York Mercantile Exchange). Both of these reference oil grades are so-called light, sweet crude oils, having low sulphur and producing above-average levels of gasoline once refined. Sulphur levels in the crude feedstock contribute directly to the smog-forming pollutant sulphur dioxide. Regulations are demanding low-sulphur fuels; therefore, lower concentrations are preferable as they obviate the necessity for additional, costly processing to remove it.[1]

Heavy crude oils with higher sulphur contents are lower in demand and thus trade at significantly lower prices than light, sweet crude. It is possible to obtain higher distillate fuels such as gasoline from heavy oils, albeit at considerably higher processing costs.[2]

Oil Refining: Converting Crude Oil into Gold

The United States had 148 operating refineries as of January 1, 2005, down from an all- time high of 324 in the early 1980s, owing to the removal of price controls and allocations, as well as tighter environmental regulations.[3] However, although no new oil refineries have been built in the past 30 years, capacity at existing facilities has increased to equal day-to-day demand, which stood at 17.1 million barrels per day in September 2004.

Canada has also experienced a drop in the number of refineries, from a high of 40 in the early 1980s to 21 as of 2005. Although refinery capacity has increased to 97% to maintain sufficient throughput to feed the markets, total volumetric capacity has dropped by nearly 18%.[4]

Wholesale Market

Since gasoline and other distillate fuels are bought and sold internationally and move across international borders, the wholesale price in Newfoundland or New York reflects international commodity market prices. Cargo ships and pipelines supply the wholesale markets with the largest quantities of gasoline, which is valued at the spot market price for that location.

Wholesale gasoline can also be purchased either directly from the refineries "at the gate rack" or from the terminals or bulk plants operated by the major oil companies at the "rack rate," excluding taxes but inclusive of transportation and other local market price factors.

Retail Market

To provide the retail markets with gasoline, distributors take delivery at the wholesale rack rate, and charge for local distribution, federal and local taxes, as well as retail profits.

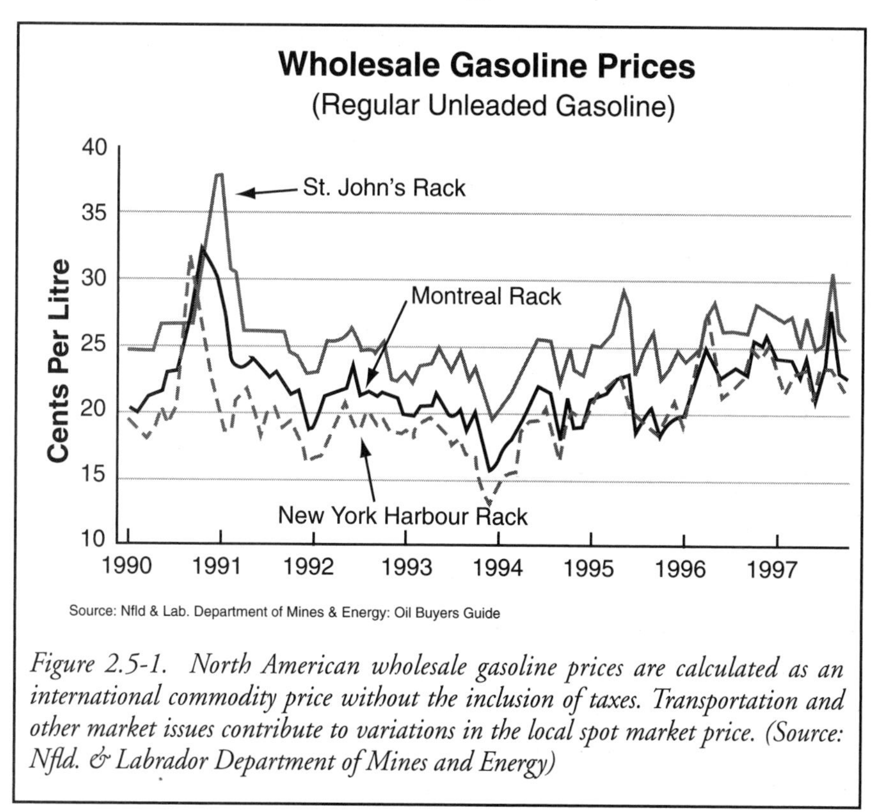

Figure 2.5-1. North American wholesale gasoline prices are calculated as an international commodity price without the inclusion of taxes. Transportation and other market issues contribute to variations in the local spot market price. (Source: Nfld. & Labrador Department of Mines and Energy)

The difference between the wholesale and retail price of transportation fuel provides governments with tax revenues, but equally importantly it provides funds for the development of fuel-efficient vehicles as well as biofuels.

Ethanol-Blended Fuels

The consumption of fossilized fuels is not the only means of extracting energy from plant life. Fermentation of grapes and apples has been fuelling binges for as long as man can remember. The naturally occurring sugars in the fruit produce wine and cider with a maximum alcohol content of approximately 14%. Applying heat to these beverages, in a process known as distillation, allows extraction of the alcohol at up to 100% concentration.

The primary source of plant sugars can vary, as automotive-grade ethanol can be produced from grains such as corn, wheat, and barley. Recent advances in enzymatic processes even allow the conversion of plant waste in the form of straw and agricultural residue into sugars which can in turn be fermented into ethanol.

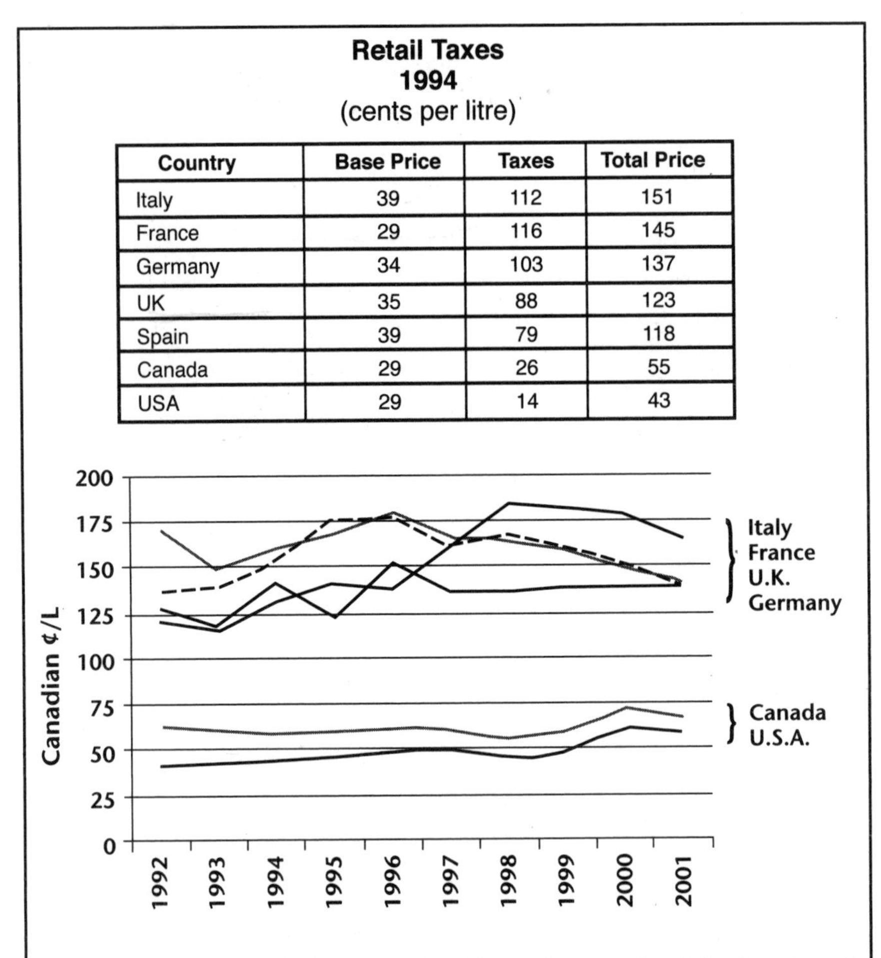

Retail Taxes
1994
(cents per litre)

Country	Base Price	Taxes	Total Price
Italy	39	112	151
France	29	116	145
Germany	34	103	137
UK	35	88	123
Spain	39	79	118
Canada	29	26	55
USA	29	14	43

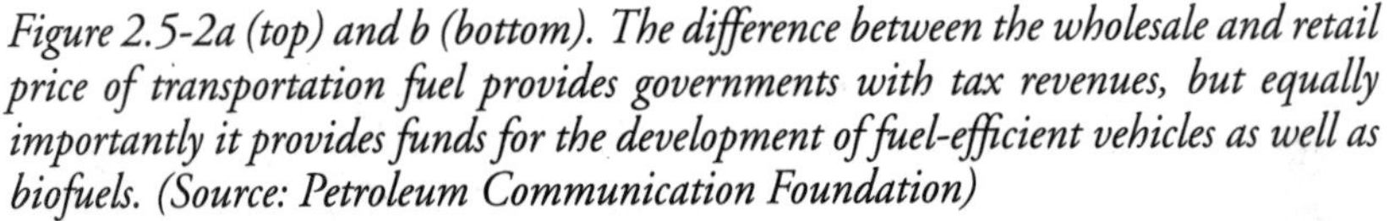

Figure 2.5-2a (top) and b (bottom). The difference between the wholesale and retail price of transportation fuel provides governments with tax revenues, but equally importantly it provides funds for the development of fuel-efficient vehicles as well as biofuels. (Source: Petroleum Communication Foundation)

Conventional grain-derived and cellulose-based ethanols (those made from farming and timber byproducts) are the same and can be easily integrated into the existing gasoline supply chain. Ethanol may be used as a blending agent or as the main fuel source. Gasoline blends with up to 10% ethanol can be used in any vehicle manufactured after 1970.[5] High-level ethanol concentrations of between 60% and 85% can be used in special "flex-fuel vehicles."[6] At the time of writing Ford, General Motors, and DaimlerChrysler warranties allowed up to 10% ethanol blends in their North American vehicles.[7]

Figure 2.5-3. Plant grains and fibre can be converted to sugar, fermented into ethyl alcohol (ethanol), and used as a blending ingredient with gasoline or as the main fuel. High concentrations of ethanol can reduce greenhouse gas emissions by up to 80% relative to gasoline. (Courtesy Iogen Corporation)

Adding ethanol to gasoline increases octane, reducing engine knock and providing cleaner and more complete combustion, which is good for the environment. Ethanol produces lower greenhouse gas emissions than gasoline: a 10% ethanol blend with gasoline (known as E10) will reduce GHG emissions by 4% for grain-produced ethanol and 8% for cellulose-based feedstocks. At concentrations of E85, GHG emissions are reduced by up to 80%.[8]

Figure 2.5-4. The majority of ethanol is derived from corn. The production of ethanol fuels strengthens agricultural regions and creates new and more stable markets for the farming community.

Greenhouse gas emission reduction was abundantly demonstrated during a recent 6,000-mile (10,000 km) driving tour conducted by Iogen Corporation. An SUV fuelled with 85% cellulose ethanol produced the same GHG emissions as a super-efficient hybrid car.

In addition to making environmental improvements, ethanol is produced from domestic renewable agricultural resources, thereby reducing our dependence on imported oil and providing a major source of economic diversity for rural farming economies.

In the United States, 77 ethanol plants produce over 3.3 billion gallons (12.5 billion liters) of ethanol per year. Canadian production currently stands at 62.6 million U.S. gallons (237 million liters) per year. Of the 77 plants in the United States, 62 use corn as the feedstock. The remainder use a variety of seed corn, corn and barley, corn and beverage waste, cheese whey, brewery waste, corn and wheat starch, sugars, corn and milo, and potato waste. In the United States there are 55 proposed new plants and 11 currently under construction. Canada currently has 6 plants with 1 under construction and 8 new proposals on the drawing board.[9]

There is no question that using domestically grown feedstocks from renewable and clean-burning grains is better than importing fossil fuel from the Middle East. However, there is some question about diverting food stocks into fuel feedstocks to continue our love affair with the automobile, thereby contributing to urban sprawl and other societal problems. A reduction in automotive usage is a key, long-term goal but one fraught with difficulty in the short term. Until urban planners can wrench the car keys from suburbanites,

Figure 2.5-5. North American ethanol production currently stands at 3.36 billion gallons (12.72 billion liters) and is expected to increase by 28% before the end of the decade. This is just a drop in the bucket compared with fossil-fuel gasoline usage, but with good demand management (through CAFE and properly managed taxation) it could go a long way toward reducing reliance on imported oil. (Courtesy Iogen Corporation)

Figure 2.5-6. Adding ethanol to gasoline increases octane, reducing engine knock and providing cleaner and more complete combustion. As well as providing performance and environmental improvements, ethanol is produced from domestic renewable agricultural resources, reducing dependence on imported oil and providing a major source of economic diversity for rural farming economies.

society might as well use all the clean "transition" fuels at its disposal.

Cellulose ethanol eliminates the diversion of food crop to fuel feedstocks and is an advanced new transportation fuel which has some advantages over grain-based ethanol:

- Unlike grain-based ethanol production, the manufacturing process does not consume fossil fuels for distillation, further reducing greenhouse gas emissions.
- Cellulose ethanol is derived from non-food renewable sources such as straw and corn stover.
- There is a potential for large-scale production since it is made from agricultural residues which are produced in large quantities and would otherwise be destroyed by burning.

Cellulose-based ethanol has, in the past, been very expensive as a result of the inefficient processes required to produce it. The Iogen Corporation, working in conjunction with the Government of Canada and Shell Corporation, has recently launched a cost-effective method for producing what the company refers to as EcoEthanol™. Once the first phase of production, using straw and corn stover, is under way, it won't be long before other feedstocks can be used, including hay, fast-growing switch grass, and wood-processing byproducts.

Figure 2.5-7. Iogen Corporation founder Brian Foody Sr. stands in a field of straw grass, which may become the twenty-first-century equivalent of a Saudi Arabian oil field. (Courtesy Iogen Corporation)

The Downside to Ethanol

There are very few downsides to ethanol fuel. Because ethanol contains oxygen it permits cleaner and more complete combustion, which helps keep fuel injection systems deposit free. Be aware, though, that ethanol acts as a solvent and can loosen residues in your car's fuel system, necessitating more frequent fuel filter changes.

Gasoline containing 10% ethanol has approximately 3% less energy than regular gasoline. However, this loss of energy is partially offset by the increased combustion efficiency resulting from the higher octane and oxygen concentration in the ethanol blend. Actual road tests show that fuel economy may be reduced by approximately 2% when using E10 concentrations. To put this in perspective, driving 12 mph (20 kph) over the 60 mph (100 kph) speed limit increases fuel consumption by an average of 20%.

Biodiesel as a Source of Green Fuel and Heat

Internal combustion engines ignite fuel using one of two methods. A spark ignition engine produces power through the combustion of the gasoline and air mixture contained within the cylinders. An electric spark which jumps across the gap of an electrode ignites this volatile mixture.

The compression ignition or diesel engine, as it is more commonly known, uses heat developed during the compression cycle. (Have you ever noticed how hot a bicycle air pump becomes after a few strokes?) With a compression ratio of 18:1 or higher, sufficient heat is developed to cause diesel fuel sprayed into the cylinder to self-ignite. The higher energy content of diesel fuel and the oxygen-rich combustion process of the compression ignition engine contribute to improved fuel economy, power, and reduced CO_2 emissions, all desirable features for fleet and other high-mileage vehicles.

Biodiesel, like its cousin ethanol, is a domestic, relatively clean-burning renewable fuel source for diesel engines and oil-based heating appliances. It is derived from virgin or recycled vegetable oils and animal fat residues from rendering and fish-processing facilities. Biodiesel is produced by chemically reacting these vegetable oils or animal fats with alcohol and a catalyst to produce compounds known as fatty acid methyl esters (FAME or biodiesel) and the co-product of the chemical production process, glycerine.

Biodiesel that is destined for use as transportation fuel must meet the requirements of the American Society for Testing and Materials (ASTM) Standard D6751 for pure or "neat" fuel graded B100.[10] Fossil fuel diesel (or petrodiesel) must meet its own similar requirements within the ASTM standard. Biodiesel and petrodiesel can be blended at any desired rate, with a blend of 5% biodiesel and 95% petrodiesel denoted as B5, for example.

Figure 2.5-8. Loading straw into an ethanol plant is a lot cleaner and much more sustainable than extracting oil from the North Sea. Waste grasses, straw, and wood products are "carbon-neutral" fuels that are available domestically and in endless supply. Think of these sources as "stored sunshine." (Courtesy Iogen Corporation)

The testing and certification of any transportation fuel is a requirement of automotive and engine manufacturers implemented to minimize the risk of damage and related warranty costs.

From an agricultural viewpoint, biodiesel offers many of the same advantages to the farming community as ethanol feedstocks. In addition, the process of making biodiesel is relatively simple and low cost, which may lead to cooperative and rural ownership of processing and production facilities, further increasing farming income and risk diversity. Even the co-product of the biodiesel manufacturing process, glycerine, has a ready market in the food, cosmetic, and other industries.

Biodiesel as a Transportation Fuel

The modern diesel engine is a far cry from the smoky, anaemic model of the 1970s. As a direct result of Mr. Nixon's oil crisis, consumers lined up to purchase Volkswagen Rabbit and Mercedes diesel cars, enticed by fuel economy claims. In addition to the lack of power and acceleration, when the mercury dipped below 32°F (0°C) and a diesel engine wasn't plugged in overnight a bus ride was a sure bet the next morning. Is it any wonder that people were sceptical about diesel-powered cars?

Figure 2.5-9. Biodiesel fuel, a renewable, clean-burning fuel source for diesel engines and oil-based heating appliances, is produced from virgin vegetable oils as well as waste vegetable or animal fats. Dr. Martin Reaney, Chair of Lipid Utilization and Quality at the University of Saskatchewan, studies the canola oil seed plant used extensively in the production of biodiesel.

Figure 2.5-10. As the price of petrodiesel continues to rise and biodiesel production costs fall, rural communities will produce their own democratic energy to fuel their part of the economy. (Courtesy Lyle Estill/Piedmont Biofuels)

Fast forward to today. Gone are the smelly, smoky, lumbering diesels of old. Witness the new Mercedes E320 family of "common rail, turbo-diesel" engines that offer no "dieseling" noise, smoke, or vibration, achieve superb mileage, and have better acceleration than the same model car equipped with a gasoline engine.[11]

Biodiesel offers some distinct advantages as an automotive fuel:[12]

- It can be substituted (according to vehicle manufacturer blending limits) for diesel fuel in all modern automobiles. Although B100 may cause failure of fuel system components such as hoses, o-rings, and gaskets that are made with natural rubber, most manufacturers stopped using natural rubber in favour of synthetic materials in the early 1990s. According to the U.S. Department of Energy, B20 blends minimize all of these problems. If in doubt, check with your vehicle manufacturer to ensure compliance with warranty and reliability issues.
- Performance is not compromised using biodiesel. According to a 3 ½-year test conducted by the U.S. Department of Energy in 1998, using low blends of canola-based biodiesel provides a small increase in fuel economy. Numerous lab and field trials have shown that biodiesel offers the same horsepower, torque, and haulage rates as petrodiesel.

Figure 2.5-11. The simplicity of a biodiesel production facility is shown in this aerial view of a continuous production plant. Soybeans are delivered to the processing plant for conversion to food-grade oil. Alcohol and a catalyst are added to the oil to produce biodiesel and glycerine. The biodiesel is stored in the tank farm. The glycerine byproduct is sold and the water and alcohol are recycled, creating an environmentally friendly processing cycle. (Courtesy West Central Soy)

- Lubricity (the capacity to reduce engine wear from friction) is considerably higher with biodiesel. Even at very low concentration levels, lubricity is markedly improved. Reductions in the sulphur levels of petrodiesel to meet new, stringent emissions regulations have, at the same time, reduced lubricity levels in petrodiesel dramatically. Biodiesel blending is currently being considered by the petrodiesel industry as a means of circumventing this problem.
- Because of biodiesel's higher cetane (ignition) rating, engine noise and ignition knocking (the broken motor sound when older diesels are idling) are reduced.

Biodiesel as a Heating Fuel

Heating fuel oil consumption in the United States is currently hovering around 7 billion gallons (26.5 billion liters) per annum according to the U.S. Department of Energy, with the vast majority supplied by imported oil.[13] Blending a mixture of 20% biodiesel with No. 2 petrodiesel would reduce fossil fuel consumption and replace that amount of petrodiesel with domestic, renewable, clean-burning biodiesel. If this strategy were adopted across the

United States, biodiesel consumption would increase by 1.4 billion gallons per year (5.3 billion liters per year), reducing carbon monoxide, hydrocarbon, and particulate emissions by approximately 20% while supporting domestic agricultural economies.[14]

Are you ready to start using biodiesel? The downside is that finding biodiesel can be challenging. In the Northeastern U.S., suppliers are starting to fill this niche market, albeit slowly. Many distributors purchase biodiesel from large producers such as World Energy Alternatives, LLC (www.worldenergy.com) or enter a purchasing (and often producing) co-operative such as Piedmont Biofuels of Chatham County, North Carolina (www.biofuels.coop). According to Piedmont partner Lyle Estill, production limitations and price can become a deterrent to many people. However, Piedmont cannot keep up with demand even though it charges $3.50 per gallon ($0.92 per liter) for B100. Of course, if you purchase a B20 blend the price difference compared to petrodiesel is quite small while the environmental improvements are considerable.

For information regarding supply of biodiesel consult the National Biodiesel Board in the United States at www.biodiesel.org (1-800-841-5849) and the Canadian Renewable Fuels Association at www.greenfuels.org (1-416-304-1324).

Figure 2.5-12. Over 200 public and private fleets in the United States and Canada currently use biodiesel, and the number is increasing rapidly. Environmental stewardship regarding climate change, urban smog, and air quality is creating mass-market acceptance of biodiesel fuel. (Courtesy Saskatchewan Canola Development Commission)

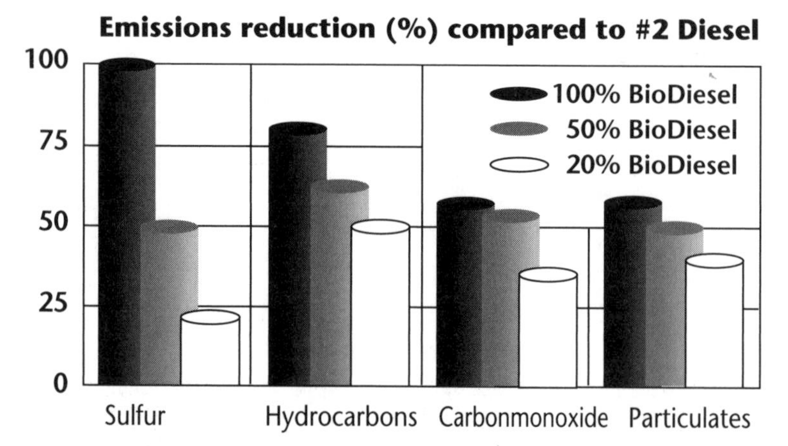

Figure 2.5-13. Blending a mixture of B20 (20% biodiesel to 80% petrodiesel) would reduce carbon monoxide, hydrocarbon, and particulate emissions by approximately 20%. In addition, biodiesel would assist rural domestic suppliers and help keep North American money at home.

Figure 2.5-14. Biodiesel can be used in virtually any oil-burning furnace or boiler provided that fuel lines and gaskets are not made of natural rubber. Because biodiesel will gel (like No. 2 heating oil), storage tanks should be installed indoors or underground. Alternatively, your fuel supplier can provide winter-blended heating oil containing both No. 1 and No. 2 heating oil with biodiesel blended to B20 to achieve the desired low-temperature storage performance.

Summary

Biodiesel fuel will not solve all of North America's transportation and heating supply issues single-handedly. Other energy options including aggressive demand-side management are also required. But since biodiesel is a direct replacement for or blending agent with regular petrodiesel or heating oil, no infrastructure or societal changes are necessary in order to start using this fuel source immediately: no waiting for low-cost fuel cells or exotic batteries. Biodiesel provides a clean, renewable, domestic, and economically diverse energy source which will help pave the way to a more sustainable future.

Figure 2.5-15. Energy prices are rising, domestic supplies are dwindling, U.S. domestic production has peaked, and the environment is taking a beating. Perhaps it's time to consider biofuels for your transportation and home heating needs. Here the author is seen whizzing around in his biodiesel-powered Smart™ car.

3 Technical Background of Diesel/Biodiesel Fuels

3.1 Petrodiesel

We learned in Chapter 2.5 that crude oil can be sold on the world markets using a variety of 30 grades, with "Brent" and "West Texas Intermediate" forming the principal benchmarks. The grade or composition of the crude oil determines its value, with lighter, low-sulphur varieties commanding higher prices. The composition of the crude oils can vary greatly in chemical and physical makeup, ranging from low-viscosity liquids to heavy tar-like materials.

Before crude oil can be sold on the market, it must undergo a complex transformation known as refining. The refining process breaks the crude oil into various constituent parts using a distillation process that takes advantage of the differing boiling points between the major hydrocarbon components or "fractions." The principal distillate fractions from lightest to heaviest are shown in the "cracking tower", Figure 3.3-1.

After distillation has separated the crude oil into the various fractions, they are pumped to the conversion facilities where hydrocarbon "cracking" converts lower-quality fractions into higher-value products such as gasoline and various grades of diesel fuel.

The final step in the refining process is to perform end treatments such as octane adjustment, the introduction of fuel additives for winterization, and the addition of lubricants. The fuel products are then graded and stored

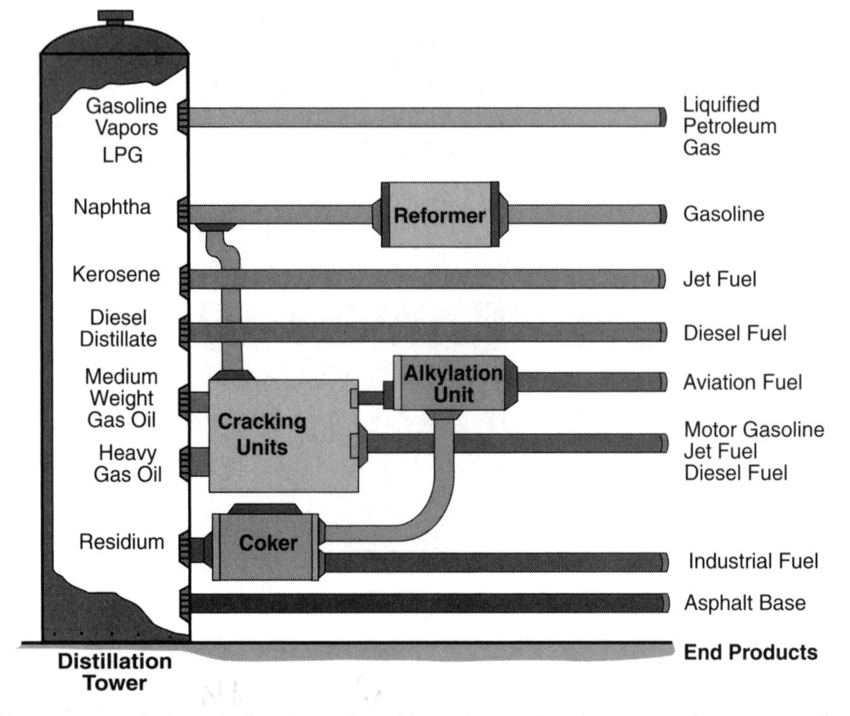

Figure 3.1-1. Before crude oil can be sold on the market, it must undergo a complex transformation process known as distillation refining and cracking. The different boiling points between the major hydrocarbon components or "fractions" are used to separate out the desired high-value products such as gasoline and diesel fuel.

in "tank farms" ready for distribution and retail sale.

It is important to provide quality testing at every step in the distribution chain, including the retail storage and sales facilities. Water, microbial growth, and other contaminants can enter the supply chain at any point, resulting in fuel quality below specification limits.

In the United States, diesel fuel is categorized into five specific grades, each of which is denoted by a test specification developed by the American Society for Testing and Materials, ASTM D975-04:

Grade Low Sulfur No. 1-D—A special-purpose, light distillate fuel for automotive diesel engines requiring low sulfur fuel and requiring higher volatility than that provided by Grade Low Sulfur No. 2-D.

Grade Low Sulfur No. 2-D—A general-purpose, middle distillate fuel for automotive diesel engines requiring low sulfur fuel. It is also suitable for use in non-automotive applications, especially in conditions of varying speed and load.

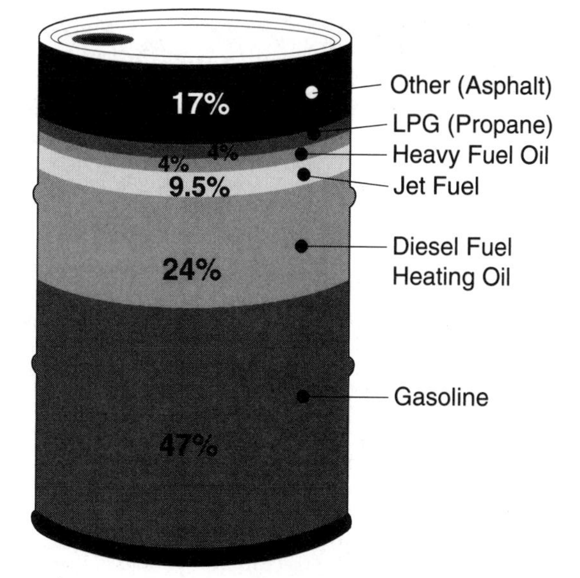

Figure 3.1-2. A 42-gallon barrel of crude oil will yield approximately 44.1 gallons of refined hydrocarbon products, an increase of 5%, as a result of a process known as "refinery swell." Shown here is the approximate percentage yield of each major product generated from a typical barrel of crude oil.

Grade No. 1-D—A special-purpose, light distillate fuel for automotive diesel engines in applications requiring higher volatility than that provided by Grade No. 2-D fuels.

Grade No. 2-D—A general-purpose, middle distillate fuel for automotive diesel engines, which is also suitable for use in non-automotive applications, especially in conditions of frequently varying speed and load.

Grade No. 4-D—A heavy distillate fuel, or a blend of distillate and residual oil, for low- and medium-speed diesel engines in non-automotive applications involving predominantly constant speed and load.

In Canada, diesel fuel quality is the responsibility of the Canadian General Standards Board (CGSB). As in the United States, there are a number of standards for diesel fuel, including:

1. CAN/CGSB-3.517 Automotive Low-Sulfur Diesel Fuel
2. CAN/CGSB-3.520 Automotive Low-Sulfur Diesel Fuel Containing Low Levels Of Biodiesel Esters (B1-B5)
3. CAN/CGSB-3.16 Mining Diesel Fuel
4. CAN/CGSB-3.6 Regular Sulfur Diesel Fuel
5. CAN/CGSB-3.18 Diesel Fuel for Locomotive-Type Medium Speed Diesel Engines
6. CGSB 3-GP-11d Naval Distillate Fuel

These standards include a number of substandards which are suitable for a variety of end uses of the fuel, as illustrated below:

TABLE 3.1-2
Fuel Types and Applications

Standard	Fuel Type	Intended Application
CAN/CGSB-3.517 Automotive Low-Sulfur Diesel Fuel	Type A-LS	High speed diesel engine applications involving frequent and relatively wide variations in loads and speeds and when ambient temperatures require better low temperature properties. Examples include urban transit buses and passenger vehicles. Maximum fuel sulfur of 500 mg/kg.
	Type B-LS	High speed diesel engines in services involving relatively high loads and uniform speeds and when ambient temperatures and fuel storage conditions allow its use. Examples include intercity trucks and construction equipment. Maximum fuel sulfur of 500 mg/kg.
	Type A-ULS	Same as Type A-LS. Maximum fuel sulfur of 15 mg/kg.
	Type B-ULS	Same as Type B-LS. Maximum fuel sulfur of 15 mg/kg.
CAN/CGSB-3.520 Automotive Low-Sulfur Diesel Fuel Containing Low Levels Of Biodiesel Esters (B1-B5)	Type A-LS, Bx 500 mg/kg.	Same as Type A-LS. Maximum fuel sulfur of Biodiesel content from 1.0 to 5% by volume.
	Type B-LS, Bx	Same as Type B-LS. Maximum fuel sulfur of 500 mg/kg. Biodiesel content from 1.0 to 5% by volume.
CAN/CGSB-3.6 Regular Sulfur Diesel Fuel	Type A	Similar to Type A-LS, but fuel use is mainly limited to off-road applications. Maximum fuel sulfur of 3000 mg/kg.
	Type B	Similar to Type B-LS, but fuel use is mainly limited to off-road applications. Maximum fuel sulfur of 5000 mg/kg.
CAN/CGSB-3.16 Mining Diesel Fuel	SPECIAL	High speed diesel engines used in underground mining equipment. Maximum fuel sulfur of 2500 mg/kg.
	SPECIAL - LS	High speed diesel engines used in underground mining equipment. Maximum fuel sulfur of 500 mg/kg.
CAN/CGSB-3.18 Diesel Fuel for Locomotive-Type Medium Speed Diesel Engines		Medium speed diesel engines in locomotive service. Other medium speed diesel engines may also use this fuel. Maximum fuel sulfur of 5000 mg/kg.
CGSB 3-GP-11D Naval Distillate Fuel	Type 11	High speed and medium speed diesel engines, gas turbines and boilers in marine service and when ambient temperatures are higher than -1°C. Maximum fuel sulfur of 5000 mg/kg.
	Type 15	High speed and medium speed diesel engines, gas turbines, and boilers in marine service and when ambient temperatures are higher than -12°C. Maximum fuel sulfur of 5000 mg/kg.

Table 3.1-2. *Fuel Types and Applications*
Source: Canadian General Standards Board

Petrodiesel Fuel Properties

The primary consideration regarding diesel fuel is its compatibility with light-, medium-, and heavy-duty diesel engines used in the transportation, farming, and construction sectors. Diesel fuel should be of consistent quality, safe to use, relatively non-polluting, and should not contribute to engine wear and premature failure.

This may seem obvious, but given the large volume of fuel production, the degree of variability in crude oil, the complexity of the refining process, and the possibility of fuel contamination and decomposition in storage, fuel companies perform a remarkable service and are to be commended for their high-quality products. Consider for a moment how few concerns people have about fuel issues (other than price) on a day-to-day basis.

In order to provide this high level of service, both the ASTM and the CGSB facilitate industry stakeholder meetings to develop fuel standards for the wide range of applications required of diesel fuel. The average car or light-, medium-, or heavy-duty truck generally uses only two types of fuel, although even this minor variation in composition is not apparent to most users.

Diesel fuel No. 2 (No. 2-D) is the preferred fuel for most applications as it provides greater energy density, offering increased fuel economy compared with Diesel No. 1 (No. 1-D), which is similar to kerosene. Given that No. 1-D has lower energy density, one would assume that it would be of lesser value. Unfortunately, No. 2-D suffers from "fuel freezing" at a relatively warm temperature. This is known as cloud point, the temperature at which wax crystals inherent in the fuel begin to form. Should these crystals form in an automotive fuel tank, they will be drawn into the fuel system filter, causing plugging. Depending on the severity of filter plugging, a condition known as fuel starvation will either rob the engine of power or cause it to stop operating completely. The condition is reversible once temperatures rise sufficiently above the cloud point, although costly fuel system degumming may be required depending on the severity of cold plugging.

If ambient temperatures continue to decrease below the cloud point of the fuel, complete solidification will result. This occurs when the fuel reaches its pour and freezing point limits.

No. 1-D has cloud, pour, and freezing points considerably lower than those of No. 2-D and is therefore mixed with No. 2-D in ratios appropriate to withstand the lowest expected temperatures in a given geographic locale. As a general rule, the farther south one travels on the North American continent during winter, the lower the ratio of No.1-D in the diesel fuel mix.

In addition to the differences between No. 1-D and No. 2-D, fuels vary in sulphur content that is either standard or low level. From a retail or fuel purchasing perspective, this is an immaterial point, as all diesel engines operate on either grade of fuel. For environmental reasons, however, smog-forming sulfur has been steadily reduced in diesel fuel from 5000 parts per million (ppm) in 1993 to 15 ppm in 2006, further helping to improve atmospheric conditions.

The lubricating properties or "lubricity" of diesel fuel is directly related to high sulfur content. With the move toward reducing sulfur concentrations in diesel fuel, alternative lubricating agents such as biodiesel must be substituted.

Property	Grade **No. 1-D (low sulfur)**	Grade **No. 2-D (low sulfur)**
Flashpoint °C, min	38	52
Water and Sediment, % vol. Max	0.05	0.05
Distillation Temp., °C, 90% Min. Max	 - 288	 282 338
Kinematic Viscosity, mm^2/s at 40°C Min. Max.	 1.3 2.4	 1.9 4.1
Ramsbottom carbon residue, on 10%, %mass, max.	0.15	0.35
Ash, %mass, max.	0.01	0.01
Sulfur, % mass, max.	0.05	0.05
Copper Strip Corrosion, Max. 3 hours at 50°C	No. 3	No. 3
Cetane Number, min.	40	40
One of the following properties must be met: 1) cetane index 2) aromaticity, % vol., max	 40 35	 40 35
Cloud Point °C, max.	TBD	TBD

Table 3.1-3
Requirements for Selected Grades of Diesel Fuel Oils
ASTM D 975-97

According to Table 3.1-3, each of No.1-D and No. 2-D fuels must meet the noted specification in order to be sold as commercial diesel fuel and more importantly to ensure compliance with diesel engine manufacturers' warranty and quality specifications. It is well known that properly serviced light-duty diesel-powered automobiles and trucks will provide in excess of 300,000 miles (500,000 km) of service, and high-quality fuel is the first step in ensuring long engine life.

Figure 3.1-3. It is well known that properly serviced diesel-powered automobiles and trucks will provide in excess of 300,000 miles (500,000 km) of service, and high-quality fuel is the first step in ensuring long engine life.

Petrodiesel Specification Overview

As petro and biodiesel fuels are designed to be compatible for blending and use in diesel-powered vehicles, several of the specification parameters are either identical or similar in nature. The following is a brief discussion of petrodiesel specifications. A more detailed discussion of petro/biodiesel specification tests will be given later in the book in Chapter 8.3. Note that bolded keywords are listed in the **glossary.**

Flashpoint: Diesel fuel flashpoint measures the ability of heated diesel fuel to ignite and records the lowest temperature at which the fuel combusts. No.1-D has a flashpoint of 38 °C, while No.2-D is rated 52 °C. (For comparative purposes, gasoline has a flashpoint rating of -40°C). When the flashpoint of a given chemical is lower than the normal ambient temperature, the risk of ignition due to a spill or other accident is greatly increased. Biodiesel has a minimum flashpoint rating of 130°C, which necessitates considerable preheating of the fuel before ignition takes place, thus rating the fuel as "nonflammable."

Water and Sediment: Water that is solubilized (dissolved) in diesel fuel will contribute to oxidation (rusting) of metal components and can contribute to microbial growth while the fuel is in storage. Water is introduced into the fuel system by leakage though storage tank and piping systems as well as by atmospheric water vapour condensation on the inside of storage tanks. Retail fuel outlets will typically use either electronic sensors or "water reactive paste" applied to tank dipsticks to determine the degree of water ingress.

Sedimentation may contribute to fuel filter plugging, or in a worst-case scenario may pass through the filter and cause serious damage to the close-fitting components of the fuel injection system.

Distillation Temperature: A sample of the diesel fuel is subjected to progressive heating, and the initial boiling point (IBP) is recorded as well as the temperature at each 10% distillation of the sample, concluding with the final boiling point (FBP). The distillation of a petrodiesel sample provides data regarding the composition of the fuel. Test results are recorded at 90% point, which is the temperature at which 90% of all the fuel components boil. No.1-D has no minimum point, but the 90% point must be lower than 288°C. No.2-D has a lower limit for the 90% point of 282°C and a maximum limit of 338°C. Distillation range

Kinematic Viscosity: This is the measure of a fluid's resistance to flow measured over time. Molasses has a very high viscosity, which can be lowered by heating. In a similar manner, diesel fuel viscosity is inversely proportional to temperature: increasing temperature lowers viscosity measurement. Kinematic viscosity plays an important role in the ability of the fuel pressure pump and injection system to atomize the fuel mixture to ensure complete burning. The measure of kinematic viscosity is mm^2 per second (of fuel draining through a calibrated cup) when measured at 40°C. mm^3?

Ash: Ash is the residue left once the fuel has been completely combusted, like wood ash remaining in a fireplace or woodstove. The ASTM test requires that a sample of fuel be heated to the combustion point. The remaining material comprises carbon and other inorganic materials which are abrasive and will contribute to engine wear. The specification indicates the amount of maximum carbon and percentage of total ash produced during the test.

Sulfur: Sulfur is a major contributor to atmospheric smog and its concentration in all petroleum fuels has been steadily regulated downwards. By 2006, the sulfur content for on-highway fuels will reach the extremely low level of 15 parts per million. Sulfur is a lubricating agent, and alternative lubricity additives such as biodiesel must be substituted as a result of the near-elimination of sulphur in petrodiesel fuels.

Copper Strip Corrosion: Diesel fuel contains hundreds of compounds, many of which can be corrosive to the metal components of the fuel injection system. Copper is particularly susceptible to chemical corrosion, thus providing a rapid means of indicating the corrosive nature of diesel fuel. Polished copper strips are placed in a sample of heated fuel for 3 hours and are then washed and compared to thirteen reference samples which have been treated to varying degrees of corrosive degradation and graded from number 1a through 4c. The limit of acceptable corrosion is established by comparing the test sample to reference sample number 3a.

Cetane Number: Air is rapidly heated during the compression phase of the diesel engine cycle. Upon completion of this cycle, diesel fuel is finely sprayed or atomized into the combustion cylinder by the fuel injection system where it ignites on contact with the heated air. The measure of the fuel's ability to ignite quickly under these conditions is known as the cetane number.

Cloud Point: Cloud point is the temperature at which wax crystals first begin to appear as a sample of diesel fuel is cooled. There is no specific value defined for cloud point, with the test report indicating the limit for the given fuel sample. As a general rule, a refiner will blend diesel fuels that will exhibit a cloud point approximately 6°C colder than the lowest temperature that will occur 90% of the time (10th percentile).

Cloud point temperature will vary for each biodiesel fuel, based on feedstock and processing techniques as well as blending levels with petrodiesel. In temperate locations, cloud point rating is considered unimportant, while northerly, winter locations may be very concerned about the rating. For these reasons, the report merely indicates the measured cloud point temperature for reference, not as a pass/fail measurement.

3.2
The Diesel Engine

There are two types of engine predominantly used to power road-based vehicles. The spark ignition engine operates on gasoline or less frequently on liquid petroleum gas (LPG). The compression ignition engine uses diesel fuel and is named after Rudolf Diesel, who patented his heavy oil engine in 1892.

Diesel engines are very popular in Europe with demand expected to rise from a current level of approximately 30% of total car sales to an anticipated 40% of the market. Surprisingly, the greatest demand for diesel technology comes not from the penny-pinching small-car market but from the luxury car buyers, with 44% of all luxury cars sold in Europe powered by diesel engines.[1] Luxury diesel sales represent a very large percentage of specific markets: Belgium – 87%; France – 82%; Austria – 77%; and Italy – 70%.[2]

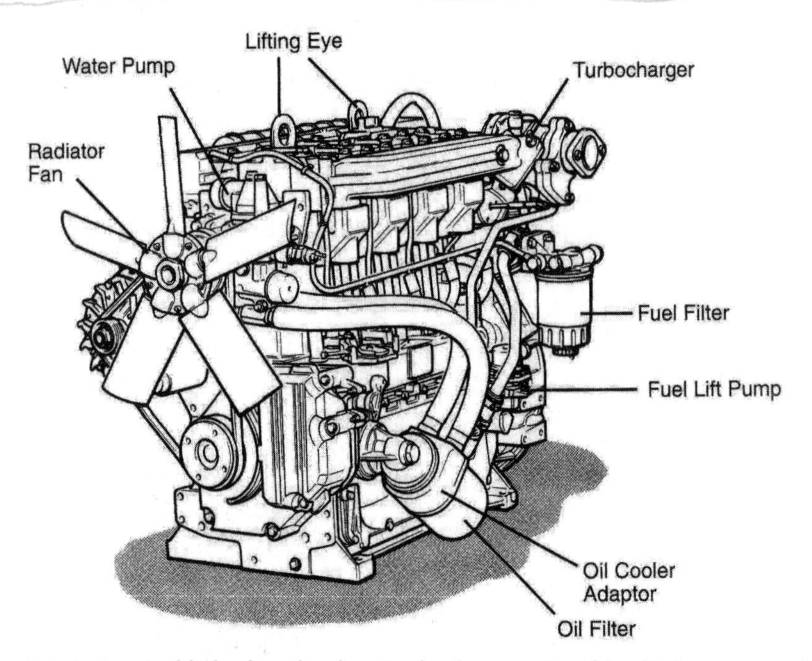

Figure 3.2-1. A typical light-duty diesel engine has between 4 and 8 cylinders, arranged with a series of intake and exhaust valves which control the air admission, compression, ignition, power, and exhaust phases. Because of their high volumetric efficiency, diesel engines use 30% to 60% less fuel than gasoline engines of similar power. (Courtesy Lister Petter Company)

Most North Americans have a complete disdain for diesel engines, thinking of them as slow, noisy, polluting, and generally uninspired. While this may have been the case with grandpa's old smoker, advances in technology have placed the diesel ahead of the gasoline-powered engine in several key areas:[3]

Fuel Economy: Because of their high volumetric efficiency, diesel engines use 30% to 60% less fuel than gasoline engines of similar power.

Power: Diesels produce more torque and power at lower engine speed than gasoline engines of similar displacement.

Durability: Diesel engines are designed to last well in excess of 300,000 miles (500,000 km) and require less maintenance than gasoline engines.

Greenhouse Gas Emissions: Diesel fuel contains more energy per gallon than gasoline, allowing a diesel engine to burn less fuel and produce significantly lower CO^2 emissions.

Noise and Smoke: Using the latest "Common Rail Direct Injection" (CDI) and lean-burn technology as well as particulate traps and catalytic converters, today's diesel vehicles have none of the smoke or noise common with older designs.

As a result, diesel-powered vehicles offer a cleaner, quieter, and more powerful alternative to identical automobiles and trucks equipped with less efficient gasoline-powered engines.[4]

The light-duty diesel market in North America is practically nonexistent, while the Europeans have created huge fuel and GHG savings by developing

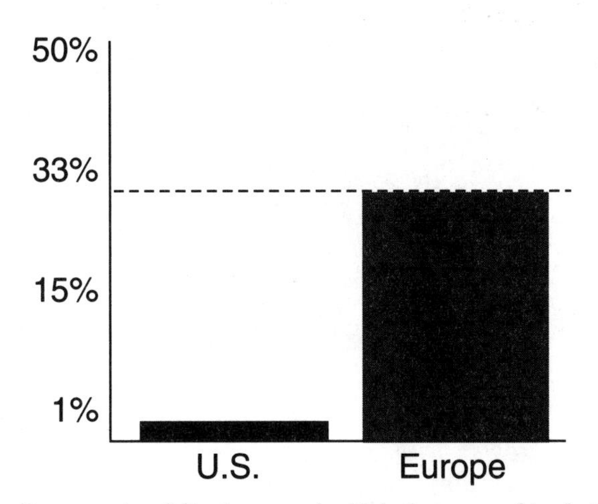

Figure 3.2-2. Current sales of diesel-powered vehicles have soared in the European Union owing to the numerous advantages diesels have over gasoline engines. In the luxury car market France has seen 82% of market share go to diesel engines.[5]

advanced engine technologies and emissions standards. Concurrently, the United States is fostering the Tier 2 emissions regulations that are a detriment to the development of diesel engine technology here, denying consumers access to the obvious benefits and national fuel consumption reductions that diesel technology would bring.

Nevertheless, progressive companies such as Mercedes-Benz will introduce to North America their line of BLUETEC diesel powertrains, which meet all emission standards. The introduction and delivery of these vehicles is being timed to occur in conjunction with the mandated release of ultra-low sulfur diesel fuels in the fall of 2006.

Diesel engines are perfectly suited to improving the fuel economy of the light trucks and SUVs that continue to fascinate consumers. Because the diesel engine is more powerful and fuel efficient than gasoline models, it could allow manufacturers to provide the vehicles people demand as well as the fuel economy so desperately needed to slow America's dependence on foreign oil.

In order to get the benefits of diesel technology in North America, regulators may have to develop a compromise and recognize that different engine technologies have different emission profiles. U.S. emissions standards

Figure 3.2-3. The prestigious Mercedes-Benz E320 CDI offers a 25% improvement in fuel efficiency compared to its gasoline-powered counterpart, while outperforming it by a full second in 0 to 60 mph acceleration, according to company data. Motor Trend's Frank Markus is even more direct: "Your eyes, ears, and nose will have trouble detecting that this is a diesel, while your backside will detect some serious pressure just at idle". In 2006 the company introduced its BLUETEC powertrain system, which meets all North American emission standards. (Courtesy Mercedes-Benz Canada)

accommodate gasoline engines which emit relatively little NOx but much more CO, HC, and CO_2 than their diesel counterparts. By placing the emphasis on smog-forming NOx and particulate matter (of greater concern in diesels) instead of CO2, the U.S. standards discourage the development and marketing of diesel engines.

A more balanced approach to matching technical developments with emissions standards would encourage the development and marketing of diesels in North America.[6]

Figure 3.2-4. With "lean-burn" technology as well as particulate traps and catalytic converters, there is no need for modern diesel-powered vehicles to have any of the smoke or noise common in older designs. The City of Ottawa, Canada "Ecobus" features these technological advances.

Environmental Benefits of Biodiesel

A study conducted by the U.S. EPA[7] found that tests conducted using a 20% blend of biodiesel with No. 2-D fuel in a highway truck, an urban bus, and a full-sized pickup truck showed the following emissions reductions compared with 100% No. 2-D fuel:

- total hydrocarbons (HC) up to 20%
- carbon monoxide (CO) up to 12%
- total particulate matter (PM) up to 12%

Biodiesel does result in increased nitrogen oxide (NOx) emissions, but this test showed an increase of less than 3%.

These reductions are impressive by any standard despite the fact that none of the test vehicles was equipped with the latest emission reduction

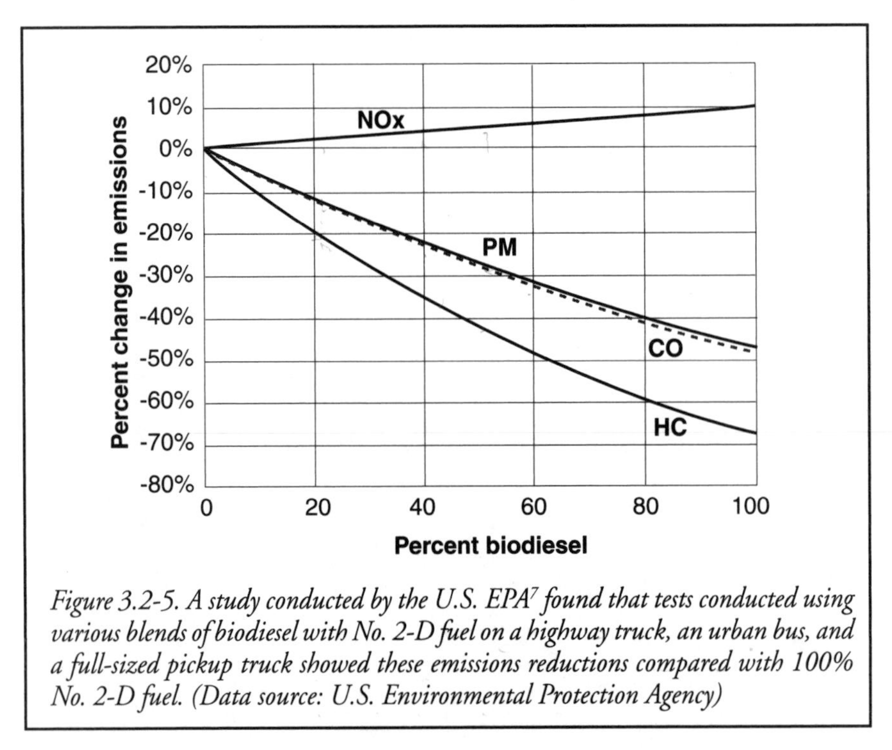

Figure 3.2-5. A study conducted by the U.S. EPA[7] found that tests conducted using various blends of biodiesel with No. 2-D fuel on a highway truck, an urban bus, and a full-sized pickup truck showed these emissions reductions compared with 100% No. 2-D fuel. (Data source: U.S. Environmental Protection Agency)

technology such as exhaust gas recirculation, NOx adsorbers, or particulate matter (PM) traps.

Engine Technology

All internal combustion engines (such as the model shown in Figure 3.2-1) include a cylinder block which houses the major components of the running gear. An eccentric crankshaft runs along the length of the engine providing support and a means of transferring the linear motion of the piston to the rotary motion required to turn the vehicle's drive wheels.

Each piston is fitted into the cylinder machined in the engine block. A series of piston rings installed on the outer diameter of the piston and placed in contact with the cylinder wall ensures a gas-tight seal.

An intake and exhaust valve (or more commonly multiple intake and exhaust valves per cylinder) are operated from the camshaft, which forms part of the running gear of the engine and is synchronized with the rotation of the crankshaft. As the crankshaft rotates, the camshaft sequentially opens and closes the intake and exhaust valves in accordance with the four-stroke operating theory described below.

In the spark ignition engine, a properly balanced mixture of fuel and air is admitted into the cylinder and compressed, creating a volatile, explosive mixture. An electric spark is activated at the correct timing sequence, igniting the mixture and providing a downward power force on the piston which is then transferred to the crankshaft and ultimately to the drive wheels.

In contrast, the diesel engine eliminates the spark plug and related components and instead uses the heat of compression to perform the ignition function. The four-stroke timing sequence diagrams shown below illustrate this process.

In the diesel engine, air alone is admitted into the cylinder during the first 180° rotation of the crankshaft, creating the intake cycle shown in Figure 3.2-6a. The intake valve (left side) is opened by the camshaft (not shown) as the crankshaft rotates in a clockwise direction, forcing the piston down. The downward motion creates a vacuum which draws air into the cylinder. At the lowest point of piston travel in the cylinder, the intake valve closes, creating an air-tight chamber.

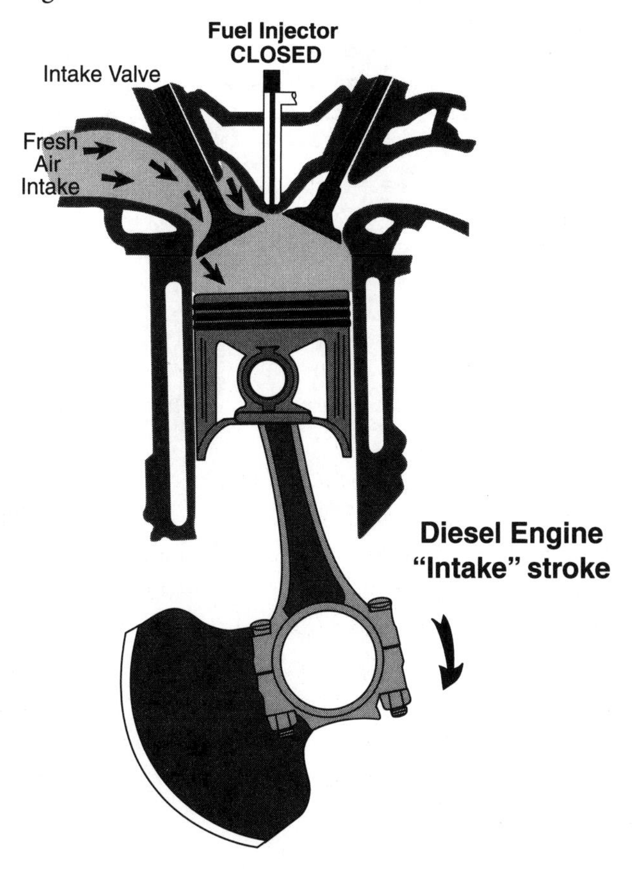

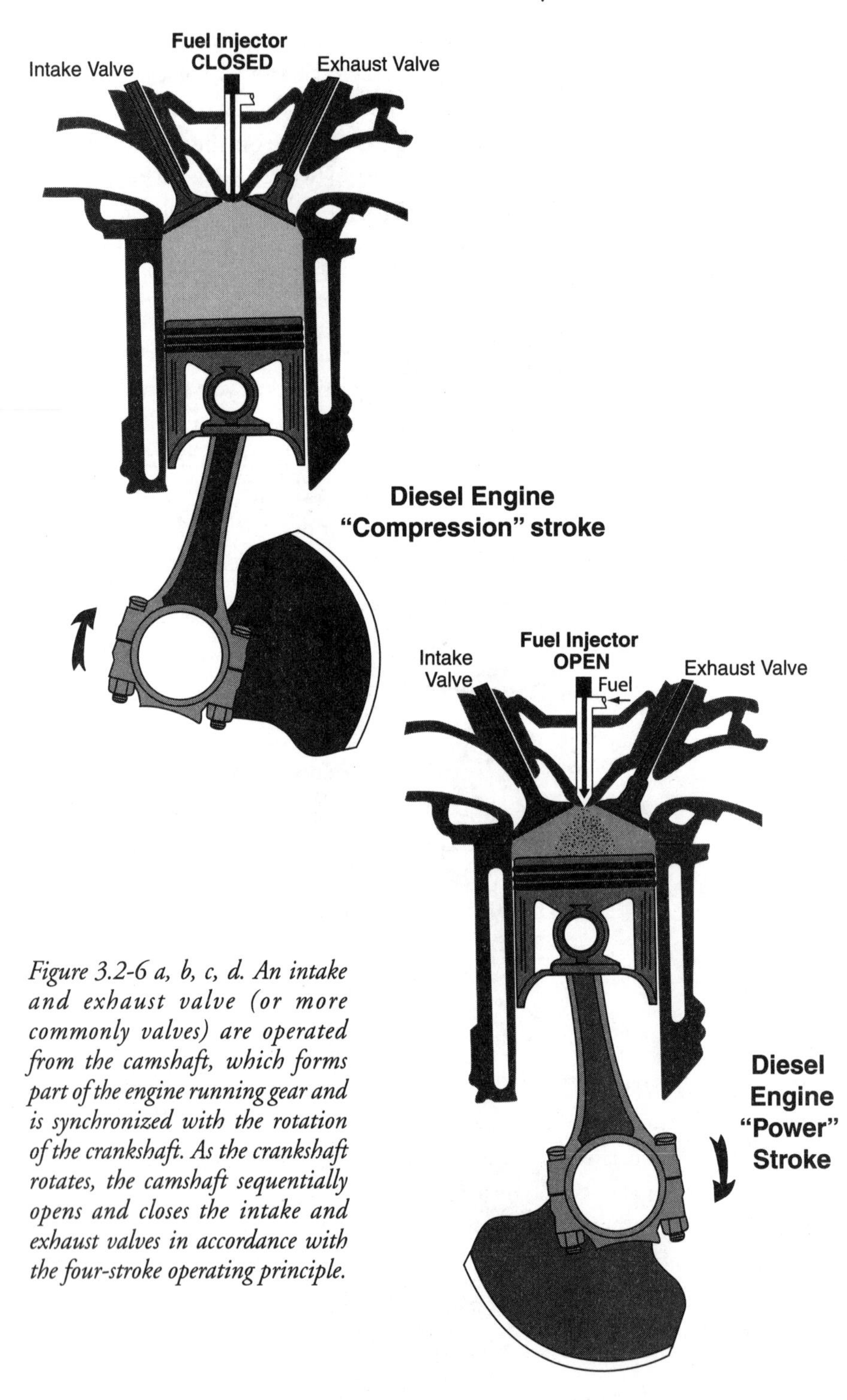

Figure 3.2-6 a, b, c, d. An intake and exhaust valve (or more commonly valves) are operated from the camshaft, which forms part of the engine running gear and is synchronized with the rotation of the crankshaft. As the crankshaft rotates, the camshaft sequentially opens and closes the intake and exhaust valves in accordance with the four-stroke operating principle.

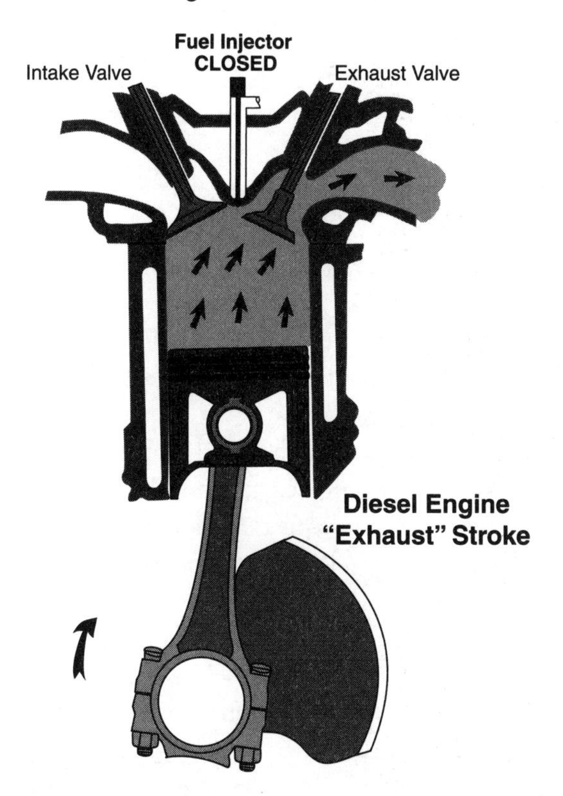

The next 180° rotation cycle is known as the compression stroke, detailed in Figure 3.2-6b. The piston is now travelling upwards, compressing the air inside the cylinder. In diesel engines, the compression ratio (ratio of cylinder volume at bottom compared to volume at top) may range as high as 21:1, causing pressures in excess of 500 pounds per square inch (3.4 MPa) and temperatures in excess of 1,000 °F (538 °C).

When the piston reaches the top of the cylinder at the conclusion of the intake cycle, the power stroke begins, as shown in Figure 3.2-6c. Diesel fuel is supplied to the fuel injector under very high pressure. Using mechanical or more commonly computer/electronic control, the injector is opened and fuel is sprayed in a fine mist (atomized) into the cylinder. A fraction of a second later, the fuel will ignite, causing massive expansion of the burning gases and forcing the piston downward, applying power to the piston rod, crankshaft, and vehicle drive train.

The delay time between the opening of the fuel injector (signifying the initial fuel spray into the cylinder) and ignition of the fuel is known as ignition delay and is determined by the cetane rating of the diesel fuel. The ability of a fuel to combust in the presence of heat is known as the auto-ignition ability of fuel and is a key property of diesel fuel.

The final 180° cycle is the exhaust stroke, whereupon the camshaft opens the exhaust valve as seen in Figure 3.2-6d. As the piston sweeps upward, combusted fuel (exhaust) components are forced out of the cylinder and sent on their way to the exhaust system, noise reduction muffler, and tailpipe. Advanced engines may also contain particulate matter traps to capture the dusty particles caused by the incomplete combustion of diesel fuel, NOx adsorbers to capture and treat nitrogen oxides in the exhaust stream, and catalytic converters which utilize the high temperatures to catalyze or neutralize exhaust gas components.

It requires two full rotations of the crankshaft to obtain one half-cycle (180°) power stroke from the engine. In order to improve the smoothness and power of the engine, multiple pistons are fitted into the cylinder block and connected to the crankshaft. Each piston is "timed" to produce its power stroke at slightly different times, allowing almost continuous overlap in the power-generation cycle of the engine, in turn reducing mechanical vibration and noise.

Automobiles such as the Mercedes-Benz Smart car have only 3 cylinders and a total engine displacement (net cylinder volume x number of cylinders) of only 52 cubic inches (850 cc). By contrast, a large diesel-powered pickup truck may have an engine with 8 cylinders and a displacement of 458 cubic inches (7,500 cc). Of course industrial engines can have much larger displacements than this.

Fuel Injection Systems

At the risk of oversimplifying diesel engine fuel injection technology, I am going to make the generalization that there are two important classes of diesel engine that are relevant to our discussion; "basic" and "common rail" fuel injection. A common thread that will be repeated over and over regardless of which type of injection system is used is the necessity of using high-quality fuels in a diesel engine to ensure long life.

Older mechanical or electronic fuel injection systems tolerate substandard fuel better than latest technology "Common Rail Direct Injection" (CDI) systems do. The issue of fuel quality and injection system design will begin to reveal a nasty problem as we start to review biodiesel fuel quality later in the book.

Basic Fuel Injection

Figure 3.2-7 shows an overview of a typical fuel injection system used in many current and all older diesel vehicles. Fuel is drawn from the vehicle's storage tank and filtered to remove any debris and water that may be present in the tank. The filter is manufactured from tightly woven cellulose material which is able to stop particles larger than 20 microns (0.0008 inches) in size, preventing abrasive material from damaging the fuel system components.

Filtered fuel enters a high-pressure injection pump which pressurizes the fuel, feeding it into the fuel injector located in the top end of the cylinder.

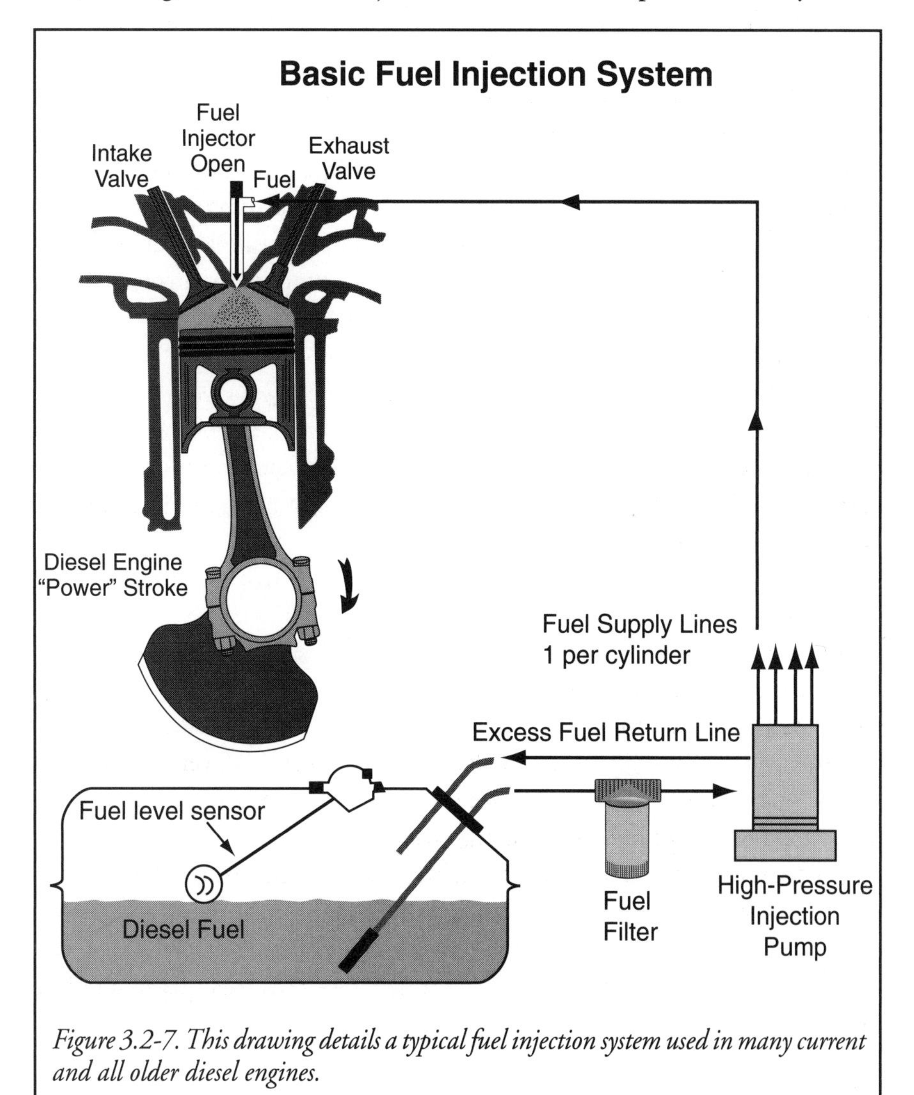

Figure 3.2-7. This drawing details a typical fuel injection system used in many current and all older diesel engines.

The injection pump is provided with multiple fuel supply lines, one for each cylinder and fuel injector. As diesel fuel is incompressible, an excess fuel return line constantly recirculates fuel when an internal pressure relief valve reaches fuel operating conditions. Pressurizing any liquid will cause its temperature to rise, thereby increasing the temperature of fuel in the tank and distribution system.

When the piston reaches the top of the cylinder during the compression stroke, air in the cylinder will be heated above the ignition temperature of the fuel. Depending on the model and age of the engine, fuel will be atomized by either mechanical or electrical control of the injector nozzle. The "fineness" of the resulting fuel spray will determine combustion completeness, engine noise, and efficiency. A disadvantage of this system is that fuel pump pressure varies according to engine speed, resulting in varying fuel spray patterns.

Researchers have found that this problem can be minimized with the use of increased and constant fuel pressure and by "timing" the spray pattern by modulating the fuel injector to create "microbursts" of fuel in the cylinder as the piston is sweeping through the end of the compression stroke and the beginning of the power stroke. In order to accomplish this, much higher pressures and faster fuel injectors are required.

Common Rail Direct Injection

Common Rail Direct Injection (CDI) systems (Figure 3.2-8) use an ultrahigh pressure injection pump and a "common rail" or pressure manifold to ensure high, even fuel pressure. CDI systems pressurize the diesel fuel to enormous pressures, often in excess of 22,000 pounds per square inch (psi) (≈150 Mpa). The common rail manifold is able to act as a pressure storage reservoir and ensure that fuel is instantly available when required by the fuel injectors. Note that higher fuel pressures generate higher excess fuel temperatures as compared with basic fuel injection systems.

Special piezoelectric fuel injectors can open and close thousands of times per second, allowing the fuel control computer to provide multiple "bursts" of fuel, offering precise control of the combustion process. Pilot injection of minute amounts of diesel fuel prior to the main combustion injection initiation virtually eliminates diesel engine "clacking" noises.

A 4-valve-per-cylinder, 2-intake and 2-exhaust arrangement increases power, fuel efficiency, and responsiveness by allowing the fuel injector to be placed in a central location, creating a symmetrical fuel spray pattern and best fuel/air mixing.

Engine and Vehicle Efficiency

Although the internal combustion engine and fossil fuel-powered vehicle have been around for a long, long time, they are not the most energy efficient members of the mechanical engineering team. Gasoline and diesel fuel contain a given amount of energy per unit of volume. No. 2-D, for example, contains approximately 129,000 BTU per gallon. Due to inefficiency of the ignition process of the fuel through the drive motion of the vehicle, most of the energy contained in a quantity of fuel is wasted:

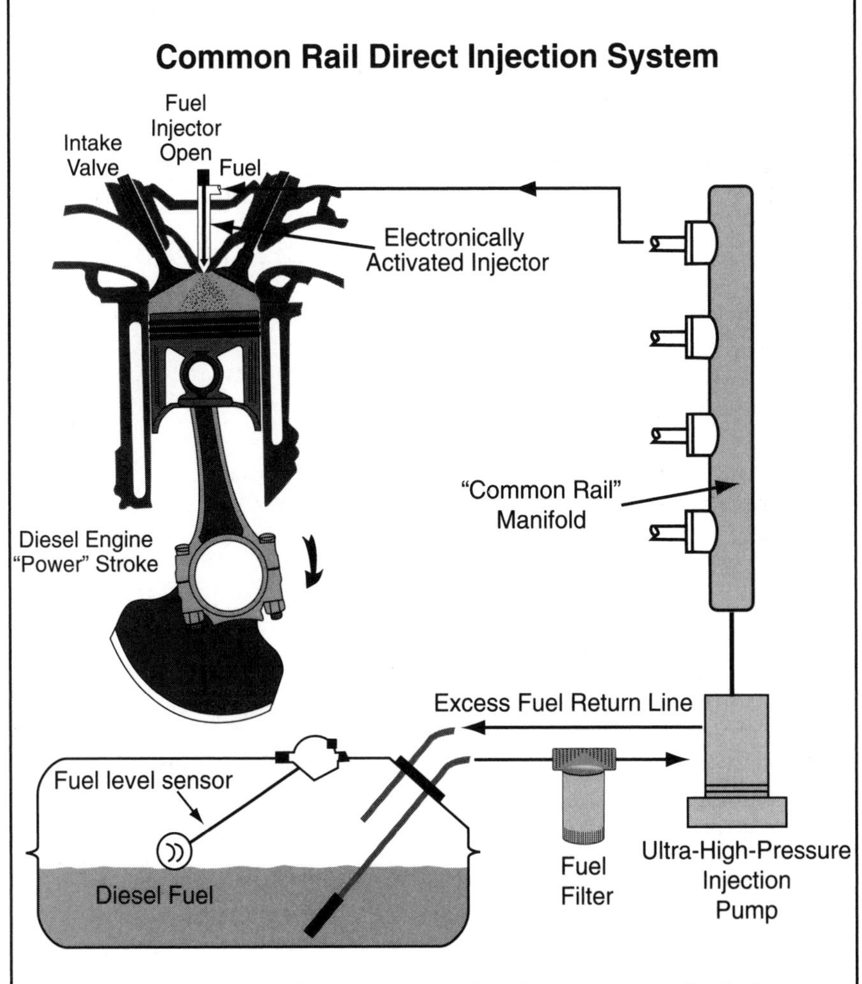

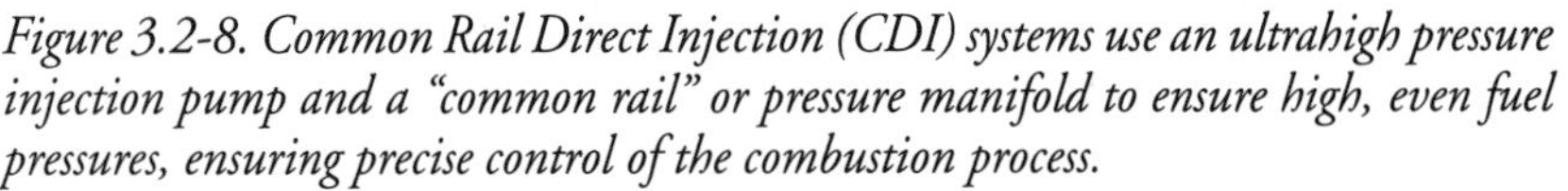
Figure 3.2-8. Common Rail Direct Injection (CDI) systems use an ultrahigh pressure injection pump and a "common rail" or pressure manifold to ensure high, even fuel pressures, ensuring precise control of the combustion process.

- 35% is lost to coolant and oil heat
- 35% is lost to exhaust gases
- 10% is lost to engine friction

Only 20% or approximately 26,000 BTU of energy remains to produce power to move the vehicle. An additional 15% is used in the rolling resistance of the tires, transmission losses, and overcoming air resistance. Movement and acceleration of the vehicle may be attributed to as little as 5% of the total energy of the fuel.

While the internal combustion engine does offer poor fuel conversion efficiency, there are numerous incremental as well as major technological improvements that can be used to improve these results provided North Americans are willing to embrace them. Engine design improvements, a move to clean-diesel technology, smaller cars, and hybrid technology are all examples of techniques that can help reduce fuel consumption and improve our environmental record at the same time.

3.3
An Introduction to Biodiesel

The concept of using plant matter to operate internal combustion engines is older than the gasoline and diesel fuels that are so ubiquitous in our lives today. Rudolf Diesel developed the compression ignition engine and demonstrated it at the Paris World's Exhibition in 1900. His fuel of choice for powering the new engine: peanut oil.

Although the concept of using plant matter to operate internal combustion engines has been revisited numerous times since Diesel's early experiments, the discovery of cheap fossil oils delayed any significant development of biofuels.

The development of the diesel engine and fuel system progressed very quickly after its first demonstrations to the public owing to its increased efficiency compared with that of the steam engine, its relative portability (paving the way for automotive, farming, and industrial uses), and access to cheap and convenient diesel fuel oil. Engine development continued for the next 80 years using low viscosity petrodiesel fuel, while the much higher

Figure 3.3-1. The discovery of cheap fossil oils delayed any significant development of biofuels. Given the massive financial subsidies and military support lavished on the fossil fuel industry as well as the environmental damage it has caused, perhaps biofuels should have been with us from the beginning.

viscosity plant oils were left behind on the grocer's shelves for baking, salad dressing, and French fries.

With the first worldwide oil "shortages" in the 1970s, researchers began working in earnest in an attempt to develop the biofuel market. The many shortcomings related to the direct use of plant oils and their **total** incompatibility with petrodiesel fuel[1] pushed researchers in the direction of chemically modified forms of plant oils and animal fats known as biodiesel.

Figure 3.3-2. Biodiesel is a renewable, relatively clean-burning, carbon-neutral fuel that can be obtained from a variety of oilseed plants, waste oils, and rendered animal fats.

Biodiesel is a renewable, relatively clean-burning, carbon-neutral fuel that can be obtained from a variety of oilseed plants, waste oils, and rendered animal fats. These unprocessed materials (collectively referred to as feedstock "oils") can be converted into a petrodiesel-compatible fuel using a process known as chemical transesterification.

The properties of rendered animal fats and plant oil vary widely from those of petroleum diesel fuel, primarily in the areas of viscosity, atomization, and the coking of engine components. All plant and animal oils have essentially the same chemical structure, consisting of triglycerides, which are chemical compounds formed from one molecule of glycerol and three fatty acids. Your next Greek salad will contain an oil and vinegar dressing, which chemists refer to as a "triglyceride and acetic acid surfactant," regardless of whether you prefer extra virgin olive oil or plain old Mazola™. Little wonder chemists are generally not invited to many dinner parties.

Glycerol (common name glycerin) is an alcohol that can combine with up to three fatty acids to form mono-, di-, and triglycerides.

Fatty acids are chains of hydrocarbons that vary in carbon length depending on the oil feedstock. If each carbon atom has 2 associated hydrogen atoms, the fatty acid is known to be saturated. If 2 carbon atoms are double bonded, having less hydrogen, the fatty acid is unsaturated. Likewise if more than 2 carbon atoms are unsaturated, the fatty acid is said to be polyunsaturated.[2]

Triglycerides are the main compounds or components of animal fat and vegetable oils. They have a lower density than water and will therefore float on it. If the oil is solid at room temperature the triglycerides are known as "fats;" if they are liquid they are called "oils." As a general rule, triglycerides that are liquid at room temperature are unsaturated, which is a desirable property for engine fuels.

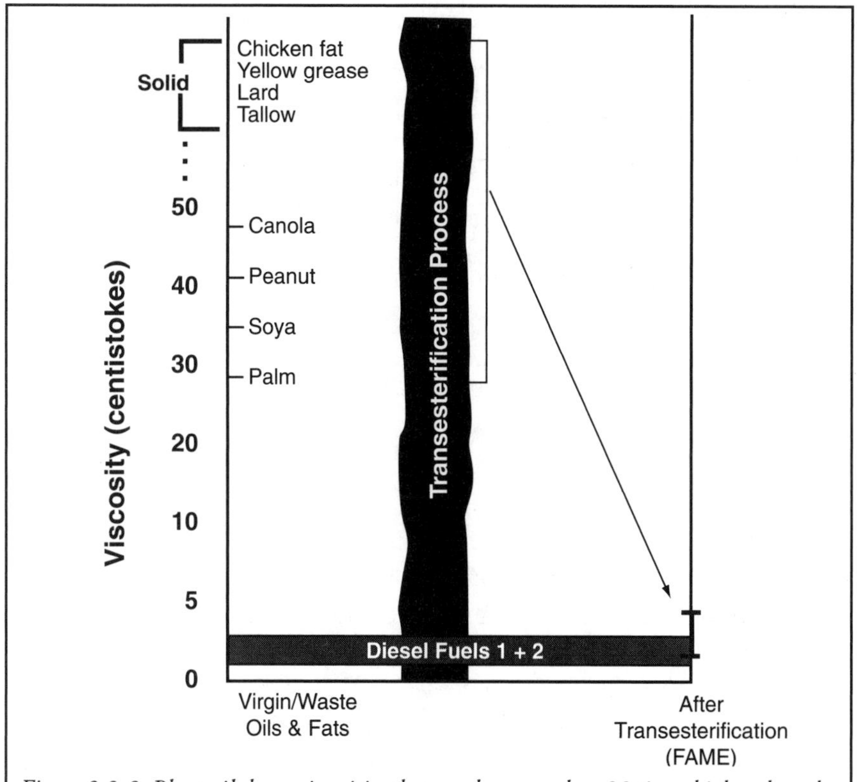

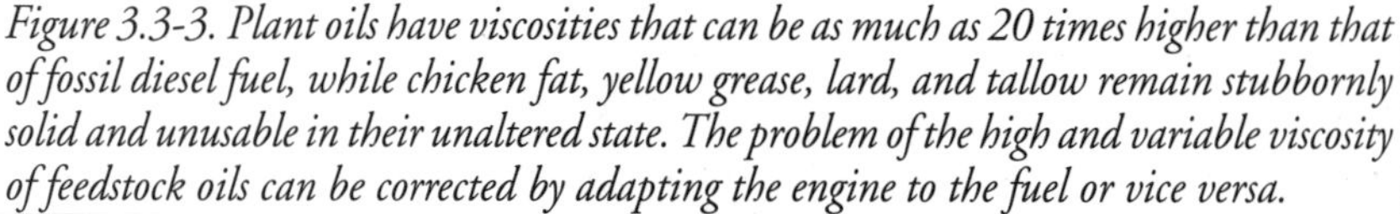
Figure 3.3-3. Plant oils have viscosities that can be as much as 20 times higher than that of fossil diesel fuel, while chicken fat, yellow grease, lard, and tallow remain stubbornly solid and unusable in their unaltered state. The problem of the high and variable viscosity of feedstock oils can be corrected by adapting the engine to the fuel or vice versa.

Plant oils have viscosities that can be as much as 20 times higher than that of fossil diesel fuel, while chicken fat, yellow grease, lard, and tallow remain stubbornly solid and unusable in their unaltered state. The problem of the high and variable viscosity of feedstock oils can be corrected by adapting the engine to the fuel or vice versa. Chapter 7.2 will debate the relative merits of the former process, while the main theme of this book focuses on the latter concept: adapting the fuel to the millions of engines that are now operating or will be produced for many years to come.

Chemical transesterification of feedstock oils is a well-known process which solves the problem of feedstock viscosity as demonstrated in Figure 3.3-3. The process was first described in 1852[3] when it was originally used as a means of producing high-quality soaps, and with a bit of retooling it was found to work wonders in the production of biodiesel. Simply stated, biodiesel is produced by the reaction of feedstock oils with an alcohol in the presence of a catalyst to produce fatty acid methyl esters (FAME) or biodiesel. The typical process is:

100 kg feedstock oil + 10 kg methanol ➜ 100 kg FAME + 10 kg glycerol

The resulting FAME is known to be chemically contaminated with numerous compounds resulting from the esterification process, requiring further downstream processing to ensure a fuel quality compatible with ASTM Standard D6751 for biodiesel, a fuel comprised of "the mono-alkyl

Figure 3.3-4. Professor Martin Mittelbach, Ph.D., University of Graz, Austria, and his team of researchers have long been acknowledged as the founders of the modern biodiesel industry. Dr. Mittelbach is the author of numerous technical papers on the subject and the author of Biodiesel: The Comprehensive Handbook.

esters of fatty acids derived from vegetable oils or animal fats." Hereinafter, FAME that is directly taken from the reaction process will be referred to as "raw FAME" and when it meets the fuel quality standard it will be referred to as FAME.

The first reference to FAME production was in 1937[3] and within the next year a bus fuelled with palm oil-based biodiesel ran between Brussels and Louvain.[4] However, at that point further scientific research and production ground to a halt.

Some forty years later, Professor Martin Mittelbach, Ph.D., University of Graz, Austria, and his team of researchers were producing rapeseed oil-based biodiesel and testing its feasibility as a diesel fuel substitute.

When prodded to discuss the origins of the modern biodiesel industry, Dr. Mittelbach modestly admits that he has been "involved since the beginning. We (University of Graz) were the first to produce biodiesel in Europe, more than 20 years ago, although I am not exactly sure what moved me into this program!...Because of my research on carbon-based compounds, a discussion ensued with an agricultural group in Austria that had some experience using straight vegetable oils mixed with 50% fossil fuels in tractors. The farmers found that after a period of time running on this mixture, total engine breakdowns would occur and they had to stop this practice. We looked into the problems, examined the prior research, and I guess you could say the rest is history."[5]

Within the next decade hundreds of research programs sprang up around the world as interest in clean and renewable fuels started to take hold.

In the United States, the demand for soybean meal (the residual husk of the bean after crushing) was greater than the demand for oil, causing an imbalance in supply and depressed soy oil prices and galvanizing the United Soybean Board into action. It voted to promote biodiesel production using soy oil, which ultimately led to the current National Biodiesel Board (NBB), headed by its executive director Joe Jobe.

The Canadian Renewable Fuels Association (CRFA), which merged with the Biodiesel Association of Canada, develops market strategy and educational data for both the ethanol and the biodiesel industries.

As a result of lobbying efforts and continued research, biodiesel has received the support of numerous federal, state, provincial, and local governments who see it as a means of reducing greenhouse gas and smog-forming emissions, supporting local agribusiness, and helping to reduce North American dependency on foreign oil.

Technical Issues

The Pros of Biodiesel

Blending

One primary advantage of biodiesel is its ability to fit almost seamlessly into the existing fuel distribution and retail sales system while other alternative "fuels" such as hydrogen[6] require the complete rebuilding of distribution technology at a cost of trillions of dollars.

Figure 3.3-5. A primary advantage of biodiesel is its ability to fit almost seamlessly into the existing fuel distribution and retail sales system, while other alternative "fuels" such as hydrogen require the complete rebuilding of distribution technology at a cost of trillions of dollars.

Biodiesel can be used in all modern diesel engines and oil-fired heating systems with minor (if any) modifications. Two notes of caution:

1. FAME may cause long-term degradation of natural rubber hoses and gaskets and some paints, and replacing natural rubber hoses, "O" rings, and gaskets with polymeric (synthetic) versions such as Viton® may be necessary. However, experience has shown that blends of 20% biodiesel or less seldom cause any problems at all, and in any event late-model vehicles seldom use natural rubber components.[7]

2. Fossil diesel fuels may develop microbial growth that forms deposits on the walls of fuel storage tanks, pipes, and other components. The solvent abilities of FAME can cause these deposits to loosen, which can lead to fuel filter plugging and a resulting loss of engine power. Therefore, it is gener-

ally suggested that fuel filters in vehicles that have been operating on fossil diesel and switch to biodiesel be replaced a few months after the change takes place.

Biodiesel fuels are in commercial use in many European countries including Austria, the Czech Republic, Germany, France, Italy, and Spain. Biodiesel can be used in either its pure or neat form or as blends mixed with fossil diesel fuels. Neat biodiesel is designated B100, while blends are marked "BXX" where "XX" represents the percentage of biodiesel in the fuel mixture. Further, because biodiesel is stable at any concentration, users are free to choose the blending level they prefer based on availability, desired operating temperature, or price[8].

Biodiesel Concentration

Germany, Austria, and Sweden market neat biodiesel, although blends of 5%-20% are the preferred concentration. The "European Directive for the Promotion of the Use of Biofuels," published in 2003, mandates that all member states ensure minimum market shares of biofuels (ethanol, ethanol derivatives, and biodiesel). Market share of biofuels is to reach 5.75% by 2010.[9]

For several reasons, such as the maximum available production of biodiesel, fuel quality and stability, and the political realities of displacing fossil diesel from the market, North American industry proponents consider a blend level of B20 to be the upper limit concentration.

The question of blend concentration and related issues of politics and fuel quality are very complex; accordingly Chapter 6 in its entirety has been reserved to debate the matter further.

Biodegradability and Nontoxicity

Biodiesel is readily biodegradable and nontoxic, making it the ideal fuel choice when used in environmentally sensitive areas such as parklands or marine habitats. It is known to be less toxic than table salt and is as biodegradable as sugar.

High Cetane Value

The cetane value is a rating of the relative ignition quality of diesel and biodiesel fuels, with higher ratings offering improved ignition performance. As the cetane value increases, fuel ignition will be smoother and more complete, improving combustion and reducing emissions from unburned fuel. Virtually all biodiesel fuels have cetane values several percentage points higher than that of petroleum diesel fuel.

Figure 3.3-6. Biodiesel is readily biodegradable and nontoxic, making it the ideal fuel choice when used in environmentally sensitive areas such as parklands, marine habitats, or ski resorts. It is known to be less toxic than table salt and is as biodegradable as sugar.

High Lubricity

Biodiesel has excellent lubricating properties, far in excess of those of petrodiesel, which help to reduce fuel system and engine wear. As petroleum diesel fuel sulfur levels continue to be legislated downwards, its lubricity will decline to the point where additives will be necessary. The addition of 1% biodiesel to low-sulfur petrodiesel will improve the fuel blend lubricity to within specification.

Low Emissions

As a renewable fuel source, biodiesel operates on a closed-carbon cycle, which reduces CO_2 production by 2.2 kg for every liter of fossil fuel displaced.[10] This is because of the regenerative (biological) nature of all energy sources that absorb CO_2 from the atmosphere during their growing phase, only to release the same compound during fuel combustion. Additionally, the FAME molecule contains 11% oxygen, which leads to improved combustion and significant reductions in Particulate Matter (PM) or soot.

As part of the Clean Air Act Amendments enacted by the U.S. Congress, the Environmental Protection Agency (EPA) was directed to ensure that any new commercially available motor vehicle fuel or fuel additive would not present an increased health risk to the public. Under this directive, EPA established a registration program and testing protocols which are outlined in CFR Title 40 Part 79 as part of Tier I and Tier II emissions testing.

The EPA completed a major study of the impact of various concentrations of soybean-based biodiesel in the operation of heavy-duty highway-based

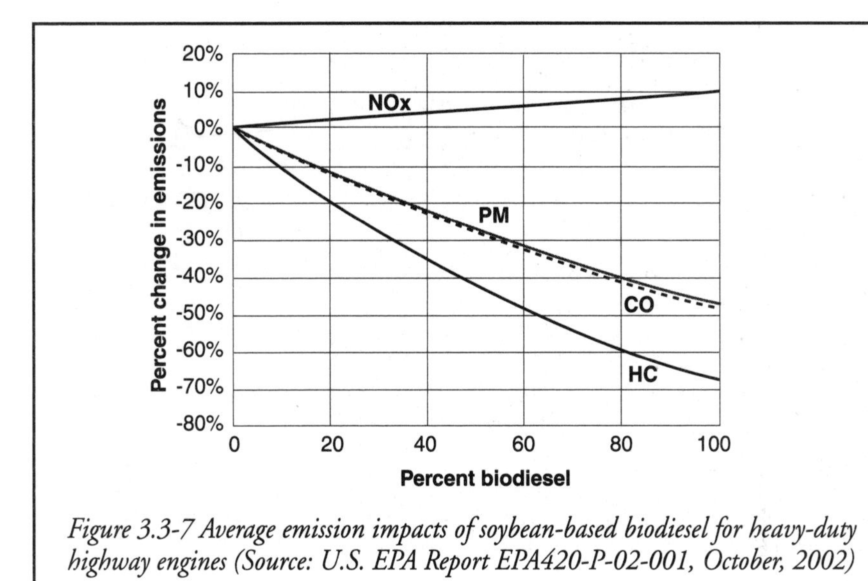

Figure 3.3-7 Average emission impacts of soybean-based biodiesel for heavy-duty highway engines (Source: U.S. EPA Report EPA420-P-02-001, October, 2002)

vehicles. The results of the study are shown in graphical form in Figure 3.3-7 and clearly demonstrate the superior emissions reductions of FAME fuels.

Nitrogen oxides (NOx) do increase as a result of high engine combustion temperatures and are discussed later in this chapter.

Renewability

Notwithstanding the findings of David Pimentel, insect specialist at Cornell University, and his associates, biodiesel has a very high net energy balance, in excess of 300%. Pimentel stirred a considerable amount of public controversy with the release of his 11-page report published in Natural Resources Research (Vol. 14:1, pp. 65-76) which concludes that soybean-based biodiesel has a negative energy balance, with an energy input 27% higher than its energy output. International media love controversy and brief, simplistic, contrary news items, thus providing Pimentel with a Warholian window of opportunity to cast doubt on the entire biofuel sector.

His comments are completely at odds with the numerous studies that have found the opposite to be true. Dr. Robert McCormick of the U.S. Department of Energy states that "the Pimentel/Patzek study uses outdated information on agricultural practices as well as unrealistic and unsubstantiated assumptions regarding energy inputs. At least eight other peer-reviewed studies have been conducted over the past 12 years and find exactly the opposite, that biodiesel has a highly positive energy balance."

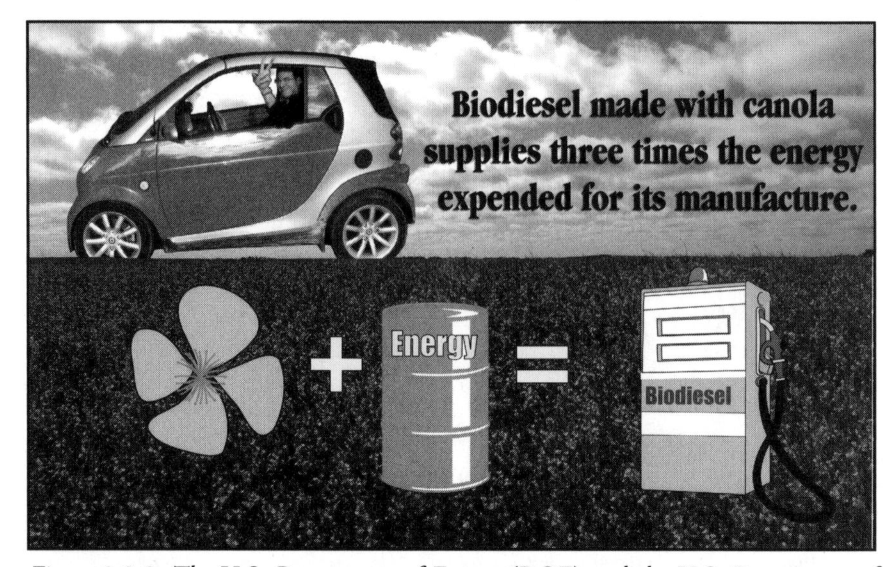

Figure 3.3-8. The U.S. Department of Energy (DOE) and the U.S. Department of Agriculture (USDA) in 1998 completed a thorough study of the energy balance of biodiesel and found that for every unit of fossil energy used in the entire biodiesel production cycle, 3.2 units of energy were delivered when the fuel was consumed.

The U.S. Department of Energy (DOE) and the U.S. Department of Agriculture (USDA) in 1998 completed a thorough study of the energy balance of biodiesel and found that for every unit of fossil energy used in the entire biodiesel production cycle, 3.2 units of energy were delivered when the fuel was consumed.[11]

Given that Pimentel's co-author, Ted Patzek, is a former oil company employee and is now a director of the University of California Oil Consortium, is it possible that the findings in the report could be skewed?

Readers who are interested in learning more about the life-cycle energy requirements for the production of soybean-based biodiesel are encouraged to read the entire 286 pages of analysis completed by the U.S. Departments of Energy and Agriculture (versus 1.5 pages of analysis completed by Pimentel) at www.biodiesel.org.

Low Sulfur

In order for petroleum diesel fuel to be given a rating of "low" or "ultra-low" sulfur content, it is necessary to subject the fuel to an energy-intensive refining process that generates additional carbon dioxide emissions. The resulting fuel will have reduced lubricity levels that must be supplemented with lubricating additives.

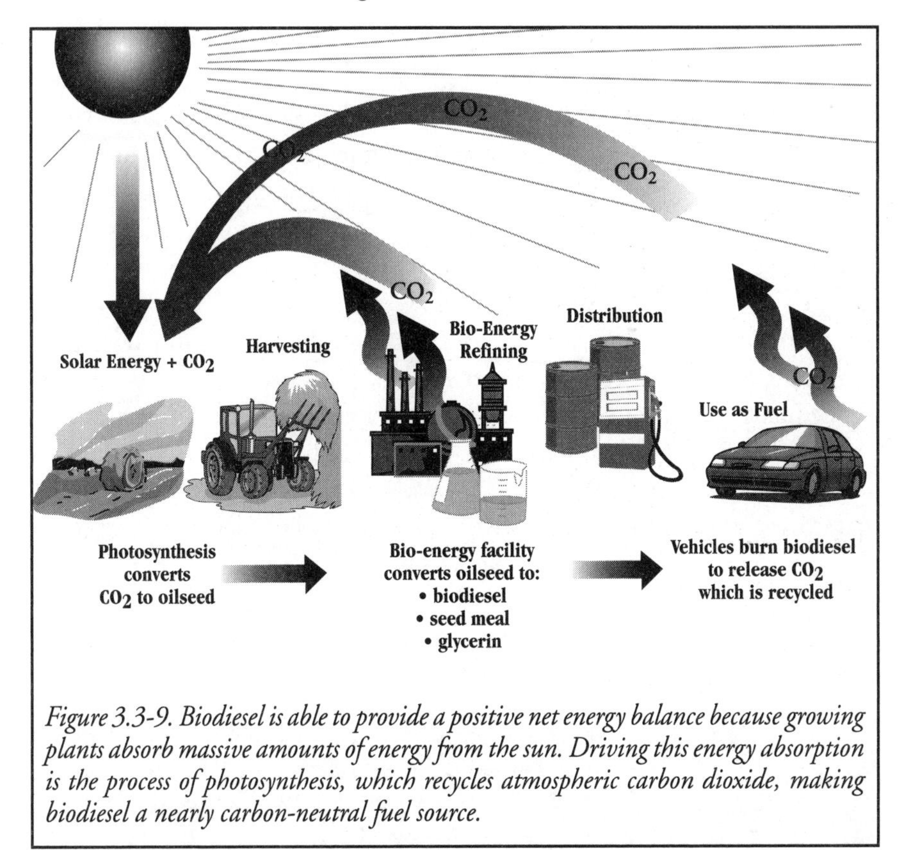

Figure 3.3-9. Biodiesel is able to provide a positive net energy balance because growing plants absorb massive amounts of energy from the sun. Driving this energy absorption is the process of photosynthesis, which recycles atmospheric carbon dioxide, making biodiesel a nearly carbon-neutral fuel source.

Biodiesel by contrast retains its excellent lubricity while being intrinsically free of sulfur. Having a virtually zero-sulfur level allows the optimum use of oxidation catalytic converters in the exhaust system.

The Cons of Biodiesel

So far, biodiesel sounds like the perfect fuel. Unfortunately, there are a number of issues which must be considered on the negative side of the balance sheet.

Oxidation and Bacteriological Stability

Biodiesel is biodegradable, which is an excellent environmental benefit, but it creates long-term storage and fuel stability issues. When it is exposed to high temperatures, oxygen, or sunlight or placed in contact with non-ferrous metals, FAME will deteriorate, resulting in polymerisation (fuel thickening) which leads to plugged filters and "glazing" within the fuel injection system. To prevent storage degradation, anti-oxidant additives may be added to extend storage life. Of course, the simplest corrective action may be to simply limit the storage time of FAME fuels through rapid consumption and rotation.

Water content can destabilize FAME as well as create an active growing medium for microorganisms. Biodiesel manufacturers produce fuels that have very low water content, but biodiesel is hygroscopic and actually attracts water much more readily than petrodiesel does. (Although according to one German report, the hydroscopic nature of biodiesel may prevent the growth of bacteria, by denying a water saturated growing media[12]).

Biodiesel will become saturated at water levels above approximately 1,000 ppm, and if water ingress continues unchecked, water no longer remains bonded and collects at the bottom of the storage tank, leading to the very condition that promotes microbial growth. To test this theory, I placed two samples of FAME and rain water (simulating a rain-induced leakage into a fuel storage unit) in small beakers which were placed in a darkened area for a period of 90 days. One of the samples was treated with Kathon® FP1.5 diesel fuel biocide. At the end of this period, a black, slimy growth had formed at the water/oil interface of the untreated sample, indicating the presence of bacteria. The second sample was bacteria free.

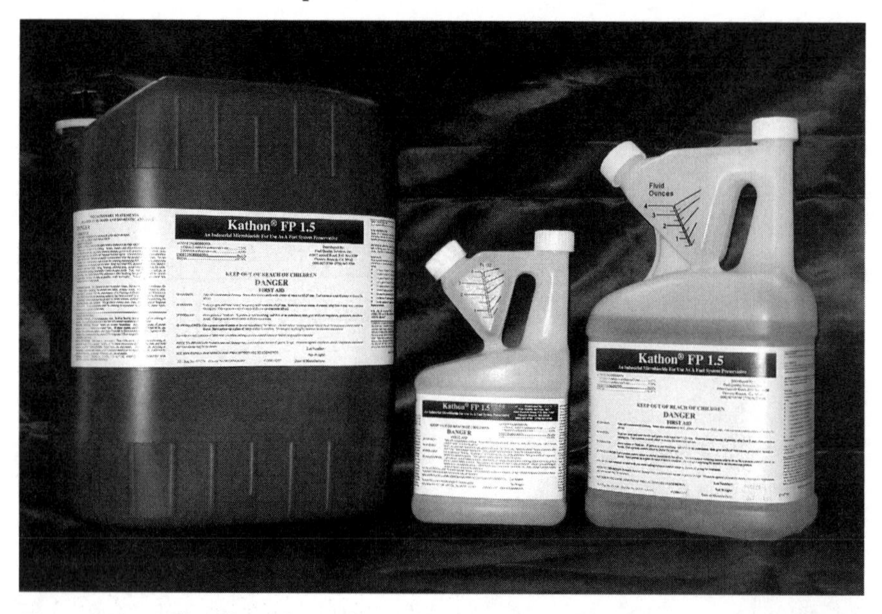

Figure 3.3-10. The effects of microbial contamination include product spoilage, corrosion on fuel-wetted surfaces, filter plugging, and engine failure. Microbes are not selective about the type of fuel product they consume and will contaminate all types of petroleum-based products, fuels, feedstock, and especially biodiesel. Kathon® is a registered trademark of Rohm & Haas Company. (Image courtesy Fuel Quality Services, Inc.)

Howard Chesneau is President of Fuel Quality Services Inc., the North American distributor of Kathon® biocide and LTSA-35A fuel stabilizer as well as a complete series of microbial test kits. According to Mr. Chesneau: "Conditions with excessive wetting of the fuel can occur where storage tank plumbing or access seals are damaged or if partially filled steel tanks are subjected to repeated thermal excursions, causing condensation of atmospheric moisture. The effects of microbial contamination include product spoilage, corrosion on fuel-wetted surfaces, filter plugging, and engine failure. Microbes are not selective about the type of fuel product they consume and will contaminate all types of petroleum-based products, fuels, feedstock, and especially biodiesel. Microbes require only droplets of condensation or a single millimeter film of water to initiate the fuel system degradation process."

Mr. Chesneau continues by explaining that biodiesel manufacturers are taking excessive risks by pushing biodiesel into markets that are not accustomed to the issues of this fuel type. "In the industry's zest to get product into the marketplace, fuel quality standards are lacking in the United States and Canada. There is no requirement for storage, oxidation, or thermal stability as is required by the European Union. This will, and already has, led to trouble in the market."

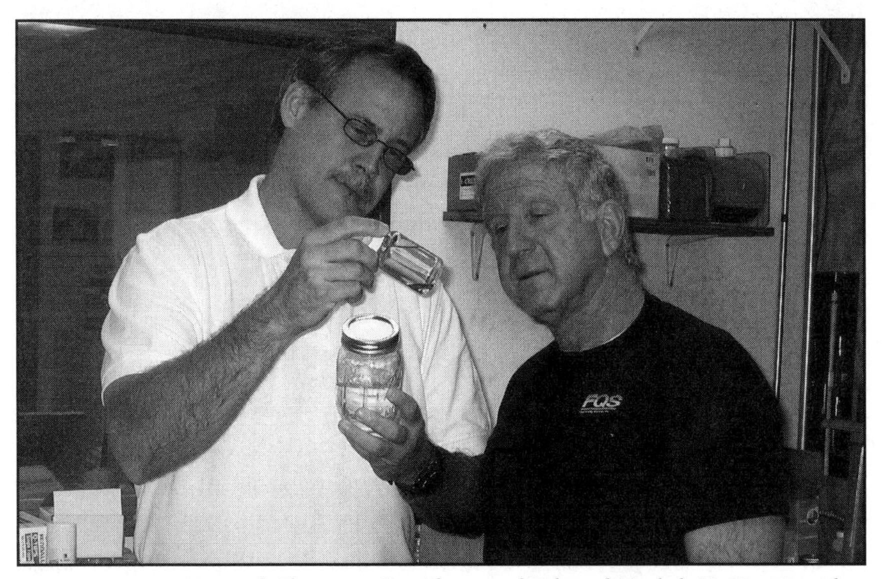

Figure 3.3-11. Howard Chesneau, President, and Edward English II, Vice President and Technical Director of Fuel Quality Services Inc., are seen inspecting biodiesel using a microbial test kit distributed by the company. (Courtesy Fuel Quality Services Inc.)

Edward English II, Vice President and Technical Director of Fuel Quality Services, concurs: "Biodiesel is a very different product from fossil diesel fuel and there needs to be better education surrounding handling and storage of these fuels. B100 may well leave the factory meeting ASTM standards, but each time it is pumped, transferred, stored, blended and dispensed, it will pick up ever-increasing amounts of water. This stuff is like a sponge, and the only way neat or blended biodiesel will remain stable and microbial free is through proper handling, monitoring and remedial procedures."

"Our detection kits allow fuel handlers and retailers a very quick 'pregnancy test' to determine if the fuel is contaminated with microbial growth," interjects Mr. Chesneau. "Unlike the food industry, we are not concerned about what type or quantity of bacteria is present in biodiesel, simply whether they are there or not. A positive result will then require remedial treatment with biocide additive and investigation as to where the water contamination entered the fuel chain, to prevent repeat problems."

The discussion continues to include bio-heat products (biodiesel blended with home heating oil), discussed in further detail in Chapter 9. "Oxidation of biodiesel can begin through contact with various non-ferrous metals such as brass, copper, lead and plumbing solders," states Mr. English. "The majority of oil-fired home heating systems contain fuel lines plumbed with copper and this will definitely create sediment and sludge, especially when fuel sits during the six-month period after the heating season. For this reason, we recommend twice-yearly treatments of oil storage tanks to prevent expensive fuel-related problems from occurring."

"As much as we would love to sell lots of biocide and fuel stabilizers, the fact of the matter is that careful handling and attention to storage environments will go a long way toward ensuring quality biodiesel," Mr. Chesneau concludes.

Nitrogen Oxide Emissions

Numerous studies have confirmed that overall emissions from the combustion of biodiesel are low but show slightly elevated levels of nitrogen oxides (NOx). This increase is regarded as a problem related to higher combustion cylinder temperatures and is not inherently a fuel-related issue. Manufacturers believe that improvements in engine sensor and management technology as well as NOx catalytic reduction are just around the corner.

Unfortunately, this issue of NOx emissions is placing a damper on the entire diesel engine industry. Diesel proponents believe that gasoline-based exhaust emission strategies are strangling the potential of the diesel market

Figure 3.3-12. The release of Mercedes-Benz BLUETEC technology coincided with the US release of low-sulfur diesel fuel in the autumn of 2006. Restricting sulfur content to a maximum of 15 ppm permits the use of particulate filters and efficient nitrogen oxide exhaust treatment. But the question remains: will the vehicles using BLUETEC technology be certified for use with biodiesel fuels? (Courtesy Mercedes-Benz Canada).

and that each engine technology should receive its own emission profile in light of the considerable fuel and greenhouse gas emissions savings of diesel engines.

Not every automotive manufacturer is worried. Mercedes-Benz is well known for its innovative, quality automobiles featuring state-of-the-art engineering. The company marked the epitome of its technological prowess by showcasing its leading edge BLUETEC technology, launching the diesel power train of the future. "Vision has therefore become reality as the extremely economical Mercedes-Benz CDI models are the cleanest diesel in the world in every category and consume between 20 and 40 percent less fuel than the gasoline counterparts," states a January 2006 corporate press release. Mercedes-Benz certainly recognizes the importance of emissions reduction technology, as over 50% of their total production volume is now captured by diesel engines.

The release of the BLUETEC technology coincided with the US release of low-sulfur diesel fuel in the autumn of 2006. Sulfur occurs naturally in mineral diesel fuel (see Chapter 3.1) in varying amounts, resulting in corrosive action which damages NOx reduction technology. Restricting sulfur content to a maximum of 15 ppm permits the use of particulate filters and efficient nitrogen oxide exhaust treatment. But the question remains: will the vehicles using BLUETEC technology be certified for use with biodiesel fuels?

Cold Flow Issues

No. 2 diesel fuel suffers from a thickening condition known as "waxing" or "gelling" when temperatures drop below the cloud point of the fuel. Should this occur within the fuel system of a vehicle, expensive cleaning becomes necessary.

By contrast, biodiesel suffers from a similar but reversible problem. Should low-temperature fuel gelling occur, causing loss of engine power or complete fuel starvation, the problem can be remedied by simply moving the vehicle to a warm location such as a parking garage until fuel temperatures moderate.

Both No. 2 diesel fuel and biodiesel can be "winterized" by the addition of so called "pour point enhancers," which may be as simple as blending No.1-D into the fuel blend.

No. 2 diesel fuel is treated by fuel refineries to meet the expected minimum temperatures within a given geographical location. To a retail consumer, the transition from "summer" to "winter" diesel fuel is completely transparent. However, adding **any amount** of biodiesel fuel into an already

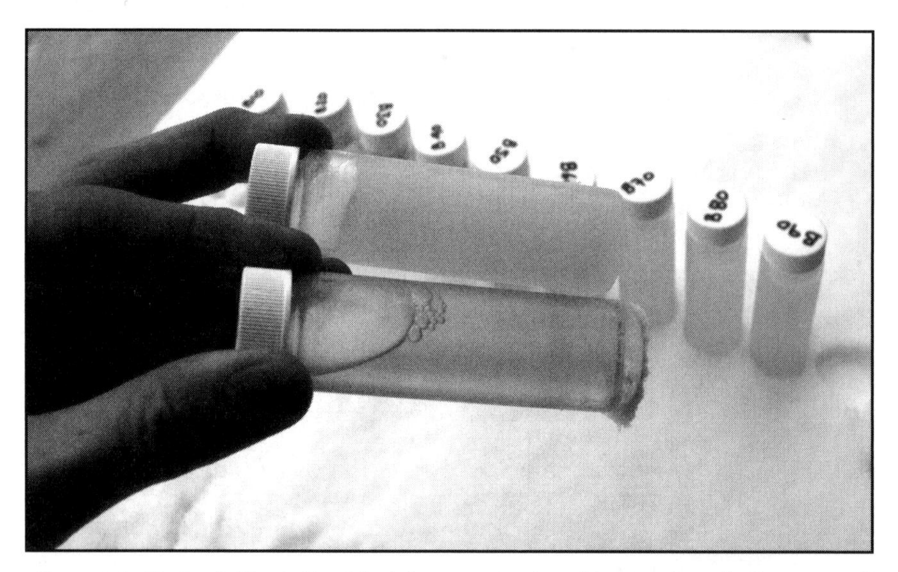

Figure 3.3-13. Both No. 2 diesel fuel (bottom view) and biodiesel can be "winterized" through the addition of so-called "pour point enhancers," which may be as simple as blending No.1-D into the fuel blend. In this image, one vial of No.2-D as well as vials of biodiesel in concentrations ranging from B10 through B100 are allowed to sit outdoors. B100 (top vial) is butter-solid at 14°F (-10°C), while No. 2-D as well as B10, 20, 30, and 40 all remain functional, above their respective cloud point ratings. Using this simple arrangement of test vials will provide the biodiesel user with a method of determining maximum winter blending limits.

winterized diesel fuel mix will degrade the minimum operating temperature by raising the cloud and pour points. This occurs because all biodiesel fuels are known to have cloud point temperatures greater than No.2-D; therefore adding biodiesel to an already winter-treated fuel dilutes the total amount of pour point enhancer present in the fuel mixture, reducing cold weather performance.

Although the addition of No.1-D will greatly improve the minimum working temperature of fuel, there is a resultant loss of approximately 3,000 BTU of energy per gallon (790 BTU per liter), resulting in higher fuel consumption and higher operating costs. In addition, the lubricity and viscosity of the No.1-D and No. 2-D fuel mixture will be reduced, which may have a detrimental effect on engine wear and engine life as well as requiring increased lubricating oil changes as a result of low-viscosity fuel leakage past piston rings.

To counter these problems, fuel-additive companies have developed a range of products to improve the cold-weather performance of petrodiesel and biodiesel fuels. Primrose Oil Company Inc. offers the Flow-Master® winter diesel fuel treatment product it claims to be more cost effective than using No.1-D blended fuels.

The reason fuel gels at cold temperatures is that waxes inherent in the fuel begin to form microscopic crystals. If untreated, these crystals will immediately agglomerate (combine) with one another to form a gel and eventually solidify, blocking fuel lines and filters. Pour point enhancers limit the ability of wax crystals to grow large enough to agglomerate. Primrose indicates that its product will improve the cold flow rating of any untreated fuel by a minimum of 20°F to 30°F (11°C to 17°C).

It is virtually impossible to determine the exact blend ratio or concentration of biodiesel that can be used at a given geographical location without knowledge of the fuel's cloud point temperature.[13] There is considerable variability in biodiesel cloud point temperature resulting from the inconsistency of the feedstock oil saturation level, with long-chain compounds displaying poor cold weather properties. Tallow, lard, palm oil, and yellow greases may remain solid or semisolid at room temperature, requiring great care in blending and storage.

Some researchers have taken a different approach to the problem by attempting to modify the FAME chemical structure through the use of alternate alcohols which have shown improved cold weather performance. Unfortunately, these alcohols have a higher cost and this process is of limited value in the current marketplace.[14]

Repeated cooling and filtration of crystal growth within the FAME has also been attempted with varying rates of effectiveness. However, this process requires considerable amounts of energy and removes valuable esters that are lost during the filtration process, lowering overall biodiesel yields.[15]

As a result of biodiesel demand growth in the northern United States and Canada, including a Minnesota bill that requires all on-highway diesel fuels to contain at least 2% biodiesel, the National Biodiesel Board commissioned the Cold Flow Blending Consortium. It studied ways to improve blending techniques in order to limit problems associated with lower temperature blending without the use of pour point enhancers.[16] The Consortium developed a test rig to simulate proportional and splash blending at a terminal, using No.1 and No. 2 diesel fuels as well as three biodiesel samples with a range of cold flow properties.

Key test results indicate that biodiesel must be at least 10°F (5.5°C) above its cloud point to successfully blend with diesel fuels in cold climates.

Chapter 8.3 will provide a direct test method that may be used to determine the cloud and pour points of any petro/biodiesel mixture.

Lower Energy Content

Biodiesel has a lower energy density or "energy content" than No. 2 diesel fuel: approximately 12.5% by weight and 8% by volume. As fuel injection systems operate on a volumetric basis, the theoretical energy loss will be the lower value. Experience has shown that actual energy losses are lower than this as a result of the slightly higher viscosity of the fuel, preventing blow past the fuel injector components and reducing leakage into the engine cylinder.

In most fleet applications, biodiesel consumption rates were between 0% and 5% higher than those of petrodiesel.

In fact, the minor change in "real world" energy content will go largely unnoticed, and if you have any ability to moderate your driving speed a

Fuel Type	BTU per Pound	BTU per U.S. Gallon
No. 1 Diesel	17,800	125,500
No. 2 Diesel	18,300	129,050
Soybean Diesel	16,000	118,170
Energy Density Difference (Biodiesel vs. No. 2-D)	-12.5%	-8%

Table 3.3-1.
Comparison of Energy Density of Petroleum and Soybean Biodiesel Fuels

mere 2 miles per hour (3.2 kph) slower at highway speeds will more than compensate for the difference.

OEM Warranty Issues

One of the popular misconceptions about biodiesel is that it will not affect engine and fuel system warranties provided the fuel meets applicable specifications.[17] Statements such as these are not only misleading; they are simply wrong.

All major engine, vehicle, and fuel injection equipment manufacturers have clearly stated guidelines regarding the use of biodiesel fuels. Without hesitation, all manufacturers state that biodiesel that is used within the blend limits of their warranty statements must meet the appropriate national and/or international fuel standards.

Volkswagen of America sums up the general OEM equipment position regarding biodiesel with the following statement:

"Volkswagen of America Inc. is proud to be the automobile industry's leader in diesel technology for passenger cars and light-duty trucks. Many customers have expressed interest to us in operating their Volkswagen TDI vehicles on 'biodiesel' fuel....Volkswagen has determined that diesel fuel containing up to 5% biodiesel meets the technical specifications for Volkswagen vehicles equipped with TDI engines imported into the United States. While

Figure 3.3-14. Volkswagen of America has determined that diesel fuel containing up to 5% biodiesel certified to ASTM standard D6751 meets the technical specifications for Volkswagen vehicles equipped with TDI engines imported into the United States.

this historic decision by Volkswagen is a first step in a renewable fuel strategy for our cars, it is not a departure from our strong recommendation that you use only high-quality fuel, nor from our long-standing fuel requirements for warranty purposes. We must stress that vehicle damage that results from misfueling or from the usage of substandard or unapproved fuels [read home brewed biodiesel] cannot be covered under our vehicle warranties."[18]

Although the policy statement indicates that high levels or off-specification biodiesel may limit OEM warranty service liability, the wording must be interpreted on a broader basis. The Volkswagen policy statement stipulates that "vehicle damage that results from misfueling or from the usage of substandard...fuels," clearly placing the burden of proof regarding **any** fuel-related damage, fossil- or FAME-based, on the customer.

Small-scale producers recognize that the cost of performing the fuel quality tests may be in excess of the value of the fuel produced, causing one of their greatest problems: lack of fuel quality. This is also the major reason that the biodiesel industry and automotive and engine manufacturers limit biodiesel concentration levels and distance themselves from the home brew community which advocates using B100. The OEMs do not want to give any credibility to fuel producers who cannot meet and guarantee minimum fuel quality standards.

Figure 3.3-15. The test procedures required to certify the quality of biodiesel are based on well established test methods that non-technical people lack both the ability and the necessary equipment to perform.

Biodiesel quality and fuel blend concentrations are some of the most critical issues facing both commercial and small-scale biodiesel producers everywhere. Given their importance, further analysis of these issues will be discussed in Chapter 6.

Biodiesel Fuel Quality

ASTM D6751 Fuel Quality Standard

As with petroleum-based diesel fuel, biodiesel must meet its own fuel quality standards. Because of the similarities between fossil and biodiesel fuels, as well as their affinity for each other when blended, several of the test parameters of each standard are closely linked.

The ASTM has created the standards and test procedures for FAME fuel at a concentration of 100%. The test procedures are based on well-established

Analysis	Method	Min.	Max
Acid Number, mg KOH/gram	ASTM D664		0.80
Ash, Sulfated, Mass %	ASTM D874		0.020
Total Sulfur, Mass %	ASTM D5453		0.05
Cetane Number	ASTM D613	47	
Cloud Point, °C	ASTM D5773		
Copper Corrosion	ASTM D130		No. 3
Vacuum Distillation End Point	ASTM D1160		360°C at 90%
Flash Point, °C	ASTM D93 Procedure A	130.0	
Method for Determination of Free and Total Glycerin	ASTM D6584		
Free Glycerin, Mass %			0.020
Total Glycerin, Mass %			0.240
Phosphorus (P), Mass %	ASTM D5185		0.001
Carbon Residue, Mass %	ASTM D4530		0.05
Kinematic Viscosity, Centistokes at 40°C	ASTM D445	1.9	6.0
Water and Sediment	ASTM D1796		0.050

Table 3.3-2 D6751-02 Test Parameters

Reprinted with permission from D 6751-03a Standard Specification for Biodiesel Fuel (B100) Blend Stock for Distillate Fuels, copyright ASTM International, 100 Bar Harbor Drive, West Conshohocken, PA, 19428.

test processes that generally exceed the ability of non-technical people working outside the laboratory to perform. However, it is important to understand the linkages between the various test criteria in order to grasp the technical issues related to FAME fuel quality.

Acid Number: The acid number describes the free fatty acids in the FAME, which are known to lead to corrosion. Water in the fuel may be symptomatic of a high reading. The value is calculated by titrating a 1-gram sample of FAME with a quantity of base, measured in milligrams of potassium hydroxide.

Ash, Sulfated: This test indicates the quantity of metallic residue left over from the catalyst used in the transesterification process. A sample of FAME is combusted and the residue is treated and weighed to determine the residual noncombustible mineral ash

Total Sulfur: The total sulfur test determines the amount of sulfur contained in the FAME. Reducing sulfur content of all fuels is required to lower atmospheric sulfur dioxide emissions. Vegetable oils, including most waste oils, typically exhibit very low sulfur levels, although contamination or processes that utilize sulfuric acid may be problematic.

Cetane Number: The cetane number of biodiesel is a direct indication of the ignitability of the fuel. FAME is known to have cetane levels that are superior to those of fossil diesel fuels. A specially developed compression ignition engine is operated using a reference fuel as well as the subject FAME, providing a relative comparison or cetane number.

Cloud Point: This is the temperature at which wax crystals begin forming as the FAME is cooled. Fuels which are operated below their cloud point are likely to cause filter plugging and subsequent fuel starvation of the engine. Cloud point measurements are recorded and reported to the client. No further analysis is provided.

Copper Corrosion: Fuels which have high levels of free fatty acids will cause specially polished strips of copper to corrode when subjected to elevated temperatures. The degree of corrosion is compared to a series of reference strips to determine a pass/fail condition.

Vacuum Distillation End Point: This is the temperature at which 90% of the fuel sample will be distilled, under condition of reduced pressure, allowing a determination of the makeup of the FAME. Petrodiesel fuels contain hundreds of compounds, resulting in much broader and variable so-called distillation curves. ASTM D6751 states that the atmospheric boiling point of biodiesel generally ranges from 330°C to 357°C. ASTM D1160 is an additional, optional specification which can be used to detect higher boiling compounds caused by fuel contamination or high levels of di- and triglycerides. When the D1160 tests are conducted, a table of 10% volume distillation levels between the initial boiling point (IBP) and final boiling point (FBP) of the FAME is provided.

Flash Point: FAME fuels are classified as nonflammable as their flash or ignition points are above the minimum level of 130°C. This test is a measure of residual alcohol which would result from incomplete methanol recovery and/or washing of the raw FAME. Methanol is highly toxic, flammable, and can be easily inhaled because of its low vapor pressure.

Free Glycerin: Free glycerin refers to suspended glycerin compounds that remain in the raw FAME as a result of improper washing. Excessive glycerin levels cause carbon deposits on fuel injection components as well as engine valves, valve seats, pistons, and rings, which leads to degraded engine performance and eventual engine failure. Free glycerin forms sludge in fuel storage tanks, resulting in plugged filters and engine starvation.

Total Glycerin: This is the sum of the free (suspended) glycerin and the bonded glycerin present in the mono-, di- and triglycerides in the FAME. Elevated levels of total glycerin result from incomplete reaction of the feedstock oils during the transesterification process and will compound the problems noted under "Free Glycerin." This is a common problem with the vast majority of small-scale and home brew processes. Later in the book we will discuss actual certified test results of home brew fuel and see that they are often out of specification by a very significant amount.

Supplementary data may be provided by the test laboratory to indicate the mono-, di-, and triglyceride levels present in the FAME. Although it is not a condition of ASTM standards, the European standard for biodiesel EN 14214 requires a minimum ester content of 96.5%. Low ester levels can be attributed to minor contaminants as well as to high levels of unreacted glycerides resulting from poor chemical procedures.

Phosphorus: This is a measure of contaminants resulting from the refining process for feedstock oils and fats as well as from the use of phosphoric acid during the production process. Vegetable oil feedstock should have very low levels of phosphorous contamination.

Carbon Residue: A fuel sample is combusted and the remains constitute the carbon residue. Excessive levels of glycerin are the likely cause of a test failure.

Kinematic Viscosity: This is a measure of a fluid's resistance to flow under the force of gravity. High viscosity fluids such as molasses will become less resistant to flow when heated; therefore this test is conducted at a specified temperature of 40°C. If the viscosity of the FAME is too low, pressurized fuel will leak past injector valve seats as well as piston rings, diluting engine lubrication oil. If the viscosity is too high, excessive amounts of strain will be placed on the fuel pump and injection components, leading to failure. Additionally, the fuel will not atomize correctly, causing incomplete combustion and rapid carbon deposit buildup. A sample of the fuel is passed through a cup which has a calibrated hole at the exit. Timing the fuel flow will provide the data necessary to determine the viscosity number measured in centistokes or mm^2 per second.

Water and Sediment: As discussed earlier, free or bonded water in FAME fuels will lead to the formation of free fatty acids and corrosion of engine and fuel storage materials and will also promote microbial growth. Sediment can plug fuel filters, and if the sediment is small enough to pass through filters it can cause abrasive damage to fuel injection components and other high-tolerance items.

The BQ-9000 Quality System

Biodiesel may meet the applicable quality standards at the manufacturer's door, but that does not guarantee that the fuel being pumped into your car will be within specification. As we have discussed earlier, fuel contamination through water ingress, condensation, microbial growth, blending contamination, contact with nonferrous metals, and a host of other sources can jeopardize quality.

These issues have little to do with testing standards but are related to management and handling procedures. Virtually all major product manufacturers and retailers have learned that it doesn't matter whether you produce and deliver bread or semiconductors; it is the quality standards and procedures that ensure the final product is delivered "on spec."

The International Standards Organization (ISO) recognized this issue nearly a generation ago with the development of the ISO 9000 quality control process, which has been adopted on a worldwide basis. In a similar manner, the NBB directed an independent committee to develop the BQ-9000 program, which mirrors the philosophy of the ISO 9000 program but is geared to the biodiesel industry.

The NBB realized that a second level of checks and balances implemented throughout the industry would further ensure that fuels meeting ASTM D6751 would arrive at the customer's location in the same condition. The program includes protocols on storage, sampling, testing, blending, shipping, distribution, and fuel management practices.

Companies are awarded accreditation following the successful completion of a series of formal reviews and audits which prove that they have the ability to deliver on their quality commitments. Essentially, a proponent develops an in-house quality policy manual which states its quality objectives and procedures, "saying what we will do" from a quality perspective. The audit team periodically monitors the company's implementation program by reviewing tracking documents, interviewing staff, and ensuring that everyone involved "does what they say they would do."

Once the BQ-9000 has been fully implemented, customers can ask for certificates of conformance for each biodiesel delivery. These certificates enable the accredited company to follow the fuel from "cradle to grave," ensuring that proper procedures have been implemented during each step of the manufacturing and distribution chain. Should a fuel quality complaint be logged with the company, it becomes a simple matter of tracing the documentation chain backwards through the system until the failure or breakdown is identi-

Figure 3.3-16. In September 2004, Peter Cremer North America became the first BQ-9000-accredited producer of biodiesel, landing the company a high-profile customer: almost immediately after receiving the accreditation, Peter Cremer was named the biodiesel supplier of choice for DaimlerChryler's B5 Jeep Liberty factory. NBB Executive Director Joe Jobe and then Biodiesel Association of Canada Executive Director Christine Paquette beam as the first units fuelled with biodiesel begin to roll off the line.

fied and then implementing a corrective action.

Ultimately, BQ-9000 is also a marketing tool, ensuring customers that the fuel supplied by an accredited company is warranteed and that any problems will be quickly identified and corrected. In September 2004, Peter Cremer North America became the first BQ-9000-accredited producer of biodiesel, landing the company a high-profile customer: almost immediately after receiving the accreditation, Peter Cremer was named the biodiesel supplier of choice for DaimlerChrysler's B5 Jeep Liberty factory.[18] Relationships like this prove that the delivery of a quality product is the key to success.

Fuel Quality Summary

The quality of any fuel can be assured only by careful chemical engineering procedures, examination by certified testing laboratories, and vigilant storage and retail handling procedures. A failure of testing or procedure will place the biodiesel at risk somewhere along the distribution chain, virtually ensuring that poor quality fuel will be delivered to the end client, causing inconvenience and financial loss.

Or, as Christine Paquette, past executive director of the Biodiesel Association of Canada, points out: "Our clients want clean, renewable fuels that work; they don't want to waste time and money while their vehicles are disabled on the side of the road. When it comes to biodiesel, the most important issue is quality."

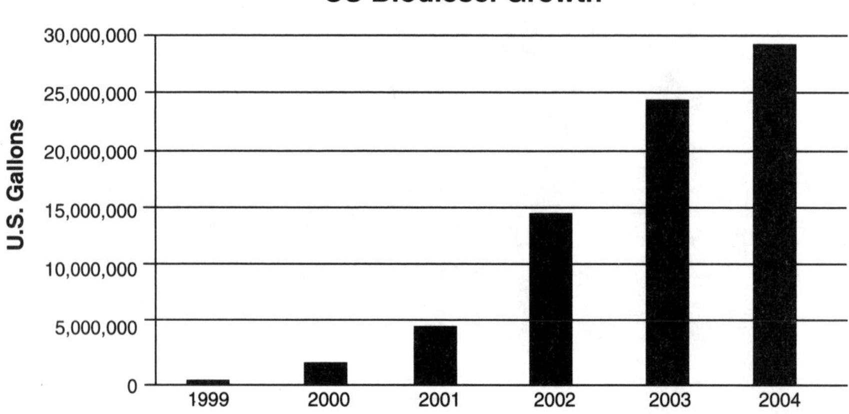

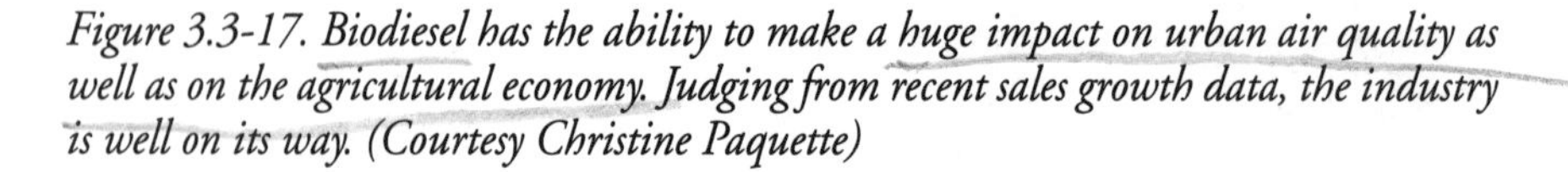
Figure 3.3-17. Biodiesel has the ability to make a huge impact on urban air quality as well as on the agricultural economy. Judging from recent sales growth data, the industry is well on its way. (Courtesy Christine Paquette)

4
Oil Feed Stocks

4.1
An Introduction to Biodiesel Feedstock

Biodiesel is the product obtained from the transesterification of animal fats or vegetable oils, yielding fatty acid methyl esters or FAME. As we discussed in Chapter 3, it is possible to use virtually any fat or vegetable oil feedstock to complete the transesterification process, including:

- vegetable oils such as soybean, canola, rapeseed, palm oil, sunflower oil, etc.
- restaurant and commercial kitchen waste frying and cooking oils
- animal fats including beef tallow, lard, and rendered materials
- yellow grease, processed waste cooking and fryer oils
- brown grease, a lower quality waste grease recovered from restaurant plumbing traps or from highly oxidized yellow grease

All plant and animal oils have essentially the same chemical structure, consisting of triglycerides, which are chemical compounds formed from one molecule of glycerol and three molecules of fatty acids. Triglycerides are the main compounds or components of animal fat and vegetable oils, which have viscosities many times greater than that of petrodiesel fuel. High viscosity fuels are not compatible with standard diesel engines unless significant mechanical modifications are undertaken to ensure proper operation with the alternative fuel source (see Chapter 7.2).

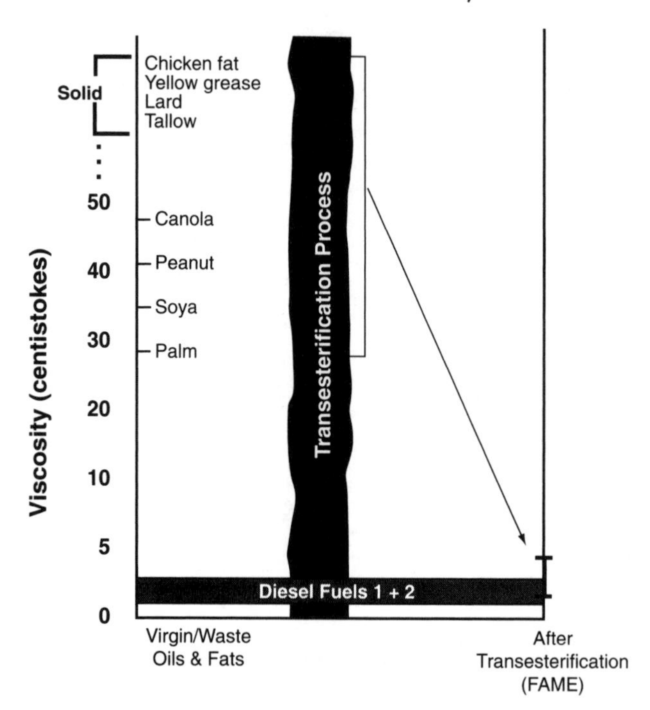

Figure 4.1-1. Triglycerides are the main compounds or components of animal fat and vegetable oils, which have viscosities many times greater than that of petrodiesel fuel. High viscosity fuels are not compatible with standard diesel engines unless significant modifications are undertaken to ensure proper operation with the alternative fuel source.

Section of Chapter 4	Feedstock Oil/Fat	Free Fatty Acid Level	Relative Cost
4.2	Refined Vegetable Oil (Soybean, Canola)	0 - 1%	Highest
4.3	Waste Fryer Oil/Fat	2 - 7%	Low
4.4	Animal Fats (Beef Tallow, Lard)	5 - 30%	Low
4.5	Yellow Grease	7 - 30%	Moderate
4.6	Brown Grease	> 30%	Very Low

Table 4.1-1
Range of Free Fatty Acids (FFAs) in various feedstock oils and fats (Source: 3M Food Services Business and Canada Challenge Inc.)

Vegetable oils also contain varying amounts of fatty acids, which are chains of hydrocarbons that differ in carbon length depending on the oil feedstock. The fatty acid composition of the feedstock will determine important characteristics such as melting point and stability in the presence of atmospheric oxygen. Oils and fats are also comprised of free fatty acids (FFAs) which are not bonded to the parent oil molecules such as triglycerides. These free fatty acids are found in increasing concentration in used or waste oils that have been subject to heavy oxidation such as deep fryer oils and yellow grease.

In general, the lower the feedstock cost the higher the free fatty acid level, which in turn complicates the biodiesel production process and resulting yield. The home-scale or microproducer will almost certainly wish to focus on feedstock oils that are either free or of very low cost. This will virtually eliminate the best feedstock, refined vegetable oils, due to cost.

On the other hand, small-scale producers such as farm and rural co-operative producers will most likely use poor-quality, off-specification or excess-production oilseeds as a value-added market stream. It is also possible to use yellow grease from a third-party supplier, provided feedstock supply and economics make sense.

Large-scale commercial biodiesel producers will use the most economic oil feedstock for their location: the Canadian prairies will use canola, Americans will concentrate on soybean, and tropical regions will use palm oil. Regardless of the source, it is important to realize that the chemical and physical properties of the biodiesel will be determined by the type and quality of the original feedstock.

Virtually any plant oil or animal fat can be used to produce biodiesel, although not all parent feedstock materials are feasible because of their economic or physical properties. The North American markets are awash in numerous potential feedstock materials; the drive to produce biodiesel is often related to the economics of the feedstock material or to filling an existing need in the marketplace. For example, low oilseed commodity prices often drive producers to sell below the cost of production to provide immediate cash flow. Increasing market demand resulting from the production of biodiesel may support prices through additional oilseed consumption channels. Similarly, fish processing plants generate vast quantities of waste oils and other byproducts that can provide complementary income through biodiesel feedstock production.

However, switching from fossil fuels to biofuels in order to meet our growing energy requirements is neither possible nor in our long-term interests. According to one research study,[1] it is only possible to grow approximately

Figure 4.1-2. In general, the lower the feedstock cost, the higher the free fatty acid level, which in turn complicates the biodiesel production process. The microproducer will almost certainly wish to focus on feedstock oils that are either free or of very low cost, such as recycled fryer oils.

15% to 20% of current North American biodiesel requirements, and this level of production would continue to fall in the face of rising energy consumption. Even if we could grow our way out of the fossil fuel hole, environmental concerns about rampant biofuel consumption due to an excessively mobile and energy-inefficient society and the ecological destruction from increased levels of energy-intensive agriculture would offset the achievement.

An article published by environmentalist George Monbiot[2] suggests that every year we use the equivalent of four centuries' worth of the potential energy of plant and animal products. If this theory is correct, it means that the idea of replacing fossil fuels with those derived from current biological supply streams is a fallacy. It also points out that we are depleting our fossil legacy account at an astonishing and unsustainable rate.

Mr. Monbiot's article argues that biodiesel creates extensive competition for land use in the developed world but is also causing rampant destruction of the developing world's ecosystem. He points out that "between 1985 and

2000, the development of oil palm plantations was responsible for an estimated 87 per cent of deforestation in Malaysia" because of the planting of oil palm trees used for biodiesel production. The report argues that destroying vast forests and replacing them with small, scrubby oil palm plants releases huge amounts of carbon into the atmosphere when the larger trees are felled and burned. "In terms of its impact on both local and global environments, palm biodiesel is more destructive than crude oil from Nigeria," Mr. Monbiot proclaims.

Attempting to meet the rising demand for fossil fuel by replacing it with biofuel is impossible. The key is to combine lower and more efficient energy consumption with biofuel production that has long-term sustainability as the central focus (see Chapter 2.1).

4.2
Refined Vegetable Oils

"Biodiesel production can be sustainable. I do a lot of work in biorefinery processing and the goal for me is to find out what am I going to do with all that other stuff that results from lopsided agricultural processes, that produce useful, yet unmarketable co-products as an offshoot of the primary product", explains Dr. Martin Reaney, when queried about the balance between agricultural production and market demand. "For example, I am always in the situation working with high producing oilseeds such as canola. Here in Saskatchewan we have a plant that can produce say, 42% oil coming off the field and future varieties that will be pushing 50%. While this is very impressive when speaking about biodiesel or food oil feedstock, it still leaves us with 15% to 20% protein byproduct as a result of the oil extraction process. To put this in perspective, our yearly canola harvest could easily produce enough protein to feed everyone in the country one and a half times over. Unfortunately, we use the protein meal in resource-wasteful ways, feeding it to animals which convert it to milk or meat at fairly low levels of efficiency. The proper trick is to find the right sustainable and economic balance between all of the agricultural components in the processing system".

Figure 4.2-1. Dr. Martin Reaney, formerly a research scientist working in bioproducts and processing at Agriculture and Agri-Food Canada, is currently Chair of Lipid Utilization and Quality at the University of Saskatchewan. His work focuses on the development of new technology for processing oilseeds as well as the creation of commercial products from both oilseed and protein meal streams. Dr. Reaney is also an expert on the canola oilseed biodiesel market.

"Soybeans are another case in point. They are a fairly low-quality crop for oil production, yet they are excellent producers of meal which can be used in a huge array of products ranging from tofu to every imaginable animal feed," Dr. Reaney explains. "The imbalance between the market for soy meal and the oil is one of the drivers moving the United States in the development of soy-based biodiesel."

Producing biodiesel from vegetable oil generates a coproduct, glycerin, at a significant rate of 10% by mass. With biodiesel production rates set to soar in the coming years, the market for glycerin is expected to become saturated, depressing prices and lowering demand. As a result, commercial as well as microscale biodiesel producers are on the lookout for increased markets for this crude glycerin (see Chapter 11).

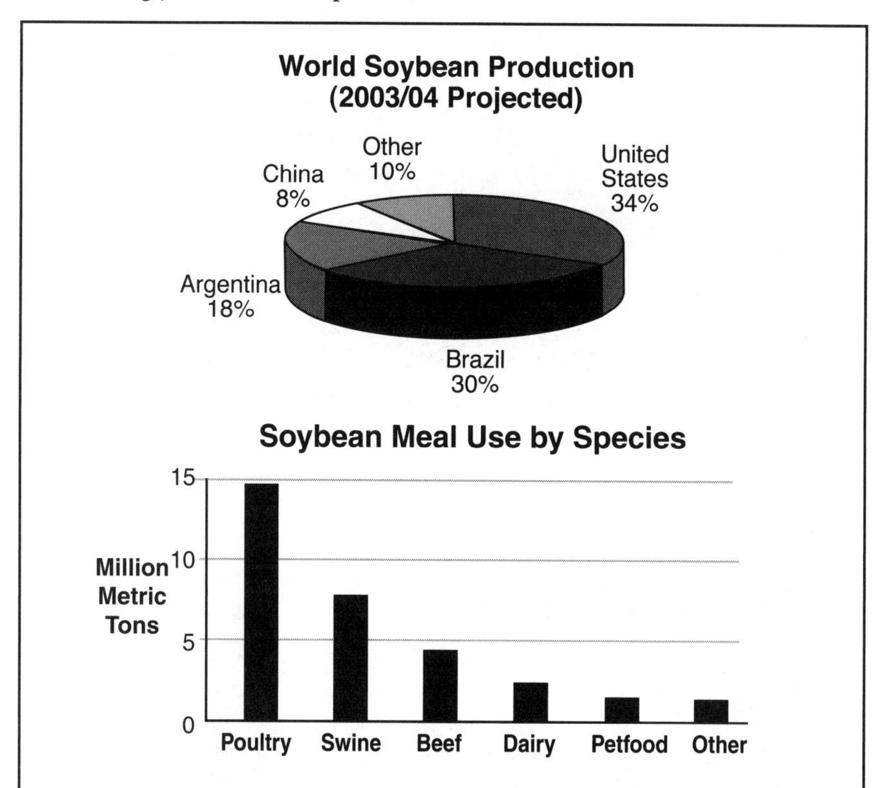

Figure 4.2-2a/b. "Soybeans are another case in point. They are a fairly low-quality crop for oil production, yet they are excellent producers of meal which can be used in a huge array of products ranging from tofu to every imaginable animal feed," Dr. Reaney explains. "The imbalance between the market for soy meal and the oil is one of the drivers moving the United States in the development of soy-based biodiesel." (Source: www.soymeal.org)

Balancing the energy, market and resource efficiency throughout the entire biolife-cycle will be the only way to ensure sustainable agribusiness models of the future.

Dr. Reaney understands the relationship between sustainable agriculture and our endless energy needs. Formerly a research scientist working in bio-products and processing at Agriculture and Agri-Food Canada, he is now Chair of Lipid Utilization and Quality at the University of Saskatchewan. His work focuses on the development of new technology for processing oilseeds and commercial products from both the oilseed and protein meal streams with a strong emphasis on the biodiesel market.

Figure 4.2-3. "If we could get the chokecherry to divert its production from sugars to oil, we could get double the oil production of canola without any need for crop rotation," Dr. Reaney explains.

As we passed by rows of bushes on the university's research grounds, Dr. Reaney grabbed a handful of chokecherries and proffered a few samples of the sweet-tasting fruit. "These chokecherries are part of a breeding program to improve chokecherries as a fruit. They are some of the most consistent producers, yielding a crop every year from the same piece of land. Contrast this with canola that must be rotated so that it produces a crop on the same land only every four years to ensure disease-free growth. If we could get the chokecherry to divert its production from sugars to oil, we would be able to get double the oil production compared to canola without the need for crop rotation."

Dr. Reaney finishes a few more chokecherries and then continues explaining the need to improve the sustainability of agricultural processes whether the crop in question is for eating or placing in your fuel tank.

Figure 4.2-4. The largest single agricultural input is nitrogen, which is currently produced using natural gas. This is not sustainable in the long term as we simply do not have sufficient natural gas to support the process, nor do we want to keep throwing carbon dioxide into the atmosphere as the by-product of the conversion process. There are a number of nitrogen-fixing plants such as sea buckthorn, shown here, that can be grown, harvested, and mulched directly into the soil adjacent to the main crop.

"The largest single agricultural input is nitrogen, which is currently produced using the Haber-Bosch process to convert natural gas into fertilizer products. This is not sustainable in the long term as we simply do not have sufficient natural gas to support the process, nor do we want to keep throwing carbon dioxide into the atmosphere as the byproduct of the conversion process. There are a number of nitrogen-fixing plants such as sea buckthorn that can be grown, harvested, and mulched directly into the soil adjacent to the main crop. This will be much more sustainable in the long run and will help improve the overall energy/nutrient cycle."

Peacock Industries

Ensuring a balance between all parts of the agricultural process may not be easy to visualize on a national agribusiness scale, but at the local level the effect is quite obvious. "Peacock Industries is a company that I am working with that has oil as the byproduct, or more correctly the coproduct, that had no value in their original product lines," Dr. Reaney explains. "Peacock has developed a natural nematocide/ pesticide and a soil stabilizer that are produced from a formulated and stabilized mustard bran. Their process is based on cold pressing mustard seed, which is a relative of the canola plant, to extract the oil and meal components. They are looking to me to figure out what to do with the oil."

Figure 4.2-5. Rob Reddekopp of Peacock Industries is shown with a sample of the company's natural nematocide/pesticide that is produced from the refining of mustard seed.

The dual process of converting mustard seed into multiple products is an excellent example of the biorefining process. Although Peacock is just beginning to ramp up its production process, the company is well on of its way to converting the relatively low-value mustard product into multiple, higher-value income streams.

Operations Manager Rob Reddekopp is very enthused about the production facility and Peacock's ability to provide value-added products from locally grown feedstock. "Our facility is now running a half-tonne-per-day press, providing a 25% yield of oil or about 150 liters per day, which is the equivalent of 2 gallons (7.6 liters) of oil per bushel (35.2 liters) of seed. With the other presses running we could easily double our oil production to 100,000 liters per year (26,400 gallons per year)," Mr. Reddekop comments. One man's waste is another man's gold, and with Dr. Reaney's assistance Peacock intends to either sell the oil directly to biodiesel processors or make fuel themselves. (With access to off-specification or low-cost oilseed and a few thousand dollars in press equipment, an enterprising farm operator or small-scale cooperative venture could produce a perfect biodiesel oil feedstock, while using the protein meal for animal feed, fertilizer or combustion fuel, for example). Without scaling up their current process, Peacock Industries could produce approximately 13,000 gallons (≈50,000 liters) of biodiesel per year.

Figure 4.2-6. Rob Reddekopp's "thousand watt smile" is powered by the knowledge that the mustard oil that was previously a waste product will become a valuable feedstock in the biodiesel production market. A "noodle" of canola meal is being ejected into Rob's hand. The twin presses are able to process one half-tonne per day of oilseed, yielding 25% oil.

Figure 4.2-7. Henrey Hanom of Peacock Industries is seen topping up the seed loader. A screw auger loads the seed into the white polyethylene hopper which in turn feeds the two presses. Canola meal is expressed from the press and is fed into the screw auger (left) which feeds the storage unit. Oil drains from the press "heads" and drips into a trough and storage container. The oil is filtered and is immediately ready for use.

Figure 4.2-8. An overview of the twin presses, electric drive motor, and seed storage hopper. Although small cold-press units such as these are not fast, they are economical to operate and will continue to run 24 hours per day.

Figure 4.2-9. A close-up view of the press "head" clearly shows the canola meal noodle being expressed (left). Oil drips from the perforated collar (right) into the trough mounted under the unit. Oil then flows via gravity feed to a storage tank.

Figure 4.2-10. Canola meal reaches the fabric sheeting storage hopper which assists air flow and helps to expedite drying. The expressed meal has a sawdust-like texture and is very dry to the touch. The meal is then forwarded into Peacock's proprietary production system where the natural soil treatments and pesticides are produced.

Figure 4.2-11. This front view of the twin presses details the canola meal being expressed. The angled tray under the press "heads" is the oil distribution trough.

Milligan Bio-Tech Inc.

The canola seed growers in the Canadian prairies produce high-grade, crystal clear canola oil for the food industry. Vagaries in the weather or during storage may cause the oilseed to produce off-color or high free fatty acid oil that is rejected by the market. Crop failures and crashing commodity prices are all too common in farming communities, resulting in financial hardship or worse. Progressive thinking in the rural Saskatchewan community of Foam Lake led to the creation of Milligan Bio-Tech Inc. as a way of diversifying crop development and reducing the economic risk inherent in the canola business. "Local businesses, farmers, and people with roots in the community came together in the creation of the company. You might say that we are a private version of a community cooperative, working to ensure the long-term survival of our industry and community." says Milligan principal Zenneth Faye.

Milligan realized that having farmers focus only on selling oilseed would not provide sufficient diversification for the community and decided on a biorefining venture as a way of adding value by converting off-specification oilseed to high-value products. Assessing its options, Milligan reviewed nu-

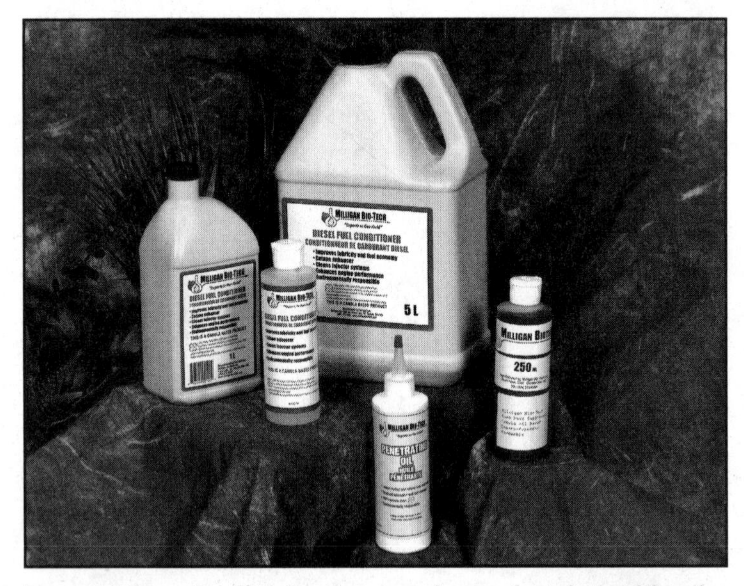

Figure 4.2-12. Milligan Bio-Tech Inc. has created a number of oilseed-based products such as penetrating and lubrication oil as well as a diesel fuel additive which increases fuel lubricity and cetane. Milligan has even developed a novel use for the glycerin byproduct, creating an ecologically friendly, biodegradable, and long-lasting dust suppressant for gravel roads which can replace salt-based calcium chloride products currently in use. (Courtesy Milligan Bio-Tech Inc.)

Figure 4.2-13. According to Milligan principal Zenneth Faye, rural communities cannot afford to be dependent on one income stream without taking unnecessary financial risks. "Diversification is the key to community growth and sustainability."

merous concepts including a mobile seed crushing/oil extraction plant that would allow the seed husk to be left at the farm site, providing oilseed meal that could be used as a feed supplement for livestock.

Virgin oils would then be graded and sold as feedstock oil in the biofuel and lubricant processing stream. Ultimately, the company decided on a semi-centralized seed crushing and biodiesel production facility at its Foam Lake location.

"After consulting with various technical experts, we decided on a hub and spoke arrangement between oilseed crushing and centralized biodiesel production," Zenneth explains. "Our model is based on having a 10-tonne-

Figure 4.2-14. Clockwise from bottom left: Zenneth Faye, Milligan Bio-Tech Inc.: Dr. Neil Westcott, Agriculture Canada; Ed Hogan, Natural Resources Canada; and Dr. Martin Reaney, University of Saskatchewan.

per-day oilseed crushing capacity and a 30-tonne-per-day biodiesel production facility located at Foam Lake (Saskatchewan). As we ramp up commercialization, other oilseed facilities throughout the prairies will develop their own crushing capacity, keep the protein meal for animal feed, and ship the oil to us for processing into biodiesel."

Milligan is now in the process of realizing this dream after having worked for a number of years on small-scale testing and processing off-specification seed into biodiesel and the various complementary products discussed earlier. "We expect to have our 9-million-liter-per-year (≈2.4 million gallons/year) biodiesel facility up and running in 2006," Mr. Faye continues. "Once we are on-line, we can fulfill 100% of our plant capacity through the use of off-specification canola, which currently rots in storage or out in the fields. Out of Canada's 8-million-tonne canola production, we lose anywhere from 5% to 10% of our production to off-spec seed, for which the farmer is not paid. Our biodiesel production system does not care about the quality of feedstock we use and can therefore pay the producer rather than see product go to waste. Being a farmer myself, I know firsthand the feeling of helplessness from losing so much income and having to rely on subsidies or crop insurance.".

Milligan is also looking at the possibility of using financial arbitrage to sell or withhold canola seed from the commodity markets in order to produce biodiesel which would remove some of the variability of oilseed prices and producer risk. "Oilseed prices can fluctuate enormously depending upon sup-

Figure 4.2-15. "Out of Canada's 8 million tonnes of canola production, we lose anywhere from 5% to 10% to off-spec seed, production for which the farmer is not paid. Our biodiesel system does not care about the quality of feedstock we use and can therefore pay the producer rather than see product go to waste. Being a farmer myself, I know first hand what it feels like to lose so much income," explains Milligan Bio-Tech's Zenneth Faye.

ply and demand issues," explains Zenneth Faye. "During a three-year period starting in 2000, canola oscillated between C$290 and C$390 per metric tonne. When these fluctuations in price are coupled with severe weather and natural spoilage, it can easily make or break the farming operation."

Figure 4.2-16. "We expect to have our 9-million-liter-per-year (≈2.4 million gallons/year) biodiesel facility up and running in 2006," Mr. Faye explains. "Once we are on-line, we can fulfill 100% of our capacity through the use of off-specification canola, which currently rots in storage or out in the fields.

Figure 4.2-17. Milligan Bio-Tech Inc. started rolling the oilseed presses into its Foam Lake facility in late 2005. Once these massive units are up and running, the local facility will be able to crush approximately 10 tonnes of canola per day.

Oil-Producing Crop Inputs

There are hundreds of oil-producing crops, although selection as a biodiesel feedstock is generally based on the plant's ability to produce the maximum amount of oil and useful protein meal for a given geographical location. There is no point in the North American biodiesel industry being excited about the fact that oil palms produce approximately five times more oil than canola since oil palm plants do not grow here. Geography, biology, and money govern the feedstock industry.

Research undertaken by Dr. Mittelbach[1], biodiesel research scientist of the University of Graz, Austria, shows worldwide vegetable oil production from nine major plant species during the 2003/2004 season (see Table 4.2-1). Soybean production tops the list as a result of its versatility as well as the fact that it produces one of the lowest-cost oils. The 2003/04 settlement price for soybean oil was 29.97 cents per pound while peanut oil was 60.84 cents per pound[2]. This doubling of cost virtually eliminates peanut oil from commercial biodiesel production except under conditions of poor oil quality or low commodity prices due to market duress.

We learned earlier that biodiesel produced from soybean oil using North American farming practices generates a net positive energy ratio of approximately 3:1. This means that in the production of soybean-based biodiesel, from seed planting, harvesting, and processing to fuelling your vehicle, there are three units of energy output for every one unit required to produce the fuel.

Crops that provide higher levels of oil and/or reduced fertilizer, insecticides, and extraction energy may be able to produce biodiesel at a lower cost. The following graph shows the relative productivity or energy yield of a range of feedstocks:

Vegetable Oil	2003/04 Production (million metric tons)	Vegetable Oil	2003/2004 Production (million metric tons)
Soybean	31.83	Cottonseed	3.90
Palm	28.13	Palm Kernel	3.50
Rapeseed	12.57	Coconut	3.33
Sunflower	9.42	Olive	2.81
Peanut	4.81	TOTAL	11.20 (JMI)

Table 4.2-1
Current worldwide production of nine major vegetable oils
(Source: Mittelbach, Biodiesel: The Comprehensive Handbook)

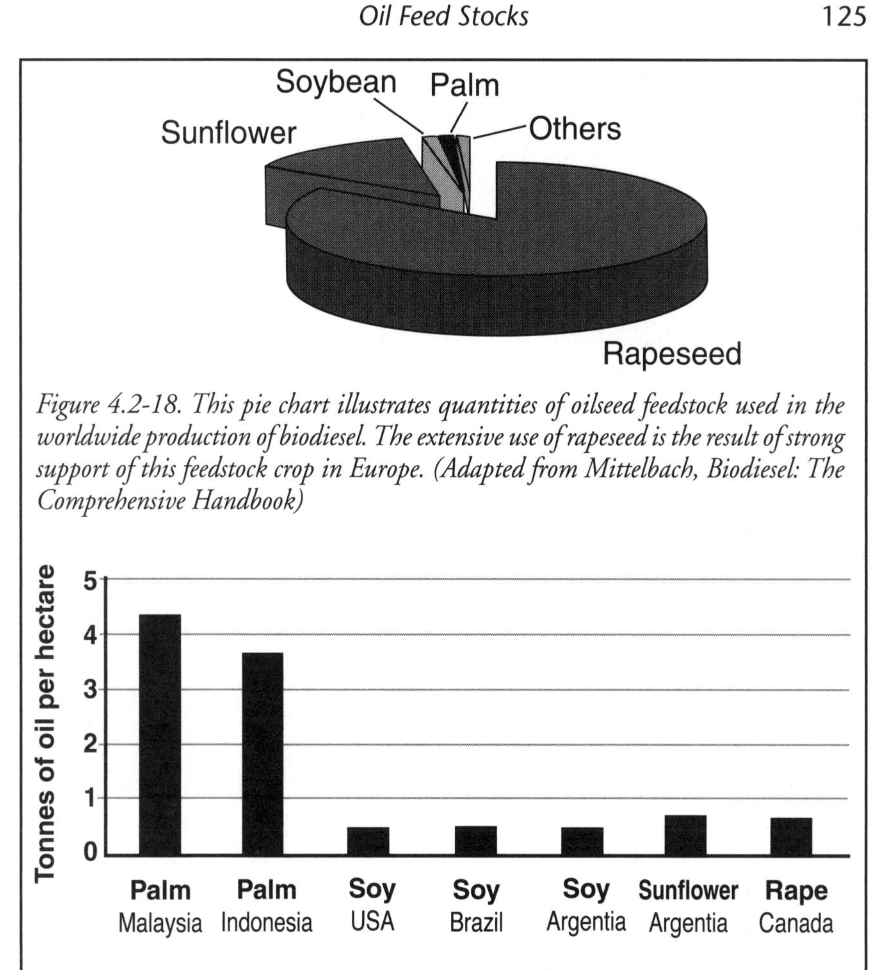

Figure 4.2-18. This pie chart illustrates quantities of oilseed feedstock used in the worldwide production of biodiesel. The extensive use of rapeseed is the result of strong support of this feedstock crop in Europe. (Adapted from Mittelbach, Biodiesel: The Comprehensive Handbook)

Figure 4.2-19. Oil production per hectare of land for a selection of oil-producing plants[2]

Figure 4.2-19 shows the oil production per hectare of land or "specific yield" for a selection of oil-producing plants. Although Malaysian and Indonesian oil palm plants provide as much as five times the specific yield of soybean, these plants grow only in tropical and subtropical areas and are of no value to North American producers. Palm oil is a moderately priced feedstock and given the desire of biodiesel producers to compete with subsidized petrodiesel, low feedstock cost is mandatory. However, many of the palm oil-producing countries have poor environmental regulations which may result in unacceptable ecological damage in the rush to maximize profits. Additionally, palm oil contains very high levels of saturated fatty acids and produce biodiesel with unacceptably high cloud and pour points of approximately 50°F (10°C), which may limit cold weather blend levels in northern climates.

Many people incorrectly assume that palm oil's tremendous productivity advantage places it in the lead for supplying the biodiesel feedstock market. Although palm oil is a good-quality, low-cost feedstock, it is not nearly as prolific in its production of valuable meal as soybean is (Figure 4.2-20). A reduction in soybean production in favour of lower-cost palm oil would create a shortfall in the world soy protein meal market. This would result in a cyclic effect, increasing soybean production to fill the meal gap, producing excess oil. Of course one cannot forget that soy is home-grown, keeping politicians and farming organizations alert to the "dumping" of palm oil. The interconnectedness of agribusiness markets has end results which are not as simple and logical as one would expect.

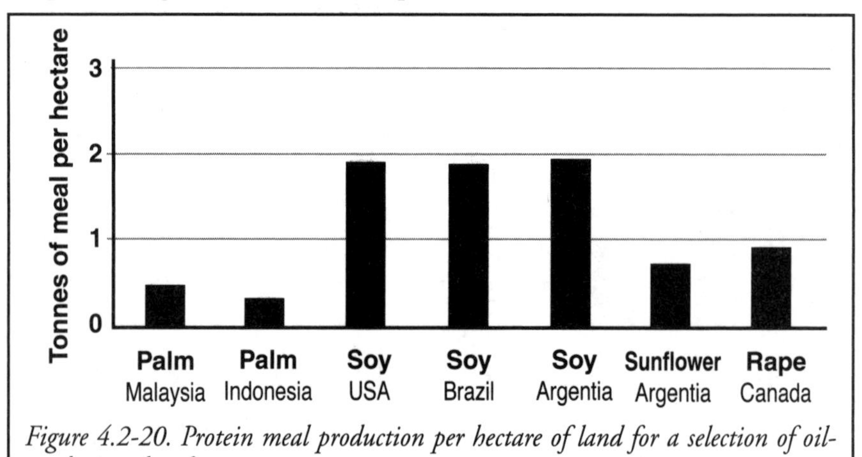

Figure 4.2-20. Protein meal production per hectare of land for a selection of oil-producing plants[2]

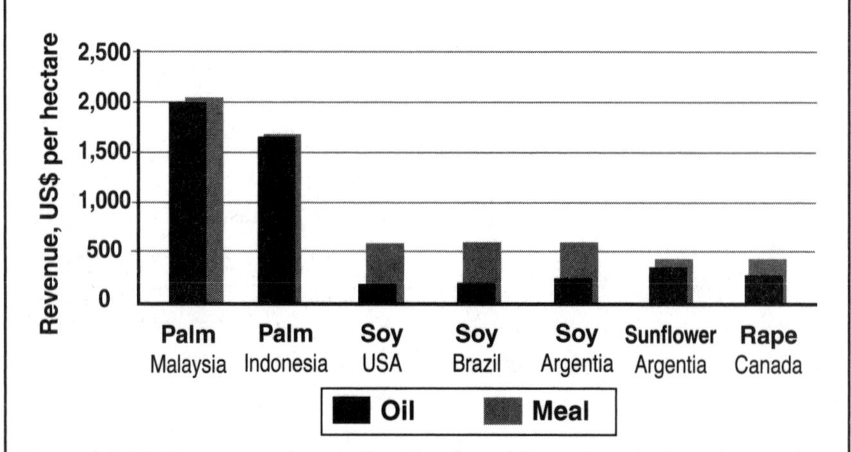

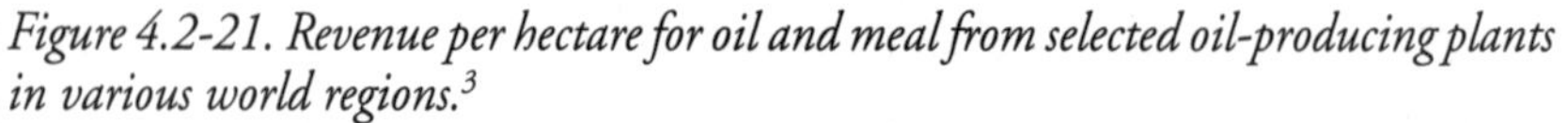

Figure 4.2-21. Revenue per hectare for oil and meal from selected oil-producing plants in various world regions.[3]

The interconnectedness of the markets is clearly demonstrated when revenues from oil and protein meal are superimposed as in Figure 4.2-21. Palm contributes an insignificant share of revenue and volume to the world's meal market, while soy producers in North and South America receive greater value and higher production from meal than from oil.

To summarize:

- Using virgin, refined, food-grade vegetable oils to replace current, inefficient fossil fuel consumption is a completely misguided and unsustainable effort when viewed in the context of the world's energy consumption. There is simply no better solution than to reduce energy consumption through efficiency measures.
- If the industry is to prosper, biofuels and biodiesel must be considered within the context of agricultural sustainability. Displacing land that could be used to grow food in order to grow fuel crops is no solution. However, developing biodiesel as a coproduct of an existing process or technology may offer one answer to the sustainability question.
- The increase in biofuel production is dwarfed by rising energy consumption in both the developed and developing world and any benefits, perceived or real, will be lost in the not-too-distant future.

Figure 4.2-22. Sunflower seed oil is a major oil feedstock for European biodiesel but isn't on the radar in North America. Sunflowers, like all food-grade oilseeds, are best used in the production of food products and only off-specification oilseeds or oil byproducts should be used to support the biodiesel industry.

4.3
An Overview of Waste Oils and Fats

There are few people who deny that they enjoy the flavour of fried foods such as fish and chips, chicken, or even doughnuts. Although they may tempt the taste buds, these foods are contributing to an epidemic of overweight and obese people around the world. Nevertheless, consumption of fast and greasy foods is growing (along with our midriffs), creating a waste stream problem for restaurant owners and, ironically, a healthy business for the disposal industry.

Figure 4.3-1. There are few people who deny they enjoy the flavour of fried foods such as fish and chips, chicken, or even doughnuts. Nevertheless, consumption of fast and greasy foods is growing (along with our midriffs), creating a waste stream problem for restaurant owners and, ironically, a healthy business for the disposal industry.

The food services industry generates several forms of waste oils and fats through the processing and cooking of raw food and from waste water streams. Animal byproducts that are handled by slaughterhouses and poultry processing plants also provide high quantities of fats that must be processed in a safe and profitable manner.

Restaurants and commercial kitchens generate waste vegetable oils and fats directly through the cooking and frying processes. Fryer oils break down after a period of use and become unsuitable for further cooking as a result of increasing free fatty acid content. Once these fryer oils and fats reach a given breakdown level, they are discarded or recycled.

Waste fryer oils and fats can be collected directly by microscale biodiesel users, providing them with a low- (or no-) cost feedstock. The type and quality of the fryer oil used by a given restaurant determines the ease of production

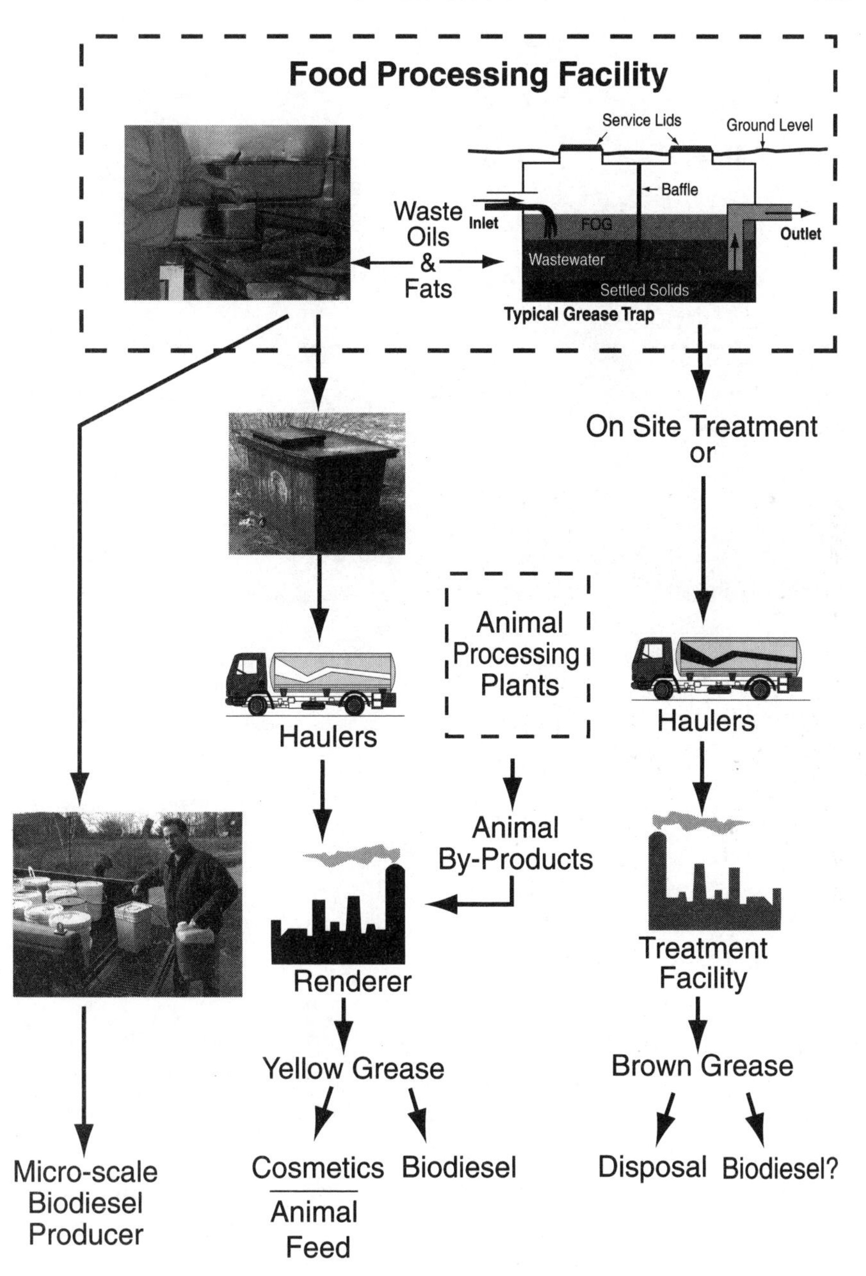

Figure 4.3-2. The food services industry generates several forms of waste oils and fats through the processing and cooking of raw food and from waste water streams. Animal byproducts that are handled by slaughterhouses and poultry processing plants also provide high quantities of fats that must be processed in a safe and profitable manner.

Figure 4.3-3. Restaurants and commercial kitchens generate waste vegetable oils and fats directly through the cooking and frying processes. Fryer oils break down after a period of use and become unsuitable for further cooking as a result of increasing free fatty acid content. Once these fryer oils and fats reach a given breakdown level, they are discarded or recycled.

as well as the quality of the biodiesel. This will be more fully discussed in Chapter 4.4.

Waste fryer oils may also be recycled, which is the most common disposal method used by the food industry. Landfill discarding is not considered an acceptable means of eliminating this product because of its high economic value.

Many rendering and recycling companies provide small oil storage tanks which hold the waste oils and fats generated by the food processor. Periodically, a hauling company will empty the tank and deliver the oil to a rendering facility for processing, which results in the production of yellow grease.

Feedstock materials for the production of yellow grease may also include animal byproducts from slaughterhouses and poultry processing plants as well as from numerous other sources such as butcher shops, supermarkets, farms, ranches, and animal shelters.[1] With meat consumption in the United States averaging 220 pounds (100 kg) per capita per year, the total volume of animal-related byproducts is enormous.

Recycling and processing of waste oils and fats now generates over 2.75 billion pounds (1.25 billion kg.) of yellow grease annually.[2] Yellow grease is a

Figure 4.3-4. Many rendering and recycling companies provide small oil storage tanks which hold the waste oils and fats generated by the food processor. Periodically a hauling company will empty the tank and deliver the oil to a rendering facility for processing, which results in the production of yellow grease.

valuable commodity, providing the livestock, poultry, pet food, and cosmetics industries with an inexpensive and high energy density product.

Many microscale biodiesel producers comment that their production of biodiesel through the use of waste oils and fats helps to save the environment by eliminating waste disposal. This is not supported by the facts: that the vast majority of waste oils and fats is recycled and reused in numerous processes which are as good as (and possibly better than) microscale biodiesel production. Nevertheless, if microscale production is done correctly and with regard to the waste streams inherent in the processing, there should be little negative impact from this activity.

Food processing industries also generate a very large amount of waste oils and fats through their waste water systems. Washing pots and pans after cooking, frying, or sautéing releases small amounts of waste oils and fats into the wash water, which would end up causing damage through buildup and the clogging of municipal sewer lines. To prevent this from occurring, special devices which capture this "trap grease" are added to the water discharge system.

Grease traps require periodic servicing which can be accomplished either onsite through the addition of chemicals or enzymes to break down the fats, oils, and grease (FOG) or by employing a vacuum removal service to evacuate the unit and haul the grease for disposal. As a result of the very high levels of water, contaminants, and free fatty acid levels, the resulting "brown grease" is nearly always disposed of by landfill means.

Figure 4.3-5. Many microscale biodiesel producers comment that their production of biodiesel through the use of waste oils and fats helps to save the environment by eliminating waste disposal. This is not supported by the facts: that the vast majority of waste oils and fats is recycled and reused in numerous processes which are as good as (and possibly better than) microscale biodiesel production.

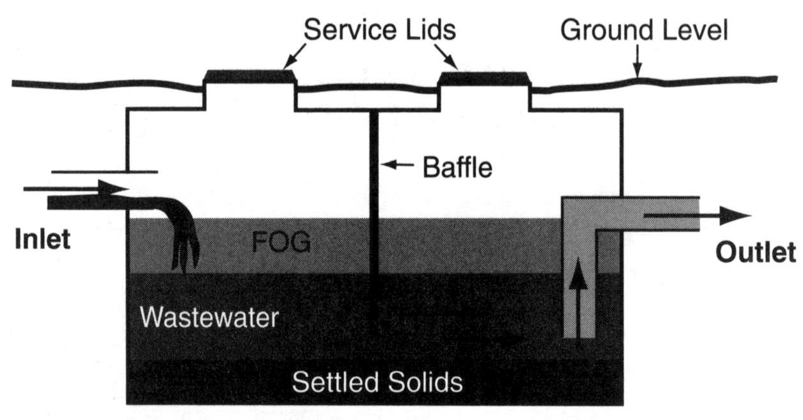

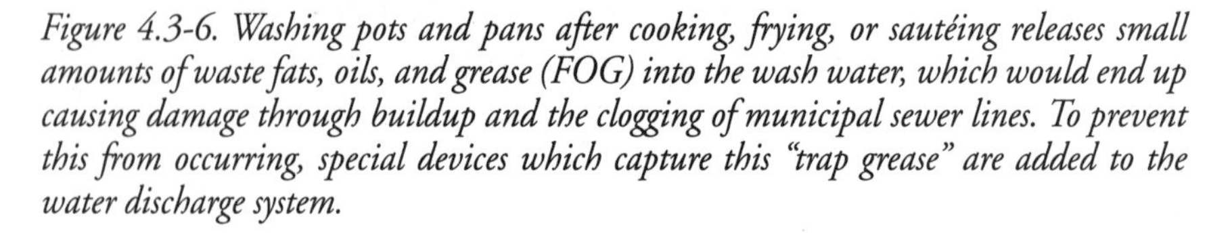

Figure 4.3-6. Washing pots and pans after cooking, frying, or sautéing releases small amounts of waste fats, oils, and grease (FOG) into the wash water, which would end up causing damage through buildup and the clogging of municipal sewer lines. To prevent this from occurring, special devices which capture this "trap grease" are added to the water discharge system.

4.4
Waste Fryer Oils and Fats

Waste fryer oils and fats are an obvious feedstock for the microproducer since it solves two problems: the food processor saves disposal or recycling costs and the producer gains a low- or no-cost feedstock.

The relative simplicity and low cost of transesterifying waste oils and fats into a "rough-quality" biodiesel have led thousands of people to produce their own fuel from this feedstock. At the commercial end of the process, numerous large-scale producers also draw on this feedstock through the purchase of controlled-specification yellow grease.

Waste fryer oils and fats comprise three basic feedstocks:

- vegetable oils such as soy, peanut, canola, and sunflower
- semi- or fully hydrogenated vegetable oil
- animal fats from lard and tallow

Figure 4.4-1. Waste fryer oils and fats are an obvious feedstock for the microproducer since it solves two problems: the food processor saves disposal or recycling costs and the producer gains a low- or no-cost feedstock.

Although each parent oil or fat produces biodiesel with differing properties and qualities, most people do not discriminate between them and give them the collective name "Waste Vegetable Oil" or WVO. This can be a mistake for microproducers as they are likely to use high-level concentrations of biodiesel in their vehicles, which may result in operational problems. In addition to the parent oil or fat, WVO contains varying amounts of water, solids (skin, bones, meat), and other contaminants resulting from the cooking process. Finally, the degradation of the WVO resulting from excessive frying time or heat increases the level of free fatty acids in the oil, causing the transesterification process to be difficult or impossible.

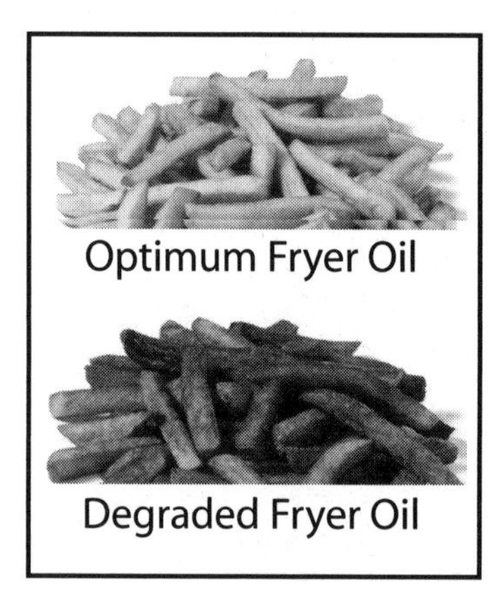

Figure 4.4-2. Fryer oils and fats that have been "overworked" not only produce poor quality fries but also make biodiesel transesterification difficult if not impossible because of excessive free fatty acid content and water contamination.

Numerous microproducers of biodiesel are often surprised when a seemingly good batch of biodiesel suddenly turns into a shimmering mass of "gravy." This can easily be avoided by paying close attention to the quality and composition of the feedstock and following the production procedures outlined in later chapters.

Vegetable Oil

Anyone who has worked with oils and fats recognizes that most (but not all) oils are liquid at room temperature while fats remain solid. After processing the feedstock into biodiesel, the cloud and pour point temperatures for the resultant biodiesel will not change appreciably. Oils that are highly saturated will exhibit cold flow problems. If the majority of the biodiesel use is in tropical or temperate areas, this may not be an issue. However, people wishing to use biodiesel in northern climates will notice that biodiesel made from saturated oil feedstock has a tendency to gel, plugging fuel filters and possibly the fuel injection system. This may cause some inconvenience and the need to tow the vehicle to a warm location until the biodiesel melts. Less severe problems will manifest themselves as fuel starvation and power loss caused by partial filter plugging.

The issue of cold weather operation can be easily corrected by blending the affected biodiesel with No. 2-D or, in severe climates, No. 1-D fuel. Cold flow performance can also be improved by obtaining WVO that is known to be produced from canola or soy oil. Alternative oils, such as, peanut, palm, and palm kernel, become buttery at the first sign of a chill in the air and therefore may not be suitable in high-blend concentrations.

Figure 4.4-3. Cold flow performance can be improved by obtaining WVO that is known to be produced from unsaturated canola or soy oil. Alternative oils, such as, peanut, palm, and palm kernel, become buttery at the first sign of a chill in the air and therefore may not be suitable in high-blend concentrations. In this image, canola-based biodiesel is shown left, while peanut oil biodiesel is turning into peanut butter even though the air temperature is above 50°F (10°C).

Semi- or Fully Hydrogenated Vegetable Oil

A quick look at the label of a tub of margarine will show that it is approximately 80% vegetable oil, yet it remains a semisolid mass at room temperature. When hydrogen gas is rapidly injected into a base vegetable oil, a process known as hydrogenation causes the oil to solidify. While this process is useful for making margarine, it introduces the same cold temperature operability issues that occur with naturally hydrogenated oils and all fats. If possible, avoid these potential feedstock oils.

Animal Fats from Lard and Tallow

Animal fats contain high levels of saturated fatty acids and produce biodiesel with severe cold weather operability issues. As with other saturated fat feedstocks, users will want to consider the liability of cold flow problems in their geographical area.

Waste Oil Quality

Regardless of which feedstock source is available, attempt to obtain it directly from the fryer to ensure minimal storage contamination. Ask the restaurant staff to stop using any water to help rinse the fryer or to ensure that rinsing is done after your containers are filled. WVO that has been stored in the recycling storage unit may pick up large quantities of water and be subjected to higher levels of free fatty acid as a result of oxidation degradation in the tank.

As the facility owner would normally pay to have the WVO recycled, the offer of free recycling in exchange for a minor amount of careful handling makes economic sense and greatly reduces the work necessary in processing the feedstock into biodiesel.

As an added suggestion, one that will benefit both you and the restaurant owner, consider purchasing a box of 3M™ Shortening Monitor test strips[1] and give them to the restaurant owner. The 3M™ Company has developed these low-cost and easy-to-use strips which indicate when fryer oils and fats have expired and ensure that oils are not discarded too early. Premature rejection is costly for the owner, while using oils beyond their useful life will reduce the quality of fried foods.

From the point of view of the microproducer, having an oil supply of consistent quality is possibly the most important step in manufacturing biodiesel and eliminating excessive waste due to rejected batches.

Feedstock oils with free fatty acid (FFA) levels exceeding 1% can become problematic. Large amounts of alkali catalyst are required to neutralize the FFAs and sufficient catalyst must remain to complete the transesterification reaction. High levels of alkali catalyst in the presence of FFAs create an unstable condition that produces soaps and emulsions, greatly complicating the biodiesel production process. (This issue will be discussed in greater detail later.)

Although the microproducer is unlikely to receive WVO that has been underused, with a desirably low level of FFAs, the use of test strips may prevent accepting oils that have been subjected to excess use and have correspondingly high FFA readings.

The Shortening Monitor strips are dipped into fryer oil at operating temperature and all four color bands are submerged for two seconds. The

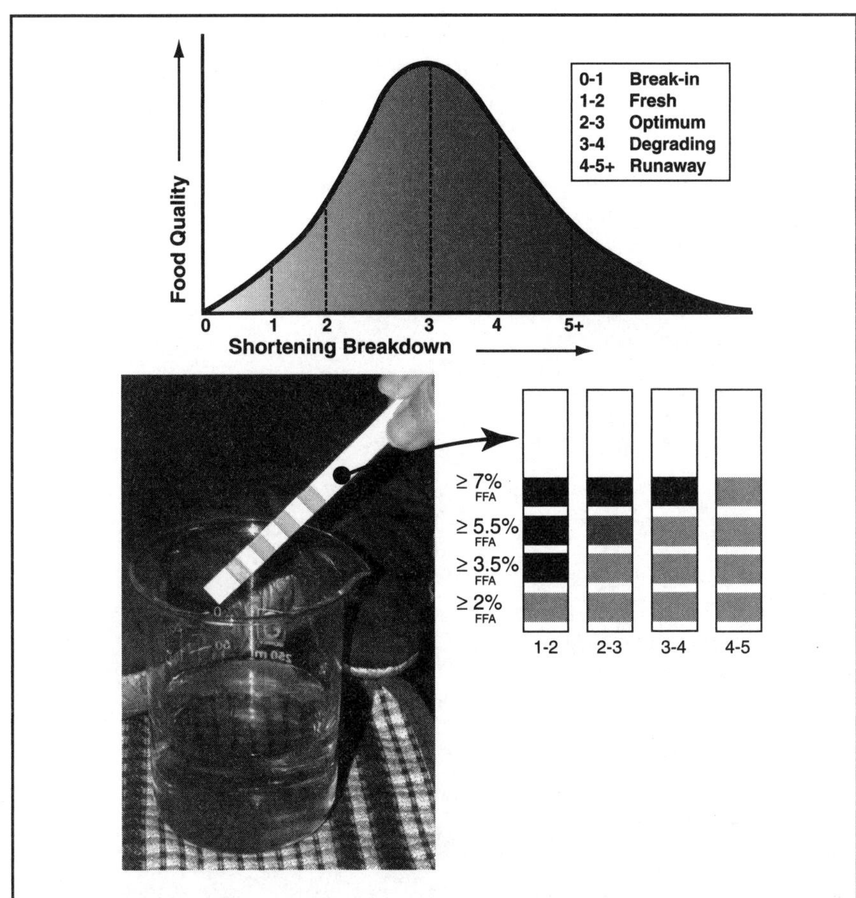

Figure 4.4-4. The 3M™ Company has developed these low-cost and easy-to-use free fatty acid test strips which indicate when fryer oils and fats have expired and ensure that oils are not discarded too early. Premature rejection is costly for the owner, while using oils beyond their useful life will reduce the quality of fried foods. Strips are dipped into a sample of hot WVO and the change in color bands indicates the level of free fatty acid. From the point of view of the microproducer, having an oil supply of consistent quality is possibly the most important step in manufacturing biodiesel and eliminating excessive waste due to rejected batches. (Adapted from 3M™ Food Services brochure for 3M™ Shortening Monitor.)

color bands are then compared to an included color chart (see Figure 4.4-4). A change of color from blue to yellow on each band indicates progressively higher levels of FFA to a maximum of 7%. The chart shown in Figure 4.4-4 indicates the approximate relationship between food quality and FFA level. Note that "food quality" is a subjective measure and it will be up to restaurant management to relate it to FFA or color band levels. Once this calibration has been completed, the fryer operator can quickly monitor quality and recycle the oil prior to runaway degradation.

Once the WVO has been transported to the microproducer's facility, the test strips can also be used as a quick way to determine the FFA level in individual batches of oil. Comparing several buckets of WVO will provide the data necessary to allow mixing of the samples with the highest and lowest FFA levels, resulting in a lower average reading.

Figure 4.4-5. Paul Zammit and Michelle Murphy, owners of The Fall River Pub and Country Gift Store in Maberly, Ontario, believe in sustainability in their business, which uses local produce in the preparation of first-rate meals and sells local Church-Key beer and a variety of community goods and crafts in the gift shop.

Figure 4.4-6. Paul is shown with a bucket of canola-based WVO before pouring it into the recycling container. Interestingly, the recycler, Rothsay/Laurenco, is Canada's first major biodiesel producer using WVO gathered from restaurants as well as animal byproducts from the rendering side of the business.

Figure 4.4-7. Once the WVO has been transported to the microproducer's facility, the test strips can also be used as a quick way to determine the FFA level in individual batches of oil. Comparing several buckets of WVO will provide the data necessary to allow mixing of the samples with the highest and lowest FFA levels, resulting in a lower average reading and easier biodiesel production.

Figure 4.4-8. One of the largest problems affecting microscale biodiesel producers is creating a big batch of "glob" rather than biodiesel. This is due largely to excess water in the WVO resulting from food particles or water used to rinse the fryer. This WVO is approximately 50% water as result of fryer rinsing. Note the phase separation: water sinking to the bottom and oil floating on top.

4.5
Yellow Grease

Yellow grease is the final edible or inedible product resulting from recovered and recycled processed fats and oils. As noted in Figure 4.3-2, inputs to the rendering and recycling process may include animal byproducts from meat packers, food service establishments, stockyards, and animal shelters. Edible rendering plants process animal tissue, generally working in conjunction with a slaughterhouse or poultry processing facility. Inedible oil rendering plants produce tallow and grease which can be used in livestock feed, soap, and the production of fatty acids.

Rendered products that contain animal fats will generally be solid or jelly-like at room temperature. In addition, the FFA profile of the input animal product sources will vary considerably, leading to a highly variable qualities of yellow grease feedstock. Yellow grease is sold using "wider" specifications than refined vegetable oils or fats, but at a lower cost, making it a good feedstock for larger-scale biodiesel producers who have the ability to monitor and adjust to this high degree of variability.

4.6
Brown Grease

Brown grease typically refers to those oils and fats that have very high levels of free fatty acids, typically above 30%, as well as other contaminants including water. The primary sources of brown grease are either yellow grease that has oxidized as a result of excessive thermal oxidation (exposure to air) stress or trap grease which is extracted from the waste water systems of restaurants and food processing facilities.

Although experiments in the production of biodiesel from brown grease have been conducted at the laboratory level, the highly variable feedstock and low yield levels of biodiesel it produces have virtually eliminated its use in this application.

Currently, brown grease has very few applications and even disposal has been largely limited to landfill methods.

4.7
Feedstock Summary

The choice of feedstock used in the biodiesel process will almost always be biased towards lowest cost per pound. Biodiesel costs are largely attributed to the feedstock material, which creates enormous pressure to locate the best price and highest yield available.

Fatty acid levels determine the yield of FAME produced from oilseed and also affect the choice of production capital equipment and process technology. Although microproducers may be willing to work around high FFA levels if the feedstock is free or essentially free, small-scale and commercial producers must weigh the costs of pre-processing these lower-quality feedstocks before determining the correct technology. FFA levels above 1% will require the commercial producer to pretreat the feedstock, while many microproducers ignore this concern and risk making "bad batches" of biodiesel.

The desired cold-weather operation characteristics also influence feedstock selection. Refined canola and soy oils have better cold flow characteristics, which may impact production and storage techniques as well as end use concentration or blending levels.

5 Commercial Biodiesel Production

5.1 Overview

For the purposes of this discussion, the definition of a commercial biodiesel plant is one that produces more than 500,000 gallons (1.9 million liters) per year. This is a very arbitrary number, but it reflects my belief that the world biodiesel industry will eventually segment itself into three layers as I discussed in the introduction to this book:

- Layer 1: commercial production (500,000+ gallons/year) (1.9 million liters)
- Layer 2: cooperative and small-scale community systems (10,000-500,000 gallons/year (38,000-1.9 million liters)
- Layer 3: micro-scale producers (0-10,000 gallons/year) (0-38,000 liters)

Although the commercial biodiesel market focuses only on the first layer, (plants operating at capacity levels of 500,000 gallons per year or higher), over time the cost of production will fall enough to allow distributed production facilities to flourish in smaller-scale agricultural areas. In due course the farming community will be able to process low-grade oil seed or other feedstock materials and convert them directly into biodiesel for self-consumption on the farm. Given that energy and fuel costs are soaring and that large farm equipment requires enormous amounts of diesel fuel to operate, the use of high-quality B100 made from zero- or near-zero-cost feedstock is only a matter of time. The slow movement to energy self-sufficiency is the only way commercial agricultural interests will be able to maintain their operations in an era of expensive, supply-constrained peak oil.

Figure 5.1-1. The West Central biodiesel plant, located in Ralston, Iowa has a rated capacity of 12 million gallons per year (45 million liters per year) and is typical of a large-scale commercial production facility. (Courtesy West Central)

Commercial biodiesel producers resist the idea of calling micro-scale producers (a.k.a. home brewers) legitimate producers of biodiesel. However, this was also the reaction of IBM and several other corporations when a handful of teenagers named Jobs, Wozniak, and Gates revolutionized the high-tech industry.

It is true that many micro-scale producers do an excellent job of making garbage-quality fuel in a vain attempt to save a dollar or two while pumping toxic chemicals into the sewer system and dumping equally toxic glycerol into the landfill while hiding behind a holy veil of "environmental sustainability."

But every dark cloud has a silver lining, and with the dissemination of some sound information on the Internet and through sources such as this book, a "disruptive technology" will emerge that will legitimize many in the micro-scale biodiesel production arena.

Many people believe that centralized production of "everything" is the only way to go. The idea that bigger is better and the belief in economies of scale have been paramount in the era of cheap energy and easy transportation. However, there are those who believe that distributed energy generation on the "right scale" is the correct way forward. If there is a profit to be made, then the technology will be created. The vast number of home brewers working away in garages are now being joined by companies that are attempting to meet the demand (see Chapter 7.1).

The following overview of large-scale biodiesel production is provided as a starting point to help define the existing technologies and ultimately to see how the process can be scaled down to service the second and third layers of the biodiesel industry, which is the theme of the balance of this book.

5.2
Technology Selection – Process Selection: Batch vs. Continuous

Overview

The target audience for this book is not chemistry professors and chemical engineers, since they have the university laboratories and technical manuals to fulfill their need for information. The intended audience is the group of people in the second and third "layers" of the market who are curious or wish to understand the technical and environmental issues related to FAME production.

I have assumed that most people will have minimal, if any, background in chemistry and chemical engineering in order to make the descriptions of techniques and equipment accessible to every reader. Those wanting more detailed information should consult the reference section of this book.

As we learned earlier, biodiesel or FAME is produced when the triglycerides contained in animal fats and vegetable oils reacts with alcohol in the presence of a catalyst. This process, known as transesterification, "knocks" the glycerine molecule from the fat or oil and replaces it with the alcohol. At the conclusion of the reaction, you are left with the desired FAME as well as unrefined glycerin that is contaminated with methanol, catalyst, and soap.

Typical proportions of feedstock and process chemicals used in the commercial production of biodiesel are:

Refined Feedstock Fat or Oil *<2% FFA)*	*100 kg (220 lbs)*
Alcohol (methanol)	*10 kg (22 lbs)*
Alkali Catalyst (sodium hydroxide)	*0.3 kg (0.66 lbs)*
Catalyst Neutralizer (sulfuric acid)	*0.25 kg (0.55 lbs)*

Table 5.2-1
Typical proportions of feedstock and process chemicals used in the commercial production of biodiesel. (Source: Building a Successful Biodiesel Business, Van Gerpen, et al)

The Batch Reaction Process

The simplest method of producing FAME is the single reaction tank batch process. An updated version using two reaction tanks is used in many smaller commercial facilities (< 4 million liters per year / ≈ 1 million gallons per year). The advantage of the two-tank system is that it ensures the completeness of the reaction process and guarantees that total glycerin levels are within the ASTM specification of less than 0.24%. An example of such a system is shown in Figure 5.2-1.

Here is a summary of what happens during the two-tank batch process:

- Vegetable oil or animal fat is loaded into a storage tank and heated.
- An alcohol and catalyst (methoxide) reactant is loaded into a second tank.
- The feedstock oil and reactants are pumped into Batch Tank #1 where they are mixed and heated until the conclusion of the reaction process.
- The materials in Batch Tank #1 separate into two layers or phases: raw FAME and glycerin.
- The glycerin is pumped to a processing section which neutralizes the catalyst, recovers and recycles the methanol, and provides unrefined glycerin (glycerol) for sale.
- The raw FAME is pumped to Batch Tank #2, where it is reprocessed with methoxide reactant to remove trace amounts of bonded glycerin.
- Excess glycerin is pumped off and recovered as in the first reaction step.
- Raw FAME is sent to a washing and catalyst-neutralizing tank. Wash waters are processed for disposal.
- The high-quality FAME is dried and readied for testing and sale.

Vegetable oil or animal fats having a free fatty acid content of less than 2% and preferably less than 1% are loaded into the feedstock tank shown upper left. A steam-heated exchanger coil heats the oil feedstock to approximately 60°C. It does not make any difference which parent feedstock oil or fat is used provided that the FAME meets all of the requirements of ASTM D 6751 at the completion of the production process. The reader is again reminded that fats and oils that are fully or partially hydrogenated as well as highly saturated will be affected by cold flow issues.

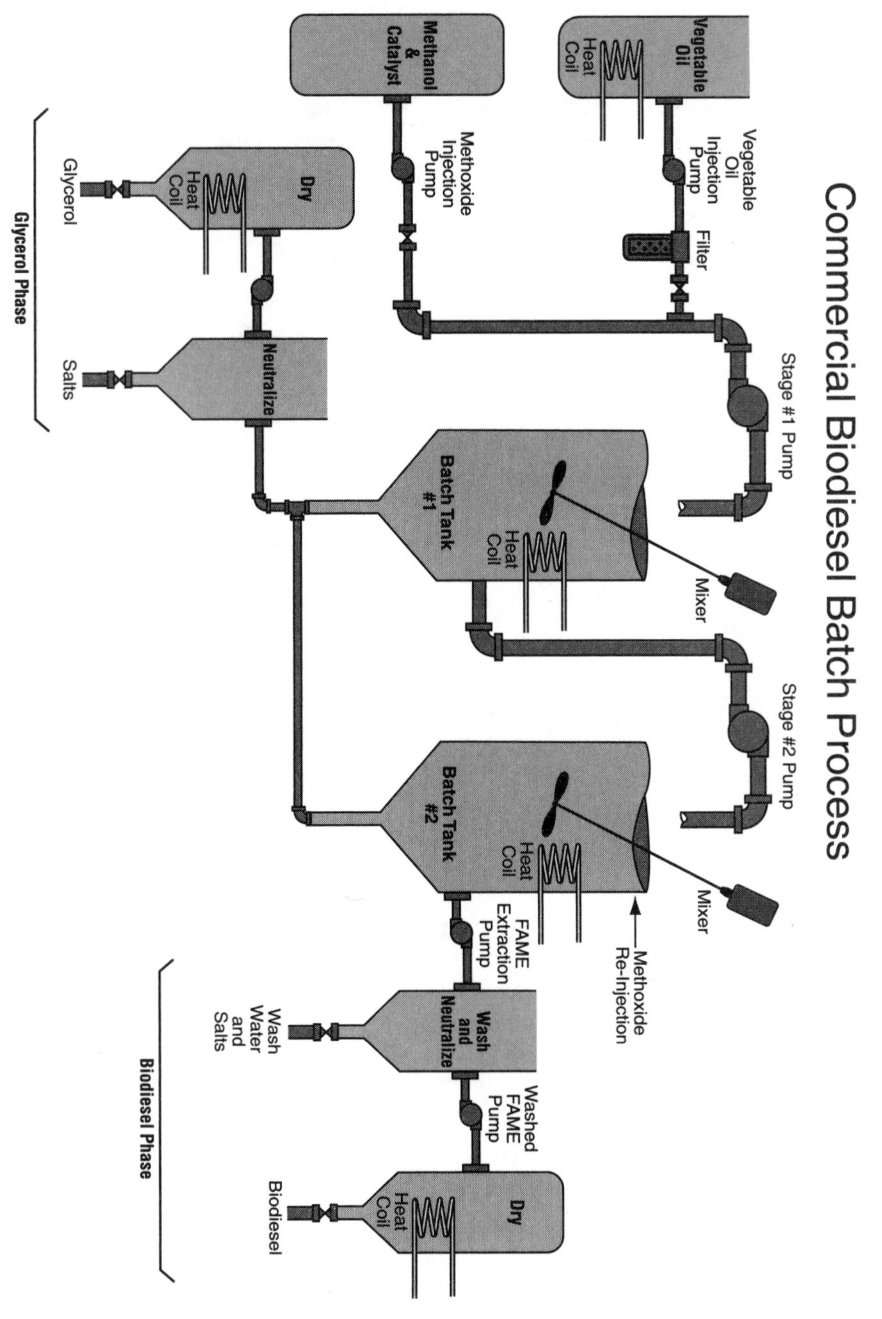

Figure 5.2-1. The simplest method of producing FAME is the single reaction tank batch process. An updated version using two reaction tanks is used to ensure the completeness of the reaction process and guarantee that total glycerin levels are within the ASTM specification of less than 0.24%.

A second storage tank is filled with a commercial premixed (methoxide) reactant solution comprising methanol (alcohol) and a catalyst of either sodium hydroxide (NaOH) or potassium hydroxide (KOH) which, when blended, is known as sodium or potassium methoxide, depending on the chosen catalyst. The sodium or potassium methoxide can also be produced at the factory.

Methanol is the most commonly used alcohol, although other alcohols may be substituted. Alcohol selection is determined by several factors including cost, toxicity, ease of recycling, and quantities required to complete the reaction process. Although methanol is principally derived from natural gas and is highly toxic in the environment, thus reducing the overall "green value" of biodiesel, it meets the litmus test of low cost, ease of use, and recycling.

Because oil and alcohol do not have an affinity for each other, a catalyst is used to initiate the transesterification reaction. A catalyst may be either alkaline (base or basic) or acidic in nature, with experience showing that acid-catalyzed reactions operate too slowly for cost-effective biodiesel production purposes.

The selection of base catalyst is also a matter of choice, although sodium hydroxide is the most commonly selected compound. The feedstock selection and free fatty acid level must be considered along with catalyst selection. The combination of base catalysts and FFA levels exceeding approximately 2% will cause the formation of soaps that can hinder the reaction process and create a contaminant in the glycerin phase.

Once the feedstock oil has reached operating temperature, it is filtered and pumped into the batch reactor tank where it is continuously heated and stirred. The methoxide reactant is then pumped into the reactor tank and vigorous mixing is continued for a period of time ranging from 30 minutes to one hour to ensure the completeness of the reaction by placing the feedstock oil, alcohol, and catalyst in close contact.

At the end of the mixing period, the mixer and heating coils are deactivated and the solution is allowed to settle, causing it to separate into two phases with the raw glycerin (glycerol) sinking to the bottom of the tank and the raw FAME floating on top as the second phase. Reaction completeness is reportedly in the range of 85% to 94%, which is below the requirements necessary to ensure that bonded glycerin meets the ASTM standard.[2]

The glycerol phase is pumped to a processing unit where methanol is captured and recycled for further use. The glycerol is also washed with acidic water to neutralize the basic catalyst. Wash water is processed and either discarded after meeting required environmental standards or recycled for further use in the plant. The glycerol is then dried and sold to a refining company.

The raw FAME that was produced in the first reaction is sent to a second reaction tank where it is reprocessed with fresh methoxide reactant in a process similar to the first reaction. Introducing a second reaction drives the transesterification to over 95% completeness, ensuring that bonded glycerin levels will now fall within the required standards.[3]

Glycerol is again pumped off and processed as in the first reaction stage.

The raw FAME is subjected to methanol capture before being washed in the same manner as the glycerol phase. A drying step ensures that the high-quality, washed FAME contains less than the prescribed amount of suspended water before it is sent to storage.

The Continuous Reaction Process

The continuously stirred tank reactor (CSTR) process is an improvement on the batch system because, as the name implies, feedstock oil and reactants are continuously added to the unit without the need to charge and discharge large, fixed-volume tanks. As a result, larger (greater than 4 million liters per year capacity) producers have migrated to this technology. The CSTR replaces the batch tanks with a series of long tubular structures containing internal "obstructions" that cause the reactants to mix as they are pumped through the unit. Varying the speed of transit through the CSTR will vary the time the reactants are in contact with each other, improving the reaction completeness.

The reaction may be completed in either one or two stages, as with the batch system. If a two-stage system is employed, approximately 80% of the methoxide reactant is mixed with the feedstock oil during the first stage. The glycerol is then removed and the remaining methoxide reactant is added to the second stage of the system.

All other aspects of the system remain essentially the same as with the batch system.

An example of a CSTR system is shown in Figure 5.2-3

Figure 5.2-2. The continuously stirred tank reactor (CSTR) process is an improvement on the batch system because, as the name implies, feedstock oil and reactants are continuously added to the unit without the need to charge and discharge large, fixed-volume tanks. As a result, larger (greater than 4 million liters per year capacity) producers have migrated to this technology. (Courtesy West Central)

High Free Fatty Acid Feedstock

In the event that high-FFA feedstocks are available at low cost, both the batch and CSTR systems can be modified to accept an additional "feedstock pretreatment unit." Using a process known as direct acid esterification, a mixture of methanol and acid (as opposed to an alkali or base catalyst) can be used to pretreat the high-FFA feedstock. Using this process, glycerol is not produced, however a mixture of methanol, chemically formed water and acid form a distinct phase which is removed. The resulting feedstock can then be sent directly to the batch or CSTR system as described above.

Numerous alternative methods of dealing with low-cost, high-FFA feedstock are used in the commercial production industry, but they are outside of the scope of this book.

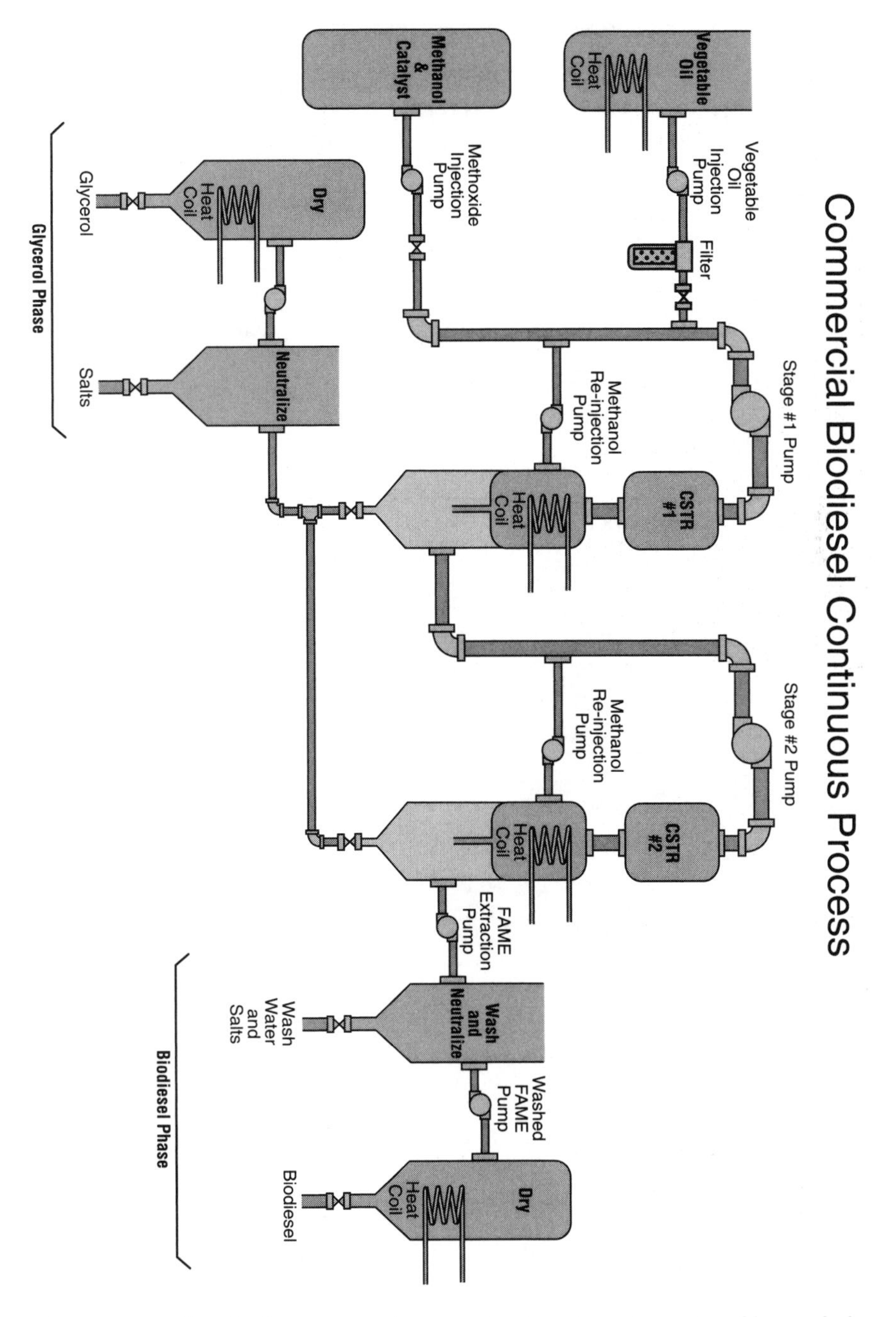

Figure 5.2-3. The CSTR reactor replaces the batch tank with a series of long tubular structures containing internal "obstructions" that cause the reactants to mix as they are pumped through the unit. Varying the speed of transit through the CSTR will vary the time the reactants are in contact with each other, improving the reaction completeness.

5.3 Economic Considerations

- **Plant Costs**

The quick rule of thumb for calculating the value of a biodiesel plant is to base it on the yearly capacity of the facility, which is the number of gallons it produces operating 24 hours a day, 350 days a year. New biodiesel production facilities cost approximately US $1.00 per gallon ($0.26 per liter) of annual capacity and have operating and maintenance costs of between $.20 and $.50 per gallon."

Figure 5.3-1. New biodiesel production facilities cost approximately US$1.00 per gallon of rated annual capacity and have operating and maintenance costs of between $0.20 and $0.50 per gallon. (Courtesy West Central)

- **Production Costs**

The primary cost in the production of biodiesel is the oil feedstock. Based on the current price for refined soy oil, feedstock costs (all figures in US$) range from $1.00 to $1.90 per gallon ($0.26 to $0.50 per liter). Production plant operating and maintenance costs are between $0.20 and $0.50 per gallon ($0.05 and $0.13 per liter) . Plants with higher operating efficiencies and greater production volumes will tend to be on the lower end of this cost range, while smaller, less efficient plants will be on the higher end.

- **Glycerol**

The value of unrefined glycerol is negligible, and it may even constitute a waste product liability unless it can be locally refined into glycerin. Small-scale plants may have the ability to refine raw glycerin; micro-scale producers will not. The price of refined glycerin is expected to become depressed in the coming years as a result of the expansion of the biodiesel market and all producers should be wary of placing too much value on this coproduct of the transesterification process. The issue of what to do with unrefined glycerol is covered in Chapter 11.

- **Subsidization**

The U.S. Department of Agriculture has created the Commodity Credit Corporation which makes direct payments to producers of ethanol and biodiesel to offset feedstock commodity costs. The subsidization program pays 40% of the cost of soybeans or other oil seeds provided they are processed into biodiesel. Readers are urged to monitor the program to ensure funding is in effect.

Figure 5.3-2. The U.S. Department of Agriculture has created the Commodity Credit Corporation which makes direct payments to producers of ethanol and biodiesel to offset feedstock commodity costs. The subsidization program pays 40% of the cost of soybeans or other oil seeds provided they are processed into biodiesel.

For more information on the CCC program contact:

> Kansas City Commodity Office
> Contract Reconciliation Division
> Financial Review Branch, Mail Stop 8758
> P. O. Box 419205
> Kansas City, Missouri 64141-6205
> Tel: (816) 926-6525
> Fax: (816) 823-1805
> E-Mail: crdfrb@kcc.usda.gov
>
> Web: www.fsa.usda.gov/Daco/kcco.htm

There are currently no direct subsidization programs available to Canadian oilseed producers for the production of biodiesel.

- **EPA Registration**

In the United States, producers of biodiesel who make fuel that is sold for on-road vehicle operation must be registered with the Environmental Protection Agency (EPA). The requirements for registration are onerous, with most producers opting to join the National Biodiesel Board (www.nbb.org) since the data required for EPA registration is provided as a benefit of joining. The regulations are described in detail in Title 40 CFR Part 79 (EPA Protection of the Environment, Registration of Fuels and Fuel Additives - www.access.gpo.gov/nara/cfr/waisidx_01/40cfr79_01.html).

- **Tax Treatment**

Tax treatment of biodiesel is subject to the ebb and flow of policies in federal, state, and provincial jurisdictions. It is not possible to provide an up-to-date listing of tax issues, incentives, and regulations in this handbook. In Canada a better approach is to visit both the National Biodiesel Board and Canadian Renewable Fuels Association websites for the latest tax information.

All U.S. producers of biodiesel are required to register with the Internal Revenue Service (IRS) by submitting Form 637. The forms are available at www.irs.gov. For those who wish to consider producing biodiesel for sale or trade, the IRS Publication 3207, Small Business Resource Guide, provides an excellent and easy-to-follow CD-ROM with all the necessary tax forms, instructions, and publications. The CD-ROM is updated yearly and a free copy is available by ordering online at the IRS website at www.irs.gov.

• Summary

Despite the economic variables listed above, market demand for biodiesel will continue to swing upwards in both Canada and the United States as a result of the Kyoto Protocol and the U.S. Energy Policy Act (EPAct) as well as new regulations regarding urban smog, greenhouse gas emissions, and minimum fuel standards.

Since federal, state, and provincial regulations fluctuate with political fortunes, readers interested in keeping abreast of these issues may wish to review policy statements and reviews at:

U.S. National Biodiesel Board: www.biodiesel.org
Canadian Renewable Fuels Association: www.greenfuels.org

The largest impediment to biodiesel production is the cost of feedstock materials. The range of feedstock and production costs discussed above suggests why waste- and low-cost feedstocks are of such value to the micro- and small-scale producer, eliminating up to 80% of the cost of producing biodiesel.

It is also true that small-scale producers, farming and community co-operative ventures that may emerge in North America and certainly in the developing world, are not concerned about return on investment as it is not the primary objective of their development strategy. On a North American scale, farming cooperatives could cold press their off-specification oil seed, produce biodiesel at well below the cost of fossil diesel fuel, and use it at high blend levels to fuel farm equipment during the growing season.

Only time will tell if a small-scale revolution will come to pass, but based on the interest and demand we have discussed so far, I think the development of this market segment is inevitable.

Therefore the next question to be considered is: Can the process for making biodiesel and treating the waste-stream materials be scaled down and still remain profitable on a smaller scale?

5.4
Scaling the Production Process Downwards

According to Biodiesel Magazine's 2006 directory (www.biodiesel-directory.com), as of the fall of 2005 there were 42 biodiesel plants in Canada and the United States with individual annual capacities of at least 1 million gallons (3.8 million liters). Their aggregate rated capacity is 316.6 million gallons per year (1.2 billion liters). In addition, there are 25 plants currently under construction or expansion which will double total rated capacity. (Note that rated capacity does not mean that plants are producing biodiesel in quantities anywhere near these manufacturing limits; the National Biodiesel Board estimates that actual production for 2005 was 75 million gallons (284 million liters).

With all of this capacity and large-scale production, is there room for a second layer of biodiesel producer who wishes to drift under the radar below 1 million gallons of annual capacity? I would argue that the commercial industry is reasonably supportive of these efforts, provided the fuel is manufactured to ASTM standards and all applicable health, safety, and environmental concerns are met in accordance with generally accepted industry principles and government regulations. In short, as long as small-scale biodiesel producers act like large-scale producers, there should be no problem legitimizing this layer of the industry.

This view appears to be supported by the fact that the National Biodiesel Board changed its membership criteria to create a new category for small-scale commercial producers with appropriately lower annual membership fees. This action has effectively opened the door for EPA registration that was previously impossible for small-scale producers to achieve.

The picture changes dramatically at the micro-scale level and this subject will be covered in detail in Chapter 7.

Biodiesel Technologies GmbH

Biodiesel Technologies GmbH is an Austrian manufacturer of high-capacity, small-size biodiesel processors. The company is active in the development, manufacturing, and sale of biodiesel processing equipment and is looking to develop international markets for its innovative CPU 1000 processor, "a FAME processor that is installed in a standard 20-foot-long transportation container," states Dr. Laszlo Kondor, Marketing Director for the firm. "The unit is delivered ready to operate after connecting the CPU to external supply and storage tanks for feedstock oil, methanol, water, compressed air, and electrical power. Although the unit is physically small and requires only one operator per shift, it has a rated capacity of over 8 million liters (2.1 million gallons) per year. Because it is competitively priced at 1.5 million Euros, the processor can be operated at lower production volumes suitable for the small-scale producer."

However, the list price of this unit at current exchange rates is approximately US$1.8 million, and there is the added cost of a facility with the necessary infrastructure for pumps, air compressors, a laboratory, and a "tank farm" necessary to handle the in- and outbound fuels and oils. Costs will certainly exceed the "benchmark" price of US$1.00 per gallon of yearly capacity.

Figure 5.4-1. Biodiesel Technologies GmbH is an Austrian manufacturer of high-capacity, small-size biodiesel processors. The list price of this unit at current exchange rates is approximately US$1.8 million, and there is the added cost of a facility with the necessary infrastructure for pumps, air compressors, a laboratory, and a "tank farm" necessary to handle the in- and outbound fuels and oils. Costs will certainly exceed the "benchmark" price of US$1.00 per gallon of yearly capacity, making even a unit of this size uneconomic.

Although the size of the processing system has been reduced, the capital cost has not, making the acquisition of such a system questionable for smaller producers. Since the yearly throughput or capacity of the system is reduced, the capital cost must be amortized into the smaller volume of biodiesel produced, making the system uneconomic.

Biodiesel Technologies has concentrated on small physical size while retaining the high-volume capability of a much larger plant. The key is to meld small physical size with correspondingly lower cost and production volume.

Olympia Green Fuels LLC

Mike Pelly is the enthusiastic developer and founder of Olympia Green Fuels LLC, located in Olympia, Washington. Mike and his business partner Jerry Cook have been concentrating on the development of a small-scale biodiesel production processor that would fit the requirements of the farming community or small business. Mike has been involved in small-scale biodiesel since 1996 after watching Fat of the Land, a humorous documentary about five women who drive from New York to San Francisco in a van powered with biodiesel. Mike has BA and BS degrees from The Evergreen State College and focused his independent studies on biodiesel production technologies. He now has several patents pending on the Pelly Model A biodiesel processor his firm is bringing to market.

Figure 5.4-2. "Fleet owners, municipalities, and farming communities can take control of fuel costs and source of supply by utilizing our economical biodiesel processor," Mike Pelly of Olympia Green Fuels explains. "The Pelly Model A unit can convert either virgin or waste vegetable oil into washed biodiesel in batches of approximately 300 to 400 gallons or larger. Our process will ensure that the fuel meets the ASTM D 6751 standard when using low FFA oil and following proper operating procedures." (Courtesy Olympia Green Fuels LLC)

"Fleet owners, municipalities, and farming communities can take control of fuel costs and source of supply by utilizing our economical biodiesel processor," Mike explains. "The Pelly Model A unit can convert either virgin or waste vegetable oil into washed biodiesel in batches of approximately 300 to 400 gallons. Our process will ensure that the fuel meets the ASTM D 6751 standard."

"We recognized that the large-scale commercial industry is well established, but there was little work being conducted on the small-scale market, specifically for those who wish to produce under 500,000 gallons (1.9 million liters) of fuel annually," Mike continues. "We see our technology serving a niche global market that is currently not well serviced."

Figure 5.4-3. "We recognized that the large-scale commercial industry is well established, but there was little work being conducted on the small-scale market, specifically for those who wish to produce under 500,000 gallons (1.9 million liters) of fuel annually," says Mike Pelly of Olympia Green Fuels. "We see our technology serving a niche global market that is currently not well serviced."

Mike goes on to explain that the list of potential clients is enormous and includes the following groups, NGOs, and other organizations that would benefit from having access to the technology:

- municipalities recycling waste vegetable oil
- farmers and farm cooperatives processing their own crops into fuel
- urban entrepreneurs making use of opportunity feedstock
- universities and technical schools (research, facilities)
- third-world-oriented non-governmental organizations (NGOs)
- vegetable-oil-intensive industries recycling their scrap into fuel
- diesel fleet operators looking for lower fuel costs
- remote fish and animal processors
- consumer fuel cooperatives making their own fuel

"We have designed the Pelly Model A processor so that it is competitively priced and safe and provides all the automation and controls necessary to produce high-quality fuels. The primary advantage of our system is its ability to provide the correctly sized system for the customer's application," Mike says.

I asked Mike if he had a checklist of conditions clients are required to meet in order to come online quickly. "As with any technical plant construction, commissioning, and operation, it is necessary to understand exactly what your requirements and expectations are before you start spending any money. We can work with clients to analyze their feedstock supply, power requirements, and other parameters required to get the system up and running in a safe and profitable manner."

Mike provided the following list of issues for potential customers to consider before they undertake the development of a biodiesel production facility:

- access to start-up capital (US$85,000 to US$150,000, depending on scale and use)
- plans to produce 50,000 to 500,000 gallons (189,000 to 1.9 million liters) of biodiesel annually
- access to a sufficient quantity and quality of feedstock for planned volumes
- markets or consumers for biodiesel fuel and glycerin at planned volumes
- access to permitted and zoned industrial site meeting environmental, fire, and building codes
- access to appropriate utilities: 220VAC, water, ventilation, climate control, sewer
- adequate fire protection and spill containment systems
- local zoning and permitting requirements

Mike concludes our discussion by saying: "The best thing for potential clients to do is speak with us regarding their specific needs. We can tailor the system exactly to current requirements while allowing the ability of the production design to grow as the client's needs change. We are confident in our product design and believe that this system will meet the requirements of anyone who wishes to produce quality biodiesel fuel at reasonable cost."

Summary

The small-scale biodiesel production market has only recently been identified and will remain an underserviced niche market for some period of time. I fully expect to see an automated, container-sized system complete with small tank farm, processing office, and equipment available in due course. It is only a matter of time before the right entrepreneur figures out how to profitably manufacture such a system, especially in view of rising world oil prices that will act as the catalyst necessary to make small-scale biodiesel production happen.

If you cannot wait and you have the mechanical aptitude to construct the equipment necessary to build your own processing facility, consider scaling up the plans provided in Chapter 8.

6 High Blend Levels of Biodiesel

6.1 The Politics of B100

The Engine Manufacturers Association (EMA) is an international association that represents manufacturers of internal combustion engines.[1] In February 2003, the EMA issued a Technical Statement on the Use of Biodiesel Fuel in Compression Ignition Engines which outlines its position on the use of biodiesel fuels in current engine technologies and recognizes the dichotomy between the U.S. Energy Policy Act (EPAct) and the "Diesel Fuel Injection Equipment Manufacturers Common Position Statement".[2]

The EPAct was amended to allow fleet vehicle owners covered by the Act to fulfill up to fifty percent of their alternative vehicle acquisition requirements by purchasing biodiesel. The Act requires a minimum of 450 gallons of B100 to be consumed per vehicle, and provided the fuel is not diluted below B20 levels prior to consumption (2,250 gallons of B20 are equal to 450 gallons of B100), this quantity of fuel constitutes one alternative fuel vehicle. Obviously, it is far less expensive to purchase and use 20% biodiesel that is compatible with diesel fuel than to purchase a new vehicle requiring an alternative or additional fuel supply infrastructure.

Figure 6.1-1. The Engine Manufacturers Association states in its February 2003 position paper that "there is limited information on the effect of neat (B100) biodiesel and biodiesel blends on engine durability under various environmental conditions. More information is needed to assess the viability of using these fuels over the mileage and operating periods typical of heavy-duty engines." These are very interesting statements given that North American biodiesel consumption represents hundreds of millions of miles of engine and vehicle use in every conceivable place, time, and temperature.

Delphi, Bosch, Siemens VDO, Denso, and Stanadyne may not be household names to most people, but they represent a majority of the fuel injection equipment manufacturers who supply the diesel engine market and have formed a consortium known as the FIE Manufacturers. Collectively, they support the development of FAME fuels but are very concerned about the risks attributed to substandard biodiesel, particularly the following:

- free methanol
- water
- free glycerine
- mono-, di-, and triglycerides (bonded glycerin)
- free fatty acids
- total solid impurity level
- alkali/alkaline earth metals
- oxidation stability

Their current position is to limit FAME concentrations to 5%. The biodiesel must meet the EN14214 standard and the unadulterated diesel fuel must meet the EN590 standard. The final B5 product must comply with EN590 (all European Standards) [3].

The FIE Manufacturers go on to state that the current ASTM D 6751 standard for FAME does not "contain a requirement for oxidation stability." Nor does it address the lack of "sufficient safeguards against blend quantity." Their collective position is that "FIE Manufacturers can accept no legal liability for failure attributable to operating their products with fuels for which

Fuel Injection Equipment - Potential Problems with FAME
(non-exhaustive list)

Fuel Characteristic	Effect	Failure Mode
Fatty acid methyl esters (general)	Softening, swelling or hardening and cracking of some elastomers including nitrile rubbers (physical effect depends upon elastomer composition) Displacement of deposits from diesel operation	Fuel leakage Filter plugging
Free methanol in FAME	Corrosion of aluminum & zinc Low flash point	Corrosion of FIE
FAME process chemicals	Entry of potassium & sodium and water hardness (alkaline earth metals) Entry of free fatty acids hastens the corrosion of non ferrous metals e.g. zinc Salt formation with organic acids (soaps) Sedimentation	Filter plugging Corrosion of FIE Filter plugging Sticking moving parts
Free water	Reversion (Hydrolysis) of FAME to fatty acid and methanol Corrosion Sustainment of bacterial growth Increase of electrical conductivity of the fuel	Corrosion of FIE Filter plugging
Free glycerine	Corrosion of non-ferrous metals Soaking of cellulose filters Sediment on moving parts and lacquering	Filter plugging Injector coking
Mono-, di- and tri-glyceride	Similar to glycerine	Injector coking
Higher modulus of elasticity	Increase of injection pressure	Potential for reduced service life
High viscosity at low temperature	Generation of excessive heat locally in rotary type distributor pumps Higher stressing of components	Fuel delivery problems Pump seizures Early life failures Poor nozzle spray atomization
Solid impurities/ particles	Potential lubricity problems	Reduced service life Nozzle seat wear Blocked nozzles
Ageing products Corrosive acids (formic & acetic)	Corrosion of all metal parts May form simple cell	Corrosion of FIE
Higher molecular organic acids	Similar to fatty acid	
Polymerisation products	Deposits, precipitation especially from fuel mixes	Filter plugging Lacquer formation by soluble polymers in hot areas

Table 6.1-1. *Source: Joint Fuel Injection Equipment Manufacturers Common Position Statement on FAME Fuels.*[5]

the products were not designed, and no warranties or representations are made to the possible effects of running these products with such fuels." The position of FIE Manufacturers (without stating it outright) is that ASTM D 6751 is a flawed standard that does not allow their collective to endorse even low B5 levels of biodiesel. This point was made very clear when the FIE Manufacturers rejected a ballot for inclusion of B5 into ASTM D 975 in December 2004 because of the absence of a fuel stability standard.[4]

The EMA is also very conservative in its endorsement of the use of certified biodiesel, recommending that blend levels "up to a maximum of 5% should not cause engine or fuel system problems, provided the B100 used in the blend meets the requirements of ASTM D 6751." It continues by stating that "if blends exceeding B5 are desired, vehicle owners and operators should consult their engine manufacturer regarding the implications of using such fuel."

The EMA also states in its February 2003 position paper that "there is limited information on the effect of neat (B100) biodiesel and biodiesel blends on engine durability during various environmental conditions. More information is needed to assess the viability of using these fuels over the mileage and operating periods typical of heavy-duty engines."

These are very interesting statements given that North American biodiesel consumption represents hundreds of millions of miles of engine and vehicle use in every conceivable place, time, and temperature. Even the U.S. military has approved biodiesel use for noncombat vehicles.

Perhaps the final word should go to the biodiesel industry as summarized by Christine Paquette of World Energy LLC: "The North American biodiesel industry is relatively new and in its infancy. If we wish it to grow to its full potential, we have to earn the trust of our customers, which will occur only if we deliver high-quality biodiesel products to them. Low-quality products and the concerns associated with high-blend applications that are outside our mandate are simply not worth pursuing on account of the potential risk of destroying the market."

Contrast the above positions with those of Dan Freeman of Seattle, Washington who owns and operates his fuel retail and distribution business, Dr. Dan's Alternative Fuel Werks, alongside his automotive repair shop. "I decided to add alternative fuels such as compressed natural gas and biodiesel to support the business as well as improve the air quality in the Seattle area," Dan explains. "Starting in 2001, I began selling ASTM certified B100, primarily from West Central and have watched sales grow up to around 15,000 gallons per month (≈57,000 liters). I have hundreds of clients who regularly

Figure 6.1-2. I have hundreds of clients who regularly fill up with B100, including operators of heavy equipment from bulldozers to tractors," states Dan Freeman of Dr. Dan's Alternative Fuel Werks. "We started a Web-based survey at the end of 2003 and have documented well over a million miles of trouble-free driving using B100."

fill up with B100, including operators of heavy equipment from bulldozers to tractors."

I asked Dan about any complaints or problems he has had with B100 sales. "I have over 2,000 happy clients, all of whom are satisfied with the fuel quality. There have been a handful of older vehicles that have required minor updates to hoses and the like, but this wasn't a surprise given the age of the vehicles, and these issues were quickly and cheaply remedied. From our experience, we have come to believe that the issue of hose and seal degradation borders on myth. I use B100 in my new Volkswagen as do approximately 700 more of my clients, all with completely satisfactory results. We started a Web-based survey at the end of 2003 and have documented well over a million miles of trouble-free driving using B100. Here in our area, we are having far more trouble with poor-quality petroleum diesel fuel than we are with biodiesel. It appears to have started when the switchover to ultralow sulfur diesel started, and it really is an epidemic."

Notwithstanding Dan's success delivering high-concentration biodiesel to the local market, his company is supplying a fuel that most manufacturers do not endorse. I asked Dan how his customers justify taking on the "risk" of using biodiesel at higher than recommended concentrations. "The Volkswagen 5% biodiesel warranty is great. It's 5% better than diesel," Dan enthuses. "But remember: the warranty isn't fuel specific. If there is a problem with the fuel it isn't covered by the warranty, period. So most of our customers don't consider this to be a problem at all. They see thousands of other users putting on millions of miles and they see that it works. What is to worry about?" It appears that Dan's reputation for supplying education as well as biodiesel fuel provides more comfort than Volkswagen's warranty policy.

"You know, Carter Volkswagen, where I purchased my car, and University Volkswagen are two local area dealers that have been generally supportive of our efforts," Dan says. "However, it is simply not to their advantage to fully support it (high blend levels of biodiesel), so they don't."

I asked Dan if he felt that OEM engine and vehicle manufacturers hesitated to give high blend biodiesel the "green light" because it might lend credibility to home brewing of poor quality fuels. "Absolutely. There is no doubt in my mind that vehicle manufacturers do not want to support any activity where there is room for bad public relations or liability. There is nothing in it for them to support these activities. I believe the North American biodiesel industry is not set up for high blend levels. They require an educated customer and a very educated retailer, whereas low levels such as B2 or B5 will provide the biggest boost for the industry, with little chance for fuel quality problems. But it doesn't matter what blend levels the industry wants to push. There will always be a customer base wanting B100."

Fair enough. The market for high blend levels is there, ready and waiting for fuel, and entrepreneurs like Dan Freeman are eager to supply the goods. But just what are the risks of using high blend biodiesel? Surely the views of engine/vehicle manufacturers and retailers like Dan and other users cannot be so diametrically opposed?

To find out, I undertook a complete engine test using B100 with the assistance of a major diesel engine/generator integrator to see if quantifiable wear or failure results could be determined when using ASTM D 6751 fuel under controlled conditions.

Manufacturer	Summarized Position Regarding Biodiesel
Cummins Engine Co.	"Cummins neither approves nor disapproves of the use of biodiesel fuel." Up to B5 should not cause serious problems.
Isuzu GM Australia Ltd	"Isuzu Japan neither approves nor prohibits using biodiesel fuel. Isuzu Japan recommends using the fuel that meets EN 14214 or ASTM D 6751 standard."
Mitsubishi Motors	Voids warranty. Have no data of testing on B100 or blends.
Premier Automotive Group (Aston Martin, Jaguar, Land Rover, Volvo)	B5 only
Scania Australia	B5 only
Volvo Truck	B4 only
Caterpillar	Neither approves nor prohibits the use of biodiesel fuels. Failures resulting from the use of any fuel are not Caterpillar factory defects and therefore the cost of repair would not be covered by Caterpillar's warranty. When using a fuel that meets Caterpillar's biodiesel specifications, ASTM PS121 or DIN 51606, and following recommendations, the use (of biodiesel) should pose no problems.
Daimler Chrysler AustraliaPacific Commercial Vehicles	Fuel must meet DIN 51606 and STIN 00.00S028. Use of B100 would not necessarily void warranties. Up to B5 suitable for all engines.
UD Nissan Trucks	"The use of biodiesel would make the vehicle's warranty conditional."
Mazda Australia Pty. Ltd.	Voids warranty.
Volkswagen America	B5 provided parent fuel (B100) is certified to ASTM D 6751

Table 6.1-2.
Biodiesel position statements of engine and equipment manufacturers from the United States and Australia.[6] Even the OEMs can't arrive at a consensus (other than to reduce concentration levels) on the issue of biodiesel usage.

6.2
Engine Analysis after Operating on B100

"My dad started the business in 1936 working primarily in the area of farm electrification prior to the days of hydro being everywhere," explains Ken Sommers, President of Sommers Motor Generator Sales in Tavistock, Ontario. "In the hydro changeover, when the frequency of the grid was modified from 25 to 60 cycles (per second), I spent a lot of my time after school working with dad rewinding motors in the shop. Around the same time we started getting into gas and diesel generators, but we really got to be well known for our PTO (power take-off) generators that were driven by tractors. In fact we ended up being the largest PTO dealer in the world, supplying product to the entire market."

"As the demand moved towards larger and more automated generators, we started to transition into the bigger systems such as the half-megawatt Cummins diesel unit you see over there in the show room or the 2-megawatt super-quiet machines we provide for emergency power systems," Ken says with a sweep of his hand across a large array of generators in their show room.

Figure 6.2-1. "As the market moved towards larger and more automated generators, we started to transition into the bigger systems such as the half-megawatt Cummins diesel unit you see in the show room or the 2-megawatt super-quiet machines we provide for emergency power systems," says Ken Sommers, President of Sommers Motor Generator Sales in Tavistock, Ontario.

"Over the years we have seen a transition to diesel engine systems so that in our medium to larger size ranges they make up 98% of the volume," interjects Neil Hevenor, system specialist at Sommers. "Diesels have traditionally been more expensive than spark ignition engines but this has changed considerably, and when you consider that diesel engines last three times as long it's no wonder so much of the business has gone this way."

I asked Ken and Neil if biodiesel was something that their customers were asking about and if there was any demand for information on the subject. Neil's reply: "Our customers have been calling. Each month we get at least a couple of calls from users and consultants asking if we have any experience in this area. Most of the calls are from people wanting to use the fuel to either sell power to the grid during peak demand/price periods or simply go off-grid themselves."

Neil went on to explain that the diesel engine manufacturers that he has spoken to are "interested" in biodiesel, but they are very uncertain about the fuel and "don't want to say anything that might come back to haunt them. They would really like to see the results of working with this fuel."

I couldn't have asked for a better segue.

Figure 6.2-2. Ken Sommers (left) and Neil Hevenor (right) are standing beside a brand-new Lister-Petter 85-cubic-inch (1.4 liter), three-cylinder, 12 kW diesel generator that will undergo a range of full-load tests using ASTM certified B100 biodiesel. At the conclusion of the test program, the engine will be disassembled and examined.

Figure 6.2-3. This Lister-Petter diesel engine drives either a single- or three-phase 12 kW Hawkpower generator. These units are some of the highest quality generators on the market and are used for both standby and prime-power (continuous) usage applications. They are routinely called into service for the operation of emergency building power, telecommunications, and computer backup systems.

Figure 6.2-4. The generator was installed in the Sommers warehouse, replacing the connection to the electrical grid. A supply of 803 liters (212 gallons) of ASTM certified biodiesel was used to power the generator on a 24-hour-per-day basis (with the exception of one oil change interval). At the conclusion of the test, the engine was fully analyzed to determine the effect, if any, of high-concentration biodiesel.

When I approached Neil about performing an engine test program, he was quite excited about the opportunity. More importantly, so was his boss, Ken. Together we developed an engine test program that would operate a diesel engine on 100% ASTM certified biodiesel, with the intention of making a number of observations during the test phase:

- analyze engine oil for wear and breakdown (two tests conducted: one after 100 hours; the second at the conclusion of the test)
- perform load transient test comparing the reaction of the engine from partial to full load and comparing the results with those of petroleum diesel fuel
- measure fuel consumption
- monitor engine and exhaust stack temperature
- monitor electrical energy generation

After the above observations were completed, the engine was returned to the Sommers factory where mechanic Robert Gingerich conducted a full internal analysis of the engine. "The Lister-Petter diesel engine is known as a high-quality workhorse and was chosen for our engine test program because the results could be transferred to any size and type of engine with a similar configuration," says Robert. "I disassembled the engine and checked every measurement and surface for abnormal signs of wear or inconsistency."

Figure 6.2-5. "The Lister-Petter diesel engine is known as a high-quality workhorse and was chosen for our engine test program because the results could be transferred to any size and type of engine with a similar configuration," says mechanic Robert Gingerich. "I disassembled the engine and checked every measurement and surface for abnormal signs of wear or inconsistency."

Robert's internal analysis was very thorough and included the following:

- cylinder walls for intact honing (boring marks)
- cylinder walls for excessive wear taper
- piston heads for carbon buildup or signs of hot or uneven combustion
- intake manifold seats and gaskets
- intake or exhaust seats for carbon buildup
- valve seats and valve faces for perfect seal
- valves and valve stems and guides
- fuel injectors
- hoses and external components

"The engine was commissioned and started on petroleum diesel fuel," Neil explains. "We used a few liters to conduct the initial baseline tests, which included the engine and exhaust stack temperature as well as the engine load variation test. This data was recorded and the fuel was changed over to the biodiesel supplied by Milligan Bio-Tech Inc. We tried to keep the electrical/engine load as high as possible while allowing some variation to occur as lights, equipment, and tools were turned on and off. This provided a fairly realistic operating condition."

Based on the engine having an average load of approximately 70%, fuel consumption was estimated to be 2.5 liters per hour (0.7 gallons per hour), providing approximately 320 hours of continuous running time. One short break was allowed in order to change the engine lubrication oil after the recommended 100-hour break-in period. This oil was sent for analysis to determine engine wear.

The engine was restarted and allowed to continue operation until the fuel supply ran dry. The crankcase oil was again changed and sent for analysis, while the engine was delivered to Robert for tear down.

"I have repaired and overhauled thousands of diesel engines of all sizes," states Robert. "It is pretty easy to tell if something is a bit out of whack, but this engine looked very good, very normal, right from the moment I popped the head off it."

We started our physical examination of the cylinder block by comparing the piston condition between each cylinder and checking for signs of excessive heat or carbon residue. The visual examination was completely normal.

The next step was the measurement of the piston wall dimensions and the taper angle. Taper angle occurs when high-pressure gases produced during the power stroke place high side-loading forces on the piston rings. This force

Figure 6.2-6. We started our physical examination of the cylinder block by checking the piston condition for each cylinder and checking for signs of excessive heat or carbon residue. The visual examination was completely normal. Measurements of piston wall dimensions, taper, and honing were all within "factory-new" specification.

Figure 6.2-7. Cylinder walls should not be smooth and glassy. A process known as "honing" lightly scores the cylinder walls to ensure proper piston ring wear and lubricating oil transfer. After 350 hours of nearly full-load operation, the honing marks are still clearly visible, with measurements indicating no change from factory-new condition.

decreases as the piston moves lower in the cylinder at the end of the power stroke, resulting in lower side forces. The variation of cylinder pressure and side force causes the cylinder wall to wear in the shape of a taper from top to bottom. Measurements of piston wall dimensions, taper, and honing were all within "factory-new" specification.

The cylinder head contains the fuel injection nozzles, valve train, and manifold connections and was the next stop on our engine assessment journey. Examining the cylinder head for carbon buildup and signs of excessive heat was Robert's next task. "The amount of carbon buildup and even color tells me that everything is operating within the normal range," Robert states. "After 350 hours, I would expect to see a bit of carbon settling on the casting surface."

Examining the cylinder head with the valves removed indicates a slight change in color and surface cleanliness between the intake valve area (right) and exhaust valve (left). Upon closer inspection, it appears as if unburned fuel is collecting on the area around the intake valve seat. "The condition you are noticing is due to "crossover," which occurs just when the exhaust and intake valves are respectively completing and starting their cycles. When this occurs a bit of pressurized hot exhaust gas will leak into the area, leaving the coating you see," Robert explains. "This is perfectly normal."

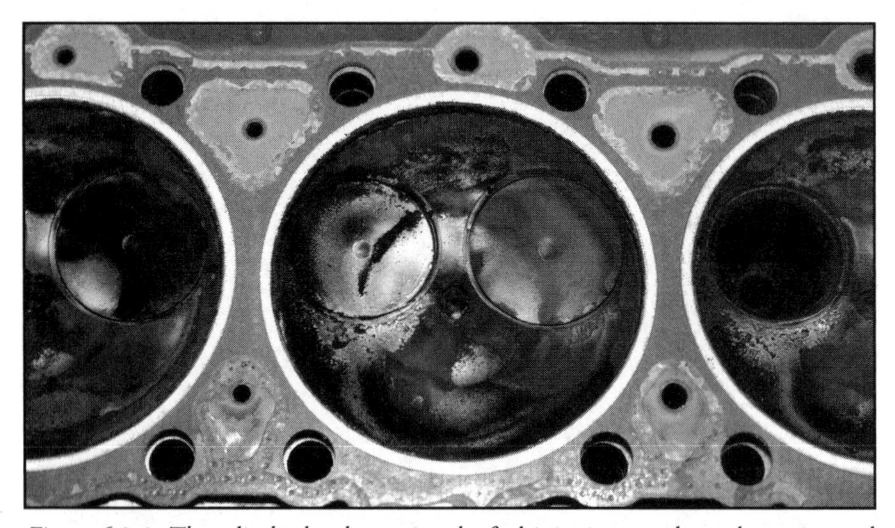

Figure 6.2-8. The cylinder head contains the fuel injection nozzles, valve train, and manifold connections and was the next stop on our engine assessment journey. "The amount of carbon buildup and even color tells me that everything is operating within the normal range," Robert states. "After 350 hours, I would expect to see a bit of carbon settling on the casting surface."

Figure 6.2-9. Examining the cylinder head with the valve removed indicates a slight change in color and surface cleanliness between the intake valve area (right) and exhaust valve (left). Upon closer inspection, it appears as if unburned fuel is collecting on the area around the intake valve seat. "The condition you are noticing is due to "crossover," which occurs just when the exhaust and intake valves are respectively completing and starting their cycles. When this occurs a bit of pressurized hot exhaust gas will leak into the area, leaving the coating you see," Robert explains. "This is perfectly normal."

The area around the exhaust valve seat does attract Roberts's attention. He is slightly concerned about the amount of carbon buildup on the surface. "I see a little bit of carbon in this area that may be the result of wet stacking and not related to the biodiesel," he explains. "Wet stacking occurs when the piston rings have not yet fully seated in a new engine and the small amount of suction that occurs during piston cycling draws lubricating oil into the cylinder, where it combusts. I am not very concerned about this, but it would be interesting to run the test for 5,000 hours and reexamine this situation."

The overall condition of the engine and generator set is well within factory new condition, with all measurements indicating that the engine has not been operated at all. There is no sign of any degradation of hoses, although this is as expected since they were in contact with B100 for a relatively short time period and the manufacturer uses biodiesel-resistant artificial (Viton™) rubber.

Figure 6.2-10. The area around the exhaust valve seat does attract Roberts's attention. He is slightly concerned about the amount of carbon buildup on the surface. "I see a little bit of carbon in this area that may be the result of wet stacking and not related to the biodiesel," he explains. "Wet stacking occurs when the piston rings have not yet fully seated in a new engine and a small amount of suction that occurs during piston cycling draws lubricating oil into the cylinder, where it combusts. I am not very concerned about this, but it would be interesting to run the test for 5,000 hours and reexamine this situation."

Figure 6.2-11. The fuel injector nozzle is located between the intake and exhaust valve seat areas and is not much larger than a pencil lead. Fuel sprays into the nozzle under tremendous pressure from the injection pump, causing it to atomize into a fine spray. Although we can barely see the injection holes, Robert is satisfied that all is normal.

Figure 6.2-12. The face stem of the intake valve is clean and measures within specification.

Figure 6.2-13. The intake valve seat area is clean and polished as if it had never been used.

Figure 6.2-14. The face stem of the exhaust valve is clean and it too measures within specification for a new engine.

Figure 6.2-15. The exhaust valve shows a small amount of carbon hazing on the seat area which Robert believes may be due to a "wet stacking" condition. He is not overly worried about this but would like to reexamine the engine after 5,000 hours of running time using B100.

Figure 6.2-16. The engine and generator set are in factory new condition, with all measurements indicating that the engine has not been operated at all. There is no sign of any degradation of hoses, although this is as expected since they were in contact with B100 for a relatively short time period.

The examination of the engine leads us to believe that there is little if any difference between operational characteristics with B100 and No. 2-D fuel. However, it is still necessary to evaluate the engine performance and lubricating oil condition to provide more conclusive results.

Engine Performance

Back in Neil's office, we start to analyze the performance of the engine generator under load conditions. The first part of the test was to establish a baseline using regular No. 2 diesel fuel. After a 30-minute warm-up time the engine was gradually loaded to maximum power. The results of the thermal and steady state electrical load test are detailed in Table 6.2-1. At the same time, a series of transient tests were conducted where the engine and generator were rapidly cycled from no load to full load and vice versa. The importance of these tests is to establish the transient power capacity of the engine, which confirms (among other issues) proper fuel combustion and energy density necessary to complete the test. The results of the transient load test using No. 2 diesel fuel are shown in Table 6.2-2.

After completion of the baseline tests the engine was stopped and the fuel supply was switched over to 100% biodiesel fuel. The engine was restarted and the generator was gradually connected to the full electrical load. Measurements of the biodiesel-fuelled baseline data were taken at several iterations, with the data from the final hour shown in Table 6.2-3 (baseline biodiesel results). Table 6.2-4 outlines the results of transient load tests that were conducted at the start and end of the test period, which confirm no degradation of performance. A summary of the entire test period is shown in Figure 6.2-5.

An examination of this data shows that there are no material differences in engine performance resulting from the use of petroleum and biodiesel fuels. The only difference noted was that exhaust stack temperature ran approximately 50°C (90°F) cooler when operating on biodiesel, although this was not unexpected.[1]

No effort was made to examine exhaust gas emissions as the there are numerous studies that have concluded the superior benefits of biodiesel in all emission gas profiles, with the exception of nitrogen oxides (NOx).[2]

Time Hrs	Temperature (Degrees C)				Oil Press Bars	Voltage			Current			Freq. (HZ)	Watts (KW)
	Ambient	Coolant	Oil	Exhaust		L1-L2	L2-L3	L3-L1	Line 1	Line 2	Line 3		
0.5	16	74	60	70	-	240	-	-	0	0	-	63.0	0
0.5	16	74	65	108	-	240	-	-	10	10	-	62.2	4.8
0.5	16	74	70	115	-	239	-	-	20	20	-	61.4	7.4
0.6	16	75	82	150	-	239	-	-	31	31	-	60.7	9.8
0.6	16	74	85	173	-	238	-	-	41	41	-	60.1	11.5
0.6	16	75	90	255	-	238	-	-	48	48	-	59.1	11.4
1.0	18	74	91	350	-	238	-	-	48	48	-	58.8	11.4
1.1	17	73	92	360	-	238	-	-	48	48	-	58.5	11.4
1.2	17	74	93	360	-	238	-	-	48	48	-	58.5	11.4
1.3	17	74	91	360	-	238	-	-	48	48	-	58.4	11.4
1.5	17	74	93	355	-	238	-	-	48	48	-	58.4	11.4
1.9	17	74	93	360	-	238	-	-	48	48	-	58.6	11.4

Table 6.2-1

Baseline engine analysis using No. 2 diesel fuel

Transients		**First Test**			**Second Test**		
		No-Load to Full-Load	**Full-Load to No-Load**	**Watts (KW)**	**No-Load to Full-Load**	**Full-Load to No-Load**	**Watts (KW)**
Frequency (Steady-State)	No-Load	62.4	62.5	0	62.5	62.5	0
	Full-Load	59.3	59.4	10.5	58.8	58.8	11.4
Voltage (Steady-State)	No-Load	240.3	240.3	0	240.2	240.3	0
	Full-Load	237.7	237.7	10.5	237.5	237.4	11.4
Voltage (V)	Min.	227.4	237.6	-	226.4	237.3	-
	Max.	240.4	250.7	-	245.3	253.5	-
Frequency (HZ)	Min.	59.1	59.3	-	55.9	58.7	-
	Max.	62.5	62.7	-	62.6	63.1	-
Transient (%) Difference	Volts	5.4	5.5	-	5.9	6.8	-
	Hertz	5.3	5.6	-	10.6	7.3	-

Table 6.2-2

Transient load test using No. 2 diesel fuel

Time Hrs	Temperature (Degrees C) Ambient	Coolant	Oil	Exhaust	Oil Press PSI	Voltage L1-L2	L2-L3	L3-L1	Current Line 1	Line 2	Line 3	Freq. (HZ)	Watts (KW)
344.7	16	47	43	53	56	240	-	-	0	0	-	62.1	0
344.7	16	66	54	86	54	239	-	-	10	10	-	61.2	2.5
344.7	16	71	54	106	52	239	-	-	21	21	-	60.5	4.9
344.8	16	71	63	146	50	238	-	-	32	32	-	59.6	7.5
344.8	16	72	67	187	46	237	-	-	42	42	-	58.9	9.9
344.9	17	74	73	248	44	237	-	-	49	49	-	58.0	11.6
345.2	18	74	73	247	42	237	-	-	49	49	-	58.0	11.6
345.5	20	75	92	358	35	237	-	-	49	49	-	57.7	11.6
345.7	24	74	91	311	35	237	-	-	49	49	-	57.9	11.6
346.2	23	74	94	291	34	237	-	-	49	49	-	57.8	11.6
346.5	23	74	94	290	34	237	-	-	49	49	-	57.8	11.6
347.2	23	74	93	296	34	237	-	-	49	49	-	57.8	11.6
347.7	23	74	94	296	34	237	-	-	49	49	-	57.8	11.6
348.2	23	75	93	305	34	237	-	-	49	49	-	57.9	11.6
348.7	23	74	93	290	34	237	-	-	49	49	-	57.9	11.6

Table 6.2-3
Baseline engine analysis using 100% biodiesel fuel

Transients		**First Test** No-Load to Full-Load	**Full-Load to No-Load**	**Watts (KW)**	**Second Test** No-Load to Full-Load	**Full-Load to No-Load**	**Watts (KW)**
Frequency (Steady-State)	No-Load	61.8	61.9	0	61.8	61.8	0
	Full-Load	58.4	58.5	10.4	57.8	57.8	11.7
Voltage (Steady-State)	No-Load	240.3	240.3	0	240.3	240.3	0
	Full-Load	237.4	237.4	10.4	237.0	237.0	11.7
Voltage (V)	Min.	226.8	237.3	-	225.1	236.8	-
	Max.	240.4	250.6	-	240.4	253.6	-
Frequency (HZ)	Min.	58.1	58.5	-	56.1	57.8	-
	Max.	61.9	62.4	-	61.7	61.9	-
Transient (%) Difference	Volts	3.6	5.6	-	6.3	8.4	-
	Hertz	6.0	6.7	-	9.2	7.1	-

Table 6.2-4
Transient load test using 100% biodiesel fuel

Notes on Baseline Analysis:

- Engine temperatures vary according to mechanical load, which is a function of generator electrical load, denoted as kilowatts (KW) in right column.
- Frequency (HZ) is a function of the rotational speed of the generator. This unit contains a 4-pole generator which requires a rotational speed of 1,800 rpm to generate alternating current at 60 Hertz or cycles per second.[3] Frequency varies as a function of generator load.
- Oil pressure will drop as a function of its temperature because of a change in viscosity.

Notes on Transient Analysis:

- Generator load directly affects engine speed as a result of a phenomenon known as "governor droop." Simply stated, governor droop will generally allow a variation in engine speed/electrical frequency of 5% between zero and full load conditions.
- The ability of the engine to respond to rapid changes in load is a good indication of engine and fuel performance.

Date	Time	Hrs	Temperature (° C)				Oil (PSI)	L1 to L2	Amps	Amps	Freq.	Kw	KwH
			Ambient	Coolant	Exhaust	Oil	Pressure	V	L1	L2	(HZ)		
21-12-05	16:00	4.1	0	47	53	43	30	240	0	0	62.1	0	0
22-12-05	10:00	6.3	12	68	120	62	30	238	49	46	58.6	10.2	9
22-12-05	14:00	9.3	13	71	204	85	30	238.6	17.7	40.4	60.2	6.9	28
22-12-05	16:30	12.1	14	72	194	86	30	239	18.5	31	60.5	5.6	45
23-12-05	08:15	27.8	17	71	195	83	31	238.6	12.8	44	60.2	6.7	146
23-12-05	15:20	34.9	18	72	211	82	31	238.4	12.7	47.4	60.3	7.5	188
28-12-05	16:30	43.0	15	72	165	82.9	31	238.8	16.4	30.7	60.2	5.6	238
29-12-05	08:00	58.5	18	72.5	193	83	32	238.6	16.8	47.7	59.6	7.4	332
30-12-05	09:00	82.9	17	72.8	177	81.7	32	238.6	15.7	28.7	59.9	5.26	479
01-01-06	17:00	138.5	17	75.6	160	83.3	42	238.5	11.8	25.2	59.9	3.84	675
03-01-06	08:30	177.6	18	72.4	225	82.4	40	238.6	14.8	47.6	59.9	7.14	796
04-01-06	08:30	201.8	19	73.1	245	91.4	36	238.5	21.4	47.2	59.3	8.13	944
05-01-06	08:30	225.9	20.5	70.5	203	83.4	38	238.9	12.5	30.1	60	5.13	1103
06-01-06	14:15	255.6	19	70.3	185	83.2	38	239	15.3	36	60.2	5.3	1309
07-01-06	10:30	275.1	19	71	195	83.3	38	238	19	32	60.1	6.6	1433
09-01-06	08:15	320.0	20	71.5	216	83.8	38	238.6	20.2	36.7	59.3	6.3	1709
10-01-06	08:15	344.0	22	73	220	89	38	238.4	42	43	58.9	9.4	1914

Table 6.2-5
Summary of 344-hour test run

Cost of Electrical Generation

Although not related to engine performance or fuel quality, the ratio of fuel consumed to the electrical energy generated can be used to calculate the cost of generation for each kilowatt hour of electricity:

803 liters of fuel ÷ 1,959 kWh of electricity generated = 0.41 l/kWh
0.41 l/kWh x $0.79 per liter biodiesel = $0.32 per kWh of electricity generated

The average price of biodiesel given here $0.79 per litre ($3.00 per gallon) is highly variable, but it indicates the cost of generating power from small generators such as the model used in these tests. This cost does not include the maintenance or depreciation of the unit, which would have to be factored into the equation before you start competing with the local electrical utility. Speaking of which, using biodiesel to generate electrical and thermal energy will be the subject of Chapter 10.

Examination of Engine Lubrication Oil

Samples of the engine oil were submitted to WearCheck International (www.wearcheck.com), a worldwide oil analysis company after the first oil change interval and at the conclusion of the 344-hour test. I asked WearCheck to provide statistical feedback on the state of the engine oil and to confirm whether there were any contaminants or other problems with the oil resulting from our use of biodiesel fuel. "Engine oil analysis is rather like a blood test," says Barry Goslin, oil diagnostician. "Our clients come to us to ensure the longevity and reliability of their equipment, and using our process is the least expensive way of finding problems before they become catastrophic."

Engine lubricating oils undergo destructive changes even in the very best and newest of engines. Oxygen, combustion gases, and high temperatures all work together over time to degrade the lubricating ability of oil. Simply changing the oil will eliminate its natural degradation but will not necessarily stop abnormal wear and contamination that might be occurring at the microscopic level. Further, without performing an oil analysis it is difficult to detect potential engine failures before they occur.

"Our typical oil analysis will provide a detailed review of contaminants that may be present in the engine's lubricating oil," continues Barry. "We are looking at three broad categories: contamination, wear, and oil condition. In each area, we have, in conjunction with engine and equipment manufacturers, determined the most likely contaminant materials and set normal, abnormal, and severe limits for their concentrations. For example, a common oil failure occurs because of the presence of glycol which leaks into the lubrication system from the cooling system due to defective gaskets or flawed metal castings. At low levels of contamination, there is unlikely to be any problem with the engine, but it provides a very accurate early warning indicator that serious problems are just around the corner."

Barry goes on to explain that with WearCheck's advanced analysis equipment it is possible to detect metal filings that are microscopic in size. "We are not only able to determine if microscopic metal "filings" are present in the oil, we are also able to identify what the base metal is. With further analysis it is possible to determine if abnormal wear is coming from a piston ring, a cylinder wall sleeve, the engine block, or friction bearings. This diagnostic test is very accurate and is used by aircraft engine manufacturers and other industries that are very concerned about quality, reliability, and the improvement of maintenance procedures." Barry explains that WearCheck seeks input from the customer which may be useful in making a diagnosis. "If I notice that oil results for the last several intervals have been fine, and then all of a sudden the soot concentration shoots up, this tells me there may be a sudden change of engine loading, for example a transport hauling light stock that suddenly switches to carrying heavy steel up steep inclines. On the other hand, a generator that has a defective bearing may increase engine load beyond normal limits and can easily contribute to indicators in the lubricating oil."

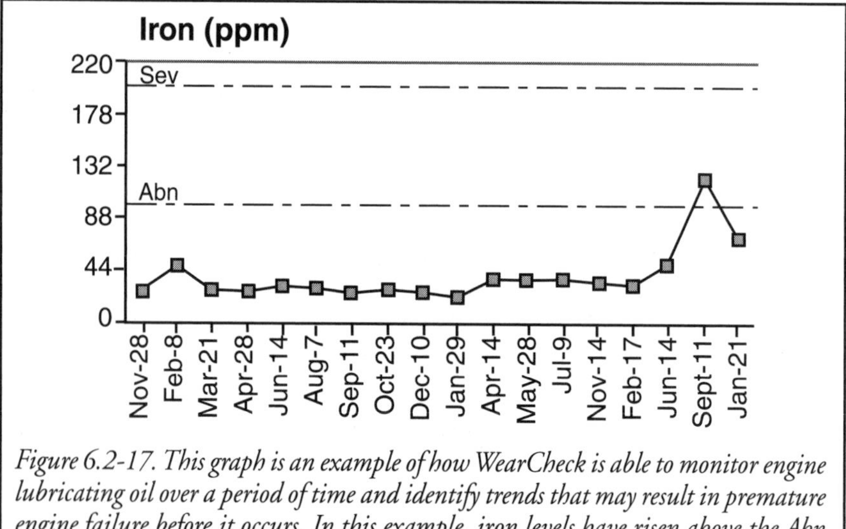

Figure 6.2-17. This graph is an example of how WearCheck is able to monitor engine lubricating oil over a period of time and identify trends that may result in premature engine failure before it occurs. In this example, iron levels have risen above the Abn (abnormal) level of 90 parts per million in a sample of engine oil, necessitating further investigation to see what caused the spike before serious and expensive engine damage results.

MOBILE OIL ANALYSIS REPORT	CONTAMINATION:	NORMAL		
	WEAR:	NORMAL		
	OIL CONDITION:	NORMAL		
RECOMMENDATION	Sample Date:	01/06/06	Current	UOM
Resample at the next service interval to monitor. Please specify the component make and model with your next sample.	Time on Unit:	255	348	hrs
	Time on Oil:	168	93	hrs
	Time on Filter:	168	93	hrs
	Oil Maint.	changed	n/a	-
	Filter Maint.	changed	n/a	-
CONTAMINATION	Sample Date	01/06/06	Current	Abn
There is no indication of any contamination in the component.	Silicon	10	6.2	
	Potassium	5.9	4.8	
	Sodium	1.0	0.2	
	Fuel (%)	< 2.0	< 2.0	
	Glycol	-	-	
	Water (%)	< 0.1	< 0.1	
	Soot (%)	0	0	
	Sulfation	85	76	
	Nitration	73	64	
WEAR	Sample Date	01/06/06	Current	Abn
All component wear rates are normal.	Iron	13	7.8	100
	Nickel	0.0	0.0	2
	Chromium	1.6	1.3	2
	Titanium	0.2	0.2	2
	Copper	1.9	1.1	330
	Aluminum	2.5	2.6	20
	Tin	0.1	0.1	15
	Lead	4.2	1.5	40
	Silver	0.2	0.3	2
OIL CONDITION	Sample Date	01/06/06	Current	Base
Oil Type: 4 Ltr. of Esso XD-3 Extra 15W40	Boron	1.7	2.0	0.0
The condition of the oil is acceptable for the time in service.	Barium	0.3	0.2	0.0
	Calcium	3221	3244	3800
	Magnesium	9.3	9.0	0.0
	Molybdenum	0.8	0.7	0.0
	Sodium	1.0	0.2	0.0
	Phosphorus	1102	1119	1350
	Sulfur	4020	4004	
	Zinc	1240	1237	1500
	Visc @ 40°C	-	-	116
	Visc @ 100°C	13.0	13.7	15.4
	Oxidation	131	110	-

Table 6.2-6

The results of the final 344-hour engine run time oil analysis show that engine wear and oil contamination levels are well within normal specifications and that using 100% biodiesel has not generated any negative data up to the time of sampling.
(Courtesy WearCheck Canada Inc.)

I asked Barry to interpret the results of our two oil analyses using 100 percent biodiesel. "Absolutely perfectly normal. This engine should have a long life provided current analysis trends continue."

Conclusion

If you are considering both sides of the B100 argument and are still uncertain, you're not alone. Engine and vehicle manufacturers doggedly insist that low or zero concentration levels of biodiesel are correct while thousands of people as well as commercial and industrial users have defiantly burned vast quantities of B100 over millions of miles, seemingly without any significant problems.

The situation is a complex blend of politics, lack of knowledge, and liability. Fossil fuel suppliers do not want to lose their control over the supply stream, while engine manufacturers and component suppliers see no upside to supporting a potentially problematic fuel source. When the rhetoric that results from lack of education is coupled with increased product warranty and liability issues, it is no wonder that commercial interests in North America shun the concept of high blend level biodiesel.

There is no simple answer. Biodiesel is not petroleum diesel fuel. It is an organic product that responds differently from fossil diesel fuel under changing conditions, and very few people in the fuel supply and engine industries are educated about what is fact and what is fiction. Rather than becoming immersed in endless debate on the matter, the commercial biodiesel industry recognizes the complexity of marketing B100 and has wisely chosen the path of supporting low blend levels. Selling 5% biodiesel blended into the entire diesel fuel supply represents far more volume than the industry can produce today and eliminates many of the issues critics and naysayers use to stop the development of the market.

As for the B100 faction, take solace in the data provided above and be thankful Dr. Dan and his crew are there to buck the trend and supply product, provided you are in their service area.

Perhaps, over time, Dan will franchise his establishment. Until then, the standoff between the various "conservative" industries and B100 proponents will continue.

7 Debunking the Home Brew Myth

7.1 Raw Biodiesel Production

While there is no doubt that a few home-scale producers are capable of making biodiesel that meets ASTM standards, there is currently no regulation or licensing of these hobbyist efforts, virtually ensuring that "off-specification" fuel is produced. Home brew operators vary in skill and production procedures from one location to the next, and **none** of the fuel samples I have examined meet the applicable standards. Ignorance, misguidance by Web peers, or lack of responsibility causes many people to engage in shortcuts or simply ignore the environmental aspects of the waste stream products, leaving those dirty little secrets hidden in the sewers, garbage dumpsters, and roadsides of the nation.

I would argue that micro-scale producers will operate in an endless vacuum of chat lines and open source forums that will propagate errors and misinformation unless this group can legitimize itself and tackle the hundreds of issues and myths that continue to circulate. I find it surprising that statements of fact are promoted and offered to people through online forums and "eco-fairs" and yet no one seems to bother testing these statements to ascertain their veracity. For instance, consider these statements from "experts" in the home brew community when they are considered in light of scientific and regulatory examination:

"The glycerin (remaining after the transesterification process) is biodegradable and nontoxic, contains no PCBs, and will not harm animal or plant life."[1]

The glycerol (raw glycerin) that is produced as a result of the transesterification process is known to contain a very high percentage of excess methanol[2]. Methanol is registered as a hazardous waste product according to the U.S. Environmental Protection Agency (40 CFR Ch. 1, Section 261.33). Methanol is also known as an "F-listed chemical" that makes each waste it contaminates hazardous at any concentration. Even one drop of an F-listed chemical on a shop rag, in an adsorbent, or in used oil is enough to create a regulated hazardous waste that must be disposed of in accordance with federal and state laws[3].

"Most of the alcohol (methanol) in the glycerin will evaporate if the glycerin is left in the sun for a week."[4]

To test this statement, a 200 ml sample (304.4 grams including beaker) of raw glycerol was weighed and placed in a warm, sunny location for a period of 10 days, after which the glycerol was reweighed to determine if any material had evaporated. The results indicated that less than 0.6 grams of methanol had evaporated in the "sunny location." The glycerol was then heated to 70°C for a period of one hour. This ensured that the water would not evaporate but the remaining alcohol would boil off through simple atmospheric distillation. The sample was again reweighed. After heating, sufficient methanol had evaporated to cause the volume to drop to 175 ml and the mass to 281.4 grams. This test clearly indicates that very little methanol evaporation occurs without the addition of excess energy, thus refuting the statement above. According to the EPA Office of Pollution Prevention and Toxics, "Workers repeatedly exposed to methanol have experienced several adverse effects. Effects range from headaches to sleep disorders and gastrointestinal problems to optic nerve damage." Given that methanol does not evaporate from raw glycerol, disposal, or downstream processing may present a severe health risk to people who are misled by such baseless information.

Figure 7.1-1. ***"Most of the alcohol (methanol) in the glycerin will evaporate if the glycerin is left in the sun for a week."***[4] *To test this statement, a 200 ml sample (304.4 grams including beaker) of raw glycerol was weighed and placed in a warm, sunny location for a period of 10 days, after which the glycerol was reweighed to determine if any material had evaporated.*

Figure 7.1-2. The sample was again reweighed. The results indicated that less than 0.6 grams of methanol had evaporated in the "sunny location."

Figure 7.1-3. The glycerol was then heated to 70°C (to ensure that water would not evaporate but remaining alcohol would) for a period of one hour.

Figure 7.1-4. After heating, sufficient methanol had evaporated to cause the volume to drop to 175 ml and the mass to 281.4 grams. This test clearly indicates that very little methanol evaporation occurs without the addition of excess energy, thus refuting the claim above.

"Most of us (home brewers) currently sewer the wash water, figuring that it is no worse than the soaps, detergents, shampoos, laundry soaps, bleach, fabric softeners, drain cleaner, shower scrubbers, tile polish, floor cleaner, car cleaners, tire cleaners, windshield bug deflector cleaner, ammonia, and a host of other "nasties" flushed by the average American household."[5]

I am not even sure if this statement needs commenting on. Surely two wrongs don't make a right? Nevertheless, this is an excellent example of the lack of responsibility that plagues unregulated processes conducted by people who are unwilling or unable to investigate the facts.

Biodiesel wash water is known to contain methanol, catalyst, and soaps and to have a high biological oxygen demand (BOD). Most cities in North America have sewer discharge bylaws that expressly forbid the disposal of chemical processing wash water that contains hazardous materials or are outside of specific limits for pH (acid/base ratio), flashpoint, etc. These regulations also forbid deliberate dilution of waste industrial process water in order to adhere to the minimum discharge limits.

Household cleaning agents are generally used in very low concentrations or small volumes. Chemical composition is carefully controlled under government guidelines and homeowners have many options to replace standard chemical-based products with biodegradable or earth-friendly options. This is not the case with biodiesel wash water (or glycerol), which is generated in much larger volumes and may be a regulated hazardous waste, depending on processing procedures.

If there is any doubt as to whether a waste product is hazardous, simply "figuring" that it is safe to dispose of is not an acceptable or responsible option. The waste material must be tested by a competent laboratory to determine the correct disposal technique.

"Anybody can make biodiesel. It's easy. You can make it in your kitchen—and it's BETTER than the petrodiesel fuel the big oil companies sell you." [6]

Biodiesel production entails a degree of risk that is not within the ability of most nontechnical people to appreciate or understand without extensive research or training. Producing biodiesel indoors with the use of large quantities of toxic, flammable methanol is not only irresponsible; it is an activity that would certainly be excluded in homeowner/tenant insurance policies.

I also take exception to the general statement that it is "BETTER" than petrodiesel.

Statements such as this are completely irresponsible and underplay the technical and personal safety risks involved in the production of home-brewed biodiesel. Lack of discussion of waste stream processing, quality, environmental controls, and the risks associated with methanol leave readers with the impression that biodiesel production is no different than making a milkshake. Such starry-eyed comments lead people astray and possibly place them in harm's way.

"Although the ASTM specifications for biodiesel allow for only 500 parts per million of water in the fuel, up to 2,500 parts per million of water in the fuel will not harm most diesel engines." [7]

This is a very interesting statement. Here, a single home-scale biodiesel producer is challenging the entire diesel engine manufacturing industry with homespun expertise and no facts to back up the statement. Barry Hertz P.Eng. is a Professor of Mechanical Engineering at the University of Saskatchewan who also acted as faculty advisor for the 1986 SMV X-Canadian World Challenge and the development of the world's most fuel-efficient car at 5,691 miles per gallon. Dr. Hertz says that such a statement is completely without merit. "Excess water in the fuel will cause severe degradation and oxidation "pit-

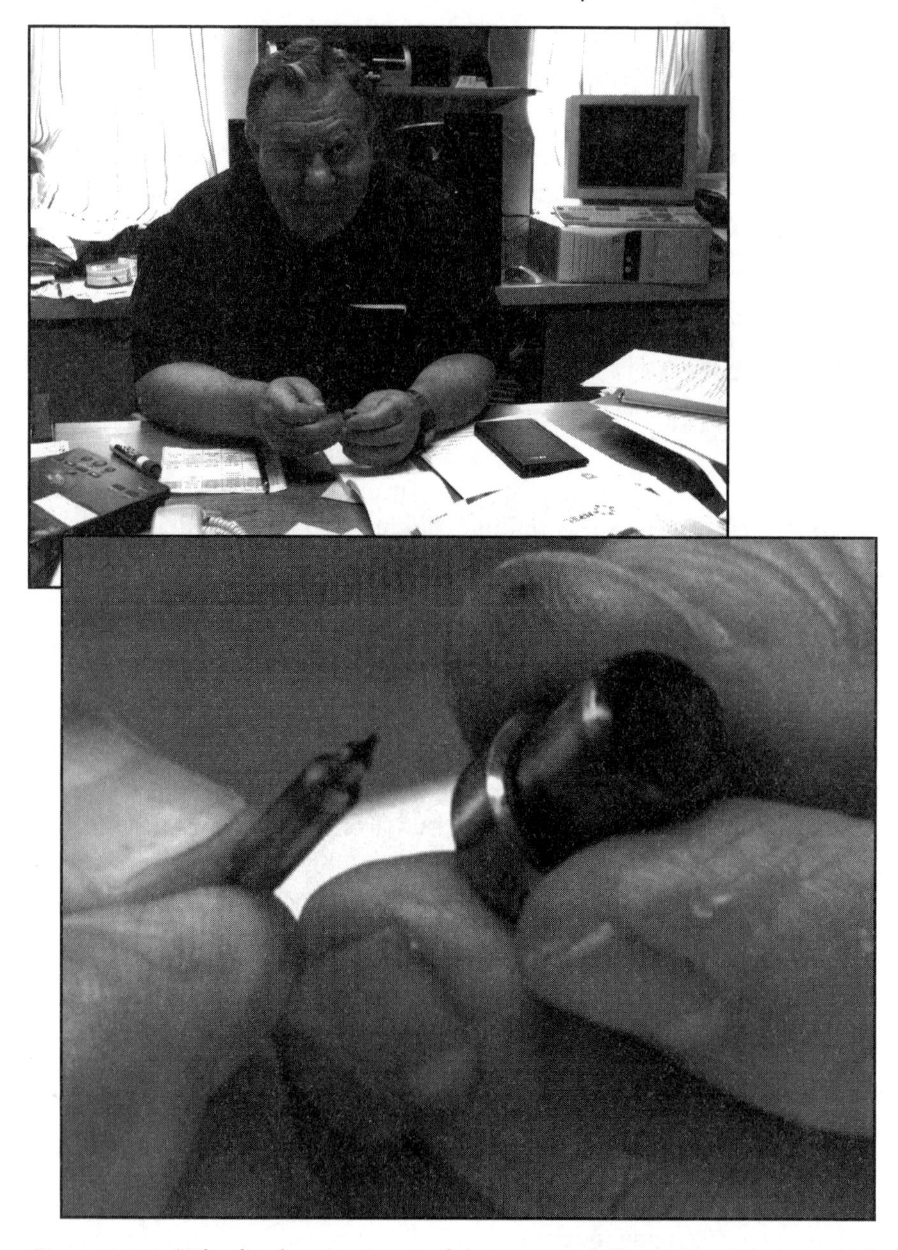

Figure 7.1-5. "The diesel engine is one of the most carefully machined devices which contains very close-tolerance parts such as this fuel injector and plunger assembly. Any minute particles of oxidized metal caused by water-contaminated fuel will damage this component or at the very least cause severe sticking, resulting in costly repairs. There is a reason the industry develops these standards, and to simply dismiss them as being excessive cannot be based on fact," states Professor of Mechanical Engineering Barry Hertz.

ting" of all ferrous metal parts that come in contact with water contaminated fuel. The diesel engine is one of the most carefully machined devices which contains very close-tolerance parts such as this fuel injector (Figure 7.1-5) and plunger assembly. Any minute particles of oxidized metal will damage this component or at the very least cause severe sticking, resulting in costly repairs. There is a reason the industry develops these standards, and to simply dismiss them as being excessive cannot be based on fact."

Online community chat rooms are not any better:

"I've had the same emulsion problems that GreaseFire [an online pet name] has experienced. The static wash on heated water goes fine. Even two times. Then, even seemingly mild bubbling causes emulsion. Same with mild misting. As for emulsion busting, I just dump in some dark glycerine byproduct and it clears up within 24 hours. Has worked great for me three times now. (thanks, Jack). I've about had my fill of washing and dealing with emulsions. Perhaps others have figured out how to avoid them, but despite my careful efforts they persist. The other great threads [online sources] going have reassured me that unwashed biodiesel won't wreck my life or the injection pumps in my 78 240D or 95 F350. After this batch I'm going to make the switch." (Source: http://biodiesel.infopop.cc/eve/ubb.x/a/tpc/f/719605551/m/990104519)

This person is describing the troubles they are having washing raw biodiesel to remove soaps and various contaminants. Others in this online community are advising that washing the biodiesel is not important and will not affect their personal health or that of the two noted vehicles.

As is typical with much of the advice given online, there is no supporting data and what is given is generally in favour of the simplest way forward. Fuel produced in this manner **will not meet** ASTM standards and ignores the volumes of evidence provided by vehicle, engine, and component manufacturers that off-specification fuel has been shown to reduce equipment life.

Summary

For the most part, home brewing of biodiesel is done under the guise of environmental stewardship, but it is really done in an effort to save money and "put the boots" to Big Oil. If home-brewed biodiesel cost $5.00 a gallon and took a week to make I would bet most people would quietly throw up a white flag, discard their alchemy aprons, and return to the fuel pumps with the rest of society.

Figure 7.1-6. If the home brew segment of the market is ever to be legitimized, an association working to develop safety, quality, and production technologies must be formed, preferably under an umbrella with its commercial cousins.

Everyone wants to save the planet; nobody wants to pay for it.

It is for economic and production labour reasons that people take shortcuts in the biodiesel production process or dispose of waste materials with the dismissive attitude that if "everyone else does it, so can I." Taking this antagonistic (or anarchistic) approach further alienates their cause. It is no wonder that legitimate commercial biodiesel producers want to distance themselves from the home brew community.

I suggest a change of attitude. Micro-scale biodiesel production is not going to go away, and given the number of people who are interested in producing their own fuel and the rising price of petrodiesel, more and more people will take part in this "revolution." Home brewers are not stupid. What they lack in finances and direct skill they make up for in energy and enthusiasm. Perhaps all that is needed is for commercial producers, universities, and industry advocacy groups to provide better direction and authority to help this movement police itself.

Biodiesel: Basics and Beyond offers a glimmer of hope in this sphere. The production methods and technologies outlined in the book will provide the prospective home brewer with an even-handed assessment of the pros and cons of producing ASTM-specification biodiesel. The book also tackles the hard questions of environmental sustainability with regard to both the input oil sources and the waste stream products. This book is not the end of the road for micro- and small-scale biodiesel production; it is the start. If the enthusiasm and energy of people who are eager to produce their own fuel are coupled with the strategies outlined in this book, perhaps a legitimate small-scale revolution will begin.

However, I would prefer to see this happen before someone dies trying to save a few dollars with the assumption that biodiesel can be produced in their kitchen.

7.2
The History of Micro-Scale Biodiesel

Converting waste vegetable fryer oil (WVO) into biodiesel has become a pop culture trend in the western world. The average person may not understand it, but those who can perform the alchemy of producing home-brewed biodiesel are granted exalted status.

As is so often the case, North Americans tend to forget that they make up only a small percent of the world's population, forgetting about the Europeans, who were the first to create biodiesel from waste vegetable oil. In 1983, Dr. Mittelbach studied the transesterification of WVO and went on to develop a commercial process to produce biodiesel from the financially attractive feedstock, earning him the World Energy Globe Award in 2001 and World Climate Star in 2003.

Figure 7.2-1. Closer to home, Dr. Thomas B. Reed is credited as the first person to roll up his sleeves and jump into a garbage dumpster to recover some precious WVO in an effort to produce biodiesel. (Courtesy Dr. Thomas B. Reed)

Closer to home, Dr. Thomas B. Reed is credited as the first person to roll up his sleeves and jump into a garbage dumpster to recover some precious WVO in an attempt to produce biodiesel. "In the summer of 1989, I learned about the conversion of animal fats and vegetable oils into their methyl ester for fuel purposes after attending some sessions at the US Department of Agriculture," states Dr. Reed. "Although they were speaking about converting refined oils at the proceedings, I wondered if the same process would work with waste oils after I discovered that there happened to be about a billion gallons a year of this stuff."

With the wonderment of a 10-year-old, Dr. Reed headed over to his local McDonald's restaurant and picked up a gallon of waste grease from the dumpster out back. "I took the icky stuff back to my laboratory at the Colorado School of Mines, where I worked at the time," Dr. Reed explains. "Wow! Just like that I produced a batch of beautiful biodiesel fuel."

"I didn't think the chemical name "transesterified waste vegetable oil" would win anyone over, so considering the source of the feedstock I decided to call my fuel McDiesel. I even applied for a copyright, but after a chat with the powers that be in the firm I decided my original name, "biodiesel," wasn't so bad!" exclaims Dr. Reed. "After these years, the new name has stuck, but I still think McDiesel was a pretty snazzy choice."

Dr. Reed continues his story by explaining that, "the Denver bus company was considering alternative fuels and we approached them to see if they would be interested in testing our alternative, clean fuel. "Yes," they said, "but we need more than a gallon for testing." So I went to our onsite laboratory at the School and made 2 drums (100 gallons) for their test program. Unfortunately, there was no political base for using WVO at the time and the concept went underground. The rest is history."

Figure 7.2-2. Dr. Thomas B. Reed is probably the first true home brewer, shown here filling up his 1989 Mercedes-Benz diesel wagon. This is definitely a picture for the history books. (Courtesy Dr. Thomas B. Reed)

Micro-Scale Production Case Studies

Micro-scale biodiesel producers (or "home brewers" if you prefer the colloquial term) are a very interesting bunch. They all have their own reasons for producing biodiesel and in many cases they have engineered their equipment from recycled materials. They definitely have a free spirit approach.

The following are case studies of a few of these determined pioneers, showcasing the engineering handiwork used in their biodiesel production equipment. I have chosen to start with Steve Anderson's raw biodiesel production system using recycled materials and conclude with a commercially available processing unit.

It should be noted that none of these systems solves the problems of fuel quality and environmental waste stream issues, but they do afford a glimpse of the diversity and effort of people who develop their own fuel refineries.

CAUTION

The manufacture and storage of biodiesel fuel as well as its component chemicals are dangerous and involve a degree of risk. The following stories and photo collection depict the production of biodiesel using methods which are known to be dangerous. They are included here for illustrative purposes only. The author, publisher, and contributors assume no liability for possible errors and omissions in the following discussion or for any personal injury, property damage, and consequential damage or loss which might occur as a result of using the information in this book, however caused.

7.3

Steve Anderson - "I'm Just a Simple Home Brewer"

Steve Anderson likes to refer to himself as a simple "home brewer." He has no pretensions about putting Big Oil out of business or getting rich making biodiesel. He just wants to save a few bucks and, for good measure, maybe even help the earth to live just a bit longer.

Figure 7.3-1. On a cool day in late winter I visited Steve Anderson and found him filling up a small bucket from a batch of biodiesel fuel that feeds the oil furnace.

Over the last few years, Steve has produced a couple of thousand gallons of raw biodiesel in his workshop/garage. It was a cool day in late winter day when I visited Steve and found him filling up a small bucket from a batch of biodiesel fuel that feeds the furnace. "I have a great deal with the owner of this garage," Steve explained. "All I have to do is keep the building warm all winter and I get to use half of the place for biodiesel production. It's a pretty good deal considering how little it costs to make the heating fuel."

Steve uses a very simple open processing unit fabricated from an old oil drum to produce his biodiesel. Without counting the "tinker factor," Steve estimates that he has invested about $500 in his setup. He explained that he has to invest some money this year to upgrade his production equipment. "I realize that I shouldn't be using an open processing system, because methanol gives off poisonous vapors. However, I have been careful to make sure the garage is open and to cover the reactor and methoxide drums when they are in use. Upgrading to a closed system is one of my next projects." Based on Dr. Tremblay's earlier comments regarding the chronic effects of methanol, this is no doubt a smart move.

Figure 7.3-2. Steve uses a very simple open processing unit fabricated from an old plastic oil drum to produce his raw biodiesel. Without counting the "tinker factor," Steve estimates that he has invested about $500 in his setup.

Figure 7.3-3. A close-up view of Steve's laboratory indicates that very little complex machinery or equipment is required to produce biodiesel.

A close-up view of the laboratory indicates that very little complex machinery or equipment is required to produce raw biodiesel. As Steve works through the process, it becomes clear that cleanliness and careful attention to detail are the main ingredients in successfully producing fuel. "I have made a couple of batches of glop," Steve informed me. "But most of the time the process works beautifully because I have worked with my waste oil suppliers to try to get a consistent product from them. They know I will be there twice a week to pick up the oil, in turn saving them thousands of dollars a year in recycling costs. In return, they ensure that the oil is not contaminated coming out of the fryer and waiting for pickup. I also check my "suppliers" to ensure that they are not using animal or hydrogenated fats, as any oil that is solid at room temperature will make a biodiesel that will gel when the weather gets cool." Feedstock oils such as these may be fine for someone living in the southern, more temperate areas of the United States, but pose considerable trouble for northerners.

Figure 7.3-4. Steve is careful not to take any hydrogenated or animal fats. These oils tend to be solid at room temperature and will quickly turn into a butter-like substance when the outside temperature begins to cool. This may not be a problem in the sunbelt of the United States, but it poses serious problems for northerners.

The Production Stages of Raw Biodiesel

The production of raw biodiesel may appear more difficult than it actually is. However, it is a smart person who produces the first batch or two on a smaller scale. After all, if you are going to make a mistake and produce a batch of glop, it might as well be a small batch.

Many beginning home brewers start out producing a 1-litre (approximately 1 quart) batch of raw biodiesel using an old blender and reducing the amount of WVO, methanol, and sodium hydroxide by a factor of 100 based on the standard recipe that follows. For example, using virgin vegetable oil, a mini batch requires 1 liter WVO, 0.2 liters (200 ml) methanol, and 3.5 grams of sodium hydroxide. All other procedures remain the same as for a large 100-liter (26.4-gallon) batch. In fact, this recipe may be scaled up or down to suit your particular needs and processing equipment requirements.

Figure 7.3-5. The first step in producing biodiesel is to filter the WVO using a filtration tank comprising an open-ended tank fitted with a pair of pantyhose. This unit filters out food particles and other debris that may be present in the oil.

Waste Oil Preparation

Figure 7.3-6. Steve is shown pouring WVO from one of his "suppliers" into the oil filtration unit.

Figure 7.3-7. While the oil is filtering through the stocking material, Steve cleans out the reactor vessel by carefully eliminating any traces of sodium hydroxide (lye) and glycerin from previous reactions. Leftover materials, especially water, increase the risk of the reaction making gravy rather than useful biodiesel.

Figure 7.3-8. Wash water is rinsed from the reaction vessel.

Figure 7.3-9. Soaps and glycerin contaminate the wash water. Unreacted sodium hydroxide may cause wash or waste water from the biodiesel process to be slightly "basic." Mixing small amounts of household vinegar (acetic acid) with the water will neutralize the base, bringing the pH level of the water to 7, or neutral. When the water has been neutralized, it may be composted or disposed of in a leeching bed. Do not pour waste water into a drain connected to a septic system.

Figure 7.3-10. The filtered WVO is now transferred to buckets...

Figure 7.3-11. ...and placed in the reaction vessel. 100 liters (26.4 gallons) of oil are processed into biodiesel at one time. (Calculations for mass and volume are greatly simplified using the metric system.)

Figure 7.3-12. Steve attaches a standard electric paint-mixing paddle to a variable speed drill to mix the oil and other chemicals. (As methanol fumes are very flammable, I explained to Steve that sparks produced by the drill motor commutator could ignite the vapors. Use Totally Enclosed Fan Cooled (TEFC) motors or preferably explosion-proof models available at any motor supply store or www.grainger.com.

Figure 7.3-13. The mixing drill is attached to a bracket, with the paddle immersed in the oil.

Figure 7.3-14. The drill is started and a slow, thorough mixing of the oil begins.

Figure 7.3-15. Steve has attached a 120-Volt water heater element to a bracket and placed it in the reaction tank oil. Heaters of this type must never be activated unless they are submersed in fluid. The WVO will require less time and energy to heat in an insulated and closed reaction vessel.

Figure 7.3-16. The heater and mixer are submersed in the WVO. Note the small thermal "bulb" just to the right of the drill. This device measures the temperature of the oil and sends a signal to a control unit.

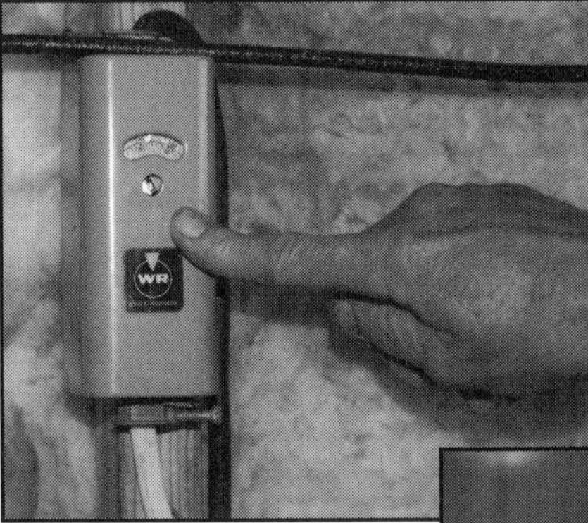

Figure 7.3-17. The heater control unit receives a signal from the temperature sensing bulb, ensuring that the oil temperature reaches and does not exceed 131°F (55°C).

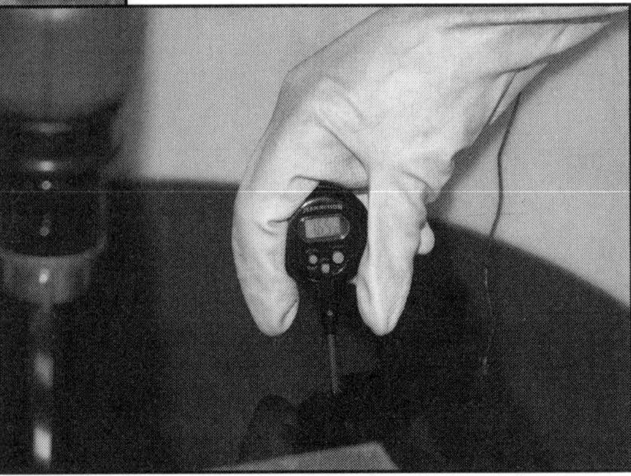

Figure 7.3-18. Methanol will be added to the WVO in subsequent steps. It gives off large amounts of explosive fumes when it boils (148.5°F/65°C), so it is important to control the temperature accurately.

Titration

If you think back to high school chemistry, you will recall the pH scale which is used to grade the strength of acids and their opposite cousin, "bases." The scale is numbered 0 through 14, with 7 being the halfway or "neutral" point on the scale. Pure water is neutral and therefore has a pH reading of 7. As the number drops below 7, the substance becomes a stronger acid. Likewise, as the reading increases beyond 7, the substance becomes more basic. Virgin vegetable oil is by nature acidic, and used fryer oils become increasingly acidic the longer and "harder" they are used.

The biodiesel transesterification process requires a base material, sodium hydroxide (lye), to act as a catalyst to cause the "exchange" of glycerin for methanol molecules. The titration process will determine the level of Free Fatty Acids (FFA) contained in the oil as well as the amount of base material required to ensure that the transesterification process is "pushed" to completion, thereby producing biodiesel and not "glop."

An acidic substance can become neutralized with the addition of a basic substance of equal but opposite strength. Using this rule and the information derived from the titration process, it is possible to calculate the amount of sodium hydroxide necessary to neutralize the FFAs and ensure that the biodiesel conversion process proceeds correctly.

In order to perform a titration process, make sure that you have a well-ventilated area. Then assemble the necessary chemicals, laboratory glassware, mass scale, and supplies:

- pH indicator known as phenolphthalein or the indicator from a swimming pool test kit known as "phenol red." Phenolphthalein is a laboratory-grade (more expensive) indicator and is slightly more accurate than swimming pool phenol red, but either indicator can be used.
- distilled water
- 100% pure sodium hydroxide or lye. Be sure to keep the lye dry, even preventing exposure to humid air.
- 99% isopropyl alcohol (Use care when purchasing alcohol as there are numerous concentrations available. If in doubt, consult with a pharmacist.)
- syringes graduated in 1 ml increments
- graduated cylinder or other accurate liquid measuring devices
- assorted small jars or beakers. At least one jar should have a sealable top to store 1 liter of fluid as a reference solution.
- triple-beam mass balance
- latex rubber gloves
- eye protection

Steve's Procedure

Note: Ensure that all beakers, containers, and syringes are clean and used for only one chemical. Syringes and glassware may be reused each time you make a new batch of biodiesel provided they are used for only one of the following steps:

1. Measure out 1 gram of sodium hydroxide on the mass balance. One gram is a very small amount; be very careful with this measurement. Alternatively, measure out 4 grams of lye and make a larger (4 liter) batch of reference solution. Measuring a larger quantity will divide any measurement error into a larger sample, improving the accuracy of the solution.
2. Measure 1 liter of the distilled water using a graduated cylinder. Place the water in the sealable bottle and add to it 1 gram of sodium hydroxide. Swirl the sealed bottle to ensure the sodium hydroxide is fully dissolved. This solution should be labeled "Reference Solution."
3. Draw 10 ml of isopropyl alcohol and place it in a small beaker.
4. Add 2 drops of phenolphthalein indicator into the alcohol and swirl until well mixed.
5. Draw a sample of WVO and add 1 ml to the phenolphthalein/alcohol solution. Use extreme care in ensuring that exactly 1 ml is dispensed. It is best to fill the syringe with, for example, 5 ml of oil and then dispense down to 4 ml rather than trying to measure a single milliliter of oil.
6. Vigorously swirl the beaker to ensure complete mixture. If you are using a sealing jar, shaking will work as well.
7. Draw 10 ml of the reference solution into a syringe. Swirling the beaker of phenolphthalein/alcohol/WVO in one hand, carefully drip 0.5 ml amounts of reference solution into the beaker at a time. The solution will be a musty yellow at the beginning of this test and will turn a uniform bright pink when the titration is complete. (The solution will turn pink when the reference solution is first injected but will return to yellow as you continue to swirl the beaker. The solution must be uniformly pink for at least 30 seconds before the titration is considered finished.)
8. Record the amount of reference solution required to complete the titration process. This will be the amount of reference solution left in the syringe subtracted from 10. For example, if 7.5 ml remain in the syringe, then 2.5 ml were used (10-7.5 ml=2.5 ml used).
9. Add 3.5 to the amount of reference solution used in Step 8 and then multiply this number by 100. This equals the mass in grams of sodium

hydroxide that will be required to complete the transesterification process for 100 litres of WVO.

In the example in Step 8, 3.5 ml would be added to 2.5 yielding 6, multiplied by 100 to give a result of 600 grams of sodium hydroxide required for every 100 liters of WVO.

Use the balance beam scales to accurately measure the required amount of sodium hydroxide.

10. It is recommended that you clean up the lab and do the test at least one more time to ensure accurate results. It is very easy to make a mistake, and accuracy is what produces a high-quality fuel.
11. Store the measured sodium hydroxide in an airtight jar and put it aside in preparation for the next stage.

Figure 7.3-19. **Steps 1 and 2**. *Measure 1 liter of the distilled water using a graduated cylinder. Place the water in the sealable bottle and add to it 1 gram of sodium hydroxide. Swirl the sealed bottle to ensure that the lye is fully dissolved. This solution should be labeled "Reference Solution."*

Figure 7.3-20. **Step 3**. *Steve is lecturing on the importance of using 99% pure isopropyl alcohol for the titration process. If you are unsure of the type of alcohol, consult a local pharmacy.*

Figure 7.3-21. **Step 3**. *Draw 10 ml of isopropyl alcohol and place it in a small beaker.*

Figure 7.3-22. **Step 4**. *Add 2 drops of phenolphthalein indicator into the alcohol and swirl until well mixed.*

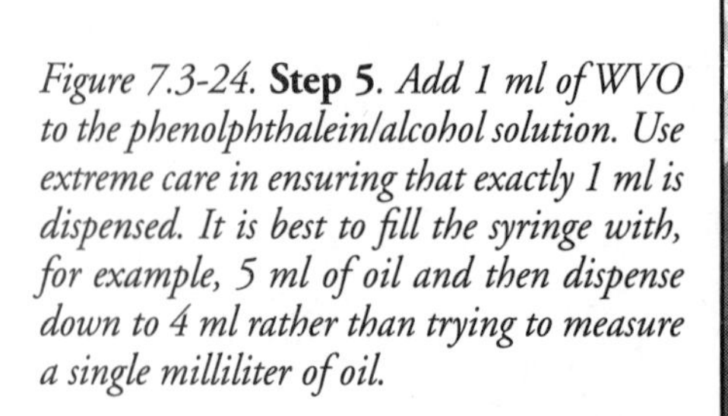

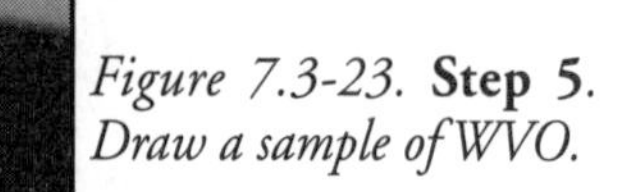

Figure 7.3-23. **Step 5**. *Draw a sample of WVO.*

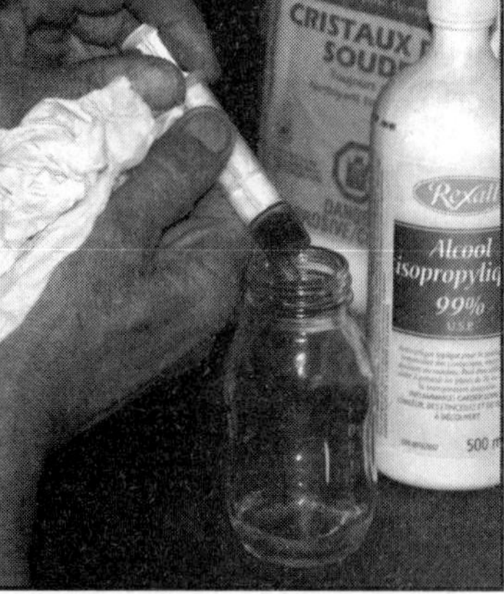

Figure 7.3-24. **Step 5**. *Add 1 ml of WVO to the phenolphthalein/alcohol solution. Use extreme care in ensuring that exactly 1 ml is dispensed. It is best to fill the syringe with, for example, 5 ml of oil and then dispense down to 4 ml rather than trying to measure a single milliliter of oil.*

Figure 7.3-25. **Step 7**. *Draw 10 ml of the reference solution into a syringe.*

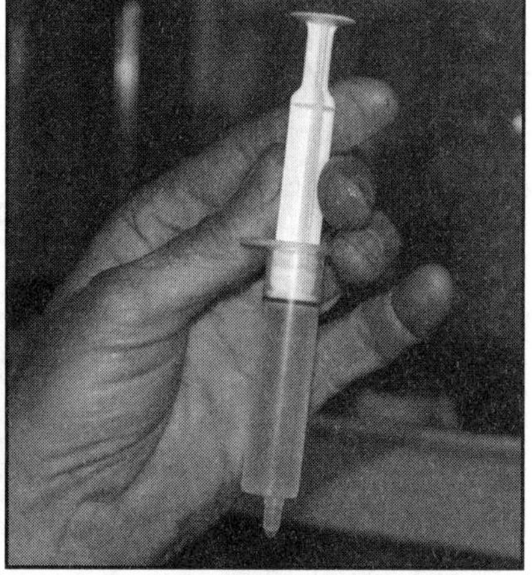

Figure 7.3-26. **Step 7**. *Swirl the beaker of phenolphthalein/alcohol/WVO in one hand while carefully dripping 0.5 ml amounts of reference solution into the beaker at a time. The solution will be a musty yellow color at the beginning of this test.*

Figure 7.3-27. **Step 7**. *The solution will turn a uniform bright pink (which is a bit hard to see in a black and white photograph) when the titration is complete. The solution will turn pink when the reference solution is first injected but will return to yellow as you continue to swirl the beaker. The solution must be uniformly pink for at least 30 seconds before the titration is considered finished.*

Figure 7.3-28. **Step 9**. *Steve is shown using a triple balance beam scale to accurately measure the amount of sodium hydroxide calculated in Step 8. It is set aside ready for the sodium methoxide steps to follow.*

Figure 7.3-29. **Step 9.** *The accuracy of the titration process and measurement of the sodium hydroxide catalyst will determine the success or failure of the biodiesel transesterification process.*

Preparation of the Sodium Methoxide Solution

Sodium methoxide is a very dangerous, flammable, and noxious solution. Given the health and safety issues involved, I do not recommend the "open vessel" method of handling of this material.

Procedure

1. Draw 20 liters of methanol (20% by volume of the WVO to be reacted) into a mixing vessel.
2. Carefully add the premeasured sodium hydroxide to the methanol. The mixing of these two components will generate a considerable amount of heat in an "exothermic" process. The fumes from this solution are deadly. Gas masks and other organic filters are useless against the sodium methoxide vapor.

Figure 7.3-30. Steve's sodium methoxide reaction vessel is made from a simple high-density polyethylene (HDPE #2) plastic 22-liter pail. A hole is drilled into the removable top to accept a paint-stirring paddle.

3. Mechanically stir the mixture to ensure complete solution, since sodium hydroxide does not actively dissolve in methanol.

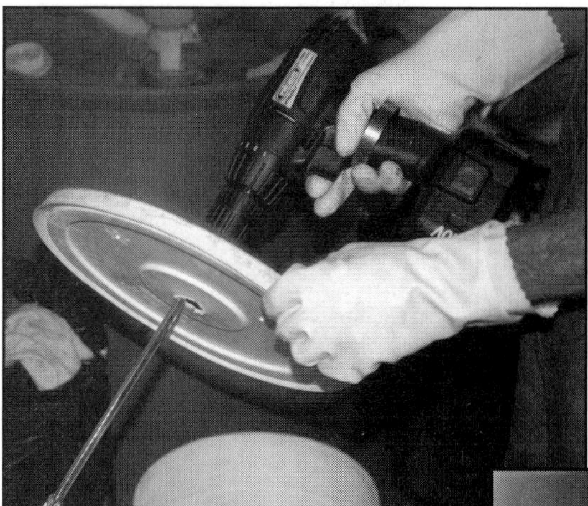

Figure 7.3-31. A cordless electric drill is fitted to the stirring paddle and the trigger is fitted with a lock, allowing the drill to operate unattended.

Figure 7.3-32. A drum of methanol "racing fuel" is fitted with a siphon pump.

Figure 7.3-33. Steve measures 20 liters of methanol (20% by volume of the WVO to be reacted) into the pail.

Figure 7.3-34. The premeasured sodium hydroxide from the titration stage is added to the methanol. The exothermic or "heat-producing" reaction of the two chemicals will produce sodium methoxide.

Figure 7.3-35. The sodium methoxide reaction vessel is allowed to sit or is mechanically stirred until all of the sodium hydroxide dissolves in the methanol.

Figure 7.3-36. Steve removes the electric heating element from the WVO reaction tank in preparation for the addition of the sodium methoxide solution.

Figure 7.3-37. The WVO mixer speed is increased to ensure complete agitation of the WVO and sodium methoxide solution.

Reaction of the Waste Vegetable Oil

Once the sodium hydroxide has been fully dissolved into the methanol, the sodium methoxide solution is ready to be added to the heated WVO to begin the transesterification process.

Transesterification involves two phases: a byproduct phase that produces glycerin and a second phase that produces "raw biodiesel." Glycerin must be refined as discussed in Chapter 8.3.10.

While some use the raw biodiesel as soon as it has been reacted and separated from the glycerin, I advocate further processing. Raw biodiesel is 100% incompatible with modern automotive engines and will cause premature failure. If you are attempting to produce a fuel with the intention of saving money, premature engine failure will quickly reverse any potential financial savings.

Procedure

1. Carefully add the sodium methoxide solution to the heated and stirred WVO.
2. Mechanically stir the solution of WVO and sodium methoxide continuously for one hour. During this time, the solution will begin to turn a "chocolate milk" color as the reaction proceeds, then a low-viscosity (thin) clear brown color. This is a good indication that the reaction has proceeded correctly.
3. Stop the mixer after one hour and allow the solution to sit for 24 hours. During this time the solution will separate into two distinct layers or "phases." The denser phase is glycerin and unreacted sodium hydroxide which, being heavier than biodiesel, will sink to the bottom. The lighter solution floating on top of the glycerin is the "raw" biodiesel or Fatty Acid Methyl Ester (FAME) phase.
4. Remove the dark brown, very viscous glycerin layer from the bottom of the reaction vessel by opening the bottom valve and allowing the glycerin to flow into a bucket. As soon as it is drained from the tank, the raw biodiesel will follow. Close the drain valve at this point.
5. You may decant the raw biodiesel into storage pails or leave it in the reaction vessel, depending on your desire to produce a high- or low-quality fuel.

Figure 7.3-38. Steve carefully pours 20 liters of sodium methoxide into the reaction vessel, which is filled with 100 liters of heated WVO.

Figure 7.3-39. As soon as the sodium methoxide touches the WVO, the reaction process begins. Note the light and dark tinges in the photograph.

Figure 7.3-40. The transesterification process is well underway. The WVO/sodium methoxide solution has turned the color of chocolate milk.

Figure 7.3-41. The reaction is nearly complete after one hour of continuous stirring. The WVO has become very clear and its viscosity (thickness) has dropped dramatically. After stirring, the solution is allowed to sit for 24 hours to allow the glycerin and raw biodiesel (FAME) to separate into two distinct layers or phases.

Figure 7.3-42. After a 24-hour settling period, the heavier, denser glycerin and unreacted sodium hydroxide are drained from the reaction vessel. The lighter, less dense raw biodiesel remains.

Figure 7.3-43. In the "simple" production method, raw biodiesel is drained from the reaction vessel and placed in buckets and allowed to sit for a further three days to one week. Although the majority of the transesterification process is completed in one hour, the reaction will continue for a longer period, allowing additional glycerin to separate from the raw biodiesel. Soap, a byproduct of the reaction process, can be seen floating on top of the raw biodiesel.

Figure 7.3-44. After allowing the raw biodiesel to sit one week, Steve wipes the floating soap layer away.

Figure 7.3-45. The raw biodiesel is decanted into storage buckets.

Figure 7.3-46. The clear, low-viscosity raw biodiesel is visible in this photograph.

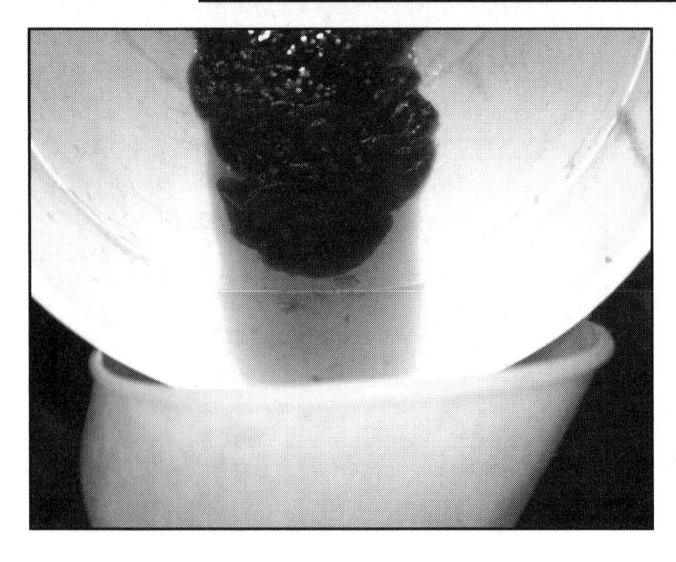

Figure 7.3-47. As the raw biodiesel is decanted, glycerin and unreacted sodium hydroxide remain behind, the result of slow, low-level reactions taking place during the seven-day settling period.

Biodiesel: Raw or Cleaned?

Some people advocate using the raw biodiesel directly. Steve Anderson estimates that he and his friends have traveled over a couple of hundred thousand miles in many different vehicles on the raw fuel he produces without any problems. Why all the fuss?

Raw biodiesel fuel is just that: raw. The reaction process will continue for a long period of time, causing glycerin to drop out of the fuel. In addition, unreacted sodium hydroxide is present in the raw fuel, causing the pH to remain very high compared to the ideal neutral pH 7 reading. High pH level is an indication that the fuel is corrosive and will, over time, damage expensive engine components.

Figure 7.3-48. Raw biodiesel fuel is just that: raw. The reaction process will continue for a long period of time, causing glycerin to drop out of the fuel, resulting in filter plugging. (Steve comments that he does not consider this to be a problem.) In addition, unreacted sodium hydroxide is present in the raw fuel, causing the pH to remain very high compared to the ideal neutral pH 7 reading. High pH level is an indication that the fuel is corrosive and will, over time, damage expensive engine components. In this view the left bottle contains Steve's raw biodiesel. The second bottle from left shows the "first wash" fuel, with milky residues of glycerin, soaps, and sodium hydroxide. The third bottle from left shows biodiesel after a second washing stage. The sample to the right is clear, washed biodiesel with a proper noncorrosive pH of 7, although the "bonded and free unreacted glycerin" measures 1.407%, which is nearly 6 times the allowable limit of 0.24%.

Analysis	Method	Min	Max	Steve Anderson's Homebrew Biodiesel	Notes
Acid Number, mg KOH/gram	ASTM D664		0.80	0.11	
Ash, Sulfated, Mass %	ASTM D874		0.020	0.0	
Total Sulfur, Mass %	ASTM D5453		0.05	0.0006	
Cetane Number	ASTM D613	47		51.6	
Cloud Point, °C	ASTM D5773			-3	
Copper Corrosion	ASTM D130		No. 3	1a	
Vacuum Distillation End Point, ° C	ASTM D1160		360°C at 90%	525	1
Flash Point, °C	ASTM D93 Procedure A	130.0		25	2
Method for Determination of Free & Total Glycerin	ASTM D6584				
Free Glycerin, Mass %			0.020	0.98	3
Total Gylcerin, Mass %			0.240	1.407	4
Phosphorus (P), Mass %	ASTM D5185		0.001	0.1	5
Carbon Residue, Mass %	ASTM D4530		0.05	0.022	
Kinematic Viscosity, Centistokes at 40°C	ASTM D445	1.9	6.0	5.828	6
Water and Sediment	ASTM D1796		0.050	0	7
Total Methanol (%)	EN 14110		0.2	2	8

Table 7.3-1

This table indicates the limits of various parameters according to ASTM D 6751 standards and compares them with Steve Anderson's raw biodiesel.[1] *Steve's home-made biodiesel fails approximately 50% of the tests with several failing by a very wide margin.*

Notes:

1. High levels of di- and trigylcerides are the likely cause of the end point distillation temperature failure. Although unreacted gylceride measurement is not contained in the ASTM data tables, it was measured and found to be 2.56%, which indicates that the biodiesel reaction was not completed.
2. Biodiesel is not considered a flammable material, and handling and storage procedures reflect this. However, here excess methanol levels have caused the flash point to drop precipitously.
3. Free glycerin levels are approximately 50 times the allowed levels. This is an indication that the fuel has not been washed, which will result in filter plugging and fuel injection system problems, particularly with newer high-pressure fuel injection systems having increased mechanical tolerances.
4. Total glycerin is the sum of free glycerin and unreacted glycerin still "bound" to the biodiesel fuel molecules. High total glycerin is indicative of unwashed fuel and incomplete reaction.
5. Phosphorous levels are 100 times greater than the allowable limits. Phosphorous can be attributed to feedstock oil, waste materials present from cooking and frying, or the introduction of phosphoric acid used in the neutralization process of free fatty acids. It is unlikely that micro-scale producers can influence this value through process techniques.
6. Here I have indicated that viscosity has failed, because the ASTM committee is currently evaluating lowering this level to 5 centistokes to more closely match diesel fuel parameters. High viscosity is known to damage fuel injection pumps, reduce atomization of fuel spray, and dilute engine lubricating oil.
7. Water and sediment is indicated as zero because the fuel has not been subjected to a water-washing process to remove contaminants. (The washing step is often skipped by home brewers in an effort to reduce the work required to produce the fuel, while producing a substandard product). Further, washing fuel without subsequent drying will virtually guarantee a failure of the water and sediment test.
8. Total methanol is not recorded in the ASTM standards but is inferred from the flash point level. I have used the equivalent levels from the European standard to indicate that total methanol has failed by a factor of 10 times.

To skip the washing, pH adjustment, and quality control steps, not to mention the waste stream recovery and disposal, is simply gambling with the longevity of your engine or furnace and the sustainability of the local environment. Older, less sensitive diesel equipment may operate on raw biodiesel for many years with apparently little problem. For newer engines with computer-controlled management systems, burning raw fuel constitutes a death sentence.

Figure 7.3-49. Older, less sensitive diesel equipment may operate on raw biodiesel for many years with apparently little problem. For newer engines with computer-controlled management systems, burning raw fuel constitutes a death sentence. This fuel filter became plugged after 12 hours of operating a high-pressure, common rail diesel engine using Steve's raw biodiesel.

Raw biodiesel has a higher viscosity than washed biodiesel because of suspended glycerin, which plugs fuel filters and affects the fuel spray and atomization pattern. Being a heavier fuel with lower vapor pressure, raw biodiesel does not burn completely, increasing atmospheric pollutants. At the same time, fuel spray condenses on the relatively cool cylinder walls and washes past the piston rings, fouling engine oil with insoluble material.

Corrosive, unreacted sodium hydroxide will attack fuel system components, particularly in late-model, high-pressure common rail direct injection (CDI) engines. Excess methanol is a solvent which will destroy polymer components used in engine fuel-system seals and hoses unless they are specifically designed to carry it.

In short, using unwashed, raw biodiesel is akin to gambling. You may save money by making raw, homemade fuel, but will the savings be depleted by expensive engine repairs? If quality, operator safety, and environmental sustainability are truly issues that concern you, the procedures and technology described in Chapter 8 will help you meet those goals.

7.4
Dave Probert – Biodiesel to Go

"A few years ago, my wife received a newsletter from the college she graduated from in Vermont," explains micro-scale biodiesel producer Dave Probert. "She pointed out an article that explained how clean and environmentally friendly biodiesel is and that it can even be made from waste vegetable oil (WVO). Being the proud owner of a Dodge Cummins diesel truck, I thought I was up on all of the latest "diesel" stuff. Clearly I was wrong."

Figure 7.4-1. Dave explains that after his biodiesel revelation, he purchased a book on the subject and started searching Websites and online discussion forums to learn more about the topic of home-scale biodiesel. "I like to get working on a winter project, fabricating things like snowshoes or a canoe. Well, a biodiesel processing system seemed like an interesting project!" (All photographs courtesy Dave Probert)

Dave explains that after this revelation, he purchased a book on the subject and started searching Websites and online discussion forums to learn more about the topic of home-scale biodiesel. "I like to get working on a winter project, fabricating things like snowshoes or a canoe. Well, a biodiesel processing system seemed like an interesting project!"

Dave's original processor was built with the obligatory water heater to heat the WVO as well as with a settling/wash tank fabricated from a 55-gallon drum, equipped with a circulating pump, filter, numerous valves, and fittings. The entire processing system was assembled in an insulated cabinet that was mounted on casters so it could be kept warm and rolled out of the way when not in use. "I have a two-car garage that I like to use for two cars," he says.

"The first batch of biodiesel rolled off the system in the spring of 2004 and it was a success," Dave says. "It is interesting to note that no two batches have acted or reacted the same, which leads me to believe that either my titration tests are inconsistent, the oil is inconsistent, or one of about a dozen process variables has changed and impacted the outcome and quality of each batch."

Dave continues: "Last winter I purchased a new single-cylinder diesel engine manufactured in India and based on a 1930s Lister engine design. This unit really caught my eye and in order to 'justify' making the purchase I decided to integrate it into the biodiesel production process by using my biodiesel fuel to power the engine which in turn drives a generator to power my biodiesel processor. What a concept. Perpetual motion perhaps?"

Figure 7.4-2a/b. "Last winter I purchased a new single-cylinder diesel engine manufactured in India. I decided to integrate it into the biodiesel production process by using my biodiesel fuel to power the engine which in turn drives a generator to power my biodiesel processor. What a concept. Perpetual motion perhaps?"

"I definitely didn't have room in my garage for any more machinery, so the next logical step was a trailer to mount all of this stuff on," Dave says. "Who knows? It might even make the unit suitable for hauling around for demonstration purposes."

Figure 7.4-3. "I definitely didn't have room in my garage for any more machinery, so the next logical step was a trailer to mount all of this stuff on," Dave says. "Who knows? It might even make the unit suitable for hauling around for demonstration purposes."

"Back to the 'obligatory' water heater for a moment. When I was reheating a batch of washed fuel while trying to speed up the drying process, the generator suddenly 'let up' and the engine sounded as if it didn't have any load on it. I suspected that the heating element had broken and after further examination discovered that was exactly what had happened. I became worried about the potential hazard of a broken element arcing in the presence of methanol vapors....I removed the water heater from the equation and bought an externally fitted drum heater to fit on the settling/wash tank, which I now use as the main reactor tank as well. Perhaps this isn't as efficient a design, but I feel a whole lot safer with the system." As you can see in the accompanying pictures, this has resulted in a very compact and simple processing system. "Simple is good," Dave says.

"I only do small batches, say 20 gallons (76 liters) of WVO, in order to keep the process manageable. This requires using 4 gallons of methanol and the proper amount of KOH (potassium hydroxide) catalyst to produce the methoxide. I found that using sodium hydroxide produced a solid glycerin which was much more difficult to work with."

Figure 7.4-4a/b/c. After having a heater element fail while processing a batch of biodiesel, Dave became worried about the safety of his original design and opted for a single processing tank with an external "band" heater. The potassium methoxide delivery tank is located on the shelf above the processing tank. Details are shown in Figures 7.4-4b and c.

"I am concerned about the issue of waste stream products generated by this process. I have been told (from online sources) that the wash water isn't a problem, but I am not 100% convinced that this is true. I know for certain that glycerin disposal is a problem almost akin to nuclear waste. We keep making more biodiesel, but we have no idea what to do with all of the leftover methanol-contaminated glycerin. I think this is one of the biggest problems facing the home-brew community," Dave says.

"I have room left in the processor cabinet to reinstall a water heater which I am going to try to use as a distiller for recovering excess methanol from the glycerin. This will not only give me the option of reusing the methanol but also make the glycerin less hazardous."

"The home brew [biodiesel] is used in my truck as well as in a 2000 VW TDI Beetle in a concentration of 50%-70% in the summer. I purchase commercially mixed B20 to fuel my vehicles in the winter as my home brew doesn't work well in the cold months. In addition, I add about 30% home brew to my furnace heating system. There is a slight reduction in horsepower when we run high-percentage biodiesel, but the decreased smoke and reduced smell more than make up for any minor loss of power."

"Although my home brew costs about $1 per gallon, excluding labor, I am not doing this less for financial reasons than for the political statement it makes. We could raise feedstock here in Maine, which would support many agricultural jobs and environmental interests. I just want to do my part," Dave concludes.

7.5
Lyle Rudensey – a.k.a. "BioLyle"

"I got into biodiesel production as a result of a meeting at an Earthday Fair here in Seattle where I was introduced to the fuel for the first time," Lyle explains. "I was amazed that this commercially available fuel was domestically produced and biodegradable. All of this was around the time when the Iraq war was starting, which seemed to me to have more to do with protecting U.S. oil interests than anything else. I felt that using biodiesel in my own car was one small statement I could make to support a sustainable fuel economy rather than the current unsustainable system our economy is now connected to. Although I purchased commercial biodiesel for a while, I really wanted to see if I could produce my own fuel."

Lyle continues our conversation by saying that he works at the University of Washington in the Department of Environmental Health and has experience in laboratory work. Although he tried to form a cooperative to share the workload, it never got off the ground and he decided to develop his own system.

"I got the information from the Web for the so-called "Appleseed" processor which incorporates an electric hot water tank. There is really nothing

Figure 7.5-1. "I got into biodiesel production as a result of a meeting at an Earthday Fair here in Seattle where I was introduced to the fuel for the first time," explains Lyle Rudensey a.k.a. "BioLyle." "I was amazed that this commercially available fuel was domestically produced and biodegradable." Lyle is shown here filtering WVO as he gets ready to produce another batch of homemade biodiesel. (All photographs courtesy of Lyle Redensey.)

Figure 7.5-2. "I got the information from the Web for the so-called "Appleseed" processor, which incorporates an electric hot water tank. I modified the plans to suit my needs, building the unit along the side of my garage. I have been producing 50-gallon batches at a time for about three years now, using 10 gallons per week of B100."

special about this concept other than that it provides an inexpensive means of creating a heated reaction tank. I modified the plans to suit my needs, building the unit along the side of my garage. I have been producing 50-gallon batches at a time for about three years now, using 10 gallons per week of B100."

"At the same time my old Toyota Tercel wagon had over 200,000 miles on it and it was time to trade it for another car. The obvious choice was a VW Jetta diesel wagon, which has a modern TDI (Turbo Direct Injection) engine which gives the vehicle a lot of pep and very little smoke. The car has been driven over 30,000 miles (48,000 km) using the fuel with no real problems to speak of."

Lyle admits that when he first started using his homemade biodiesel, he was not washing fuel and found that free glycerin plugged his fuel filter at an awkward time. "I was driving on a classic dark, rainy night with my elderly mother-in-law when the engine stalled out," Lyle explains. "At the time, I was going on the advice of others that all of the glycerin would "drop out" of the fuel within 48 hours. That turned out not to be true. And during the first year, when I didn't wash the fuel, I found that for whatever reason the fuel degraded a hose that returns excess fuel from the injector pump to the tank. After I started washing the fuel these problems never recurred, and the car has run fine ever since."

I asked Lyle if his local VW dealer had any comments on his using biodiesel and especially non-commercial-grade fuel. "Yes, our dealer held a workshop for biodiesel and TDI customers. One of the items they discussed was that after about 70,000 miles (113,000 km) the EGR (Exhaust Gas Recirculation) valve would require cleaning as a result of soot buildup. A few of the long-term biodiesel users offered our cars to the dealer to look at EGR, and he was surprised to find that the valves were very clean when compared to regular TDIs operating on petrodiesel fuel. Although VW only endorses a B5 limit, the dealership was nonetheless impressed with the how well high-concentration biodiesel worked."

We continued our discussion around the issue of the waste streams starting first with glycerin. "I have been very lucky being able to deal with

Figure 7.5-3. Lyle summarizes by saying that he wants to broaden the impact and use of biodiesel not just in the United States but in the developing world as well. "I am lucky that my hobby of home brewing biodiesel is really an extension of my day job working in environmental health education. If this bit of work I am doing helps clean even a bit of the air, then it will be worth the effort."

a commercial biodiesel producer who has agreed to take my glycerin," Lyle states. "This is an ideal situation for me, although it is one that is not open to all home brewers because of lack of proximity or willingness of commercial producers to accept this waste. As for wash water, I have been told that it does not pose a problem and have been putting it down the toilet, allowing the sewer treatment facilities to process it."

Lyle goes on to explain that he has attempted to eliminate the water washing process by using a commercially available product known as Magnesol® (www.dallasgrp.com) which is produced from magnesium silicate. Magnesol is known as a selective absorbent for hydrophillic materials such as glycerol and mono- and diglycerides. Although this process eliminates water washing and reduces total glycerin content of the biodiesel, extensive filtration is required. "This Magnesol stuff is like crushed rock and I found that even after progressive stages of filtering I could see residue in the biodiesel. I am pretty sure that this residual material would wreak havoc on my fuel system and engine, so I decided against using it," Lyle explains. "I have stuck to using water washing and we have plenty of rain on the west coast, so I use rain barrels to capture the water necessary to process the biodiesel."

Lyle summarizes by saying that he wants to broaden the impact and use of biodiesel not just in the United States but in the developing world as well. "I have been able to help promote the use of biodiesel and educate people about it through my Website www.biolyle.com as well as offering demonstrations and educational materials. Recently, I've helped to form a new group called the Breathable Bus Coalition, which aims to influence the school districts in Seattle and surrounding area to use biodiesel in its buses. Several studies have shown that children riding the bus are exposed to significant levels of diesel exhaust, which can trigger asthma and also contains numerous carcinogenic compounds."

"I am lucky that my hobby of home brewing biodiesel is really an extension of my day job working in environmental health education. If this bit of work I am doing helps clean even a bit of the air, then it will be worth the effort," Lyle concludes.

7.6
A Quick Stop at Church

Perhaps one of the most interesting aspects of researching this book has been discovering the diversity of people who enjoy bucking The System, all the while managing to make a living and trying to help the environment at the same time. On a beautiful day in early spring, I decided to visit one such group of friends who fit into this category and stop by church at the same time.

After a wonderful drive through central Ontario farm country, you come across the tiny hamlet of Pethericks Corners, just northeast of Campbellford. It is here, in an 1878 Methodist church, that John Graham and his wife

Figure 7.6-1. After a wonderful drive through central Ontario farm country, you will come across the tiny hamlet of Pethericks Corners, just east of Campbellford.

Figure 7.6-2. Here, in an 1878 Methodist church, John Graham and his wife Cherie decided to set up their business, the Church-Key Brewing Company, producing a high-quality product and selling into the premium beer market.

decided to set up their business, the Church-Key Brewing Company.

John explained that his facility is pretty small compared to the commercial "factory" brewers. Church-Key focuses on specialty beers and a quality product. "There is no way on earth that we can compete with the big boys. Our cost base is much higher, so we have to work within the niche area of the market. We have an excellent team producing a very high-quality beer, which allows us to make a living; however, I knew we could do more." I immediately assumed that an expansion or takeover by a competitor was in the works, but John just smiled and brought me in for a tour.

To the untrained eye, the brewery looks pretty much exactly the way one would expect. The big tanks dominate the room and the pungent smell of roasted hops and yeast fills the air, while refrigeration and heating equipment broods over the workers. In an industrial sort of way, it all seems rather normal. "Since we have to run a tight ship to stay in business, we decided to take it all the way," John continued. "We have adopted energy efficiency and environmental sustainability as key goals throughout the brewing process and in the operation of the building. We started with the obvious things first, such as compact fluorescent lighting, and moved on to more advanced ideas. Water must be heated as part of the brewing process, which requires fossil fuels, but we also realized that the commercial fridges in the building gave off

Figure 7.6-3. To the untrained eye, the brewery looks pretty much exactly the way one would expect. The big tanks dominate the room and the pungent smell of roasted hops and yeast fills the air, while refrigeration and heating equipment broods over the workers. It all seems rather normal.

a lot of heat, so we installed a heat recovery unit that captures this waste heat and transfers it to water used for brewing and cleaning. The propane-fired water boiler requires less energy because it's supplied with preheated water. It probably took less than one week to pay back the cost of the unit and we now have a source of clean, "free" energy."

Figure 7.6-4. John realized that the commercial fridges in the building gave off a lot of heat, so he installed a heat recovery unit that captures this waste heat and transfers it to the brewing water before it enters the boiler unit. It took less than one week to pay back the cost of the unit, and the brewery now has a source of clean, "free" energy.

John went on to explain that the reverse problem occurs at the end of the brewing cycle. "Once the brewing process is completed, the next step is to pump the hot (151°F/66°C) wort into a refrigeration unit. With all of the cold water flowing into the building, we are able to use the incoming cold water to "pre-cool" the wort before it enters the fridge. At the same time, this cold water is warmed up a bit before it enters the heat recovery unit I mentioned earlier. With a bit of thinking like this, you can reuse what is typically waste energy just by understanding your processes and using a bit of creative plumbing." The resulting energy savings help to make the microbrewery viable, while at the same time reducing the greenhouse gases and other atmospheric emissions. Saving a kilowatt of electricity is always cheaper and more environmentally sustainable than using a kilowatt, regardless of where the energy comes from.

There is more. John pointed out that all of the trucks at the brewery are diesels, and judging by the GMC 7500 sitting out front they're pretty big as well. "Aside from our retail outlet here in the church, we deliver and sell our beer only in draft kegs to bars and restaurants over a wide area of Ontario," John explained. "Fuel costs for beer delivery are a big part of our operating budget, so we decided to take a look at the feasibility of biodiesel production from waste vegetable oil (WVO)."

John and some of his adventurous friends did some research and decided that the easiest way to get into the biodiesel production game was simply

Figure 7.6-5. John Graham: "Fuel costs for beer delivery are a big part of our operating budget, so we decided to take a look at the feasibility of biodiesel production from waste vegetable oil (WVO). We quickly realized that we have a zero-cost base for collecting the WVO, as we are delivering beer to the same clients that have this waste product to get rid of."

to roll up their sleeves and give it a try. "Together we pooled our money and purchased a Fuel Meister™ system from Biodiesel Solutions, as well as assorted pumps and storage tanks to help with the collection and handling of the fuel. We quickly realized that we have a zero-cost base for collecting the WVO, as we are delivering beer to the same clients that have this waste product to get rid of. It works out well for both of us, because we are helping them get rid of a product that costs money to dispose of. The restaurants return the favor by keeping the oil clean, eliminating dish water and bacon grease for example, and having it ready for pickup. Everyone wins."

After loading the WVO into the storage tanks on board the trucks, they return to a small rented facility in the area and convert the WVO into biodiesel. "We have made well over thirty 40-gallon (150-liter) batches of mostly good biodiesel, as well as a couple of barrels of gravy!" John exclaimed. "We are pretty happy with the process and have learned a few tricks along the way to make the work easier, but all in all, it has been a very positive experience and has certainly helped to reduce our operating costs even further."

As we toured the facility, John went on to explain that this past winter they started to test biodiesel for home heating. They are pleased to report having no problems whatsoever with a B50 (50% biodiesel with petrodiesel) blend in a standard oil furnace. "We will move cautiously in making the transition, but because my buddies and I are quite handy, I can't see this being any problem in the future."

Speaking of problems, I asked John if they had any trouble using the homemade biodiesel in the delivery trucks. "No, not really. We did push it one winter with B50 and the weather got too cold and we plugged the fuel filter, but that was just being cocky."

In this part of Ontario, winter temperatures can get pretty darn cold, requiring low concentrations (less than B33 is recommended) of biodiesel blend. "Once you can plant your tomatoes, it's time to switch back to B100," John mused.

"The old church is a bit tough to heat, but we have a solution for that as well," John explained. "Later this year we intend to convert the oil boiler to run on biodiesel, which is just a natural extension of what we are already doing. The limiting factor may be our biodiesel production capacity."

Figure 7.6-6. The old church is a bit tough to heat, but John plans to convert the boiler to run on biodiesel, further reducing operating costs.

John also went on to explain that the brewing business has a huge swing in production, with beer sales going through the roof during the hot summer months. The ever-resourceful team picked up an old diesel-powered Thermo King refrigeration unit that would normally be mounted on a rail car. "Because we require so much more refrigeration room in the summer, we added the Thermo King and will run it on B100 biodiesel in the summer months. Then when our sales drop in the fall, our indoor coolers can handle the demand. This also has the effect of leveling out our biodiesel consumption and production." John has clearly thought through his business model, all the while keeping environmental sustainability front and center in his planning.

Figure 7.6-7. Because the brewery requires so much more refrigeration room in the summer, John added a Thermo King diesel-powered trailer unit that will run on B100 biodiesel in the summer months, with the indoor coolers taking care of off-peak demand.

Figure 7.6- 8. Of course the last stop on our tour is to drop by the retail store upstairs to pick up a sample of Church-Key's excellent product—in the name of research, of course.

Of course the last stop on our tour is to drop by the retail store upstairs to pick up a sample of Church-Key's excellent product—in the name of research, of course.

Figure 7.6-9. John Graham (right) and his biodiesel brewing friends Eric Dickinson (center) and Helmut Klein (left) (Paul Moring is missing) are smiling. They save money and the environment and literally get to drink the fruits of their labor.

7.7
Sean McAdam – Veggie Gas

I shudder a bit, climbing into Sean McAdam's Hummer H1 as we depart for a trip around his ecologically friendly land development project known as Chelsea Park. I see the irony of me, the environmentalist, arriving in a tiny Smart™ car only to be whisked away in Sean's big bad Hummer, the antithesis of my beliefs. As if reading my mind, Sean interjects: "Relax. I drive a Hummer for environmental reasons, notwithstanding the myriad of people who give me the finger as I drive by!"

"I know this sounds kind of crazy, but hear me out," Sean continues. "The Hummer is to me unspeakably cool, but the two main reasons for purchasing it were that it is built to last and runs on diesel fuel."

Sean goes on to explain that he is becoming increasingly aggravated by products that are designed to be disposable and have built-in obsolescence. If one considers the amount of embedded energy it takes to make a product like the typical automobile and how short a life span it will have, the Hummer begins to look sensible given that it may last 30 or more years if it is not subjected to land mines and the like. "Most of the people I know are likely

Figure 7.7-1. With his characteristic wide smile and gentle demeanour, Sean McAdam hops into his gargantuan Hummer H1 (fuelled with biodiesel, of course) to take the author for a tour of his ecologically friendly land development project called Chelsea Park (www.chelseapark.net).

to keep a car under five years," Sean explains. "When you factor in that I am using renewable biodiesel fuel to operate this vehicle, you see savings of thousands of pounds of greenhouse gas emissions. When considered in this light, it is a much more environmentally friendly choice than the average hybrid vehicle."

Sean's perspective is a provocative one, although I am still pleased to see that the windows are tinted as he pulls away for our tour.

Figure 7.7-2. "I am very proud of this project, as we have gone to great lengths not to destroy the natural setting of the property and to ensure that woodlands and wildlife remain an integral part of the project," Sean says. "For example, the more than two kilometers of roads were made entirely from stone harvested from rock cuts such as this one."

Before we start discussing Sean's involvement in biodiesel, we take a quick spin to tour his main vocation in life, ecologically friendly land development projects. As President of Greystone Developments, Sean is now wrapping up the final sales of his latest 200-acre project located near Gatineau, Quebec called Chelsea Park (www.chelseapark.net). "I am very proud of this project as we have gone to great lengths not to destroy the natural setting of the property and to ensure that woodlands and wildlife remain an integral part of the project," Sean says. "For example, the more than two kilometers of roads were made entirely from stone harvested on site and all wood that was cut was used for lumber or firewood or was recycled at a local pulp mill. We have even done our best to ensure that homes are constructed using straw bale and other sustainable technologies and we try to encourage the owners to avoid the temptation to build 'bloat houses.' Greystone Developments is proof positive that concern for the environment and good business can go hand in hand."

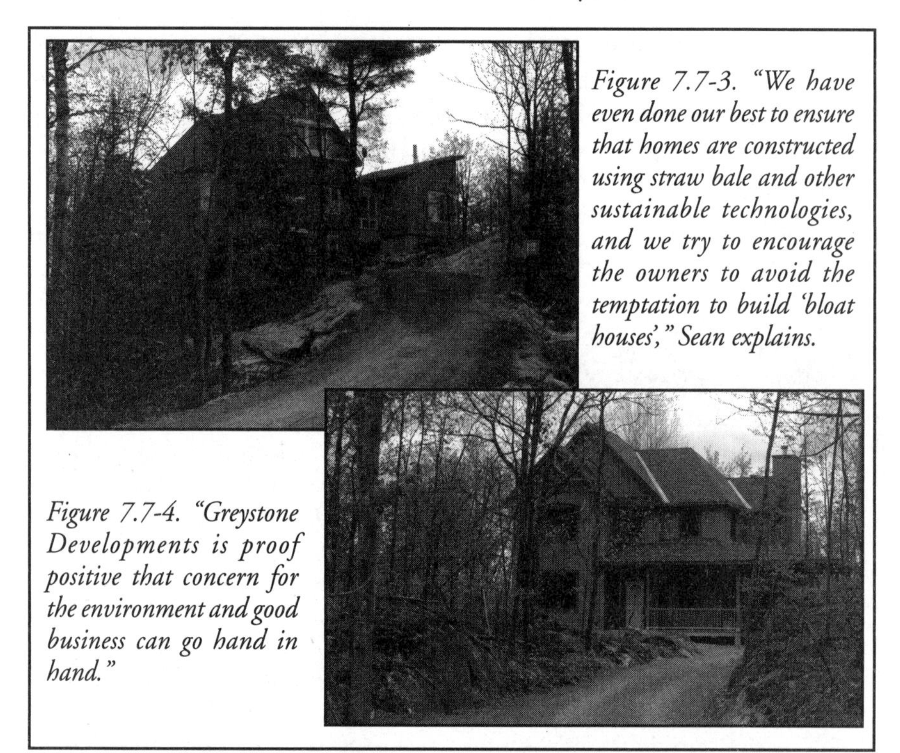

Figure 7.7-3. "We have even done our best to ensure that homes are constructed using straw bale and other sustainable technologies, and we try to encourage the owners to avoid the temptation to build 'bloat houses'," Sean explains.

Figure 7.7-4. "Greystone Developments is proof positive that concern for the environment and good business can go hand in hand."

Accompanied by Christine Paquette, past Executive Director of the Canadian Biodiesel Association, we arrive at Sean's biodiesel production facility, which he refers to as the "Veggie Gas research and development" plant (www.veggiegas.ca).

Figure 7.7-5. Accompanied by Christine Paquette, past Executive Director of the Canadian Biodiesel Association, we arrive at Sean's biodiesel production facility, which he refers to as the "Veggie Gas research and development" plant (www.veggiegas.ca).

"Veggie Gas is a pilot project to determine the feasibility of using waste fryer oil collected from local restaurants to create biodiesel for people who are interested in an ecologically sustainable alternative fuel for their vehicles," Sean explains. "I am interested in scaling the process up from this modified Fuel Meister™ system to a real commercial production facility."

Figure 7.7-6. The Veggie Gas facility is equipped with a small-scale production system manufactured by Biodiesel Solutions (www.makebiodiesel.com). "We decided to purchase this unit as a quick starting point in understanding the complexities of the technology and developing biodiesel," Sean explains.

The Veggie Gas facility is equipped with a small-scale production system manufactured by Biodiesel Solutions (www.makebiodiesel.com), as shown in Figure 7.7-6. "We decided to purchase this unit as a quick starting point in understanding the complexities of the technology and developing biodiesel," Sean explains. "For an initial investment of around US$6,000 we were able to produce biodiesel using a single-stage reaction system and a mist washer that comes with the unit."

Closer examination of the Fuel Meister™ technology reveals that it is simply a reconfiguration of the homemade processing systems developed by other biodiesel home brewers, including their shortcomings. Sean admits that the system is not truly suited to full-scale production, but says it has served him well as he has worked to understand the basic organic chemistry involved in biodiesel production and prepared to move into the million-gallon pro-

Figure 7.7-7. "I have enjoyed making biodiesel and have tested the production processes with both waste and virgin oils," Sean explains as he holds up a sample of his home brew for us to look at.

duction capacity he yearns for. The Fuel Meister™ has also allowed him to concentrate on process rather than spend time trying to source components and build a system from scratch.

"I have enjoyed making biodiesel and have tested the production processes with both waste and virgin oils," Sean continues, as he holds up a sample of his home brew for us to look at. "I have been using B100 in the Hummer during the warm months and then switch back to lower blend levels during winter. The dealer has been watching the condition of the vehicle and has seen no degradation of fuel or engine systems and I am fairly confident in the quality of fuel we have produced."

Figure 7.7-10. It may look paradoxical, but having the Kyoto Protocol endorsement painted on the side of a biodiesel-powered Hummer may actually make environmental sense.

Figure 7.7-8. Sean McAdam is shown transferring a fresh batch of his biodiesel from the washing tank to the storage rack at his Veggie Gas facility. The biodiesel produced in the facility is used to fuel his Hummer H1 with B100 during the summer and lower-level blends during the winter.

Figure 7.7-9. Sean's business partner Peter Schneider is seen here filling up the Hummer with B100 biodiesel produced earlier. The fuelling "kit" is supplied by Biodiesel Solutions and comprises a barrel-mounted fuel pump, filter assembly, and dispensing meter (detail inset).

7.8 Summary

All of the micro-scale biodiesel producers discussed above have unique motivations, methods, financial budgets, and skill levels. What they have in common is a desire to do the "right thing" for the environment and help develop a sustainable energy industry while possibly saving a few dollars in the process.

Also common among biodiesel producers is the lack of a process to ensure that the fuel they produce meets the applicable quality standards as well as a lack of understanding about how to deal with the waste stream or coproducts that are an inevitable part of the production process.

This statement is meant not to criticize responsible micro-scale biodiesel producers but to point out that because of a lack of technical knowledge, financial resources, or manpower, this group as a whole is destined to be marginalized unless something is done to change course.

To summarize this chapter, I would like to identify the key areas of failure of the micro-scale producers and use these as reference points in an attempt to correct these problems in **Chapter 8 – The Small-Scale Solution**.

Figure 7.8-1. Steve Anderson explains that he has travelled thousands of miles using raw biodiesel to power his ancient Toyota truck with nary a problem. When pressed a bit further, he admits that he keeps a few fuel filters in the glove compartment to change when necessary. I suppose whether this is a problem for you or not depends on your point of view and ability or willingness to work under your vehicle on the side of the road.

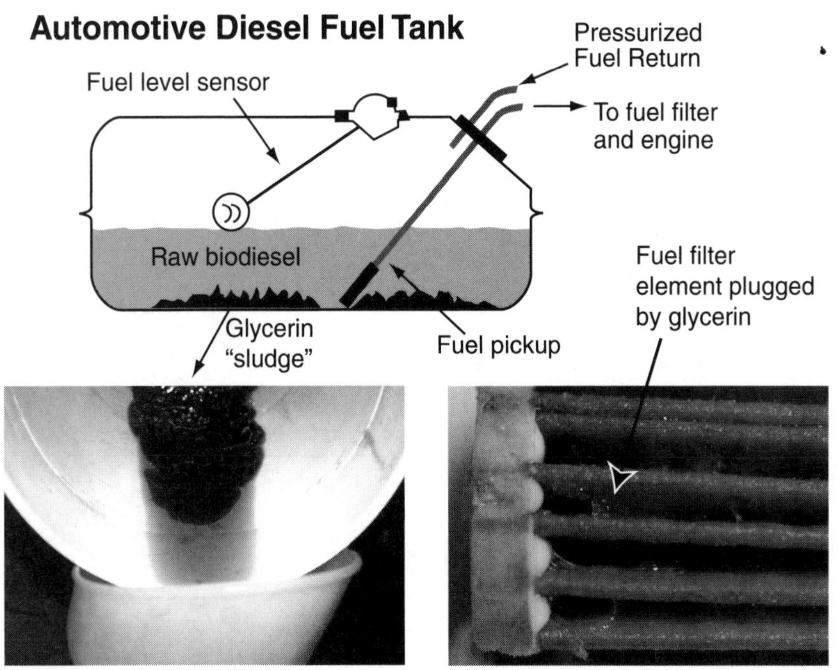

Figure 7.8-2. One problem with unwashed homemade biodiesel is the presence of free glycerine which separates out of the fuel over a period of time, due to slow, low-level reactions continuing. This photograph shows raw, unwashed biodiesel that has been allowed to sit for a period of two weeks, allowing glycerin to "drop out" of the fuel. When this occurs in your fuel tank, glycerin is drawn into the fuel filter as shown in this actual filter sample, causing engine starvation and filter plugging.

Figure 7.8-3. Dr. Andre Tremblay, Professor and Chair of the University of Ottawa Chemical Engineering Department, is seen comparing ASTM and homemade biodiesel. He comments that fuel specifications are not made to exclude anyone from producing fuel but rather to ensure a quality product that does not harm the equipment it is intended to operate.

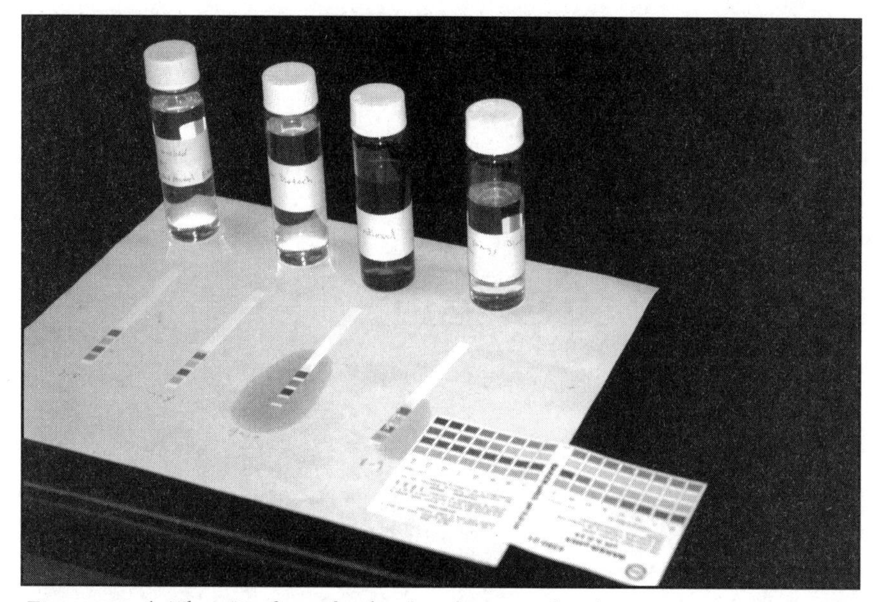

Figure 7.8-4. The pH of raw biodiesel is determined to be very caustic, which **will** *damage metallic fuel system components over time.*

Figure 7.8-5. This photograph shows a sample of homemade biodiesel after testing for "bonded glycerine." The layer of glycerin on the bottom of the container indicates that the conversion of waste oil to biodiesel was not complete, even though the producer was certain that the fuel met specification.

Figure 7.8-6. Homemade biodiesel varies considerably from one producer and one batch to another. In this photograph, "finished" biodiesel from three sources was subjected to a simple emulsion test. The vial to the left shows a very large amount of contaminants remaining after the reaction process. The middle vial shows biodiesel that was not properly washed, with soaps, catalysts, and unsaponifiable matter collecting at the interface between the biodiesel and wash water layers. The vial to the right shows the author's biodiesel subjected to the same test. Note that the wash water is clear and there are no contaminants present at the interface layer.

Figure 7.8-7. Dr. Tremblay is seen testing a sample of wash water for the presence of methanol. Improper processing techniques will result in large amounts of wash water that is considered hazardous waste because it contains methanol as well as having a high biological oxygen demand (BOD) and pH levels beyond the limits imposed by municipal sewer and disposal regulations.

The Common Results of Home-Scale Biodiesel Production

It is impossible to identify each and every issue related to micro-scale production processes and technology; nor do I profess to have all of the answers. The summary chapter at the end of the book will identify those areas where I believe additional work must be carried out in order to ensure the successful development of both micro- and small-scale biodiesel "industries."

However, after observing the difficulties, machinations, and musings that occur at biodiesel workshops and in the online community, a few problems come to the forefront:

1. lack of fixed process
2. lack of access to suitable and safe technology
3. abundance of "experts" offering advice without supporting scientific data or results
4. failure to test processes, fuels, and waste streams to applicable standards
5. lack of environmental controls for disposal of waste stream emissions into air, water, and ecosystem
6. tendency to work "underground" to avoid environmental regulations
7. disregard for authority, resulting in an anarchistic approach rather than a desire to develop legitimacy for "the cause"

Disregard for authority and environmental regulations stems from fear or contempt. The fear of micro-scale operations being shut down because of heavy-handed fire safety or environmental regulations is not altogether unfounded. There are reasons for regulations, but they may appear too onerous for the small operator to comply with.

However, working as a collective association, preferably under an umbrella national group, would allow the creation and continued development of standards that could be adopted by state, provincial, or national authorities. Consider that every local service garage or repair shop generates large amounts of federally regulated hazardous waste and handles numerous toxic chemicals yet is able to comply with applicable regulations because of small-scale waste generator rules specifically developed for this purpose. Adapting these "small-scale waste generator" rules might be a useful first step toward legitimacy.

If all of the online "experts" were corralled under the guidance of one agency, access to credible information and regulations might just be possible.

There is nothing inherently wrong with making a comment provided it is not stated as fact. The scientific process requires open access to information and test procedures and healthy debate based on the results, no matter what the scale of investigation being undertaken.

Until this happens, well-meaning people will produce off-specification fuel while continuing to wonder if what they are doing is honestly the right thing?

7.9
Straight Vegetable Oil vs. Biodiesel

Now that I have deconstructed the mythology propagated by home brewers of raw and poor-quality biodiesel, it is time to expose the bias of the North American research and technical community against Straight Vegetable Oil (SVO) systems.

It is well known amongst the petro- and biodiesel cognoscente that Rudolf Diesel first demonstrated his compression ignition engine using raw peanut oil as his fuel of choice during the Paris Exhibition at the turn of the 20th century. According to many modern biodiesel and engine fuel researchers, the pursuit of an SVO alternative cannot work and must be ignored. (Perhaps the problem can be partially linked to the tenet that "SVO systems are of no interest to capitalists because they do not require much capital.")

As I explained early on in this book, vegetable oil can be used as a source of fuel, but one of two things must occur prior to its use: either the fuel must be adapted to the engine or vice versa. When conversions to the engine and fuel system are undertaken, there is no reason why raw plant oil cannot be used as engine fuel.

However, it is important to understand that the rule of absolutes applies with plant oil fuels. The plant oil **must** be modified to work with existing engines or the engine **must** be modified to work with plant oils. There is no in-between. Simply stated, it is not practical to mix raw plant oil with either bio- or petrodiesel as a fuel extender because the high viscosity of a plant oil mix will lead to serious engine damage. Numerous reports have shown that blended plant oil and diesel fuel will cause lubricating oil dilution, resulting in thickening and sludge formation, piston ring damage due to excessive carbon deposits, and fuel injector varnishing.[1] Unfortunately, these reports often overstep their scientific boundaries by dismissing all raw plant oil engine fuels with such cavalier statements as "this approach has been used with some success although long-term durability has not been proven."[2]

This is unfortunate, since such statements are not supported by the facts.

Combined Heat and Power Systems

The Wartsila Company is a well-known power generation and ship power company that has installed over 8,600 power plants totalling over 34,000

MW of capacity as of 2005.[3] One of the hallmarks of this firm is a commitment to quality, customer support, and highly energy-efficient combined heat and power (CHP) systems using numerous solid and liquid biofuels. According to the company, vegetable-based bio-oils have been accepted for commercial operation in Wartsila power plants since 1996 and have been used as replacements for diesel fuels in numerous applications.

Olive, palm, soybean, rape, and other plant oils have been used in their operations and are all "as usable as diesel fuel." The company is also quick to point out that biodiesel may also be used in any of their applications, albeit at higher operating costs because of the additional processing necessary to convert plant oil into biodiesel. Wartsila also states that "bio-oils will be a growing group of fuels for power production as long as there is a surplus of these oils."[4]

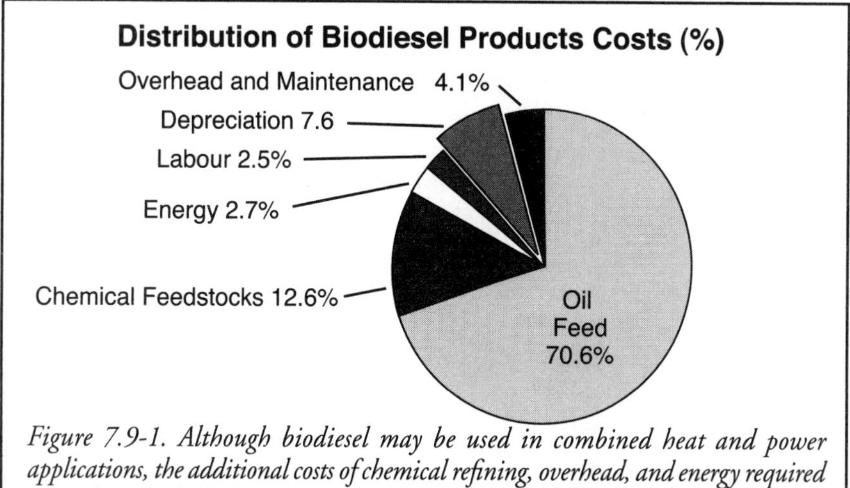

Figure 7.9-1. Although biodiesel may be used in combined heat and power applications, the additional costs of chemical refining, overhead, and energy required to produce the fuel place it at a financial disadvantage compared to SVO. (Source: Building a Successful Biodiesel Business)

By way of example, Wartsila has been hired by ItalGreen Energy, part of food producer Casa Olearia Italiana Group, to extend ItalGreen's liquid biofuels plant in Monopoli, Italy.[5] The two previously installed liquid-fuelled combined heat and power systems have a total energy output of 16 MW of electrical capacity, fuelled from excess and waste vegetable oil. The existing power systems have demonstrated a superior capacity factor or availability for operation in the range of 95%. With the additional engine system added to the mix, total electrical output will reach 24 MW of capacity, and this energy will be sold to the Italian national grid through a fixed-price contract.

Thermal energy in the form of steam is supplied to the processing plant for day-to-day operations.

A firm with the financial power and reputation for quality products enjoyed by Wartsila would not produce such a product and make these statements if they were unfounded. Clearly, the long-term durability of these systems has been proven.

As we learned in Chapter 3.3, "An Introduction to Biodiesel," the viscosity of the parent plant oil or animal fat compared to petrodiesel is the primary reason that fuel or engine modification is necessary in order to bridge the compatibility gap. If the engine in question is to remain unmodified, the plant oil must be subjected to a viscosity transformation in the form of chemical transesterification; that is, it must be converted into biodiesel.

On the other hand, the plant oil can be used directly through another means of viscosity transformation: heat. As shown in the graph in Figure 7.9-2, applying heat to plant oils and animal fats will reduce their viscosity to a level where they can be used directly in compression ignition engines, mimicking the effects of the transesterification process. This concept is quite simple to follow: butter, lard, or vegetable shortening remains solid or semisolid at room temperature, but if it is placed in a pan and heated it will quickly melt, achieving a waterlike fluidity.

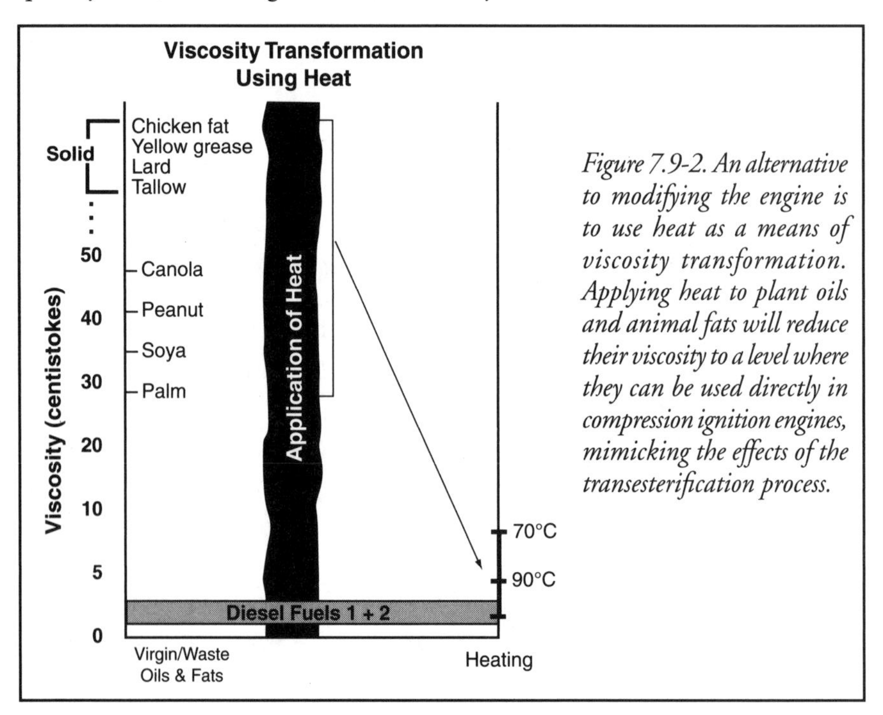

Figure 7.9-2. An alternative to modifying the engine is to use heat as a means of viscosity transformation. Applying heat to plant oils and animal fats will reduce their viscosity to a level where they can be used directly in compression ignition engines, mimicking the effects of the transesterification process.

The process of viscosity transformation in a stationary combined heat and power system is an ideal one, as the plant fuel oil may be kept hot using energy from the engine exhaust or coolant system. During idle periods, heat may also be supplied from the utility electrical supply. The concept of CHP is fairly easy to understand and a simple system is shown in overview in Figure 7.9-3.

A stationary generator comprising a compression ignition (diesel) engine is coupled to an electrical generating unit. The generator is in turn connected to the electrical distribution grid through an electricity meter which records energy sold to the grid. Although diesel engines are considered to be fuel efficient, approximately two-thirds of the energy content supplied by the fuel

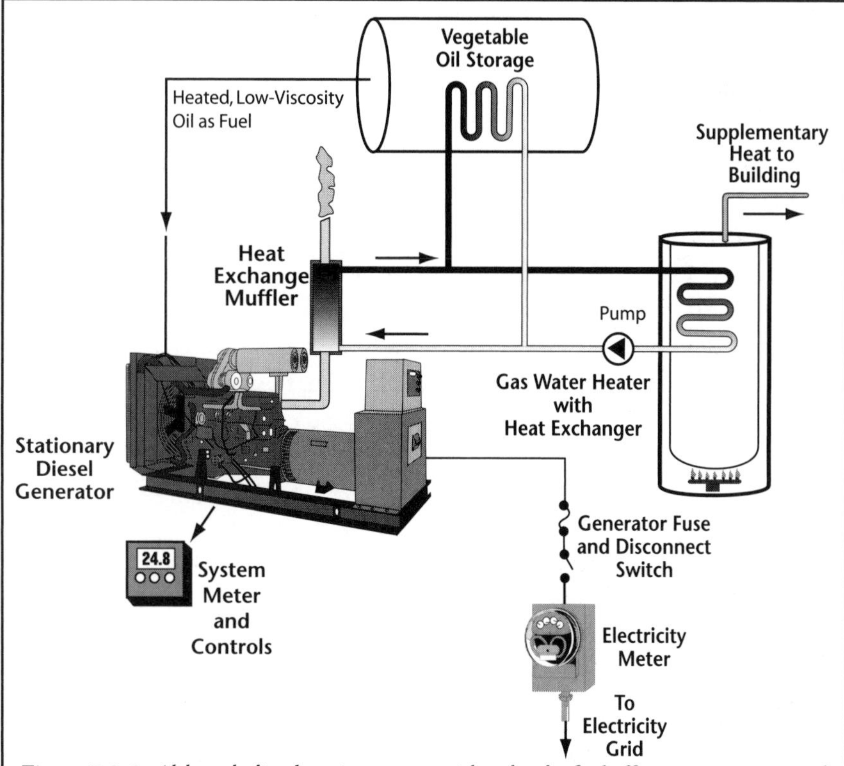

Figure 7.9-3. Although diesel engines are considered to be fuel efficient, approximately two-thirds of the energy content supplied by the fuel is wasted in the form of heat. Using a special heat recovery system attached to the exhaust gas outlet and engine coolant system, it is possible to capture this waste heat and transfer it to the building facility where it can be used for space and water heating as well as energy for various industrial processes such as food drying and cooking. When the total energy contained in the waste heat and electrical energy is calculated, CHP systems can have efficiencies of 80% or greater.

is wasted in the form of heat. Using a special heat recovery system attached to the exhaust gas outlet and engine coolant system, it is possible to capture this waste heat and transfer it to the building facility where it can be used for space and water heating as well as energy for various industrial processes such as food drying and cooking. When the total energy contained in the waste heat and electrical energy is tabulated, CHP systems can have efficiencies of 80% or greater.

Vegetable oil fuel may be stored in an insulated container which is in turn heated using utility power, waste heat from the CHP system, or a combination of both. When plant oils are heated to 80°C (176°F) or higher, their viscosity will approach that of petroleum diesel fuel, allowing the engine to operate. (Additional modifications to engine timing, fuel drainback, and other subsystems are not discussed for the sake of brevity.)

Traditional CHP systems use natural gas or diesel fuel as the input energy source, both of which are becoming more costly and are damaging to the environment. Plant or bio-oils may be used in place of fossil fuels provided the engine system is adapted to operate on the high-viscosity plant oil. Using plant oils directly results in lower operating costs than refined biodiesel fuel and may allow the use of off-specification oil seed to be cold-pressed directly into fuel for the system. The environmental and financial benefits can be very significant.

Consider companies such as Peacock Industries (Chapter 4.2), which produces approximately 50,000 liters (13,000 gallons) of "waste" oil per year. Combusting this clean oil source could reduce their electrical and heating costs to near zero, save for the initial capital cost and ongoing maintenance expenses.

Going Mobile

If stationary CHP systems make sense using plant oils as the fuel source, what about the transportation sector? To find out more about these applications I spoke to Ed Beggs, who is the President and Cofounder of Neoteric Fuels and its SVO division, PlantDrive (www.plantdrive.com). "I originally thought that I wanted an electric vehicle a number of years ago, something that was obvious, sustainable technology. After taking one model out for a test drive and having to turn off the lights and play with the clutch to nurse the car home as the battery died one evening, I knew that it wasn't right for me," Ed reminisces as he explains how he became involved in SVO transportation systems. "After the humbling experience with the electric vehicle, my search landed on biodiesel technology and the offshoot of SVO as a fuel source. I realized that SVO had several important advantages over biodiesel, including lower cost and fewer chemicals required in the processing, but found it had not been well researched and was practically ignored by academia and industry in favour of biodiesel. I decided that I would complete my Masters degree in Environment and Management, with my 1997 thesis focusing on the technological development of renewable oil fuels."

"At that time there was very little information on the subject of renewable oil fuels in North America, although the Europeans were developing the biodiesel market with a flourish and had spent time experimenting with SVO/petrodiesel fuel mixes," Ed explains. "As part of my research for the thesis, I developed an early version of the technology we now sell through our PlantDrive Website. During the research phase, hundreds of tests were conducted on a 1982 VW Jetta diesel, including lubricating oil analysis, exhaust emission profile, and engine wear. It was an exciting time for me as most of this research was fairly new and the outcome of the various tests was completely unknown."

Since the completion of his studies, Ed's Neoteric Fuels and its competitors have seen tens of thousands of mechanics experiment with converting their automobiles to operate on SVO, with an increasing number of these users having accumulated significant hours of use and total mileage. "Our poster child for the SVO movement is a fellow from British Columbia, Canada, who has accumulated 250,000 kilometers (155,000 miles) using recycled cooking oil as a fuel source," Ed states. "He drives a 1986 Ford Navistar that has not succumbed even though the engine has logged over 400,000 kilometres (250,000 miles). Even more interesting is the fact that these engines came equipped with a none-too-robust rotary fuel pump that is still forcing the SVO through. Even after all this time, the engine and fuel system are running very well, with few hints of the oil and smoke emissions that one would expect from a vehicle of that age."

Ed continues his explanation of the evolution of the diesel engine by discussing his initial concern about newer high-technology designs. "When vehicles with the modern TDI engines started appearing on our door in 1997/98, we proceeded fairly cautiously, but have found that they work just as well as the older mechanically operated, indirect fuel injection systems installed in diesel engines of bygone years. With our experience, number of installed units, and total mileage accumulated on the vehicles we have modified, I am quite happy to state that this technology works—period."

Ed says that the automobile conversion process is well known and that the technology will work summer and winter. "Our engine heat recovery

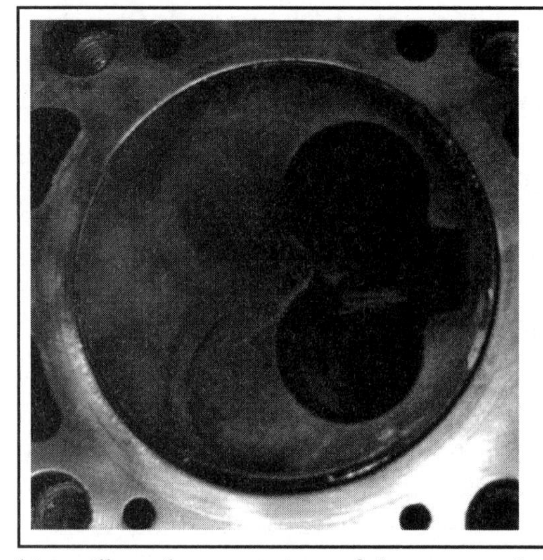

Figure 7.9-4. After operating for 20,000 km (12,400 miles) on SVO, the upper cylinder head of this VW Jetta TDI was removed and the valve train, valve seats, and engine assembly were thoroughly inspected by a VW mechanic. The examination turned up no noticeable differences between this engine and one operated on petrodiesel. (Photographs 7.9-4 through 7.9-15 courtesy Neoteric Fuels)

kits will work at any time of the year, and supplementary electric heating systems can be installed for even the coldest of climates. The main thing is to ensure that we are aware of any special applications so that the system can be fitted to match."

The system technology is deceivingly simple and very neat and tidy when installed in a vehicle. A typical installation comprises several components starting with a choice of heat recovery unit and filter. Figure 7.9-6 shows a VW TDI engine bay fitted with an electric inline SVO fuel heater which is mounted just in front of the engine oil dipstick. Located top right is the Vormax SVO oil filter assembly which ensures that the recycled oil is properly filtered and does not cause any contaminants to enter the fuel system.

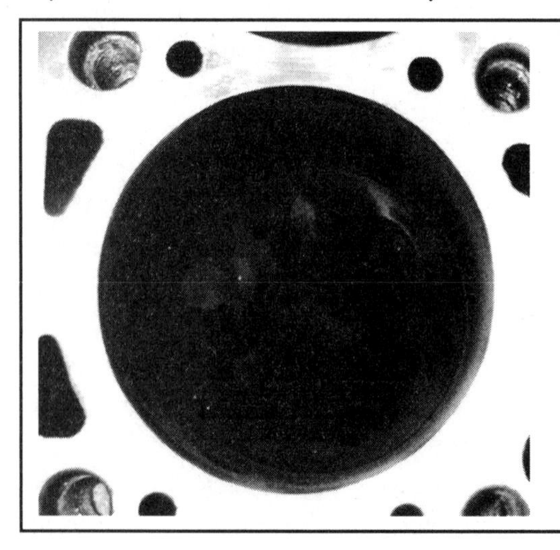

Figure 7.9-5. This photograph shows the engine block with the piston seated in the lower area of the cylinder. The top of the piston has a normal light coating of carbon which has been smudged with the impression of a fingerprint.

Ed indicates that off-vehicle prefiltering is best but that the system can filter waste oils onboard if necessary.

For vehicles fitted with a conversion kit, the engine must start on diesel (or biodiesel) fuel to allow sufficient time for the SVO to warm up. Once the SVO reaches operating temperature a switch is activated which transfers the fuel source from a diesel fuel tank to the heated SVO in a second tank. Just prior to stopping the engine, the operator switches the car back to regular diesel fuel to purge the system of SVO and ready the car for the next start.

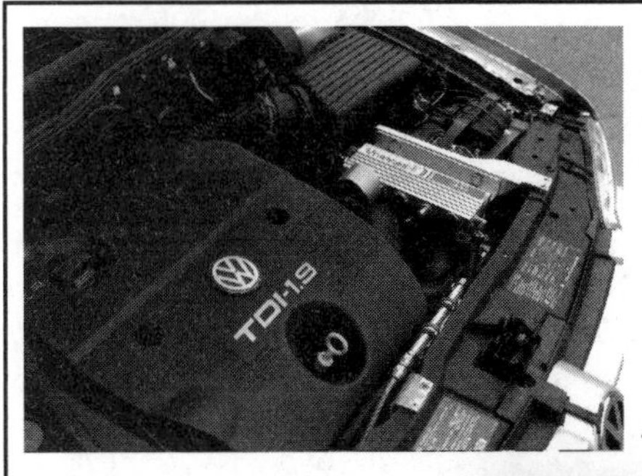

Figure 7.9-6. This photograph shows a VW TDI engine bay fitted with an electric inline SVO fuel heater which is mounted just in front of the engine oil dipstick. Located top right is the Vormax SVO oil filter assembly which ensures that the recycled oil is properly filtered and does not cause any contaminants to enter the fuel system.

Figure 7.9-7 (above) shows a clever design of a start/purge fuel tank that mounts in the spare wheel well. This small tank is used for fossil diesel fuel while the original, larger fuel tank is used for SVO storage in temperate climates.

Figure 7.9-8. (left) The VW start/purge fuel tank can be removed and placed on the ground for easy and clean filling.

Figure 7.9-9. In cold climates or where saturated or hydrogenated oils are used, it is necessary to store the oil in insulated tanks equipped with an internal heat exchanger which derives its heat from the engine cooling and/or electrical system. In this example, the owner of a VW Jetta TDI removed the spare tire and replaced it with the SVO fuel tank with an integral heat exchanger. The owner now carries a cell phone, an automobile association card, and a can of tire foam, just in case of a flat!

Figure 7.9-10. This view shows the Vormax two-stage SVO fuel filter assembly mounted to the sidewall of the vehicle trunk.

"The main thing to keep in mind when operating an SVO-powered vehicle is to perform engine lubricating oil changes on time," Ed continues. "It is possible that lubricating oil thickening (polymerization) can be an issue, since vegetable oil as a fuel may migrate past the piston rings and into the lube oil, where, unlike diesel fuel, it does not evaporate and therefore accumulates. Changing the oil on schedule will eliminate any risk from this potential problem."

Exhaust Emission Profile of SVO

Exhaust emission profiling was conducted at the Pacific Vehicle Testing Technologies Ltd. laboratories as part of Ed's Master's thesis. The tests compared the standard four EPA exhaust gases (HC, CO, NOx and CO_2) as well as opacity, which is a determinant in calculating particulate matter.

Each test was conducted for "premium" petrodiesel, waste vegetable oil, ASTM-certified biodiesel, and refined food-grade canola oil. The results are tabulated in the following graphs with the data fields recorded in "grams per kilometer" of vehicle travel. The tests were conducted in accordance with government procedures in a qualified laboratory.

Although there are no emission profile standards for SVO systems, the recorded data are within the expected averages and would likely be further reduced if the engine emission control system were "tuned" to the SVO fuel.

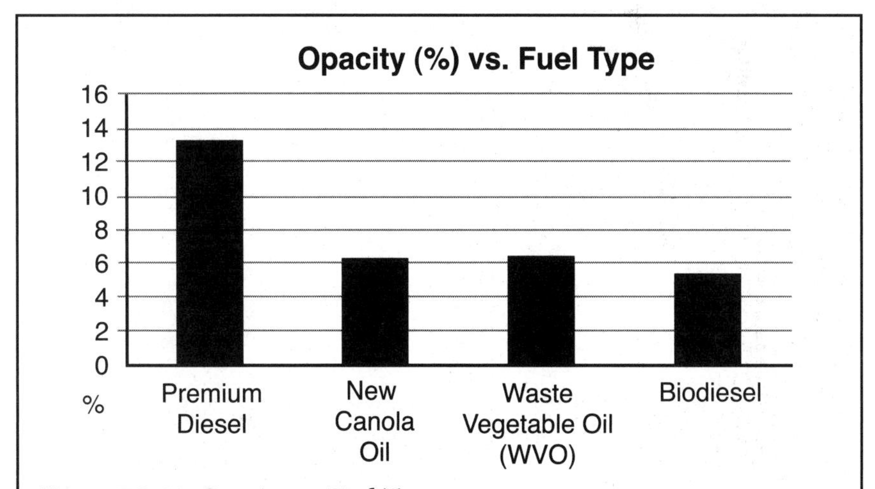

Figure 7.9-11. Opacity vs. Fuel Type

Opacity is the determination of the amount of soot present in the exhaust stream which is calculated from the reduction in light transmission through the gases. Using the opacity data, particulate matter (PM) levels can be estimated, due to their ratiometric balance.

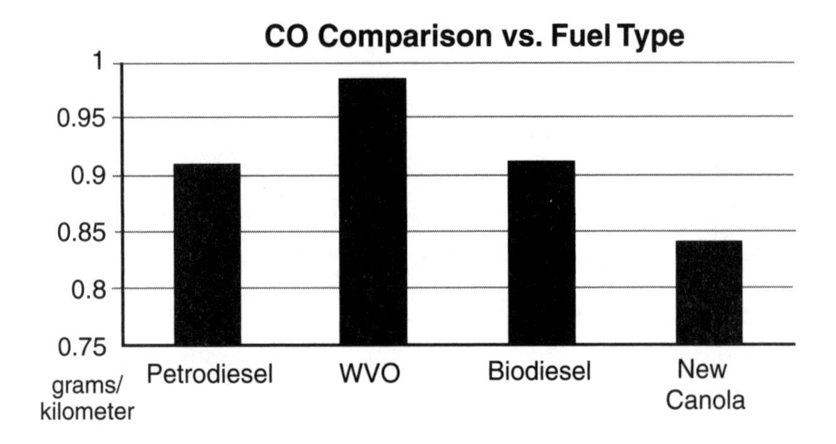

Figure 7.9-12. Carbon Monoxide vs. Fuel Type

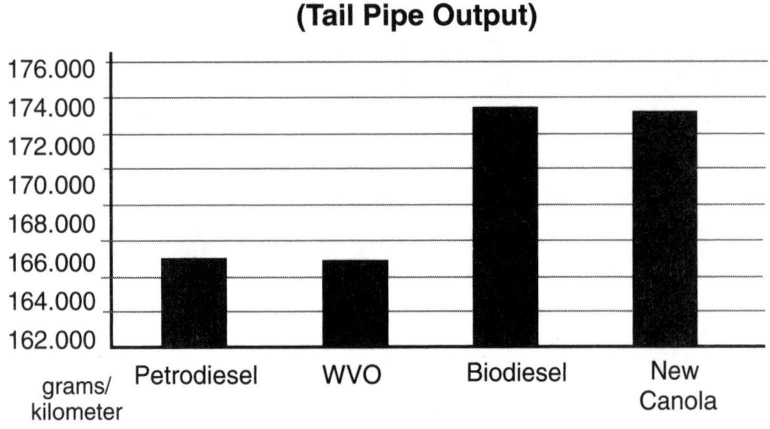

Figure 7.9-13. Carbon Dioxide vs. Fuel Type

Carbon dioxide emissions are measured at the tail pipe and do not reflect the lifecycle emissions, which would be considerably lower (see Chapter 3.3).

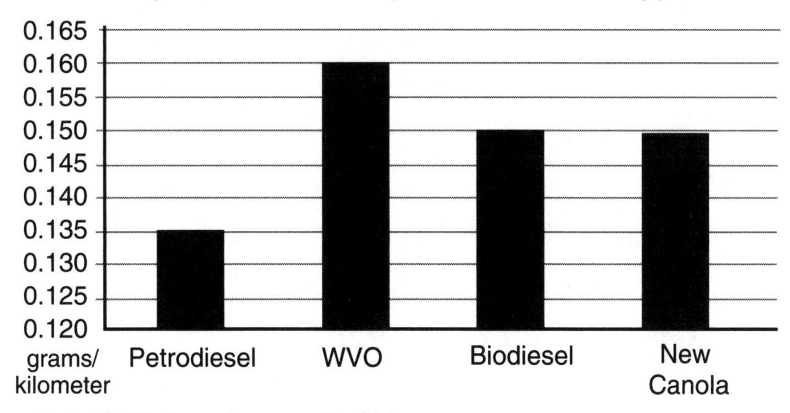

Figure 7.9-14. Hydrocarbons vs. Fuel Type

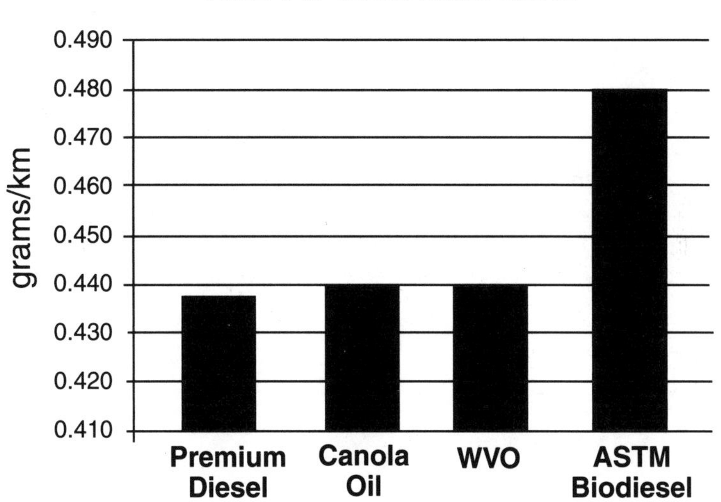

Figure 7.9-15. Nitrogen Oxides vs. Fuel Type

Summary

As the search for cleaner, less expensive fuel continues, SVO systems will become much more common in combined heat and power plants. The advantages of SVO over refined biodiesel are reduced chemical engineering costs as well as improved carbon lifecycle emissions and lower overall fuel cost.

Whether SVO systems become mainstream in the transportation sector remains to be seen. However, based on the work conducted by Ed Beggs and his associates, there is no reason not to explore the SVO option. When diesel fuel exceeds a couple of dollars a gallon, there will be a huge incentive to get on the bandwagon. With the geopolitics of oil changing rapidly, there are many who simply don't want to wait for the rush and have started on their own SVO fuel odyssey.

8 The Small-Scale Solution

8.1 Overview of the Technology

As we have seen in examples in Chapter 7, there are numerous ways of producing biodiesel from WVO or cold-pressed oilseed feedstock. Systems can range from Steve Anderson's $500 solution using an old oil barrel to Sean McAdam's financially sophisticated $6,000 commercial unit from Biodiesel Solutions.

While each of these systems can claim to partially transesterify waste oil into biodiesel, none of the examples outlined can ensure that the biodiesel produced will meet the fuel quality requirements of ASTM standard D 6751. Nor do these systems address the issues of toxic waste streams that are a byproduct of the fuel production process, chemical handling safety, or environmental sustainability.

The small-scale technology solution that I have developed represents improvements in the areas of fuel production, quality control, chemical and process safety, procedural development, and waste stream processing that have to date been sadly lacking in systems currently in use.

In keeping with the mandate of this book, which is to rigorously avoid the propagation of any myths or misinformation, statements of fact or results from specific tests have been analysed using proper scientific methods or third-party test equipment or laboratories to verify results and improvements over past efforts. Any areas of concern that remain have been catalogued and placed in the summary section, inviting further discussion and development.

The Production Facility

When Lorraine and I designed our off-the-electrical-grid home, we decided to add a small workshop area in the detached garage to house the biodiesel production facility. This extra space is located behind the door shown in Figure 8.1-1 and consists of an area of approximately 6 x 18 feet (1.8 x 5.5 meters). The room is well insulated and tightly sealed with standard construction-grade vapour barrier and acoustical gap sealant and is fitted with a ventilation fan that is capable of drawing approximately 500 cubic feet (14 cubic meters) of air from one end of the room to the other per minute of operation. The high-volume turnover air rate is required to keep methanol vapors to an absolute minimum, certainly below the lower explosive limit (LEL), as well to ensure that air quality remains high and nonpoisonous.

The production lab is shown in an overview image in Figure 8.1-2. All of the production equipment is mounted along the left wall, while the "wet

Figure 8.1-1. When Lorraine and I designed our off-the-electrical-grid home, we decided to add a small workshop area in the detached garage to house the biodiesel production facility.

Figure 8.1-2. The production lab is shown in an overview. All of the production equipment is mounted along the left wall, while the "wet laboratory," safety equipment, and storage area are located along the right wall.

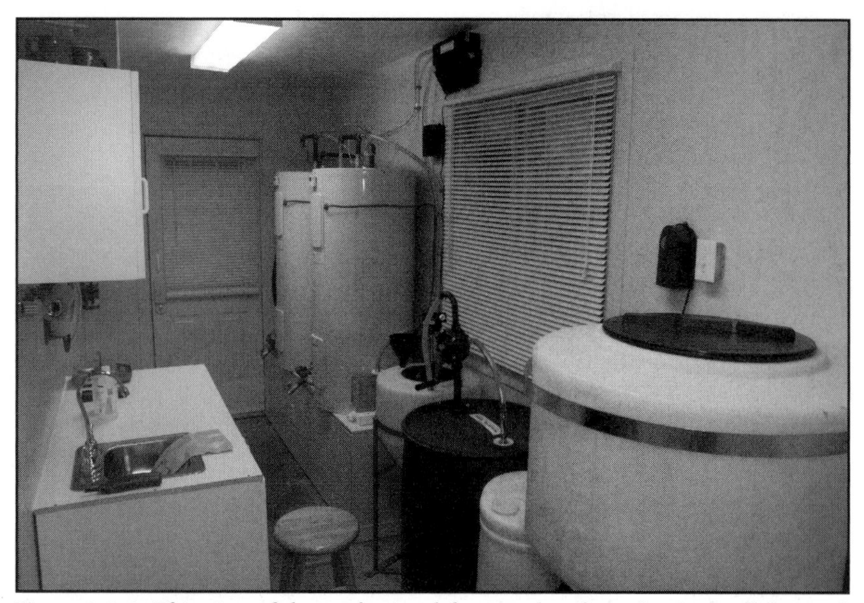

Figure 8.1-3. This view of the production lab was taken from the rear wall facing the front door. Note that the WVO and main reaction tanks are mounted on a cabinet which raises them off the floor. This arrangement provides additional storage space and makes accessing the valves and controls of the reaction tanks much easier on one's back.

laboratory," safety equipment, and storage area are located along the right wall. This image shows, starting at the bottom left corner, the first tank which receives the WVO and is used to filter, dry, and deacidify the oil to desired standards. The second tank is the main reaction tank which performs the chemical conversion (transesterification) of WVO into biodiesel. It is also used to separate the raw biodiesel from glycerol and to recycle the excess methanol prior to washing the biodiesel.

The next small white conical tank is the chemical mixer which combines the methanol stored in the adjacent tank with a powdered catalyst forming sodium (or potassium) methoxide, which is pumped into the main reaction tank to start the conversion of WVO into biodiesel.

The small white tank beside the methanol is the biodiesel wash water storage tank. It contains a submersible heater which warms the wash water to 60°C prior to spraying it into the raw biodiesel which is stored in the large white conical wash tank. The wash tank is fitted with both water spray nozzles and an air bubbling system which are used to remove contaminants from the raw biodiesel.

The last tank visible at the far rear of the picture (located behind the large conical wash tank) is the biodiesel dryer. This unit removes any remaining water from the biodiesel before final filtration and storage.

Also visible in this image is the ventilation fan located along the rear wall and the electrical subpanel which provides electrical circuit control and protection for each of the process tanks, pumps, and heaters used in the system.

Running along the right sidewall is the wet lab area which is used to analyze both the WVO and the biodiesel produced by the system. The storage cabinets include all of the necessary process chemicals as well as test equipment, scales, and measuring beakers.

Safety is of primary importance when handling any chemicals and a variety of smocks, eye protection, rubber gloves, and spill cleanup materials as well as an eyewash station and multiple fire extinguishers are provided.

The photograph in Figure 8.1-3 was taken from the rear of the production lab, showing how all of the equipment and lab facilities fit neatly into this compact area. Note that the WVO receiver tank and the main reaction tank are lifted off the floor on a horizontal storage cabinet. In addition to providing additional storage space, the cabinet also houses a small reverse osmosis water filtration unit which feeds mineral-free water to the biodiesel wash tank.

The small black box mounted above the main reaction tank is the air-to-liquid heat exchanger which is used to condense methanol vapors driven

Figure 8.1-4. This view of the wet lab shows the ample storage space for process chemicals, gloves, safety equipment, and other materials required in the biodiesel production process.

Figure 8.1-5. Safety is of primary importance when working with any chemicals. The wet lab area has storage for work smocks, gloves, and face shields as well as a telephone, eyewash station, and Class A, B, and C and foam fire extinguishers. Antistatic, ground-connected wrist straps are also provided to prevent the accidental ignition of methanol. The safe production of biodiesel does not happen by chance!

off from the biodiesel reaction process. The unit operates as a fan-driven cooling unit, causing methanol vapor to condense and drip into the storage tank located next to the reaction tank. Condensed methanol is returned to the methanol storage tank for future use.

Because of the inline design of the production system, it is possible to simultaneously process approximately four 40-gallon (≈ 150 liter) batches of biodiesel at one time, with one batch in each stage:

- drying and deacidification of WVO
- transesterification of WVO into biodiesel along with methanol recovery
- washing of raw biodiesel
- drying of biodiesel prior to testing and storage

Processing of waste stream glycerol and wash water are handled "offline" and will be discussed in later chapters.

The WVO Receiver /Dryer

WVO is delivered to the facility and immediately transferred into the receiver/dryer tank shown in Figure 8.1-6. This tank is made from a 60-gallon (227-liter) electric water heater, although larger, commercially available process tanks can be used if the system is to be scaled up for cooperative or small-scale commercial use.

Figure 8.1-6. WVO is delivered to the facility and immediately transferred into the receiver/dryer tank.

Figure 8.1-7. A load of WVO recently delivered from a restaurant will be transferred immediately to the receiver/dryer tank.

Figure 8.1-8. Geoff Shewfelt is shown inserting the suction hose of the receiver/dryer tank into a pail of WVO. The suction hose is equipped with a particle strainer at one end and a 20-micron filter at the outlet of the transfer pump, ensuring that food particles are filtered out and only WVO is sent to the receiver/dryer tank.

Figure 8.1-9. This detail view shows the suction hose and particle strainer inserted into a pail of cold-pressed canola oil that was recovered from a batch of off-specification oil seeds.

Fellow biodiesel enthusiast Geoff Shewfelt is shown in Figures 8.1-7 and 8.1-8 with a load of WVO recently received from a local restaurant that is transferred immediately into the receiver/dryer tank, where a fuel transfer pump and suction line draw the WVO from the storage pails (Figure 8.1-9). The WVO is analysed to determine its free fatty acid composition and may be subjected to deacidification and heating to remove excess water. WVO absorbs water from the foods that are fried in it, and if sufficient water remains in high-FFA oil, the transesterification process may fail, producing a jelly-like gravy rather than biodiesel. This is an important step that most biodiesel processors tend to skip.

Upon completion of drying and/or deacidification, the WVO is transferred to the main biodiesel reaction tank using the circulation pump fitted to the receiver/dryer unit.

The Biodiesel Reaction Tank

The biodiesel reaction tank is configured in a similar manner to the receiver/dryer tank as shown in Figure 8.1-10. The reaction tank is fitted with a circulation pump and a "sight glass" (Figure 8.1-11) created from reinforced braided plastic tubing. This sight glass permits the filling of the reaction tank with an exact amount of WVO and reaction chemicals and also provides a way of monitoring the completeness of the reaction process.

A vapor recovery unit is also installed above the tank to capture the excess methanol driven off during the transesterification process. Excess methanol is used to ensure the conversion of waste oil to biodiesel is driven to completion, although a large volume of methanol is not required for the transesterification

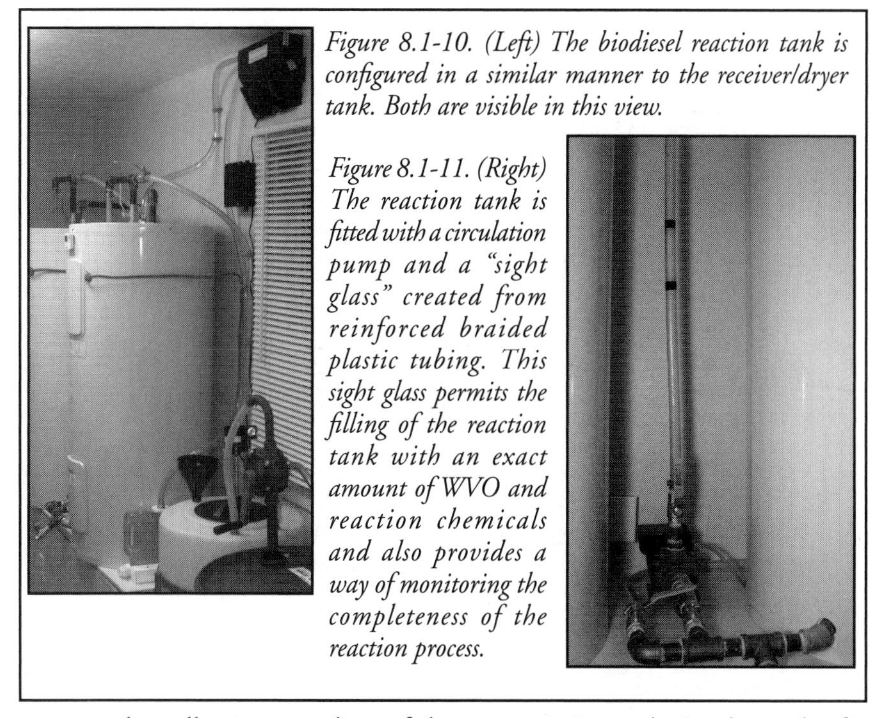

Figure 8.1-10. (Left) The biodiesel reaction tank is configured in a similar manner to the receiver/dryer tank. Both are visible in this view.

Figure 8.1-11. (Right) The reaction tank is fitted with a circulation pump and a "sight glass" created from reinforced braided plastic tubing. This sight glass permits the filling of the reaction tank with an exact amount of WVO and reaction chemicals and also provides a way of monitoring the completeness of the reaction process.

process, thus allowing recycling of the excess. It is much simpler and safer and requires less energy to capture the methanol at the reaction stage than to try and recover it from the wash water.

Leaving methanol in the biodiesel is simply not an option.

Sodium Methoxide System

In order to "crack" WVO into biodiesel and its coproduct glycerol, it is necessary to use an alcohol and catalyst solution such as sodium (or potassium) methoxide. This solution is created by the careful measurement and mixing of methyl alcohol (methanol) and sodium (or potassium) hydroxide. Figure 8.1-12 shows a 55-gallon (208-liter) drum of methanol that has been delivered by a local fuel supply company. The drum is fitted with a hand-operated chemical pump suitable for methyl alcohol and a vapor recovery tube, both of which are fitted into the bung connections of the drum.

The outlet of the hand pump is connected to the white conical bottom tank (left) and the vapor recovery line is fitted to ensure that methanol vapors are returned to the storage tank. The conical tank is fitted with a screw top sealing lid and a small funnel and stopper. The funnel is used as a hopper which allows the addition of sodium hydroxide catalyst to the previously added methanol.

Figure 8.1-12. In order to "crack" WVO into biodiesel and its coproduct glycerol, it is necessary to use an alcohol and catalyst solution such as sodium methoxide. This solution is created by the careful measurement and mixing of methyl alcohol (methanol) and sodium hydroxide in the mixing tank shown above. The barrel to the right is filled with methyl alcohol.

Figure 8.1-13. A small, spark-proof mixer is fitted to the tank and is also connected to an electrically operated mechanical timer that is used to control the mix timing of the solution, ensuring that the sodium hydroxide completely dissolves in the methanol.

A small, spark-proof mixer is fitted to the tank as shown in Figures 8.1-12 and 8.1-13. The mixer is also connected to an electrically operated mechanical timer that is used to control the mix timing of the solution, ensuring that the sodium hydroxide completely dissolves in the methanol.

The outlet at the bottom end of the conical tank is connected to the suction side of the reaction tank through a shutoff valve. Opening this valve causes the sodium methoxide solution to be drawn into the reaction tank containing the WVO, starting the transesterification process.

Upon completion of the transesterification process, two liquid components or phases are created. A glycerol phase sinks to the bottom of the reaction tank, while the lower-density raw biodiesel phase floats on top. The glycerol is removed by draining it from the tank and transferring it to a separate refining station.

Upon removal of the glycerol phase, the raw biodiesel is re-heated to cause excess methanol to boil off. The methanol may be vented into the atmosphere, or preferably directed to a reflux condenser that converts the vapors to liquid. The captured methanol may then be reused, lowering production costs.

After glycerol and methanol have been recovered from the reaction tank, the raw biodiesel is transferred to the wash tank for final processing.

Biodiesel Washing System

Raw biodiesel is transferred from the main reaction tank to the large white conical-bottom washing tank shown in Figure 8.1-14. This tank is fitted with two washing systems known as mist and bubble wash technologies. Regardless of which washing procedure is used, a small water storage tank located to the immediate left of the washing tank is required. This 10.5-gallon (40-liter) tank receives potable water from a reverse osmosis filtration unit located in the storage cabinet under the reaction tank, which is in turn fed by the household potable water supply. The purpose of the reverse osmosis system is to remove dissolved minerals such as calcium and iron that are contained in the well water in our geographical location. Others, such as Lyle Rudensey (see Chapter 7.5), use naturally soft rain water. The water storage tank is fitted with a submersible water heating element that heats the wash water to between 120°F and 140°F (50°C and 60 °C), greatly improving the wash speed and quality. Prior to starting the wash cycle, a small amount of acetic acid is added to the wash water and a submersible pump is activated.

If the mist washing process is utilized, the wash water is pumped to a series of mist heads mounted around the perimeter of the wash tank lid which causes a gentle shower of slightly acidic (softened) water to spray over

Figure 8.1-14. Raw biodiesel is transferred from the main reaction tank to the large white conical-bottom washing tank shown above. This tank is fitted with two washing systems known as mist and bubble wash technologies. The small white tank to the left of the wash tank is the wash water storage tank.

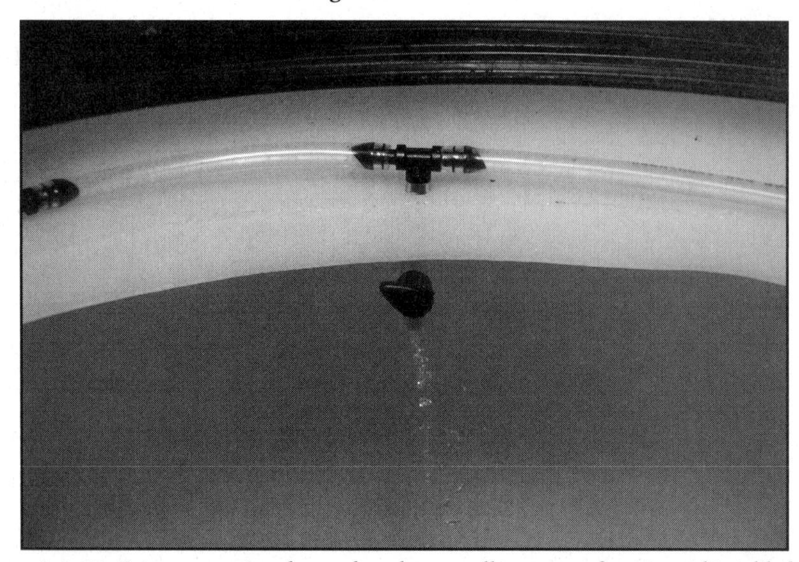

Figure 8.1-15. Prior to starting the wash cycle, a small amount of acetic acid is added to the biodiesel wash water and a submersible pump is activated, sending water to a series of mist heads mounted around the perimeter of the wash tank lid which causes a mist of slightly acidic (softened) water to spray over the biodiesel surface.

the biodiesel surface (see Figure 8.1-15). Water has a higher density than oil or biodiesel and therefore falls to the bottom of the tank, absorbing any free glycerin and excess catalyst (sodium hydroxide) along the way.

Washing may be completed using the spray mist method described above or using a "bubble washing" technique (see Figure 8.1-16) in which water is added to the biodiesel directly from the wash tank. A small air compressor (Figure 8.1-17) is connected to a heavy aquarium air stone which is lowered into the biodiesel wash tank (Figure 8.1-18). When the compressor is activated, small air bubbles carry water up through the biodiesel, absorbing impurities. The bubbles break at the surface of the biodiesel, causing the contaminated water to fall back to the bottom of the tank, where the process is repeated. After several hours, the wash water is considered to be fully saturated with contaminants and is drained off. (The merits of each washing technique will be discussed in Chapter 8.3.)

It is necessary to wash biodiesel a number of times in order to ensure compliance with fuel quality standards. At each successive washing stage the wash water contains fewer contaminants, allowing it to be reused for earlier, more heavily contaminated wash stages. This process is known as counter-current washing and greatly reduces the amount of water used in the

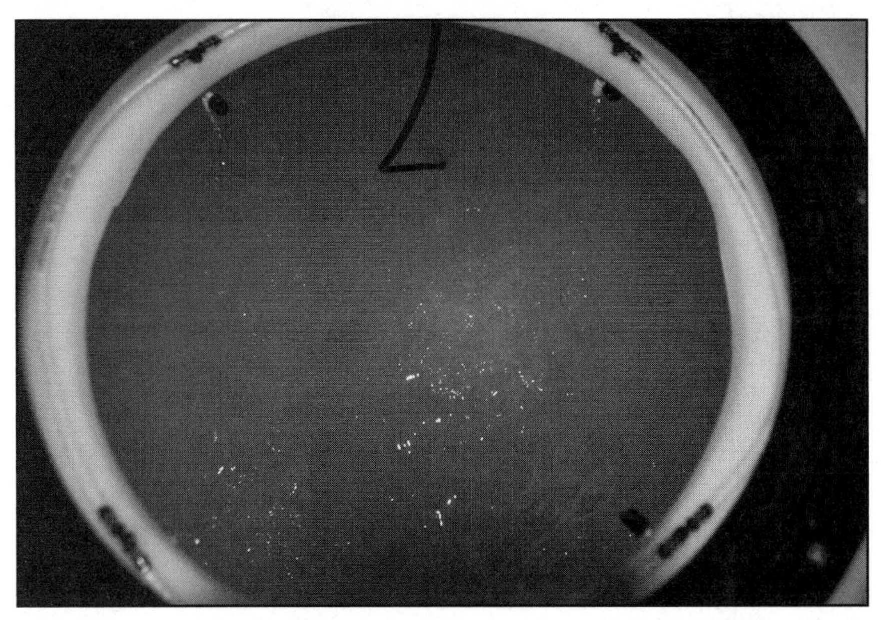

Figure 8.1-16. Biodiesel washing may use the spray mist method or a "bubble washing" technique (shown here) in which water is added to the biodiesel and tiny air bubbles are blown through the water/biodiesel mixture.

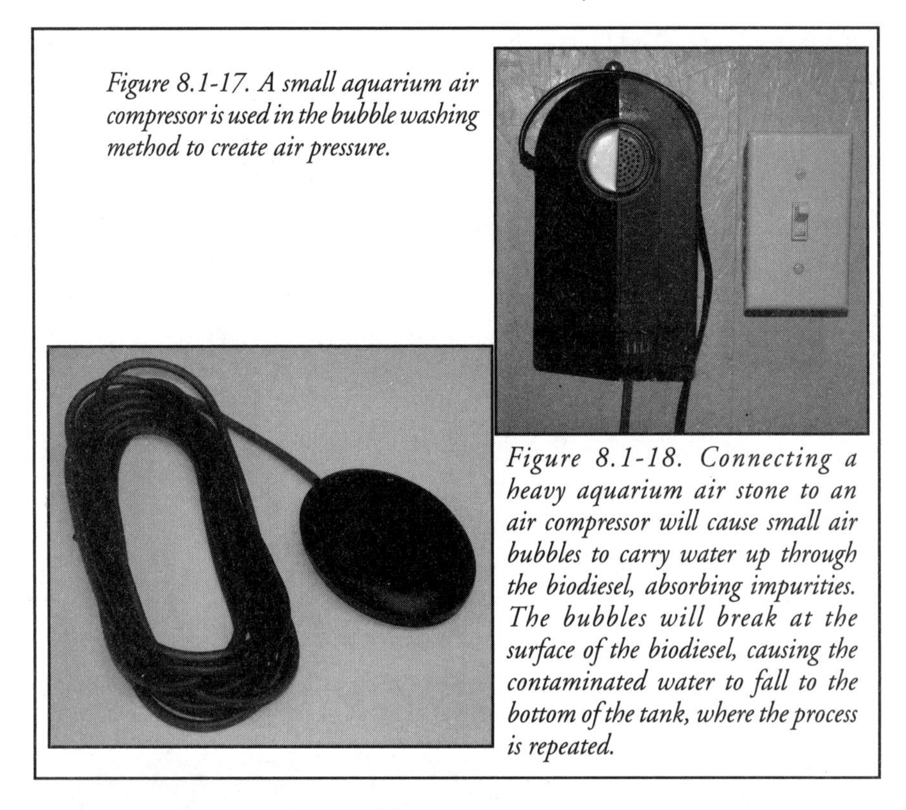

Figure 8.1-17. A small aquarium air compressor is used in the bubble washing method to create air pressure.

Figure 8.1-18. Connecting a heavy aquarium air stone to an air compressor will cause small air bubbles to carry water up through the biodiesel, absorbing impurities. The bubbles will break at the surface of the biodiesel, causing the contaminated water to fall to the bottom of the tank, where the process is repeated.

production process. When the wash water is saturated with contaminants, it is drained off and temporarily stored before final treatment and drainage into the environment. This contaminated wash water is known to be toxic and must be treated prior to release.

Biodiesel Drying and Final Filtration

After the final wash water has been removed from the washing tank, the biodiesel is pumped into the dryer tank shown in Figure 8.1-19. The dryer uses an electric water heater arrangement and is equipped with a circulating pump. It is also fitted with a heated air blower and filter arrangement which blows hot air through the biodiesel (Figure 8.1-20) before venting outside to the atmosphere. This arrangement removes excess water from the fuel, aids in ensuring compliance with the ASTM limits for water and sediment, improves the biodiesel oxidation stability, and reduces the chances of microbial growth while in storage.

Once the biodiesel has been heat-treated, it is pumped through a 20-micron fuel filter and is ready for quality testing and storage (Figure 8.1-21).

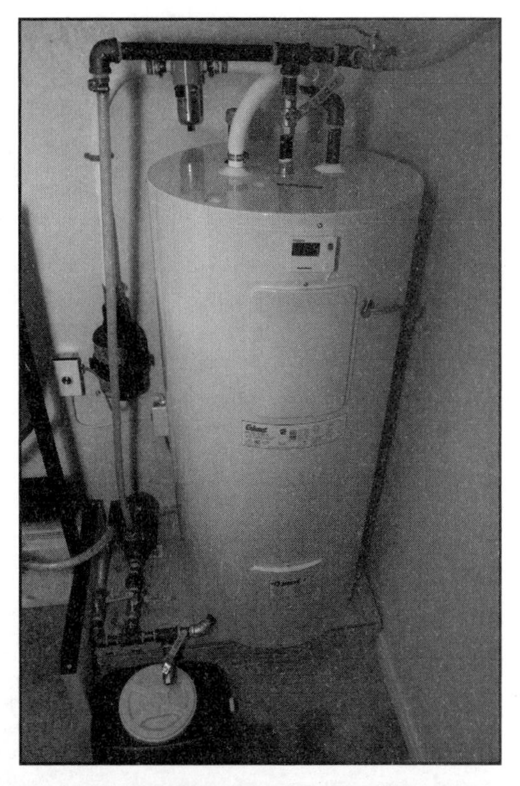

Figure 8.1-19. After the final wash water has been removed from the washing tank, the biodiesel is pumped into the dryer tank shown above.

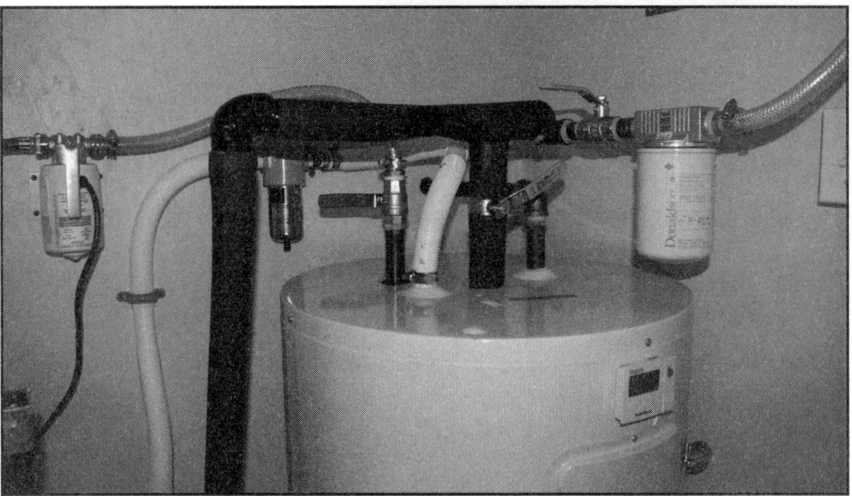

Figure 8.1-20. The fuel dryer is fitted with a heated air blower and filter arrangement which blows hot air through the biodiesel before venting outside to the atmosphere. This arrangement removes excess water from the fuel, ensures compliance with the ASTM limits for water and sediment, improves the biodiesel oxidation stability, and reduces the chances of microbial growth while in storage.

Figure 8.1-21. Once the biodiesel has been heat-treated, it is pumped through a 20-micron fuel filter and is ready for quality testing and storage.

Fuel Quality and Environmental Testing

There is no point in going to the trouble of producing biodiesel if it does not meet appropriate fuel quality standards. Likewise, the waste streams of glycerol and wash water must be dealt with in a responsible and environmentally friendly way.

Glycerol that is extracted from the biodiesel reaction is known to be a toxic waste material unless it undergoes further processing. Chapter 8.3 will discuss how to neutralize the toxicity of this material, allowing traditional means of disposal and possibly providing the biodiesel producer with a raw material feedstock that may be of some value. Chapter 8.3 deals with the detoxification process while Chapter 11 discusses some possible uses for this coproduct of the biodiesel production process.

Wash water processing may be completed offline from the biodiesel production process provided that methanol is fully recovered from the biodiesel ***before*** water washing takes place. Assuming that methanol recovery procedures are properly implemented, wash water treatment may be as simple as neutralizing the alkaline catalyst present through the addition of acetic acid (Figure 8.1-22).

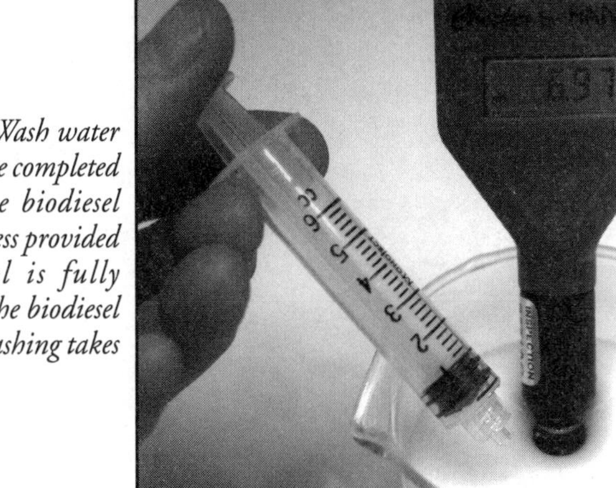

Figure 8.1-22. Wash water processing may be completed offline from the biodiesel production process provided that methanol is fully recovered from the biodiesel ***before*** *water washing takes place.*

Fuel testing comprises numerous steps to ensure compliance with appropriate standards. Chapter 8.3 outlines test procedures adapted from the ASTM that allow micro- and small-scale producers to self-monitor their production quality (Figure 8.1-23).

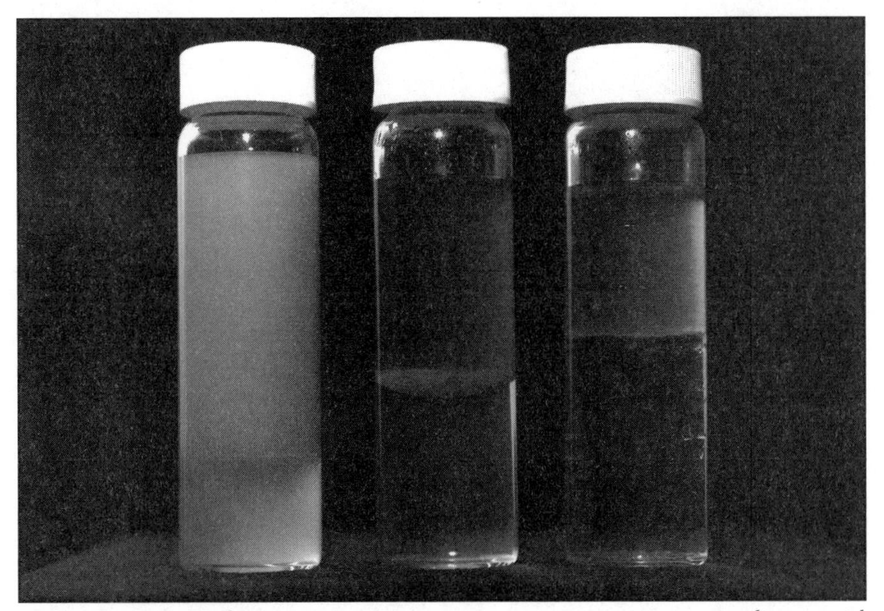

Figure 8.1-23. Fuel testing comprises numerous steps to ensure compliance with appropriate standards. Chapter 8.3 outlines test procedures adapted from the ASTM that allow micro- and small-scale producers to self-monitor their production quality.

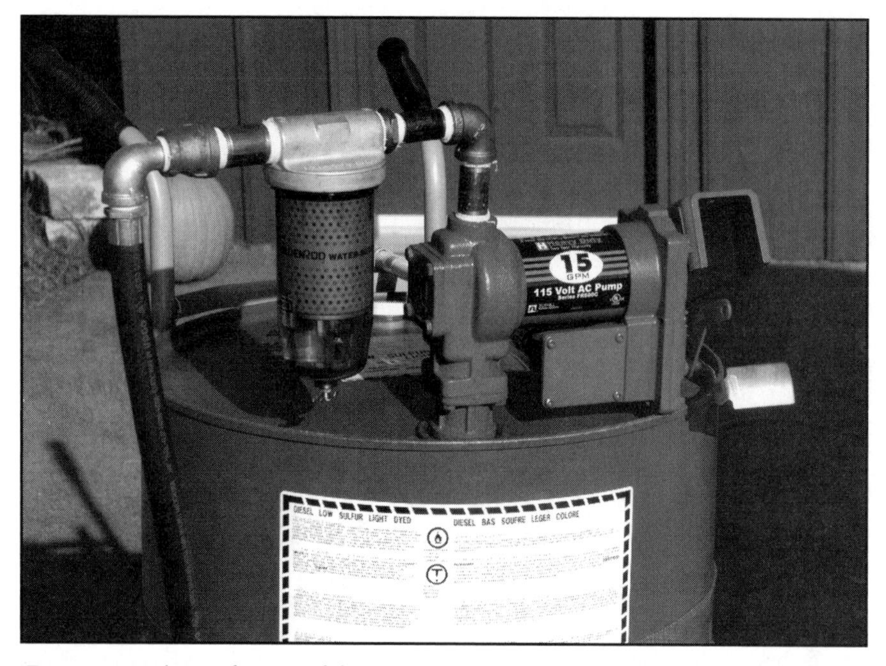

Figure 8.1-24. Quality-tested biodiesel is ready for storage and/or blending to the concentration level desired by the user. But fuel quality measures do not stop once the biodiesel is produced. It is necessary to ensure that proper storage and handling processes are continued right through the fuel consumption cycle.

Quality-tested biodiesel is ready for storage and/or blending to the concentration level desired by the user. But fuel quality measures do not stop once the biodiesel is produced. It is necessary to ensure that proper storage and handling processes are continued right through the fuel consumption cycle (Figure 8.1-24).

Biodiesel for the Off-Grid Producer

Biodiesel production is a relatively low energy-consuming process when compared with other chemical engineering procedures. However, any off-grid readers who are contemplating the production of biodiesel as part of a carbon fuel reduction program will find the energy requirements fairly high, possibly outside of the reach of your finely tuned home-energy balance. Depending on several variables, it may require between 10 and 20 kWh of electrical energy to fully process 40 gallons (150 liters) of biodiesel. At current retail rates for electricity this equates to less than 5 cents per gallon (1.3 cents per liter) for energy. For off-grid systems this amount of energy may only be available when spread over several sunny days and may require a solar thermal heating system for the chemical process heating energy. It may also be necessary to utilize some form of fossil fuel (or biodiesel-powered) generator as the motive power to operate the biodiesel production facility. For those who wish to lower energy consumption even further, consider the addition of a heat recovery muffler on the generator as a further means of improving overall thermal and electrical energy efficiency, by transferring this normally wasted thermal energy into the biodiesel production process.

With the addition of a small diesel engine (self-powered with biodiesel of course) you can create a system that will be able to generate as much energy as required burning low-cost, homemade fuel. Figures 8.1-25 through 8.1-29 detail the small 8-kW diesel engine/generator that I use for biodiesel production. The unit is mounted in a small, soundproof, insulated box that is fitted with a small ventilation fan and auxiliary exhaust muffler and fittings. The unit provides the electrical muscle for the heavy-duty work such as tank heating and circulation pump operation as well as electrical space heating during the cooler periods of the year.

Power from the off-grid system is used to operate the lower-energy items such as the room lights, room ventilation fan, wet lab work area electrical plugs, bubble and spray washer, and WVO receiver pump.

Figure 8.1-25. Geoff Shewfelt is shown opening the side door of the biodiesel-operated generator which provides the electrical muscle for the author's off-the-electrical-grid biodiesel production facility. Note the air intake located in the lower section of the door.

Alternating between these various tasks reduces the operating time of the generator to those periods when maximum electrical energy is required, while with off-grid power I can still receive WVO, wash biodiesel, and perform quality control tests while ensuring power consumption is kept to a minimum.

Figure 8.1-26. The biodiesel generator housing is fitted with fibreglass insulation that has both fireproof and acoustic qualities. The housing is also large enough to store generator fuel.

Figure 8.1-27. The top of the housing is hinged to allow access for servicing and fuel filling.

Figure 8.1-28. The generator neatly fits into the housing, which is equipped with cross-flow ventilation to ensure that the unit does not overheat during operation.

Figure 8.1-29. The "downwind" side of the generator ventilation system faces away from the house, ensuring that engine noise and smell is not a problem. Note that the exhaust system also vents away from the house for the same reason.

8.2
The Small-Scale Plant

Design Plans

8.2.1
General Issues

Prior to reviewing the design plans for the small-scale biodiesel production facility, let's discuss the construction and safety of plumbing and electrical systems. Even if you are not familiar with electrical wiring, conduit, and general construction work, it's worthwhile having a look at this section in order to understand what your electrician and plumber are talking about before you "throw the switch."

Figures 8.2.1-1 and 8.2.1-2 show an overview of the entire biodiesel laboratory system, from waste or cold-pressed oil intake through final biodiesel drying and filtration. Each stage of the production process will be discussed as follows:

- 8.2.2 Waste Oil Treatment Unit
- 8.2.3 The Biodiesel Reactor Unit with Methanol Recovery
- 8.2.4 Sodium (or Potassium) Methoxide Processor
- 8.2.5 Fuel Washing Unit (Misting and Bubble Wash)
- 8.2.6 Fuel Dryer and Final Processor
- 8.2.7 Fuel Dispensing Unit
- 8.2.8 Glycerine Processing Unit

Electrical Codes and Regulatory Issues

In North America, electrical installation work is authorized by local electrical safety inspection offices that issue work permits and review the work in accordance with national standards. In the United States, the National Electrical Code (NEC) has been developed over the last century to include almost all aspects of electrical wiring. In Canada, the Canadian Electrical Code (CEC) performs the same function as the NEC.

When you or your electrician is ready to begin wiring, it will be necessary to apply for an electrical permit. This permit will authorize you to:

- perform all electrical wiring according to NEC/CEC codes and any local ordinances in effect at the time of installation

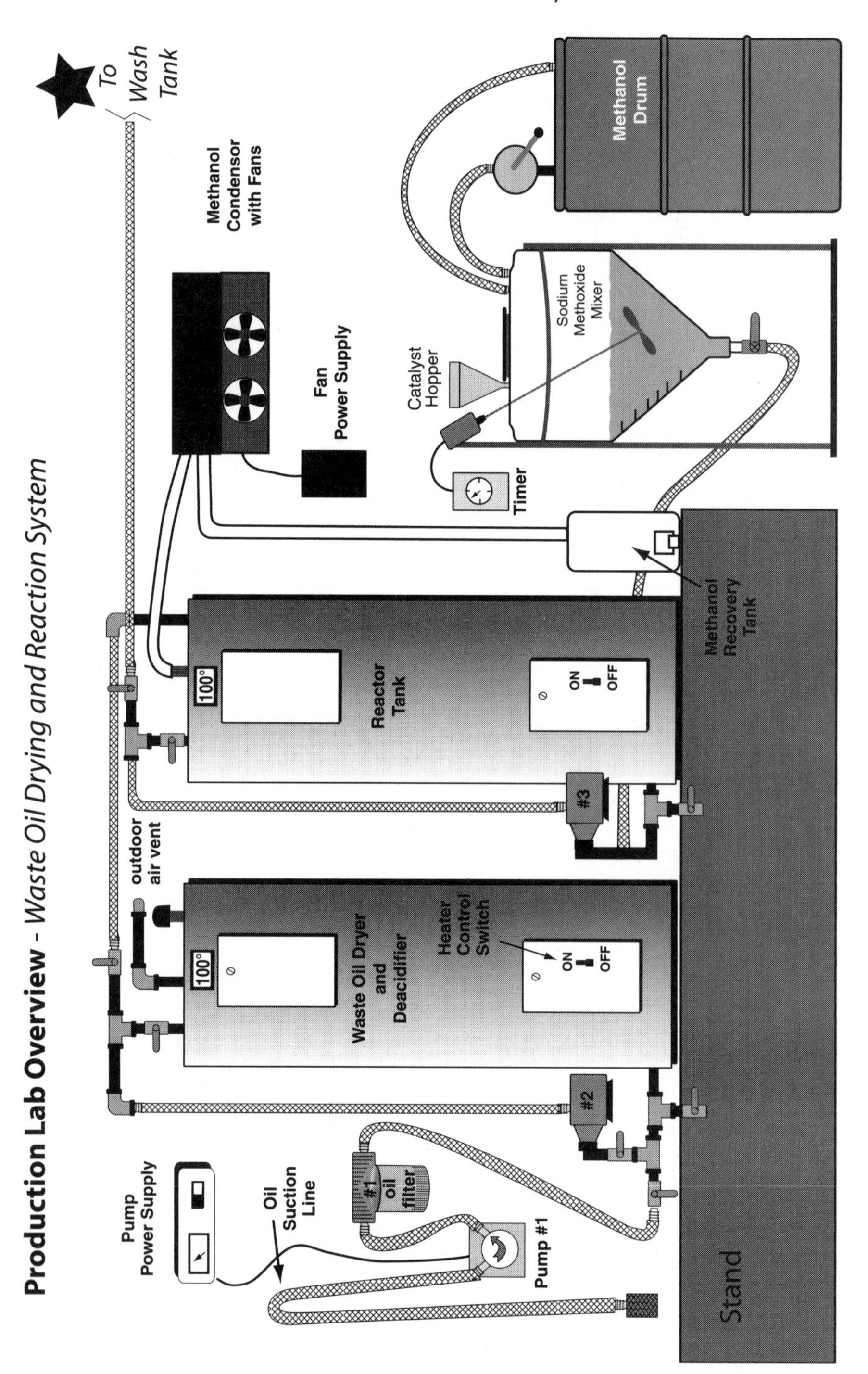

Figure 8.2.1-1. Biodiesel Production Lab Overview – Waste Oil Dryer and Reaction System.

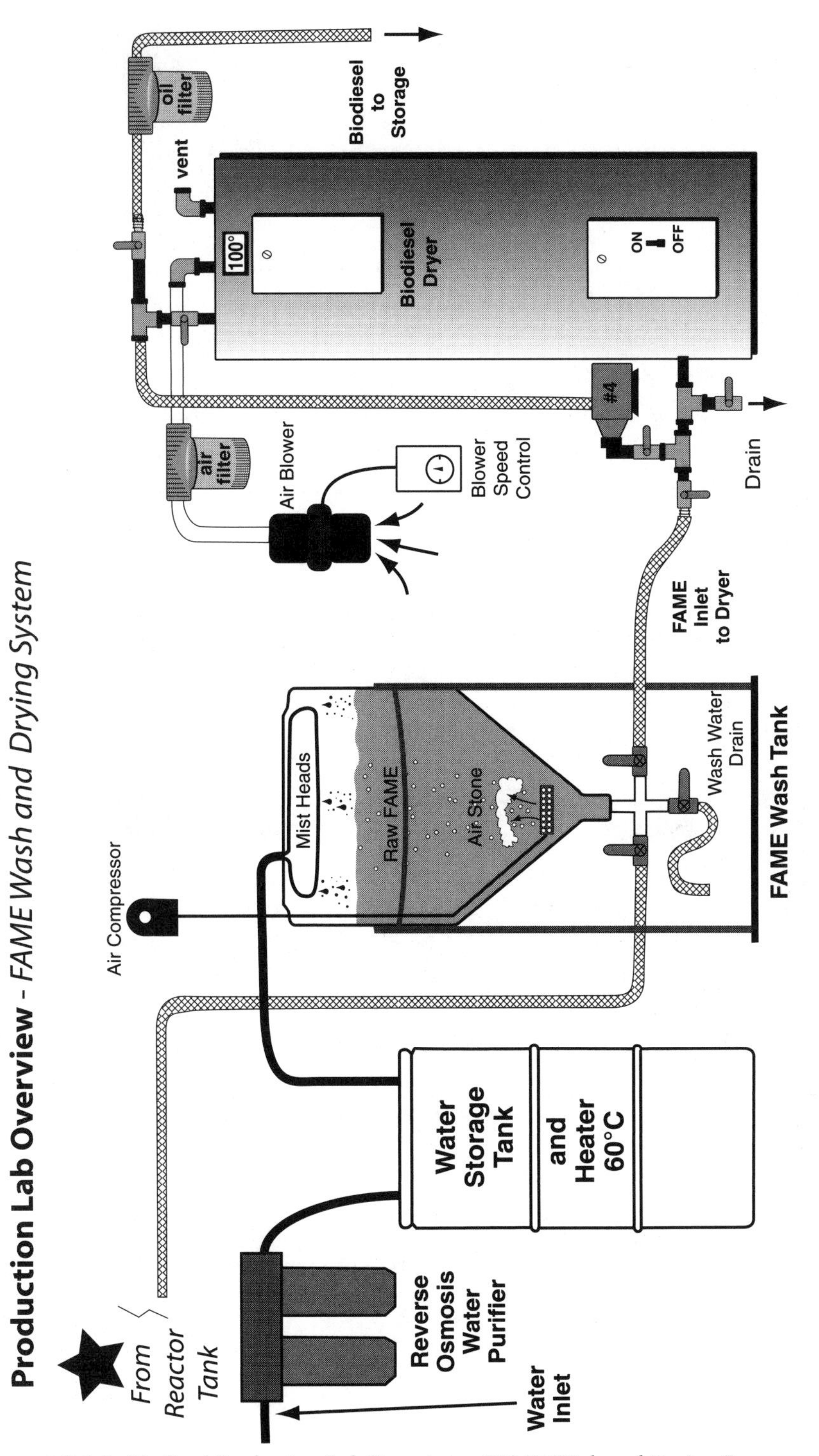

Figure 8.2.1-2. Biodiesel Production Lab Overview – FAME Wash and Drying System

- install only electrical equipment that is properly certified. Each device must have a UL, CSA, or other approval agency certification "mark."
- provide copies of wiring plans, proof of certification, or other engineering or technical documentation to help the inspector understand the renewable energy system.

You must give the inspector written notification that the work is ready for inspection, no part of the wiring work, including backfilling of trenches, can be hidden or covered until the inspector has completed the inspection and provided an authorization certificate.

Electrical inspectors will review the design and installation work and check for certification marks on the various appliances. Some system components on the market, such as foreign-made pumps, do not have test certification markings, but all electrical code rules require that products *must* have them. It is highly recommended that you check any products before you purchase them to ensure proper compliance. If you require a product and the manufacturer has not had it tested, discuss this with your inspector before you buy. A certified product may be available, albeit at a higher cost. Alternatively, field inspection of uncertified devices may be allowed in your jurisdiction for an additional fee.

CAUTION!

As electrical safety and installation code rules are updated on a regular basis and may have subtle differences from one locale to another, use the information in this chapter as a guide but discuss the details with your electrician and inspector before proceeding with installation work.

If you wish to review applicable electrical certification standards, you can purchase a copy of the NEC or CEC from:

NEC: National Fire Protection Association
1 Batterymarch Park
Quincy, Massachusetts
USA 02169-7471
www.nfpa.org

CEC: Canadian Standards Association
5060 Spectrum Way
Mississauga Ontario L4W 5N6
www.csa.ca

Wire Type

It would be wise to discuss the production room layout with your local electrical inspector prior to purchasing any materials for wiring the biodiesel system. Depending on such variables as air flow, room construction, and local ordinances, system wiring could be treated as an extension of regular house wiring or the lab could be considered to be industrial in nature, requiring cables to be installed in conduit or in an explosion-proof manner.

Wiring Color Codes

Wiring color codes are an important part of keeping the interconnection circuits straight when installing or troubleshooting the system or upgrading it at a later date. The standard color schemes are discussed below.

Bare Copper, Green, or Green with a Yellow Stripe

This wire is used to bond exposed bare metal in pumps, reaction (water heater) tanks, control cabinets, and circuit breakers to a common ground connection (discussed later). The ground wire does not carry any electrical current except during times of electrical fault.

White

This wire carries current and is normally the return or neutral conductor of a 120-volt system. The voltage measured from either the black or the red ungrounded conductor to the neutral connection is 120 volts. The white wire is also bonded to the system ground connection and the main electrical panel.

Red and Black

In a 2-wire, 120-volt system, the cable wire colors are black, white, and bare, with the black wire designated "Line 1." The voltage between the black wire and the neutral or white wire is 120 volts. Note that the bare or ground wire is not counted as a conductor when determining a 2, 3, or other wire conductor count.

In a 2-wire, 240-volt system, the cable wire colors are black, red, and bare, with the black wire designated "Line 1" and the red wire designated "Line

2." The voltage between the black and red wires measures 240 volts.

In a 3-wire, 240-volt system, the cable wire colors are black, red, white, and bare wires. The voltage measured between various cable pairs is as follows:

Black to White:	120 volts
Red to White:	120 volts
Black to Red:	240 volts

Three-wire circuits are useful where both 120 and 240 volt devices are contained within the same appliance or system, such as a clothes dryer or cook stove. Electric hot water heaters are generally powered using a two-wire, 240 volt circuit.

System Grounding

Grounding provides a method of safely dissipating electrical energy in a fault condition. Yes, that third pin you cut from your extension cord really does do something. It provides a path for electrical energy to safely dissipate when the insulation system fails within the electrical heating element of the reaction tank.

Imagine a teakettle for a moment. Two wires from the house supply enter the teakettle, plus a ground wire. During normal operation, the electricity flows from the house electrical panel via the ungrounded, black "line 1" conductor to the kettle. Current flows through the heater element and back to the panel via the white "neutral" conductor that is grounded. A separate ground wire connects the metal housing of the kettle (via the pesky third prong) to a large conductive stake driven into the earth just outside the house.

If the insulation or hot wire were to be damaged inside the kettle, it could touch the metal chassis. Because the chassis is bonded to ground (assuming you didn't cut the pin), electrical energy will travel from the chassis through the ground wire to the conductive stake. This flow of current is unrestricted due to the bypassing of the heater element, causing overheating of the electrical wires. If it were not for the circuit breaker or fuse limiting this excessive current flow, a fire could start.

Electrical energy has an affinity for a grounded or "zero potential" object and will do whatever it takes to get there. If there were no ground connection on the defective kettle's chassis, electricity would simply stay there until an opportunity arose to jump to ground. If you were to touch the kettle and simultaneously touch the sink or be standing on a wet surface, the electricity would find its path through your body. This is not a good situation.

Figure 8.2.1-3. This photograph shows a typical electrical power panel that is a staple device in virtually every home. The panel accepts 240-volt, 3-wire electrical service from the utility supply and passes it through the main disconnection circuit breaker located at the top of the panel.

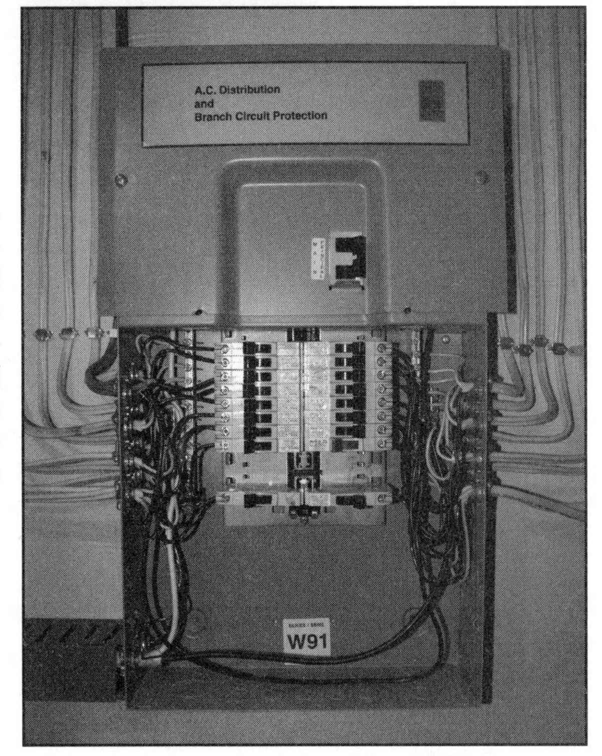

Figure 8.2.1-4. In this view the protective cover is removed from the electrical power panel, showing the electrical service power from the utility connected to a series of branch circuit breakers located along the right and left sides of the panel.

Figure 8.2.1-5. This photograph details the individual branch circuit feeder wires that deliver either 120- or 240-volt power to household circuits.

Figure 8.2.1-3 shows a typical electrical power panel that is a staple device in virtually every home. This panel accepts 240-volt, 3-wire electrical service from the utility supply and passes it through the main disconnection circuit breaker located at the top of the panel (Figure 8.2.1-4). The electrical service power is then connected to a series of branch circuit breakers located along the right and left sides of the panel. Individual branch circuit feeder wires then deliver either 120- or 240-volt power to household circuits (Figure 8.2.1-5).

The detail view of an electrical service panel is shown in Figure 8.2.1-5. In this view, the 2-wire, 120-volt circuits located on the right side of the panel are clearly visible. Although this view may appear rather daunting, an easier-to-understand schematic drawing is shown in Figure 8.2.1-6. Electrical cable strain relief bushings are detailed in Figure 8.2.1-7. A single circuit breaker will power a single water heater unit of the reaction system.

Assuming that the biodiesel production facility will be built in a separate building, it is possible to connect a single large electrical feed wire to a separate "service subpanel" located within the outbuilding/workshop. Assuming that this is the case, and basing the room design on the system layout shown in Figures 8.1-1 and 8.1-2 and the overview drawings shown in Figures 8.2.1-1 and 8.2.1-2, a typical electrical distribution system might have the ratings shown in Table 8.2.1-1.

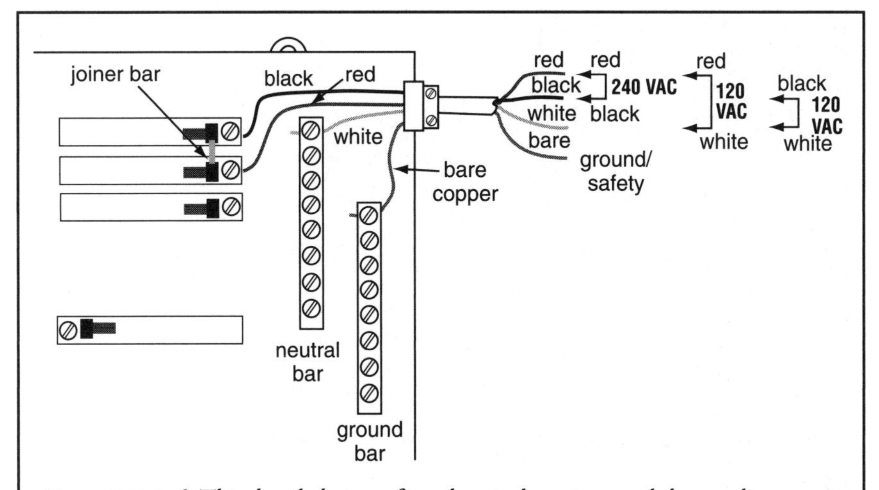

Figure 8.2.1-6. This detailed view of an electrical service panel shows a large group of 120-volt circuit breakers feeding a typical 2-wire, 120-volt household electrical system. Depending on the power rating of the various devices, each circuit path may power only one energy-demanding device or an entire roomful of small appliances and lights. A single circuit breaker will power a single water heater unit of the reaction system.

Although it is not necessary for the system to consume energy at the maximum rating, installing a panel that is slightly larger than required allows for upgrades and improvements in the design without costly rewiring charges. In addition, the main circuit breaker located in the subpanel can be used as an emergency power off switch in the event of a unit failure, chemical spill, or other mishap.

On the other hand, having a sufficiently large service panel permits simultaneous operation of each stage of the biodiesel production process. The above example allows the processing of approximately 50 gallons (189 liters) of oil/biodiesel at each stage of the system, providing 200 gallons (757 liters) of processing for the same amount of total labour required to process a single 50 gallon (189 liter) batch.

Tank Electrical Preparation

A typical 40- or 60-gallon (151/227-liter) water tank makes an ideal processing tank for each of the steps in the biodiesel production process owing to its self-draining (conical bottom) profile, low cost, energy efficiency, and integral heating system design. All models are equipped with both upper and lower electrical thermostats which are accessible through access panels as shown in Figure 8.2.1-8.

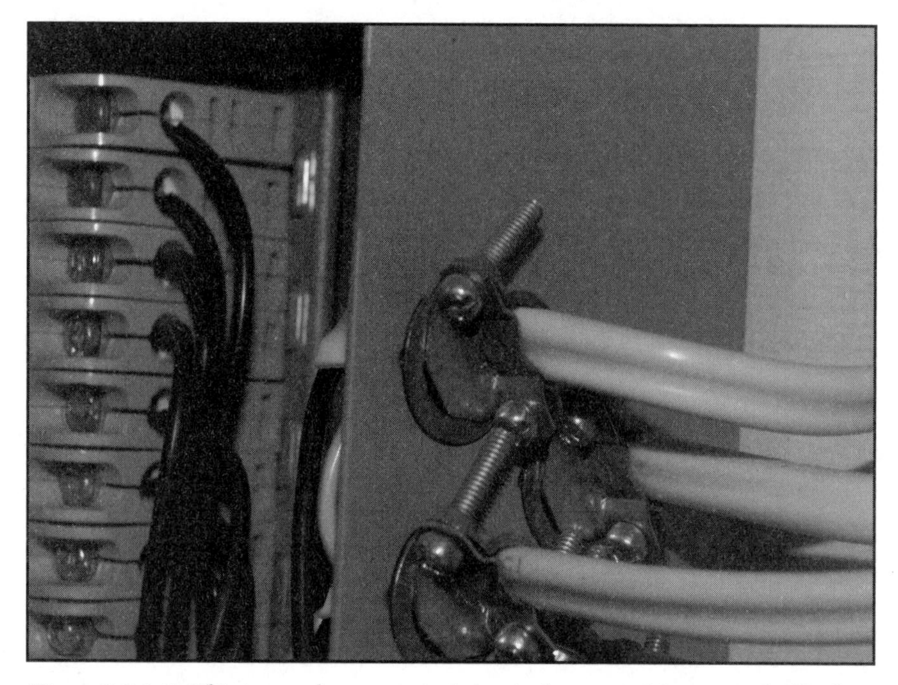

Figure 8.2.1-7. This image shows a typical electrical power cable strain relief bushing that prevents chaffing of the wire jacket and stops mechanical strain on the wire from traversing to the electrical connection points.

Circuit Function	Voltage	Current Rating (amps)	Circuit Breaker Rating (amps)
Room Lights and Outlets	120	15	15
Oil Receiver/Dryer	240	15	15
Reactor Tank	240	15	5
Wash Water Heater	120	10	15
Biodiesel Dryer	240	15	15
Tank Circulation Pump	120	8	15
Sub-Panel Rating			**100**

Table 8.2.1-1.
Typical electrical service rating for the biodiesel production facility outlined in Figures 8.2.1-1 and 8.2.1-2.

Figure 8.2.1-8. A typical 40-or 60-gallon (151/227-liter) water tank makes an ideal processing tank for each of the steps in the biodiesel production process owing to its self-draining (conical bottom) profile, low-cost, energy efficiency, and integral heating system design. All models are equipped with both upper and lower electrical thermostats which are accessible through an access panel.

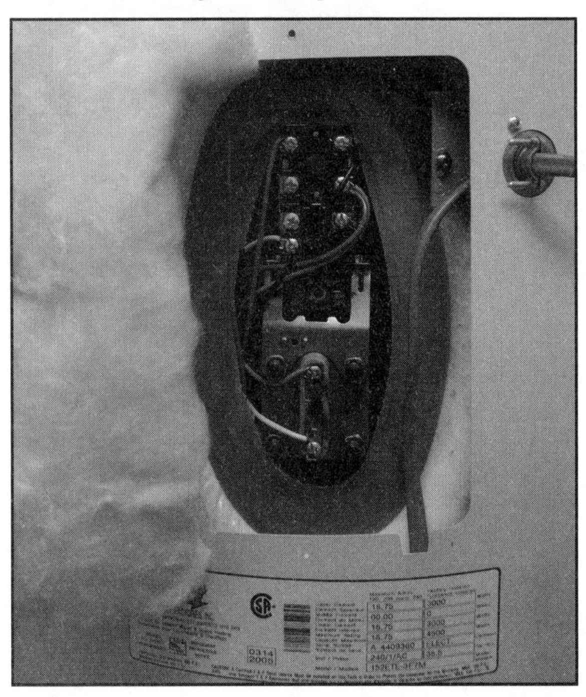

Figure 8.2.1-9. The processing tanks will not be completely filled during operation, necessitating the disconnection of the upper heating element.

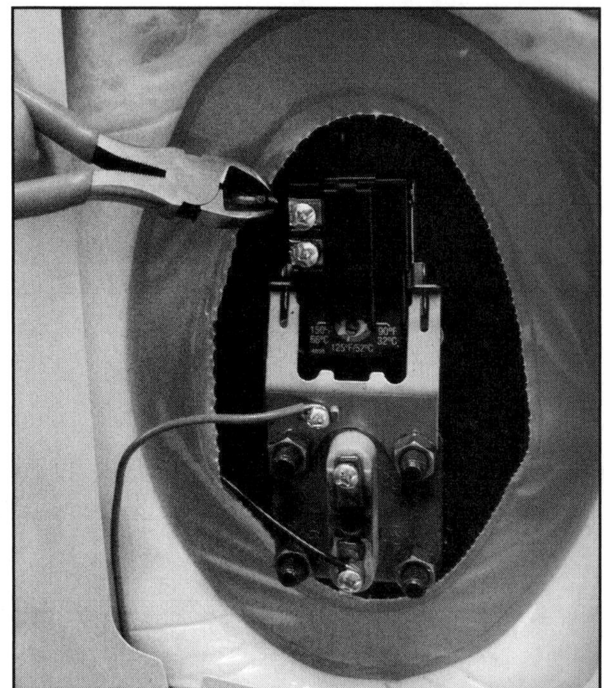

Figure 8.2.1-10. Ensure that any electrical feed wires used by the upper heating element are either completely removed or suitably capped with insulated electrical wire nuts.

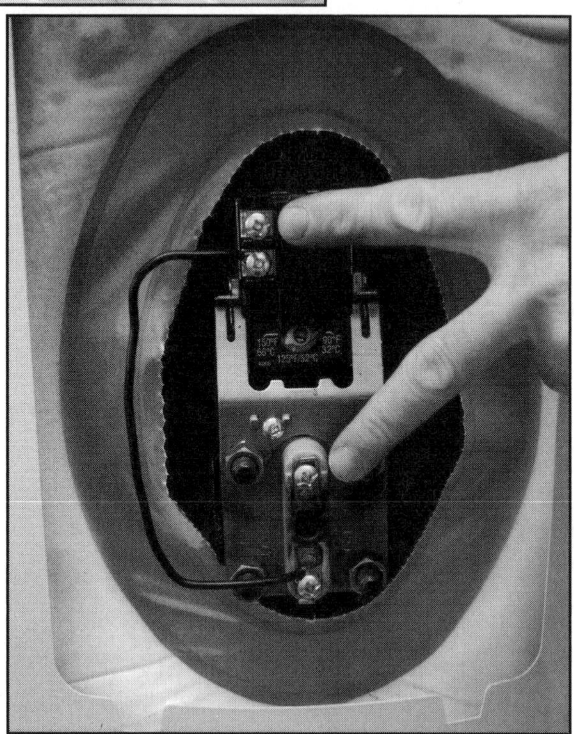

Figure 8.2.1-11. A No. 12 AWG black jumper wire is connected between one terminal of the lower thermostat and the heater assembly.

Figure 8.2.1-12. The power supply cable is then connected to the remaining terminal on the thermostat and heater element, respectively. Be certain to bend the power supply wire into a small "hook" positioned so that the terminal screws rotate towards the open end of the hook, ensuring a tight and reliable connection. The bare ground wire is connected to the green-colored screw secured to the tank chassis.

As the processing tanks will not be completely filled during the operation stage, it is necessary to disconnect the upper heating element by cutting the electrical feed wires and capping the bare ends with an insulated wire nut (Figures 8.2.1-9 and 8.2.1-10). Alternatively, the power wire may be completely removed from the heater chassis.

Once the upper heater element is disconnected, a power supply feed cable from the electrical subpanel is fed to the lower heating element. A No. 12 AWG black jumper wire is connected between one terminal of the lower thermostat and the heater assembly as shown in Figure 8.2.1-11.

The power supply cable is then connected to the remaining terminal on the thermostat and heater element, respectively. Be certain to bend the power supply wire into a small "hook" positioned so that the terminal screws rotate towards the open end of the hook, ensuring a tight and reliable connection (Figure 8.2.1-12). It does not matter which supply wire color is selected for either connection point. The bare ground wire is connected to the green-colored screw secured to the tank chassis. The finished connection should look similar to that shown in Figure 8.2.1-13.

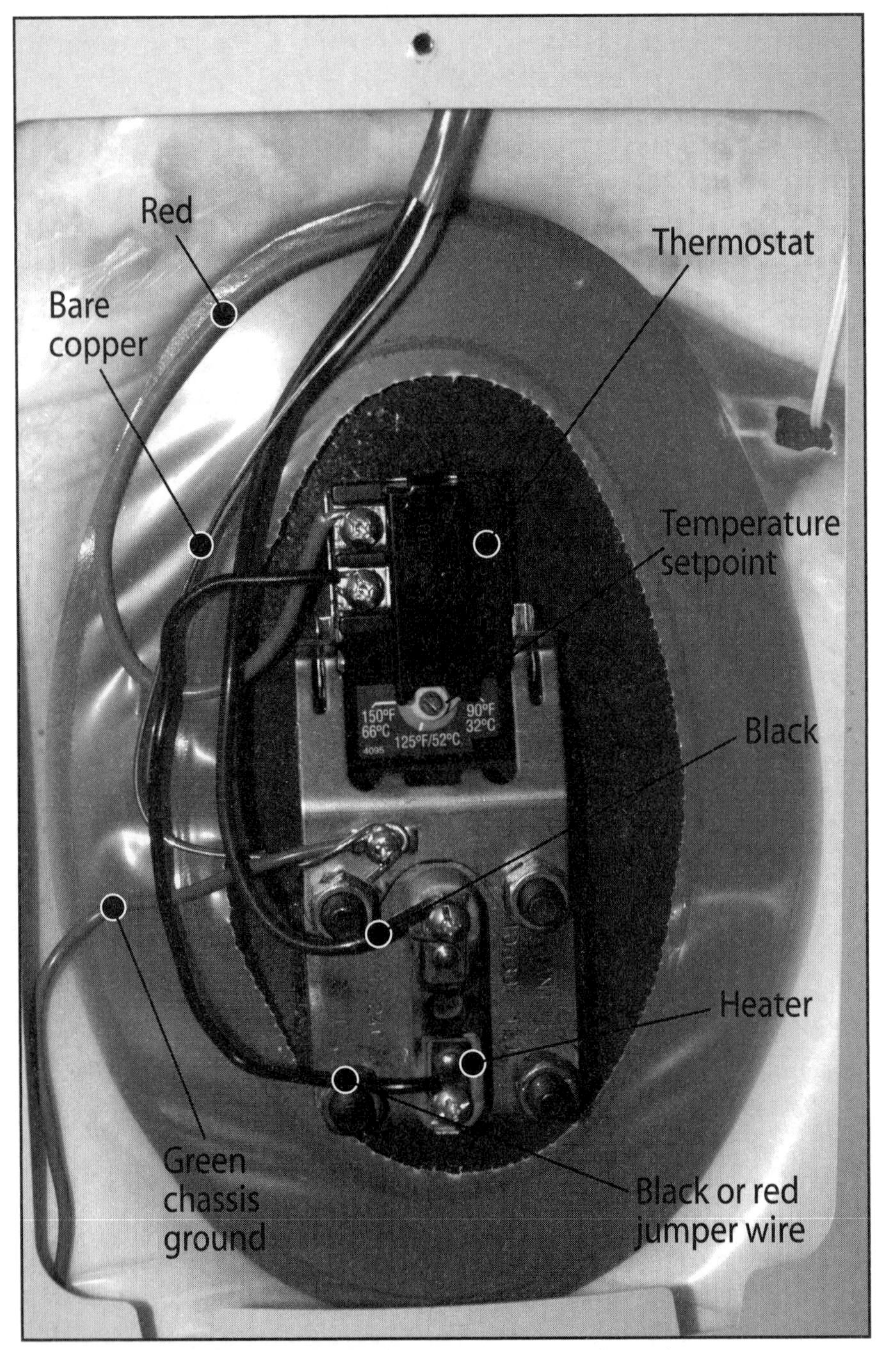

Figure 8.2.1-13. The finished connection should look similar to this.

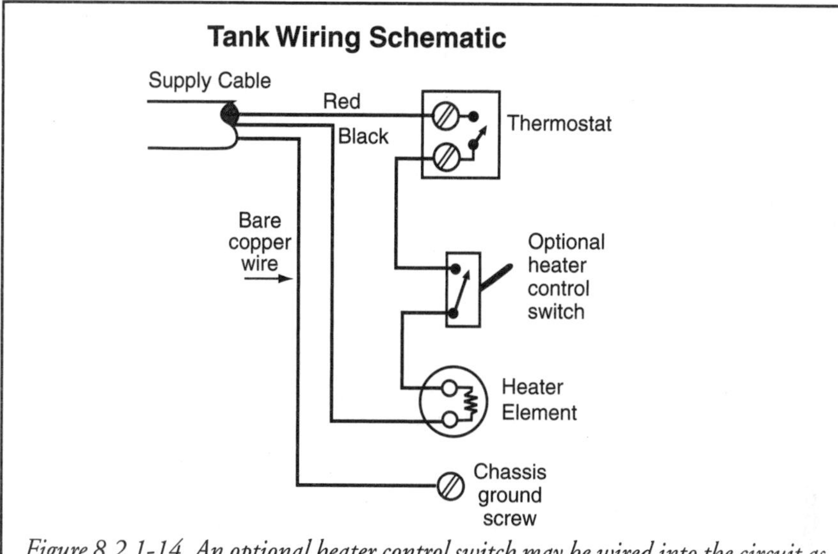

Figure 8.2.1-14. An optional heater control switch may be wired into the circuit as shown in this schematic diagram. This arrangement allows the heater to be turned off without having to access the electrical subpanel. It also provides visual feedback of heater status.

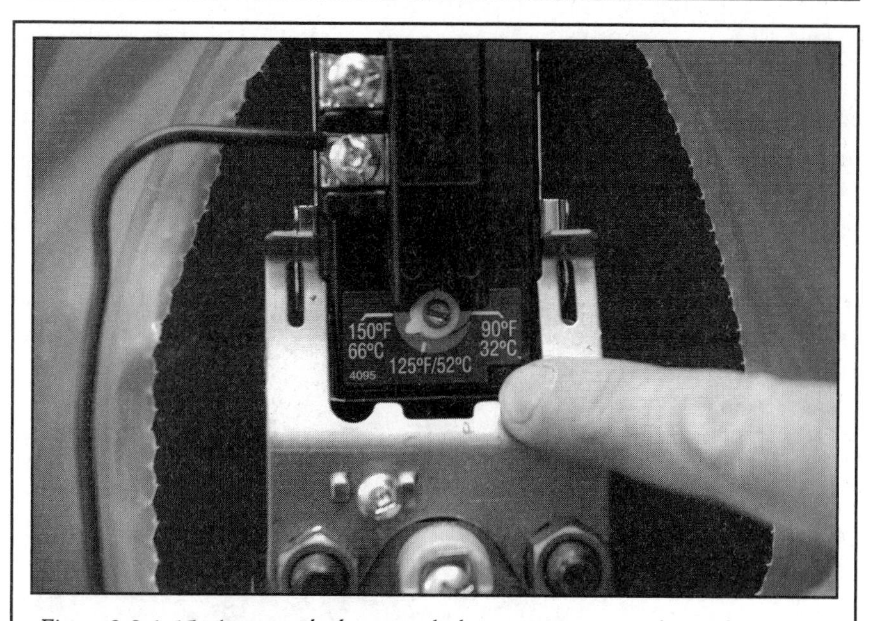

Figure 8.2.1-15. Access to the lower tank thermostat is required in order to adjust the liquid temperature from "reacting" mode to "methanol recovery" mode. This can be accomplished by drilling a small hole in the access cover and using an **insulated shaft** *screwdriver to make the adjustment.*

As an alternative, an optional heater control switch may be wired into the circuit as shown in Figure 8.2.1-14. This arrangement allows the heater to be turned off without having to access the electrical sub-panel. It also provides visual feedback of heater status.

The lower thermostat unit (Figure 8.2.1-15) has a temperature setting range that can be adjusted using a small screwdriver. The temperature range of the thermostat can be extended beyond the normal markings by removing the small mechanical detent or stop-guide which is installed to prevent hot water scalding. Access to this thermostat is required in order to adjust the liquid temperature from "reacting" mode to "methanol recovery" mode. This can be accomplished by drilling a small hole in the access cover and using an **insulated shaft** screwdriver to make the adjustment. It is also possible to set the thermostat to the maximum desired operating temperature and manually turn off the heater during normal processing.

Liquid Temperature Display

An important step in the production of biodiesel is the careful monitoring of process temperature. One of the simplest and most accurate means of measuring internal fluid temperature is to use a digital indoor/outdoor thermometer such as the model shown in Figure 8.2.1-16. These devices are available at most electronic and hardware stores and are reliable, inexpensive, and accurate. When purchasing a model, be certain the device is able to display temperatures above 175°F (80°C).

Figure 8.2.1-16. One of the simplest and most accurate means of measuring internal fluid temperature is to use a digital indoor/outdoor thermometer such as this.

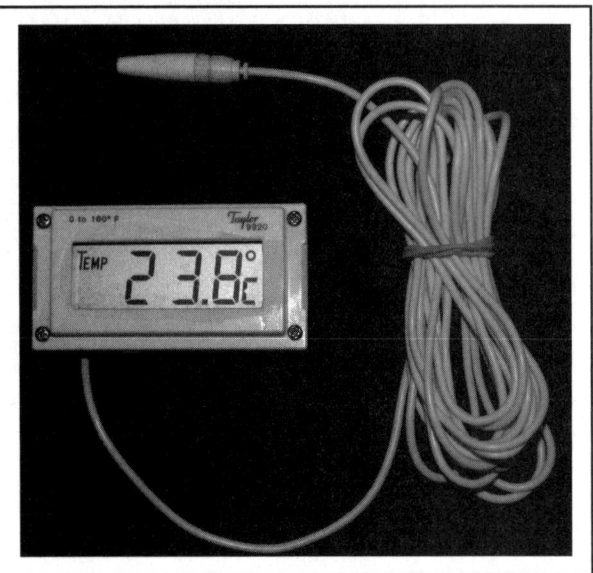

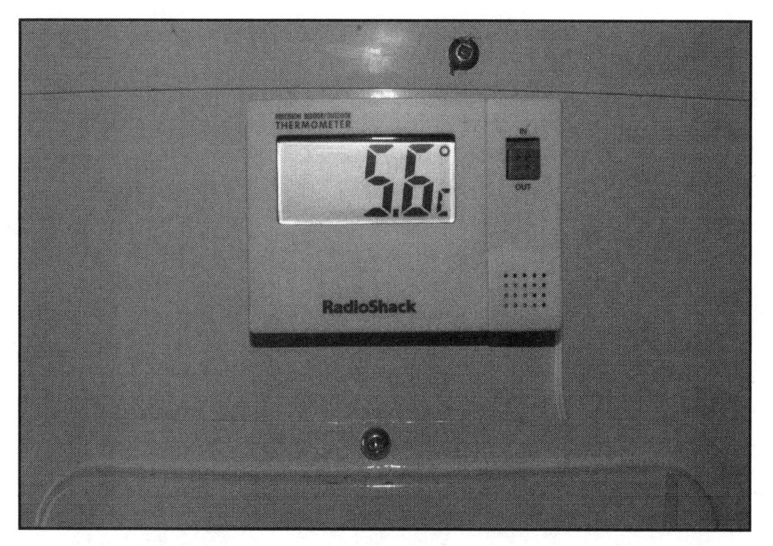

Figure 8.2.1-17. Mount the thermometer in any convenient location, keeping in mind the sensor cable routing. In this view, the thermometer has been mounted to the side of the upper section of the tank, making it visible at all times.

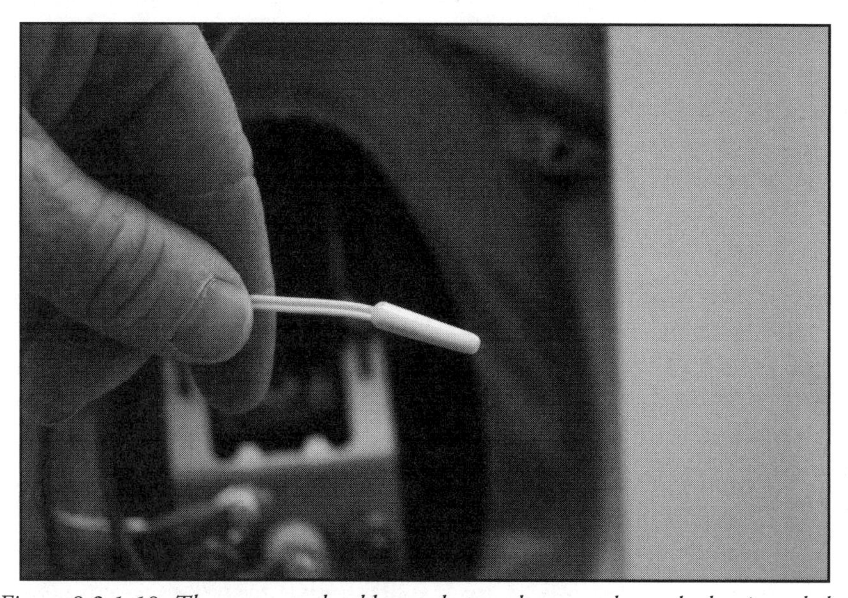

Figure 8.2.1-18. The sensor and cable can be run between the tank chassis and the insulation into the lower access cover area. A piece of stiff wire may be required in order to pull the sensor and cable from the top access area to the bottom. Ensure that the sensor wire is located well away from all electrical power cables and connections and that the wire cannot move from its intended location over time or as a result of vibration.

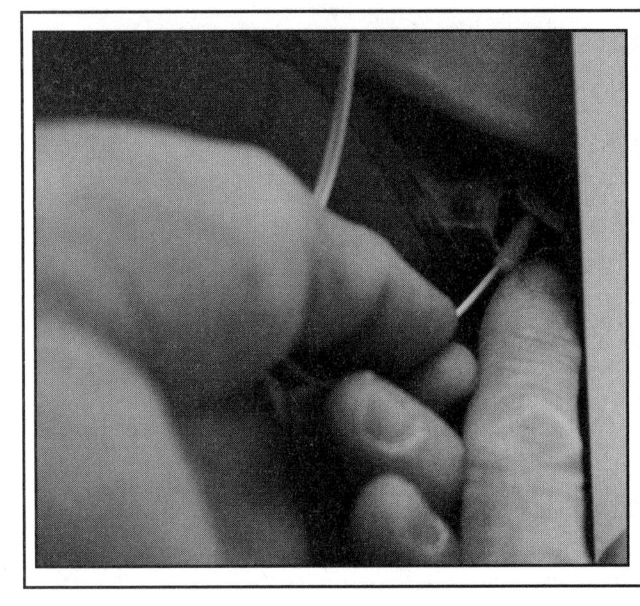

Figure 8.2.1-19. The sensor should be firmly placed between the tank wall and the insulation surrounding it to ensure that it is mounted tightly against the tank. Locating the sensor in this manner guarantees maximum accuracy when the liquid temperature is measured.

Mount the thermometer in any convenient location (Figure 8.2.1-17), keeping in mind the sensor cable routing. The sensor and cable can be run between the tank chassis and the insulation into the lower access cover area (Figure 8.2.1-18). **The sensor should be firmly placed between the tank wall and the insulation surrounding it to ensure that it is mounted tightly against the tank** (Figure 8.2.1-19). Locating the sensor in this manner guarantees maximum accuracy the liquid temperature is measured.

Black Pipe Plumbing Fittings

The majority of the plumbing connections are made with black pipe, which has threaded ends and receptacles and is available in standard incremental lengths, allowing custom plumbing designs to be developed.

A typical straight plumbing section is shown in the lower item in Figure 8.2.1-20. This section is a standard ¾" diameter, the size most commonly used throughout the biodiesel production facility. Although the ends of each pipe section are threaded, a supplementary means of sealing connections is required. The threaded ends of the upper pipe shown in this image are wrapped with four turns of "gas fitters" yellow tape, which is designed specifically for black pipe components. Alternatively, a brush-on compound known as "pipe dope" or "plumbing dope" can be used. Some pipe fitters prefer to use both, noting that a leaking fitting will result in profit-busting "callbacks."

In addition to straight pipe sections, all caps (Figure 8.2.1-21), 90° elbows (Figure 8.2.1-22), valves and other fittings are available in black pipe

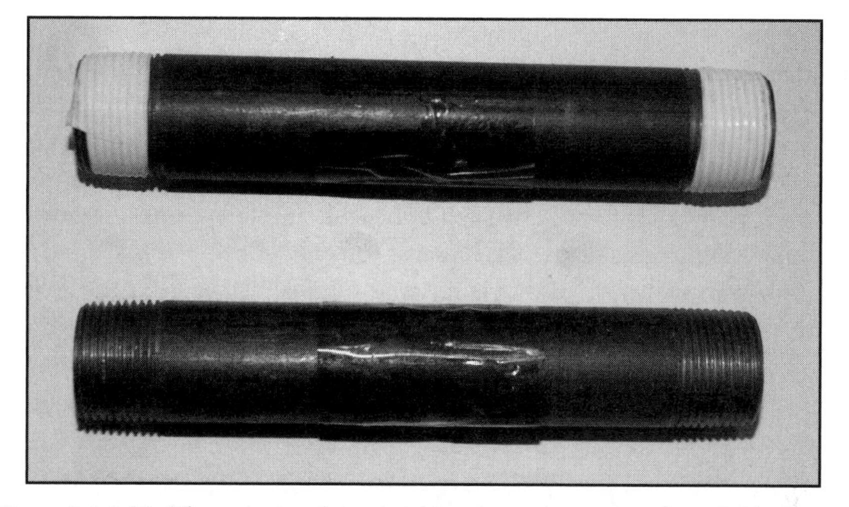

Figure 8.2.1-20. The majority of the plumbing connections are made with black pipe, which has threaded ends and receptacles and is available in standard incremental lengths, allowing custom plumbing designs to be developed.

Figures 8.2.1-21 and 8.2.1-22. In addition to straight pipe sections, all caps, 90° elbows, valves, and other fittings are available in black pipe or compatible fitting configurations.

or compatible fitting configurations. Note that some specialized fittings may be available only in zinc-plated components such as the cap shown in Figure 8.2.1-21.

Sacrificial Anode Removal

The effect of galvanic corrosion causes a low-level, slow degradation of any metallic components submersed in water, including boats, gas pipelines, and water heaters. All water heater tanks are equipped with a device known as a "sacrificial anode" which is designed to attract corrosion-causing electrical energy to it and away from the tank structure. Unfortunately the anode may also causes the release of metals into the biodiesel, which cannot be tolerated.

Figure 8.2.1-23. Locate the anode by removing the access cover on the top side of the water heater tank.

Locate the anode by removing the access cover on the top side of the water heater tank (Figure 8.2.1-23). The anode is equipped with a threaded fitting which can be removed with the assistance of a socket wrench and some judicious elbow grease (Figure 8.2.1-24). Remove the anode from the tank (Figure 8.2.1-25) and seal the fitting with a 3" section of long black pipe and a plumbing cap as shown in Figure 8.2.1-21).

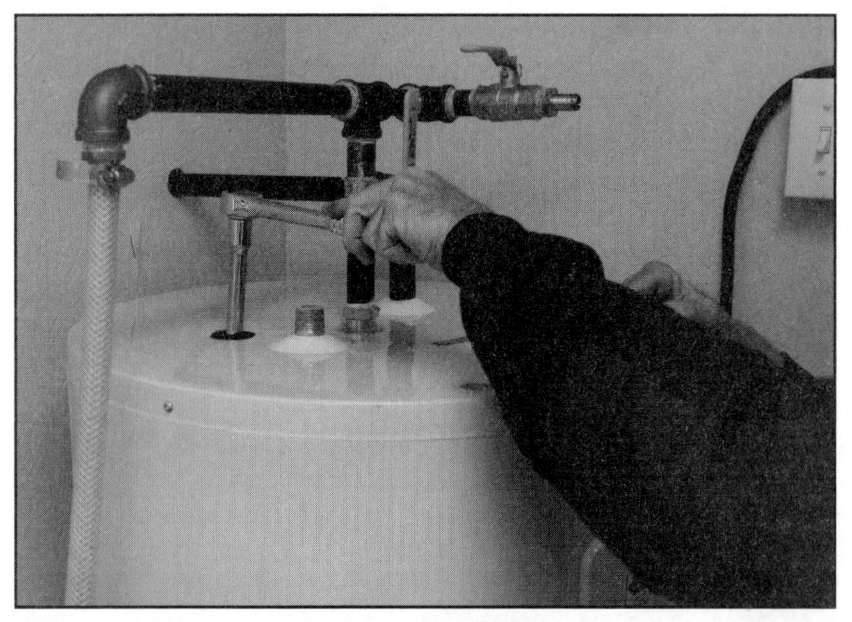

Figure 8.2.1-24. The anode is equipped with a threaded fitting which can be removed with the assistance of a socket wrench and some judicious elbow grease.

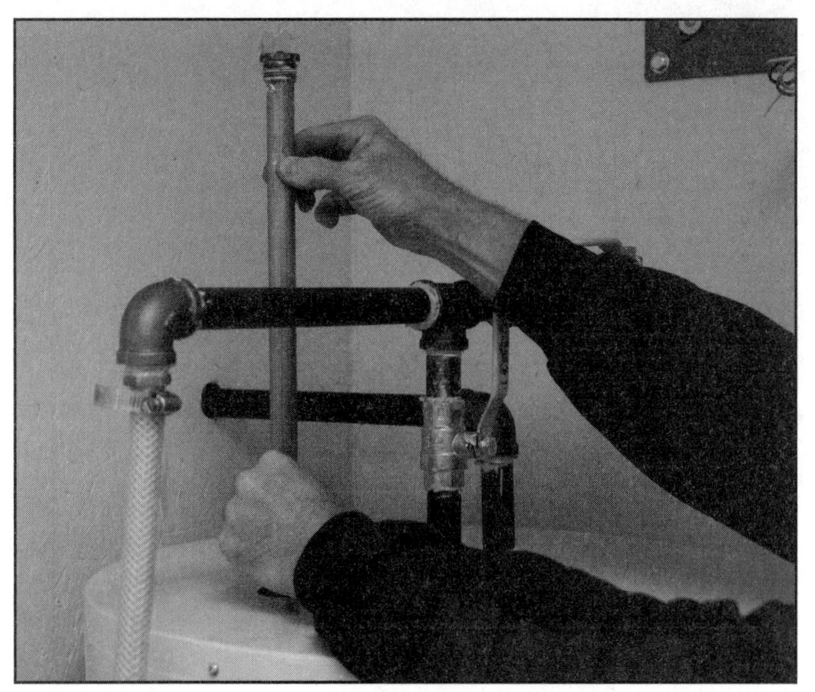

Figure 8.2.1-25. Remove the anode from the tank and seal the fitting with a 3" section of black pipe and a plumbing cap (see Figure 8.2.1-21).

Flexible Plumbing Sections

Some plumbing interconnections, including the "sight glass" connecting the vertical discharge of the pressure circulating pumps to the top of the three water heater tanks, are best fabricated with see-through, reinforced, braided, flexible plastic pipe which is available at most hardware stores.

With a micro-scale processing system, it is difficult to justify the relatively high cost of liquid level sensors or mass strain gauges necessary to detect and record the level of liquids introduced into the tanks. With clear, see-through plastic pipe, level measures can be made at a glance.

However, there is a downside to using plastic piping since it will degrade over time, and if the pipe is not verified for strength and reliability it is possible that serious leakage may occur during processing. For this reason, it is recommended that all flexible pipes be inspected at each batch run and that all plastic pipe and associated fittings be changed at least every year.

Figure 8.2.1-26. Some plumbing interconnections are best fabricated with see-through, reinforced, braided, flexible plastic pipe which is available at most hardware stores.

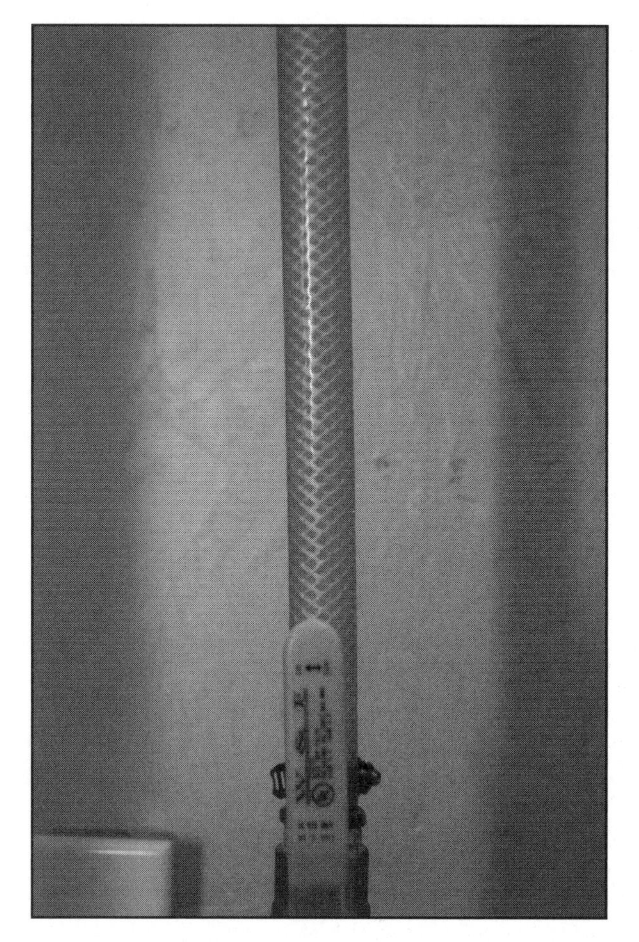

Figure 8.2.1-27. The vertical "sight glass" flexible pipe is used to determine the level of fluids added to the tanks. During commissioning of the system, measurements are made to convert liquid level to volume.

8.2.2 WVO Treatment Unit

The waste oil treatment unit is the first step in the processing of biodiesel. The tank and its associated plumbing components receive the WVO from the supply buckets or tanks, filter it to remove large particulate matter, and act as a large storage area for the oil before the transesterification process begins. Chapter 8.3 explains in detail how the unit functions.

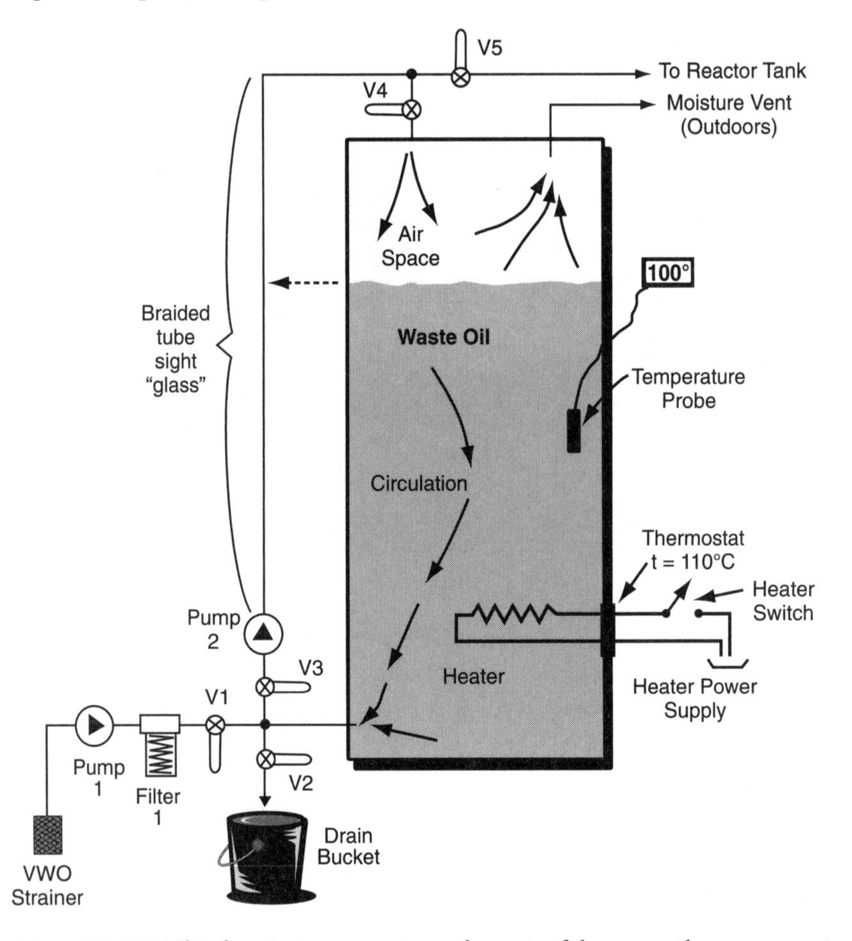

Figure 8.2.2-1. This drawing is an overview schematic of the waste oil treatment unit. The unit is also used for the drying and storage of cold-pressed oilseed oil prior to the transesterification process. Larger-scale biodiesel producers can scale the tank sizes up by using commercially available boiler tanks equipped with either direct or indirect (hot water/steam) systems.

Figure 8.2.2-1 shows the overview schematic of the treatment unit and Figure 8.2.2-2 shows a plumbing schematic of the actual assembly together with photographs that illustrate this basic design. The legend details the components used while the Resource Guide in the Appendices provides specific component supply details.

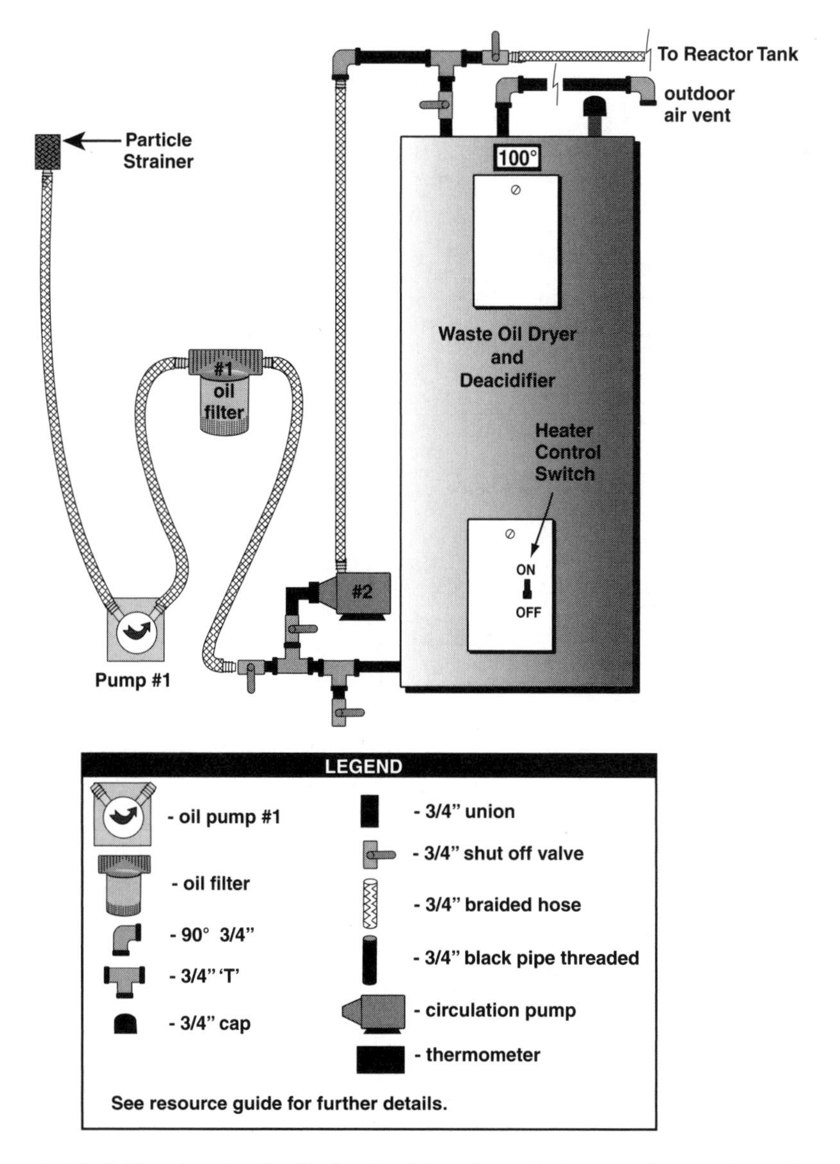

Figure 8.2.2-2. This drawing details the plumbing schematic showing the actual assembly of the WVO treatment unit.

Pump 1 draws WVO through a suction line and strainer assembly. The strainer assembly is sized to filter large waste byproducts such as bone, paper, and other contaminants. Pump 1 is a diesel fuel or oil transfer pump which may be powered directly from the 120-volt utility supply or, more commonly, from a 12-volt direct current source. 12-volt pumps are intended as fuel or oil transfer pumps in boats and transportation facilities. These models can easily be powered from a 40-ampere battery charger supply that steps the 120-volt utility mains to 12 V DC at the proper capacity.

A spin-on 20-micron oil filter is plumbed into the pressurized output pipe of the pump and to filter the majority of suspended particles from the WVO. A large number of ¾" ball valves are used to direct the flow of oil.

Pump 2 is a centrifugal circulation pump that provides sufficient fluid pressure to rapidly circulate fluid through the treatment unit as noted. Although virtually any pump may be used, sealed capacitor-start, capacitor-run models are preferred as they do not contain any centrifugal starting switches that create an electric spark when the motor is starting and stopping.

The tank is fitted with a moisture vent which is plumbed through the wall of the processing facility to the outdoors. This pipe should be angled slightly downward as it exits from the tank, ending in a 90° elbow to prevent rain from entering the vent.

The tank is fitted with a thermometer as described in section 8.2.1.

The lower heating element is connected to the utility mains supply and circuit breaker as described in section 8.2.1. An optional heater switch may be fitted to the tank to help ensure that the heater element is deactivated when filling, draining, and transferring the contents of the tank. The heater element must never be operated when the circulation pump is turned off or when the element is not completely submersed in liquid.

The thermostat of the dryer/treatment tank should be set to just above the boiling point of water (approximately 212°F/110°C) or to the maximum limit of the device, which will vary depending on the manufacturer and the addition of "stop detents" which prevent the dial from exceeding a preset maximum level. Bypassing the internal thermostat is not recommended as this may increase the risk of fire if the system is left unattended.

Figure 8.2.2-3 illustrates the assembly of the treatment tank and shows the 40-ampere battery charger/power supply mounted on the shelf above Pump 1. WVO spin-on Filter 1 is located just below the power supply shelf. The WVO suction pipe is looped around a holder on the left wall. The thermometer can be seen mounted to the top section of the tank at eye

Figure 8.2.2-3. This image illustrates the assembly of the WVO treatment tank and shows the 40-ampere battery charger/power supply mounted on the shelf above Pump 1.

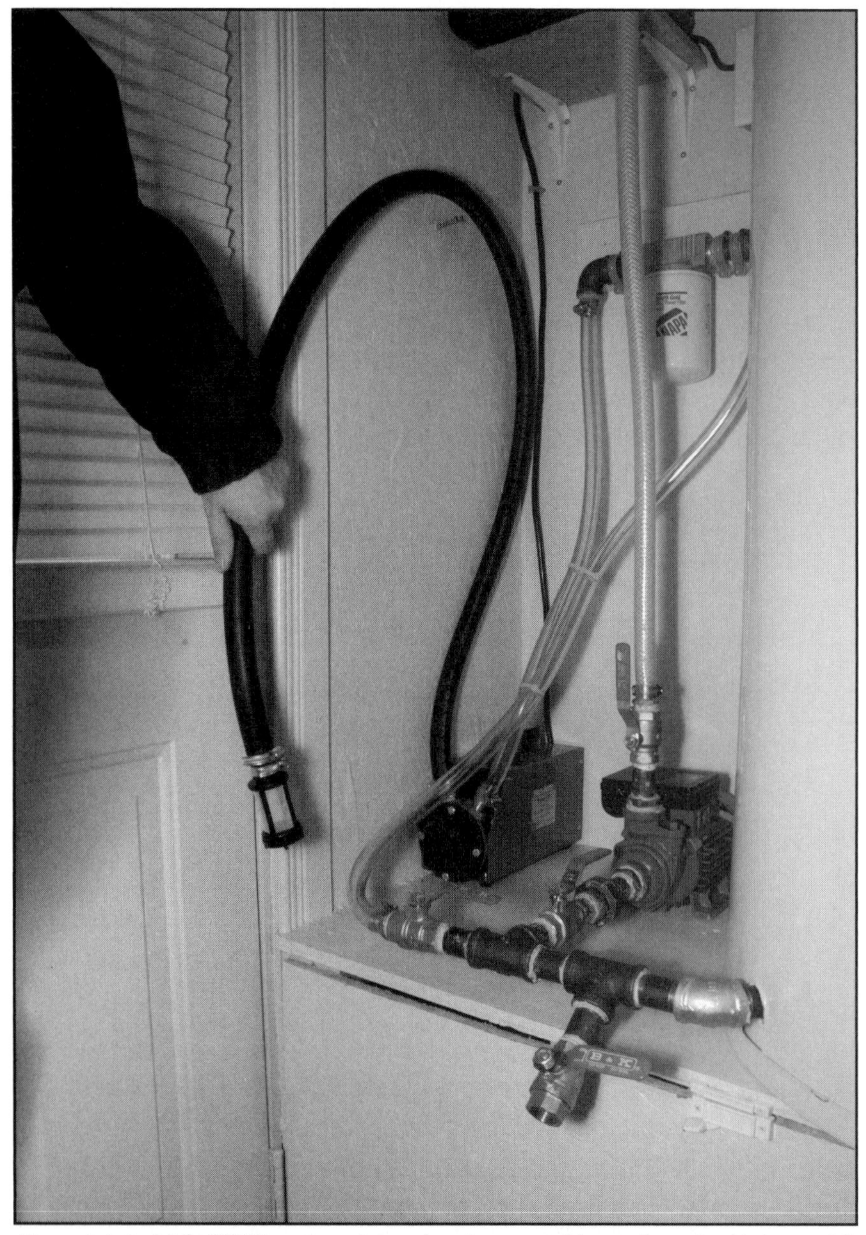

Figure 8.2.2-4. The WVO suction pipe and strainer assembly are shown in this image. The suction line is made long enough to allow the draining of oil from any tanks or buckets that may be provided by the oil supplier. The strainer assembly is intended to filter out only large debris such as napkins, bone, and other large contaminants.

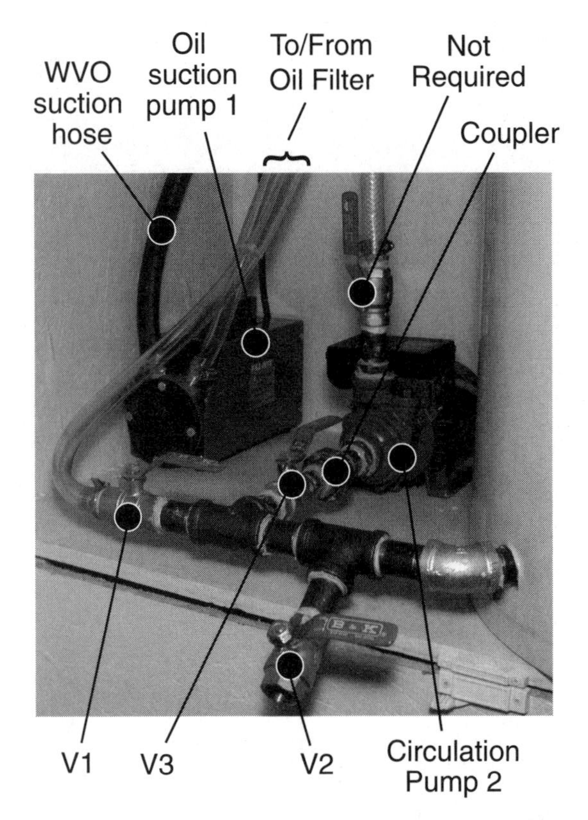

Figure 8.2.2-5. This detailed photograph with legend identifies the various components used in the assembly of the WVO treatment tank. The specific layout of the system will depend on physical space and availability of specific components. Black pipe of ¾" diameter is used extensively in this assembly. Either yellow pipe tape, thread plumbing dope, or both must be used to ensure leak-free joints. Note the use of a coupling fitting at the inlet of the circulation pump. This device facilitates the removal of Pump 2 in the event of failure.

level. The moisture vent is at the top right side of the tank, penetrating the wall of the processing facility. All vent pipes are sealed with latex, silicone, or other suitable outdoor caulking to prevent moisture ingress from rain or snow entering the building cavity.

Figure 8.2.2-4 shows the WVO suction pipe and strainer being lifted from its support. The suction line is made long enough to allow the draining of oil from any tanks or buckets that may be provided by the oil supplier. The strainer assembly is intended to filter out only large debris such as napkins, bone, and other large contaminants.

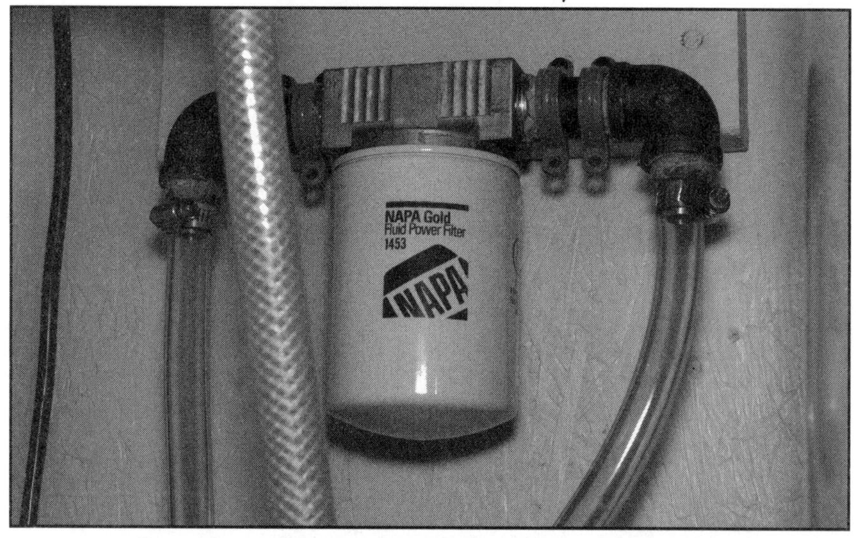

Figure 8.2.2-6. The WVO Filter 1 is shown in this image. The outlet of the WVO Pump 1 is fed into the inlet of a spin-on oil filter casting unit. The outlet of the filter is then fed into the treatment unit. Spin-on filter hubs are available from most fuel supply or auto parts stores. Keep a supply of Filter 1 on hand as heavy accumulation of particulate matter may slow Pump 1 operation and necessitate frequent changing.

Figure 8.2.2-7. After removing the suction pipe and strainer assembly from a container of WVO, a certain amount of "drainback" will occur, resulting in oil dripping in an uncontrolled manner. A small bucket placed under the strainer end of the pipe will ensure that this oil is captured, keeping slippery spills to a minimum.

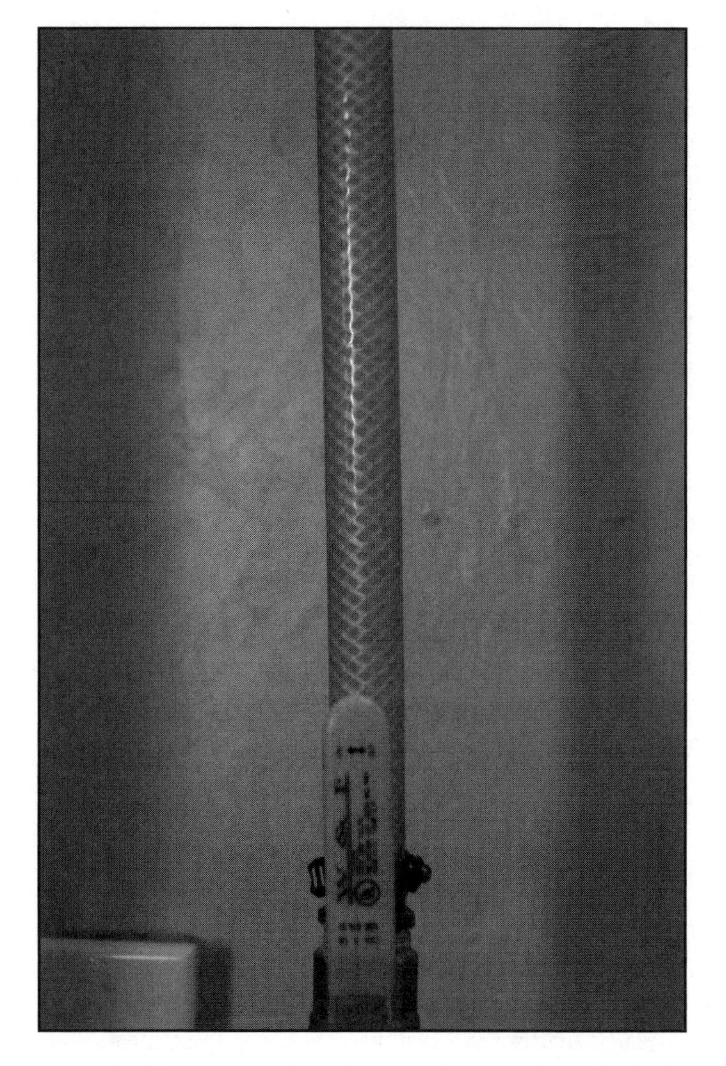

Figure 8.2.2-8. The braided plastic hose shown in this image forms the "sight glass" used to determine the vertical level of liquid in the treatment tank. Both dark-colored WVO and light-colored, cold-pressed oils are visible in the hose, making tank fill measurements very easy. A calibration process completed during system commissioning will allow the conversion of vertical fluid measurement into tank volume.

8.2.3 Biodiesel Reactor Unit with Methanol Recovery

The biodiesel reaction tank uses a construction format similar to that of the waste oil treatment unit described in the previous chapter. Figure 8.2.3-1 illustrates a schematic view of the reaction tank while Figure 8.2.3-2 shows the general assembly of the lower section of the unit and Figure 8.2.3-3 provides a general view of the top section.

The reaction tank is a 60-gallon (227-liter) electric hot water tank fitted with a series of manual ball valves to control the flow of incoming waste vegetable oil (WVO) and sodium (or potassium) methoxide. The tank is configured to allow draining of the raw, contaminated glycerol that is a coproduct of the reaction process and to direct raw biodiesel to the next processing station, the FAME washer. The reaction tank is also equipped with an air-to-liquid heat exchanger configured as a reflux condenser which captures excess methanol vapor, returning it to the methanol storage tank for future use.

The reaction tank is also fitted with a thermometer display and thermostat assembly, allowing the reactants to be heated to the desired operating temperature. The tank requires that the operating temperature be adjusted because the transesterification process operates at a nominal 55°C while the methanol recovery mode operates at a higher 80°C, to ensure complete boiling and vaporization of any excess methanol trapped in the raw biodiesel. The temperature setpoint may be adjusted by rotating the internal thermostat using an insulated shaft screwdriver inserted though a small access hole in the heater element cover plate or by setting the thermostat to the higher "methanol recovery" temperature setpoint and manually switching the heater on and off at the lower setpoint temperature during the transesterification process.

A series of valves designated V1, V2, V3, and V4 controls the flow of liquids in the lower section of the tank. Valve V1 controls the inward flow of sodium methoxide solution into the tank. Valve V2 is the tank drain and is used primarily to drain away contaminated glycerol. Valve V3 is the tank isolation valve. Valve V4 is the sight glass backflow control valve used to stop the flow of fluids trapped in the pump and sight glass during tank draining.

As with the WVO treatment unit, a braided plastic pipe is used to form the sight glass, allowing the rapid calculation of liquid volume from vertical

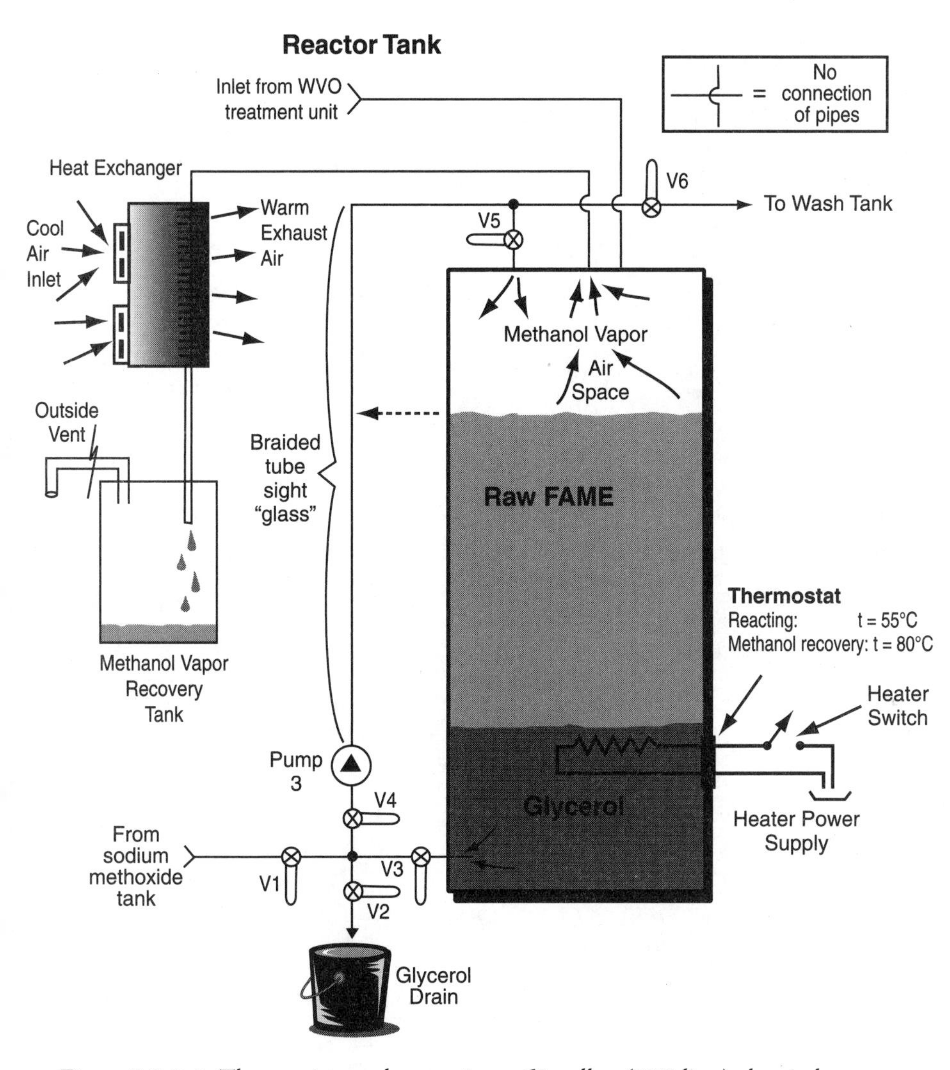

Figure 8.2.3-1. The reaction tank comprises a 60-gallon (227-liter) electric hot water tank fitted with a series of manual ball valves to control the flow of incoming waste vegetable oil (WVO) and sodium (or potassium) methoxide. The tank is configured to allow draining of the raw, contaminated glycerol that is a coproduct of the reaction process and to direct raw biodiesel to the next processing station, the FAME washer.

liquid level in the tank. (A calibration procedure will be described later.) Standard graduation marks are added to the sight glass (Figure 8.2.3-2) to determine appropriate tank fill levels required to charge the tank with WVO and sodium methoxide.

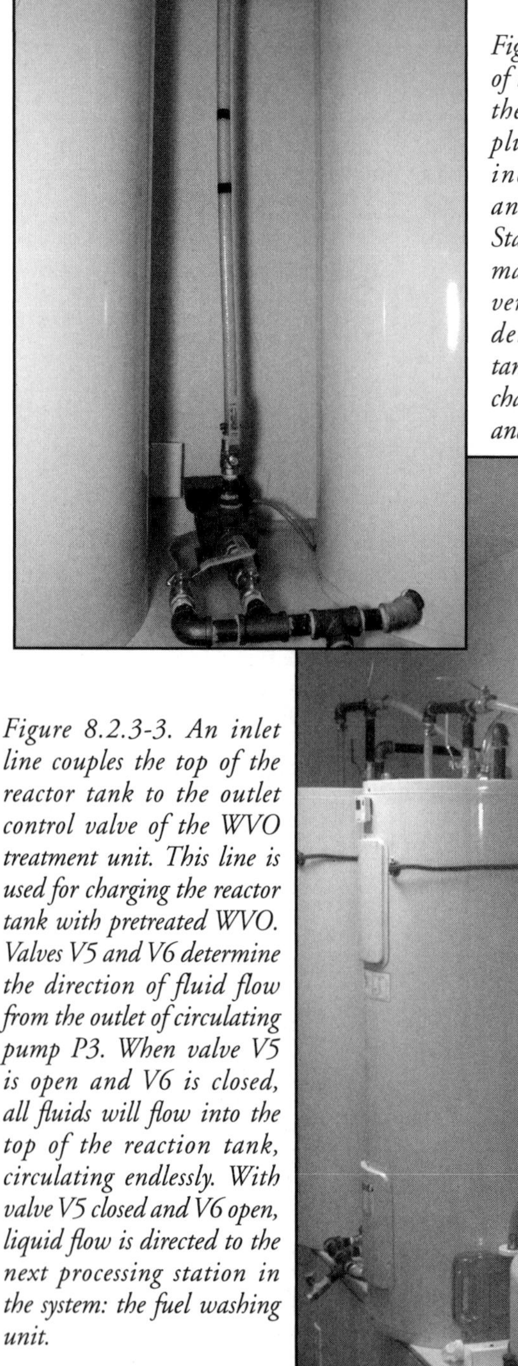

Figure 8.2.3-2. This view of the reaction tank shows the lower section of the plumbing configuration including tank drain and circulation pump. Standardized graduation marks are added to the vertical "sight glass" to determine appropriate tank fill levels required to charge the tank with WVO and sodium methoxide.

Figure 8.2.3-3. An inlet line couples the top of the reactor tank to the outlet control valve of the WVO treatment unit. This line is used for charging the reactor tank with pretreated WVO. Valves V5 and V6 determine the direction of fluid flow from the outlet of circulating pump P3. When valve V5 is open and V6 is closed, all fluids will flow into the top of the reaction tank, circulating endlessly. With valve V5 closed and V6 open, liquid flow is directed to the next processing station in the system: the fuel washing unit.

An inlet line couples the top of the reactor tank to the outlet control valve of the WVO treatment unit. This line is used for charging the reactor tank with pretreated WVO.

Valves V5 and V6 determine the direction of fluid flow from the outlet of circulating pump P3. When valve V5 is open and V6 is closed, all fluids will flow into the top of the reaction tank, circulating endlessly. With valve V5 closed and V6 open, liquid flow is directed to the next processing station in the system: the fuel washing unit.

In order to ensure proper transesterification of WVO into biodiesel, a quantity of methanol is used beyond that required by the basic chemical equilibrium. Excess methanol suspended in the FAME must be recovered to ensure compliance with fuel quality standards and to reduce the complexity of the washing process at the next station.

The simplest, lowest-energy method of recovering excess methanol from the FAME is a simple distillation process undertaken after transesterification of the WVO and removal of the glycerol. At this point the raw biodiesel is likely to be warm from the earlier transesterification process, requiring relatively little energy to "boil off" any excess methanol. Methanol vapor is directed to a heat exchange unit where it is cooled and condenses into a liquid. The liquid methanol then drips into a recovery tank fitted with an outside breather vent. At the end of the recovery process, the liquid methanol is carefully returned to the methanol storage tank.

Figure 8.2.3-4 details the plumbing configuration for the lower side of the reaction tank. Note that the plumbing is angled slightly downward towards the glycerol and tank drain fitting so as to ensure complete draining of the reaction tank as well as the pump cavity and "sight glass." The glycerol and tank drain outlet must be positioned so that a bucket or flexible hose can be added to facilitate draining.

The pipe connected to the "methoxide supply tank" is routed so as to ensure that all liquids drawn from this line are drawn completely into the reaction tank. Depending on the type of circulation pump chosen, it may be necessary to ensure that the exit fitting of the methoxide tank is mounted above the intake level of pump 3 to ensure a flooded suction inlet. Positive displacement pumps such as flexible impeller or gear types that have adequate suction head or the ability to draw liquids upward do not require a flooded suction.

Figure 8.2.3-5 details the plumbing connections at the top of the reaction tank. A short black pipe nipple and 90° fitting connect the WVO treatment unit to the reaction tank.

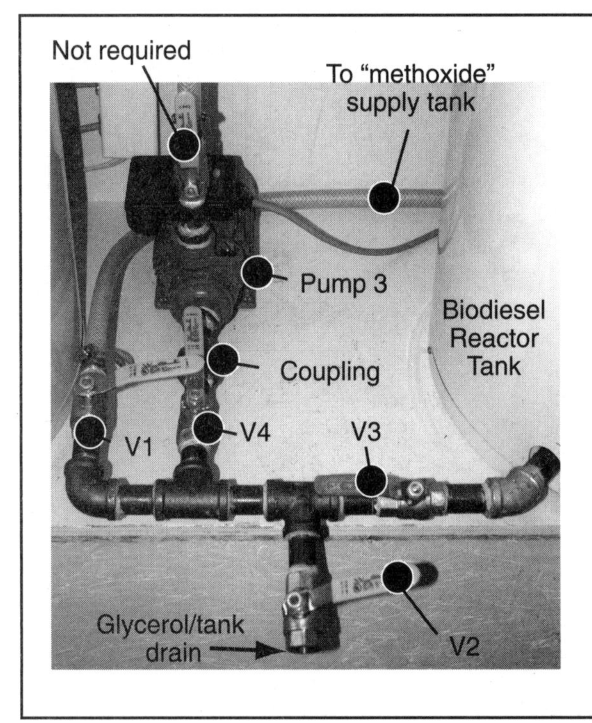

Figure 8.2.3-4. This image details the plumbing configuration for the lower side of the reaction tank. Note that the plumbing is angled slightly downward towards the glycerol and tank drain fitting so as to ensure complete draining of the reaction tank as well as the pump cavity and "sight glass." The glycerol and tank drain outlet must be positioned so that a bucket or flexible hose can be added to facilitate draining.

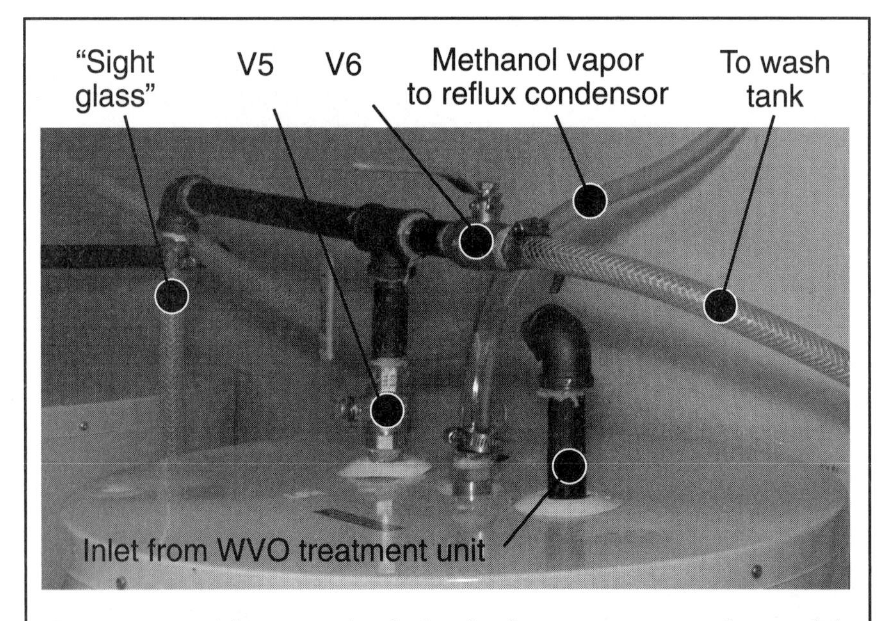

Figure 8.2.3-5. This image details the plumbing connections at the top of the reaction tank.

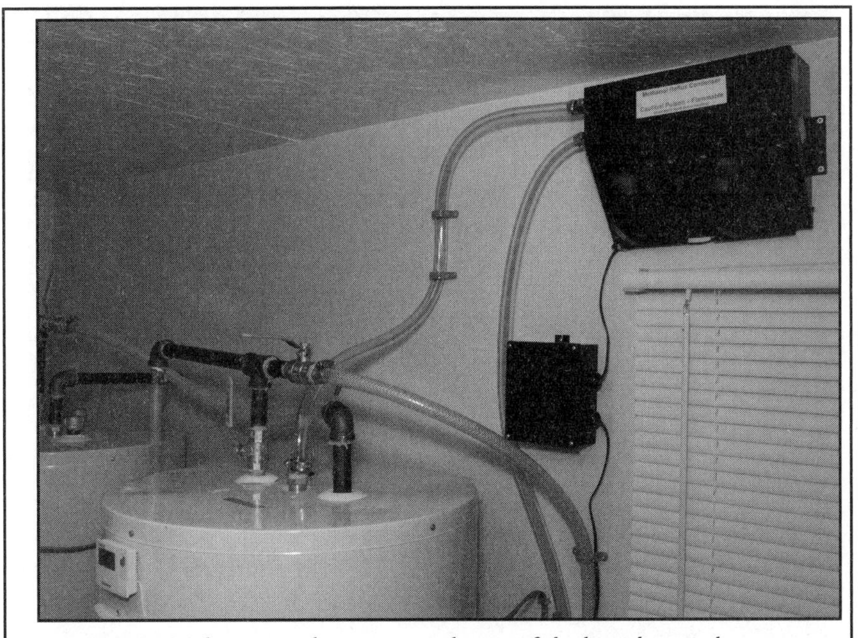

Figure 8.2.3-6. This image shows a general view of the liquid-to-air heat recovery assembly. The heat exchanger is shown mounted above the reaction tank. This reduces space requirements and allows the condensed methanol to drip as a result of the influence of gravity into the recovery tank (not seen in this view).

The "sight glass" flexible pipe is connected to valves V5 and V6 which are used to direct the flow of liquid into the tank during circulation/heating/ transesterification mode of operation or when sending raw FAME to the washing station.

An outlet "barbed" hose fitting connects the air space above the partially filled tank to the inlet of the heat exchange unit which is used for methanol recovery.

The fourth connection on the top of the tank is the sacrificial anode access fitting which is not required and is capped with a short nipple and galvanized cap.

Figure 8.2.3-6 shows a general view of the liquid-to-air heat recovery assembly. The heat exchanger is shown mounted above the reaction tank. This reduces space requirements and allows the condensed methanol to drip as a result of the influence of gravity into the recovery tank (not seen in this view).

The heat exchanger is angled slightly towards the inlet/outlet pipes to ensure that any condensate leaves the heat exchanger and does not pool inside

Figure 8.2.3-7. The heat exchanger unit shown in this view is available from most automotive and recreational vehicle supply stores. Units such as this are often used as auxiliary heating units in many larger vehicles

Figure 8.2.3-8. The heat recovery unit is a radiator core comprising dozens of aluminium fins that are bonded to a copper tube which forms the inlet and outlet connections. Ambient air is pressurized by the two fans and blown through the radiator core which removes heat from the methanol vapor flowing in the copper pipe. As the vapor cools below its boiling point of 64.7°C it condenses, converting back into a liquid before dripping into the recovery tank

the unit. Heated methanol vapor enters the exchanger at the top fitting and exits as liquid at the bottom fitting.

A low-voltage (12 volt at 1 ampere) direct current power supply is mounted directly below the heat exchanger, providing electrical power for the two low-voltage, brushless motor fans mounted to the heat exchanger core (Figure 8.2.3-7). The power supply is connected directly to the same electrical supply as circulation pump P3. This connection ensures that any time pump P3 is activated the heat recovery fans are also enabled. A low-voltage power supply and brushless motor fans are used in order to prevent the possibility of any electrical sparks occurring within the processing facility and in close proximity to any flammable materials, including methanol.

The heat exchanger unit shown in Figures 8.2.3-7 and 8.2.3-8 is available from most automotive and recreational vehicle supply stores. Units such as this are often used as auxiliary heating units in many larger vehicles. The device is a radiator core comprising dozens of aluminum fins that are bonded to a copper tube which forms the inlet and outlet connections. Ambient air is pressurized by the two fans and blown through the radiator core which removes heat from the methanol vapor flowing in the copper pipe. As the vapor cools below its boiling point of 64.7°C it condenses, converting back into a liquid before dripping into the recovery tank (Figure 8.2.3-9).

The methanol recovery tank may have to be fitted with an outdoor vent depending on the effectiveness of the heat recovery system in your application. At high ambient air temperatures methanol vapor recovery is not 100% complete, causing low-pressure buildup that should be vented outdoors.

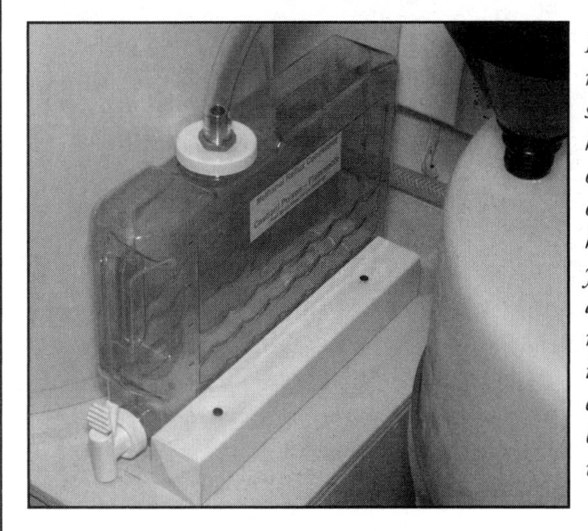

Figure 8.2.3-9. The methanol recovery tank shown in this view may have to be fitted with an outdoor vent depending on the effectiveness of the heat recovery system in your application. At high ambient air temperatures methanol vapor recovery is not 100% complete, causing low-pressure buildup that should be vented outdoors.

An alternative vapor recovery system which is more efficient but more complex is a liquid-to-liquid unit similar to the design used in stills. These units consist of a coiled copper tube which is fitted inside a larger pipe in which cool/cold water is allowed to circulate, cooling and condensing the methanol vapour in the copper tube. Although these systems are more efficient, they require a constant source of running water from either the potable supply in the building or a storage tank and circulating pump.

Larger, high-volume commercial biodiesel production systems require a water-based recirculation cooling system for methanol vapor recovery. This design is fitted with a cooling radiator and fan assembly, similar to that used in an automobile, which is plumbed in series with the recirculating, cooling water. In this configuration, ambient air is forced through the radiator, cooling the circulating water before it enters the water/methanol heat exchanger.

Figure 8.2.3-10 details the air-to-methanol heat exchanger assembly with low-voltage power supply. The inlet and outlet pipes of the heat exchanger have been shortened to simplify the photograph.

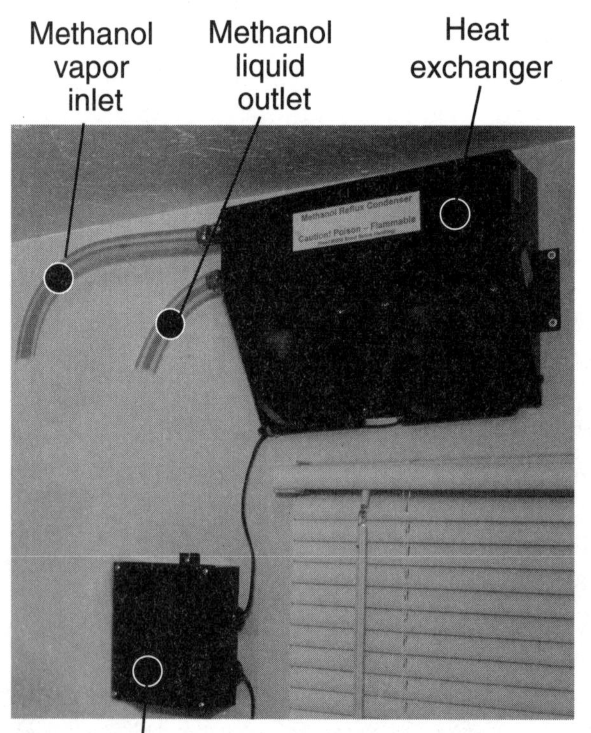

Figure 8.2.3-10. This image details the air-to-methanol heat exchanger assembly with low-voltage power supply. The inlet and outlet pipes of the heat exchanger have been shortened to simplify the photograph.

8.2.4
Sodium Methoxide Processor

In order to start the chemical transesterification of waste vegetable oil into biodiesel, a catalyst solution such as sodium methoxide or potassium methoxide must be added to the oil feedstock. As explained further in Chapter 8.3, a catalyst of either sodium hydroxide or potassium hydroxide may be mixed with an alcohol such as methanol to produce the reactant solution. For simplicity's sake, let's assume that sodium hydroxide is the catalyst of choice until we discuss the pros and cons of using alternative compounds.

Although sodium methoxide can be purchased commercially, micro-scale producers shy away from purchasing the premixed solution, preferring the

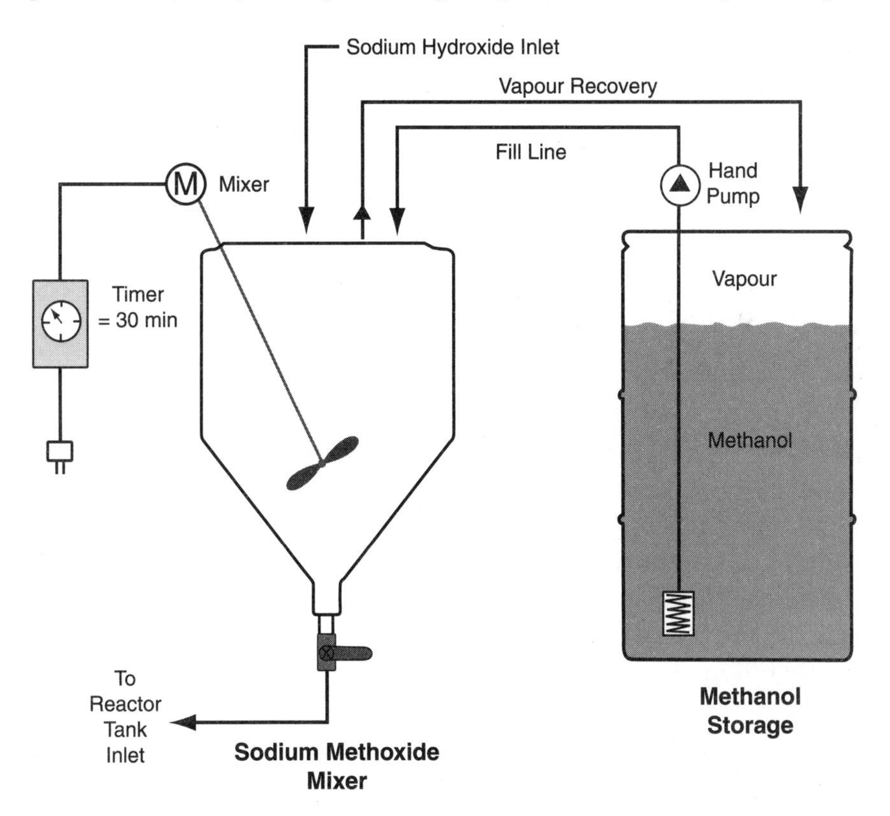

Figure 8.2.4-1. Shown above is a schematic of a relatively simple sodium methoxide mixing system using a 55-gallon (208-liter) drum of methanol supplied by a local fuel company. A second taper chemical mixing tank is equipped with a mixer, a catalyst inlet, and connections for methanol fill and vapor recovery hoses.

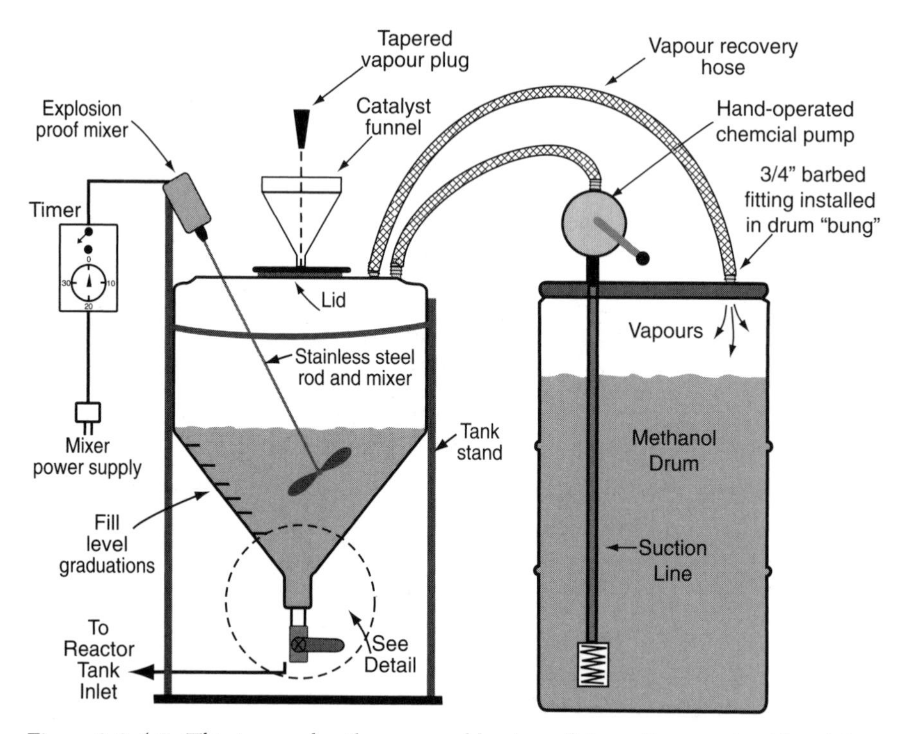

Figure 8.2.4-2. This image details an assembly view of the sodium methoxide mixing system.

less expensive self-mixing method. Purchasing a premixed concentration of sodium methoxide will not be of much assistance in any event, since it's necessary to calculate the amount of catalyst concentration required for each batch of WVO. Commercial biodiesel producers who have access to refined oil feedstock containing very low levels of FFA (preferably below 0.5%, but certainly below 1%) can use the same catalyst concentration at all times.

Varying the concentration of catalyst used in the sodium methoxide solution is made necessary because of the varying level of free fatty acid (FFA) contained in the WVO. Higher concentrations of catalyst are required to neutralize the FFAs and still have sufficient catalyst remaining to complete the transesterification process. If too low a concentration of catalyst is used with high-FFA oils, a failed reaction will result.

A relatively simple sodium methoxide mixing system is shown in a schematic view in Figure 8.2.4-1 and in a detailed construction view in Figure 8.2.4-2. The system comprises a conical mixing tank and a 55-gallon (208-liter) drum of methanol supplied by a local fuel company. Methanol (also known as methyl hydrate) is used in numerous applications such as racing and in high-performance engine fuels and industrial applications. When selecting a methanol supplier, ensure that the methanol is technical grade: 100% pure. Having the methanol delivered in 55-gallon drums is less expensive than buying it in smaller quantities and limits the amount of handling and exposure to this highly toxic, flammable substance.

CAUTION:

Methanol is a highly flammable substance. Methanol as well as sodium hydroxide and potassium hydroxide are very toxic substances. Before handling any of these materials or their containers, plumbing, and fittings, ensure that all necessary safety equipment including eye and face protection, antistatic wrist straps, fire protection, and all other appropriate safety prevention measures are in place and are followed. The reader is urged to review the safety section of this chapter before handling these substances.

The methanol supplier can deliver directly to your site and will charge a nominal barrel deposit fee for the first drum delivered. Each barrel is equipped with two treaded connection ports or "bungs" for the installation of a manually operated chemical hand pump (Figure 8.2.4-9) and a vapor recovery connection.

A 15-gallon (≈60-liter) plastic, conical-bottom chemical mixing tank is mounted in a frame and fitted with several inlet/outlet connections. Two barbed hose fittings are threaded into the shell of the conical tank to provide connection to the methanol storage tank, one for methanol intake and one for vapor recovery. The chemical tank is drilled slightly smaller than the required thread size and a ¾" tapered pipe thread tap is used to create the thread. Teflon pipe tape is wrapped around the threads of the barbed pipe fitting to ensure a gas-tight seal. Stainless steel hose clamps are then used to secure the flexible pipe to the fitting as shown in Figure 8.2.4-15. The methanol tank is equipped with threaded ports when it is delivered by the fuel supplier.

A small hole is drilled in the "top deck" of the conical tank to accept a tapered funnel which will be used as the catalyst hopper, as shown lower left

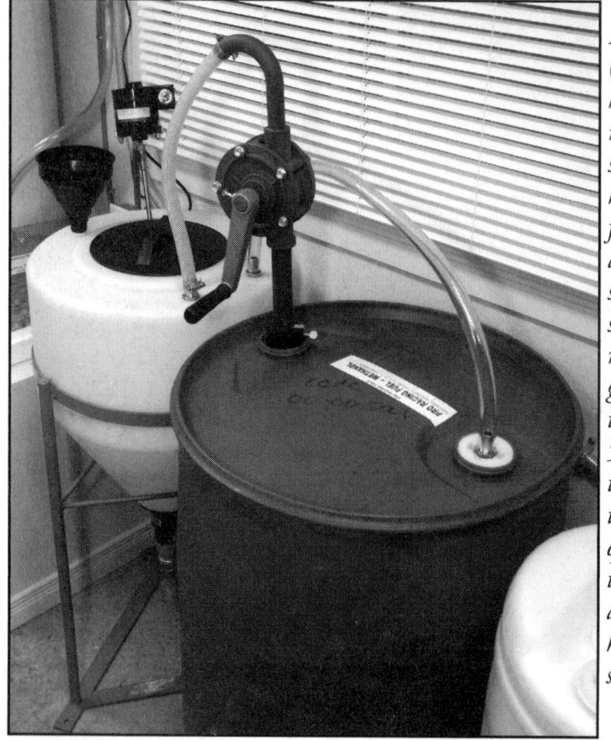

Figure 8.2.4-3. Methanol (also known as methyl hydrate) is used in numerous applications such as racing and in high-performance engine fuels and industrial applications. When selecting a methanol supplier, ensure that the methanol is technical grade: 100% pure. Having the methanol delivered in 55-gallon drums (shown at the right) is less expensive than buying it in smaller quantities and limits the amount of handling and exposure to this highly toxic, flammable substance.

Figure 8.2.4-4. The sodium methoxide chemical mixing station comprises a conical plastic chemical tank (left) and a 55-gallon drum of methanol (right).

Figure 8.2.4-5. This top view of the sodium methoxide mixing unit details the chemical mixer (top left), catalyst hopper (bottom left), methanol fill pipe (bottom right), and methanol vapor recovery pipe (top right). The screw-on tank lid has been removed in this view.

Figure 8.2.4-6. Geoff Shewfelt is shown adding a premeasured amount of catalyst to the sodium methoxide mixing hopper. Note that Geoff is wearing a face shield and rubber gloves for protection. He is also wearing an antistatic electricity suppression strap to prevent accidental ignition of methanol vapors. Safety details will be explored further in section 8.2.10.

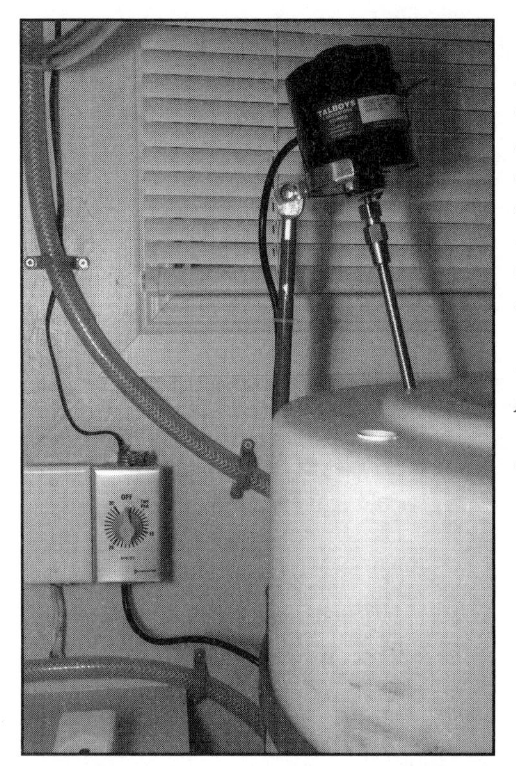

Figure 8.2.4-7. A small chemical mixing unit such as the model shown is installed to facilitate the blending of the catalyst and methanol into a homogenous solution. There are numerous chemical mixing units commercially available through laboratory supply houses. It is very important to ensure that the mixing system is spark free and preferably explosion proof. Using a variable speed drill or similar device equipped with a universal motor (with spark-creating electrical brushes) should not be considered because of the potential ignition of methanol vapors.

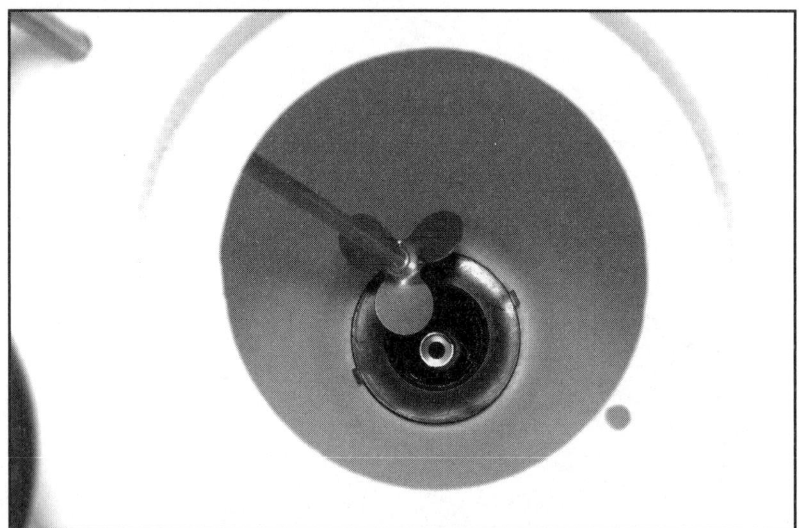

Figure 8.2.4-8. Have a local welder extend one of the chemical tank stand legs to mate with the mixing motor bracket as shown in Figure 8.2.4-7. The mixer can then be angled so that the drive shaft will extend into the tank through a small hole drilled in the tank decking as shown in the detail view above. The mixer rod should extend so that the propeller is centered and as low in the tank as possible without hitting the side walls.

in Figure 8.2.4-5. The funnel fill tube should have an inside diameter of approximately ½" (12.7 mm). The outside of the tube is wrapped with teflon pipe sealing tape and forcefully jammed and silicone sealed into the drilled hole, ensuring a vapor-tight fit. A small rubber stopper is fitted into the inside of the funnel, acting as a vapor stop. The plug is only removed when catalyst is loaded into the mixing tank, after which it is promptly replaced (Figure 8.2.4-6).

A small chemical mixing unit is also installed to facilitate the blending of the catalyst and methanol into a homogenous solution. There are numerous chemical mixing units commercially available through laboratory supply houses. It is very important to ensure that the mixing system is spark free and preferably explosion proof. Using a variable speed drill or similar device equipped with a universal motor (with spark-creating electrical brushes) should not be considered because of the potential ignition of methanol vapors. The model shown in Figure 8.2.4-7 is known as a shaded pole motor. It does not contain any brushes and is the standard model used on commercially available chemical mixing units. This particular model is equipped with an on/off switch on the motor assembly. It is strongly advised that you disable the switch either by placing a jumper wire across the switch terminals or by removing the switch and securing the switch wires together and insulating them with an appropriate size of electrical wire nut. This will prevent the inadvertent use of the switch and eliminate the potential switch spark from igniting methanol vapors that may accumulate in the area.

The chemical mixer motor will come equipped with a bracket for mounting it on a laboratory counter stand. Have a local welder extend one of the chemical tank stand legs to mate with the motor bracket as shown in Figure 8.2.4-7. The mixer can then be angled so that the drive shaft will extend into the tank through a small hole drilled in the tank decking as shown in the detail view in Figure 8.2.4-8. The mixer rod should extend so that the propeller is centered and as low in the tank as possible without hitting the side walls.

It is very difficult to make the shaft hole in the tank tight enough to prevent vapor leakage while still allowing the shaft to rotate. The shaft hole should therefore be carefully drilled on the same angle as the final position of the motor shaft, minimizing the diameter so that it just accommodates the shaft. The hole can be lubricated with a small amount of lithium grease with care being taken to ensure that none enters the tank.

The power supply cable of the mixer motor is connected to a 30-minute mechanical wall timer available at any hardware store. The wall timer is in

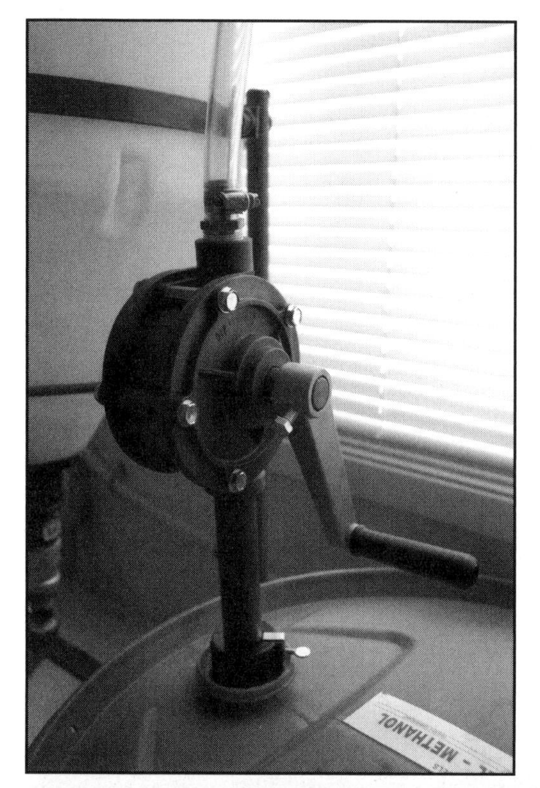

Figure 8.2.4-9. Methanol barrels are equipped with two treaded connection ports or "bungs" for the installation of a manually operated chemical hand pump such as the model shown above.

Figure 8.2.4-10. The second bung on the methanol barrel is fitted with a ¾" barbed pipe fitting and a flexible plastic pipe which connects the mixing tank and methanol storage tank, creating one part of the methanol vapor recovery system.

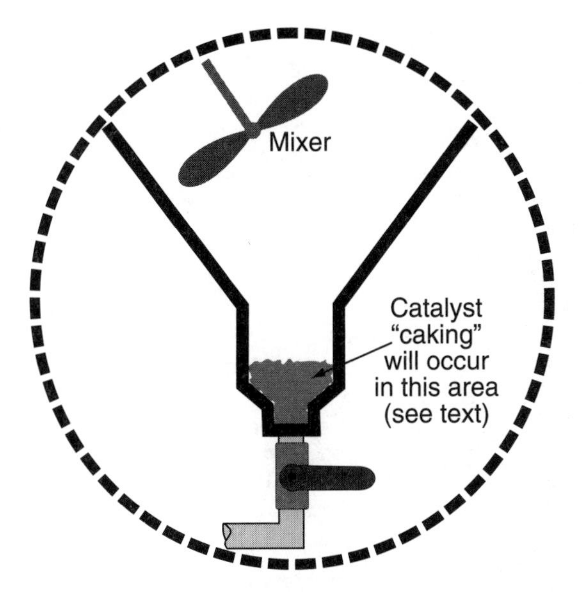

Figure 8.2.4-11. A major problem in attempting to create sodium methoxide results from the fact that sodium hydroxide catalyst does not readily dissolve in methanol and will tend to "cake" in the bottom of the mixing tank.

turn connected to a source of electrical power either from a wall socket connection or directly from the electrical supply panel.

A major problem in attempting to create sodium methoxide results from the fact that sodium hydroxide catalyst does not readily dissolve in methanol and will tend to "cake" in the bottom of the mixing tank (Figure 8.2.4-11). This problem is common in all mixing systems and may result in plugging of the plumbing lines. (One suggestion by home biodiesel producers is to put catalyst and methanol in plastic jugs and roll them on the ground until the mixture is homogenous. For obvious safety and environmental spill reasons this is not recommended.)

One solution to this problem is to obtain a rubber stopper that will fit into the bottom drain assembly of the conical tank. Depending on the tank dimensions, stoppers can be purchased from a laboratory supply company or hardware store or fabricated by a local machine shop from rubber or EPDM rod stock material. The stopper is fitted with a stainless steel "aircraft" cable and pulled through a small access hole in the tank top deck. A loop of cable held in place with a cable compression clamp (Figure 8.2.4-14) forms a release handle. A dowel can be placed through the loop to facilitate additional grip to remove stoppers that must be inserted with excess vigour.

Figure 8.2.4-12. One solution to having sodium hydroxide catalyst "cake" in the bottom of the mixing tank is to obtain a rubber stopper that will fit into the bottom drain assembly. Depending on the tank dimensions, stoppers can be purchased from a laboratory supply company or hardware store or fabricated by a local machine shop from rubber or EPDM rod stock material.

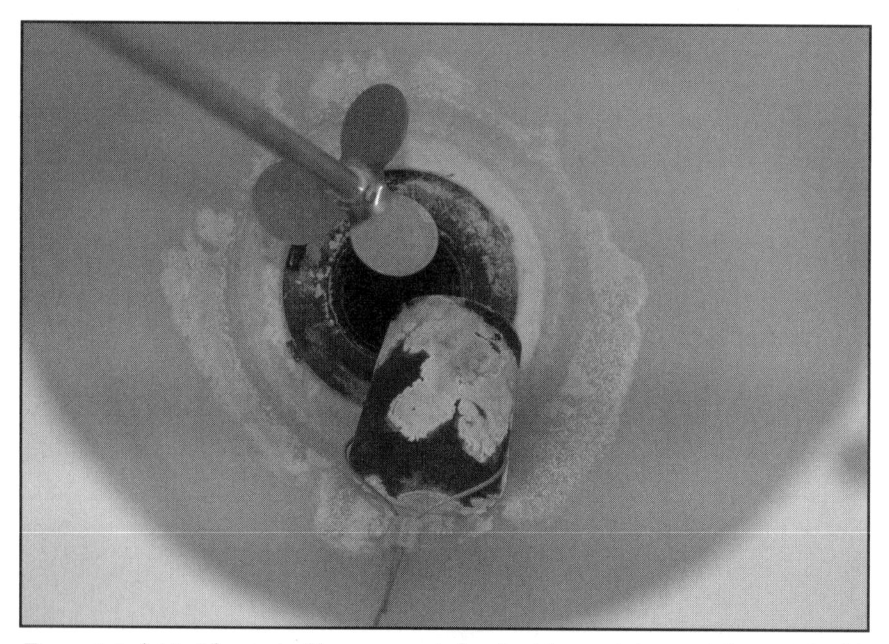

Figure 8.2.4-13. The anticaking stopper is fitted with a stainless steel "aircraft" cable and pulled through a small access hole in the tank top deck. Note the crystalline sodium methoxide residue in the bottom of the tank. ***See MSDS Saftey Data regarding this substance.***

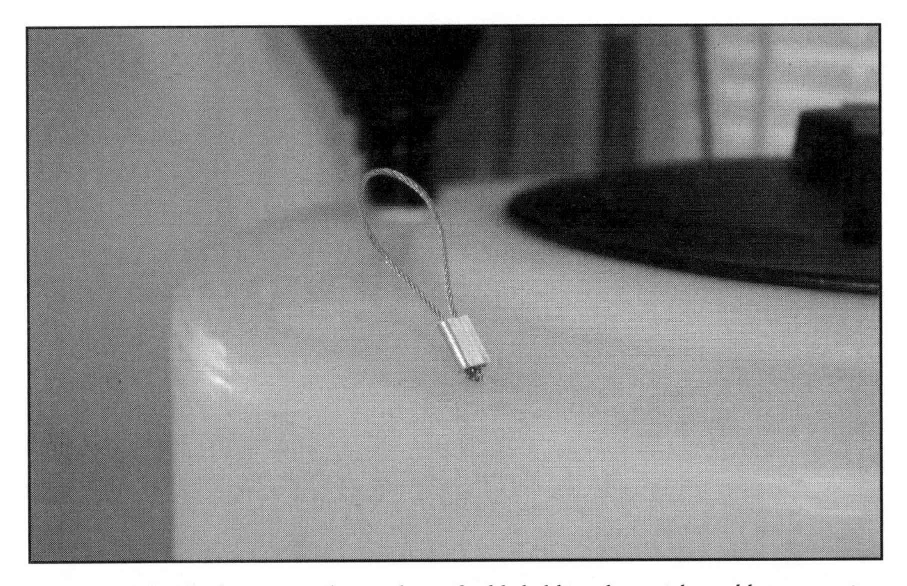

Figure 8.2.4-14. This picture shows a loop of cable held in place with a cable compression clamp, forming a release handle for the anticaking stopper. A dowel can be placed through the loop to facilitate additional grip to remove stoppers that must be inserted with excess vigour.

Care should be taken to ensure that the steel cable cannot become entrapped in the mixer propeller when the drain is fitted in place. This can be done by ensuring that the cable remains fairly taut when the stopper is in the plugged position.

An alternative method **may be** to use a large-diameter ball valve at the bottom of the tank plumbing to create the same effect as the rubber stopper. However, my experimentation has shown that this does not work as the neck of the tank discharge area tends to be very narrow and the catalyst will cake here regardless of any measures I have developed thus far. A further concern is that sodium methoxide is milk-like in color, making it very difficult to see if any caking has occurred. Once caking starts it is nearly impossible to remove the plug of catalyst without severe risk through exposure to methanol, catalyst, and sodium methoxide solution.

Examining Figure 8.2.4-13 you will see a white powder on the bottom of the tank and stopper, which is crystallized sodium methoxide. The crystallized material is just as dangerous as the aqueous compound and should be treated with care, with all necessary safety equipment in use during handling.

Testing the Mixing Tank

Prior to connecting the sodium methoxide mixing tank to the reactor tank it is advisable to perform an assembly test to ensure that there are no fluid leaks, that the mixing unit performs as expected, and that the liquid level graduation marks embedded into the tank are accurate (Figure 8.2.4-2).

Vacuum the entire tank and internal components and wipe them down with a damp cloth, ensuring that there are no traces of plastic, pipe tape, or materials from the fabrication stages.

Insert the tank stopper into the drain fitting and fill a bucket with a known amount of water and pour it into the tank. Verify that the fill level graduations located on the outside of the tank match the volume of water poured into the tank. If they are incorrect, make new graduation marks by scribing marks into the tank and filling them with black permanent marker to improve visibility. Graduation marks should occur for every 0.5 gallons (≈2 liters).

Fill the tank to a minimum of 5 gallons (≈20 liters). Start the mixing pump by operating the timer and verify that the mixer operates smoothly without undue rubbing on the tank top and that the water mixing is vigorous throughout the entire volume of the liquid. If it is not, upgrade the mixer propeller and/or motor as necessary.

When the tests have been completed, drain the tank and carefully dry all internal components. It is advisable to remove the tank lid and allow the system to "air out" for a couple of days to ensure that all traces of water have evaporated. Even small amounts of water in the alcohol can ruin the biodiesel reaction process.

Final Assembly

When fabrication and testing of the chemical mixing tank are completed, the outlet port of the tank can be fitted with a barbed hose fitting to allow connection to the inlet port of the reaction tank as described in Chapter 8.2.3. It is important to complete this connection last in order to ensure that any materials and water from the preliminary tests do not remain in the system.

As noted in Chapter 8.2.3, it may be necessary to elevate the sodium methoxide mixer so that the discharge port at the bottom of the mixing tank is above the inlet port of the reaction tank circulation pump. If there is any doubt as to the ability of the circulation pump to draw the sodium methoxide solution into the reaction tank, extend the WVO treatment unit and reaction tank frame to include the sodium methoxide mixer unit.

Figure 8.2.4-15. Two barbed hose fittings are threaded into the shell of the conical tank to provide connection to the methanol storage tank, one for methanol intake and one for vapor recovery. Stainless steel hose clamps are then used to secure the flexible pipe to the fitting as shown above. The methanol tank is equipped with threaded ports when it is delivered by the fuel supplier.

8.2.5 Fuel Washing Unit (Misting and Bubble Washing)

CAUTION!
Methanol recovery must be completed before water washing can take place.

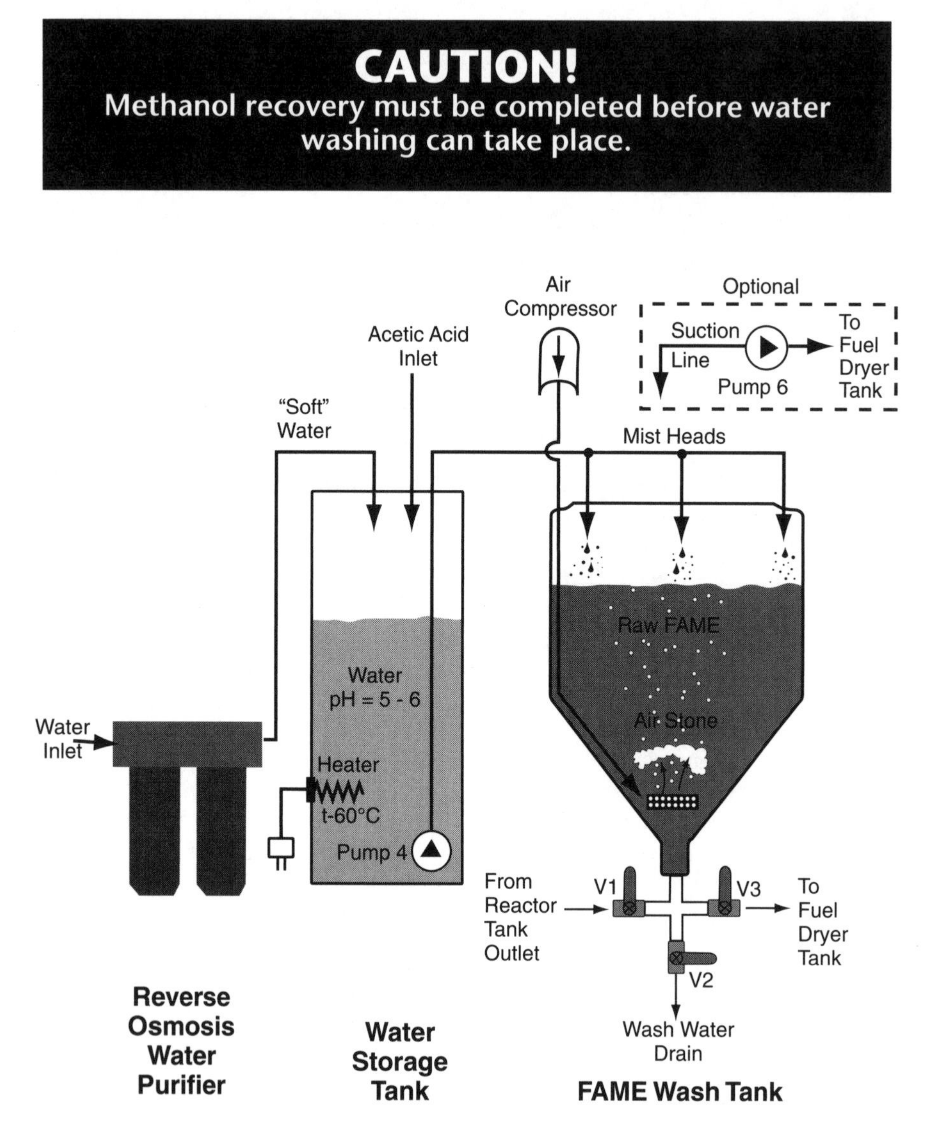

Figure 8.2.5-1. Upon completion of glycerol/biodiesel phase separation and methanol recovery, the raw biodiesel is pumped to the fuel washing unit shown in this schematic diagram.

Upon completion of glycerol/biodiesel phase separation and methanol recovery, the raw biodiesel is pumped to the fuel washing unit shown in the schematic diagram in Figure 8.2.5-1.

A chemical mixing tank is fitted with a series of drain valves equipped to allow the inflow of raw biodiesel (V1) as well as the outflow of wash water (V2) and washed biodiesel (V3). The large-capacity (80-gallon/≈300-liter) chemical mixing tank is mounted in a support frame and elevated on a 12" (30 cm) high pedestal as shown in Figure 8.2.5-3. The elevation of the wash tank facilitates the draining of wash water from valve V2.

The washing process begins with an optional reverse osmosis water softener/purifier which supplies water to a small (20-gallon/75-liter) water heater tank which is seen to the left of the washing tank in Figure 8.2.5-2. (If local water conditions supply relatively soft water with few dissolved minerals the water softener may not be required. A simple test is to look at the inside of the household teakettle to see if minerals have built up on the kettle walls. If none are present the household water supply is acceptable for biodiesel washing without a water softener). In the configuration shown, a small HDPE plastic tank has been used, although a traditional cottage-size (or larger) electric water heater tank can also be used.

The water heater tank is equipped with an internal heating element that raises the water temperature to 120°F (50°C). An inlet port on the water heater tank is used to inject acetic acid (white vinegar) into the wash water to increase wash water acidity. A submersible or external pressure pump is provided that supplies water at a minimum pressure of 20 psi (≈140 kPa).

Popular texts typically describe two washing procedures: mist and bubble washing. The tank of biodiesel shown in Figure 8.2.5-4 is shown undergoing both procedures simultaneously, although for reasons that will be described in Chapter 8.3 mist washing is the preferred method.

As the name implies, mist washing requires the application of a mist of wash water onto the biodiesel surface. Figure 8.2.5-4 shows a series of four spray heads that have been installed around the perimeter of the wash tank lid area. These spray nozzles are fed by a submersible wash water pressure pump (Figure 8.2.5-5) which is placed in the water tank. An externally mounted pressure pump could also be installed with a standard electric water heater.

Figure 8.2.5-2. The fuel washing station is equipped with a wash water/heater tank shown left and a chemical mixing tank fitted with mist and/or bubble washing technologies.

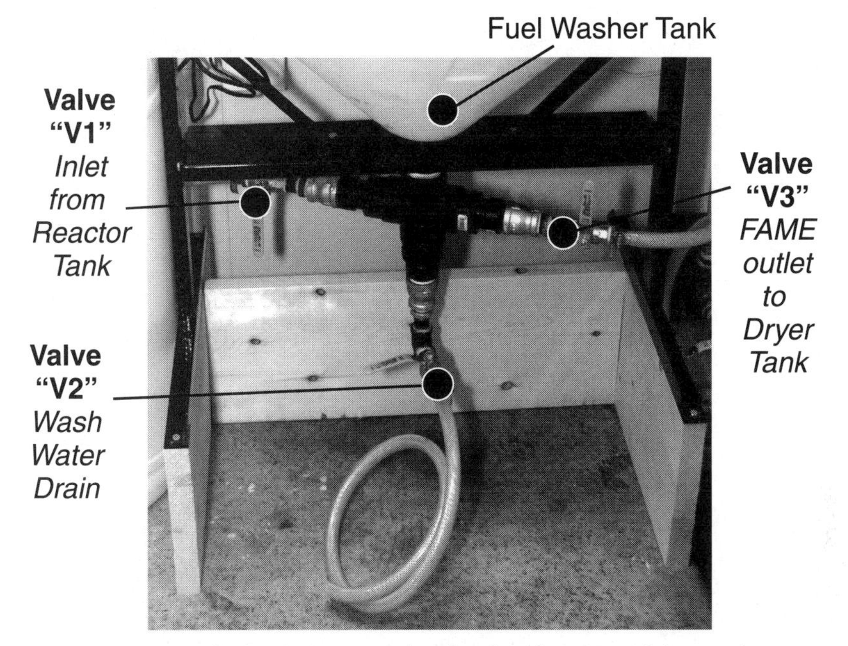

Figure 8.2.5-3. A chemical mixing tank is fitted with a series of drain valves equipped to allow the inflow of raw biodiesel (V1) as well as the outflow of wash water (V2) and washed biodiesel (V3).

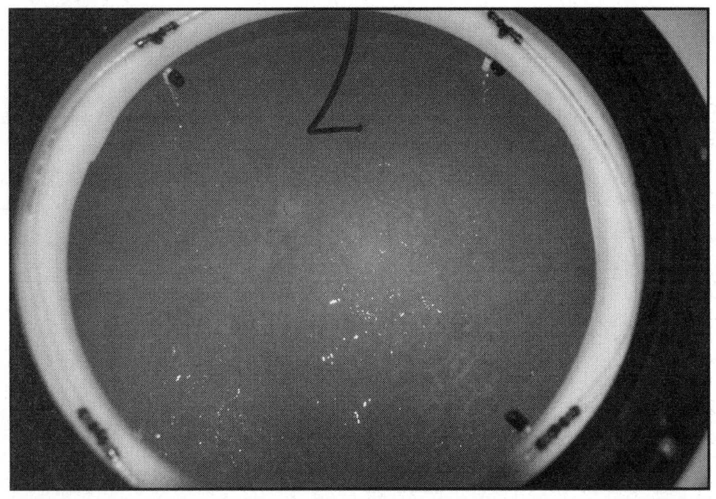

Figure 8.2.5-4. In this view, biodiesel is shown being both mist and bubble washed simultaneously. In either case, the concept is to use slightly softened (chemically acidic) water to wash impurities out of the biodiesel. As water has a higher density than biodiesel it sinks to the bottom of the wash tank where it can be drained off and processed for disposal. **CAUTION! Methanol recovery must be completed before water washing can take place.**

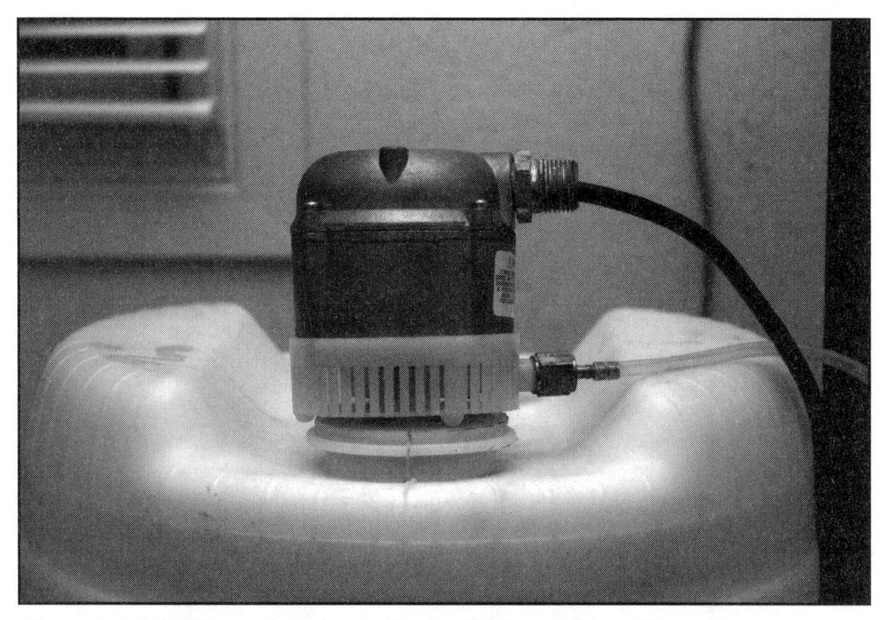

Figure 8.2.5-5. The wash spray nozzles are fed by a submersible wash water pressure pump which is placed in the water tank. An externally mounted pressure pump could also be installed with a standard electric water heater.

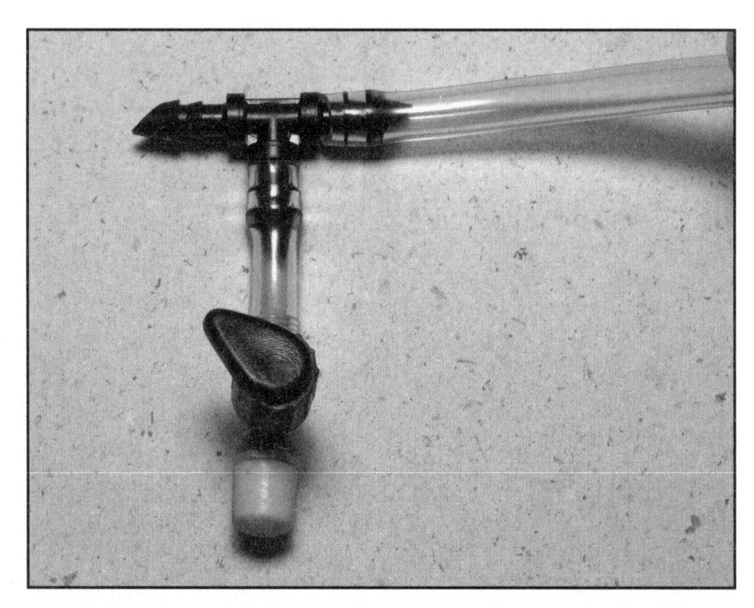

Figure 8.2.5-6. This photograph shows a close-up view of a spray nozzle of the type used for automatic plant watering. These nozzles and the associated fittings and pipe are available from most hardware stores including Home Depot.

As mentioned earlier, the only important considerations for the wash water station are that it:

- contain a minimum of 20 gallons (75 liters) of softened water for each 50 gallons (190 liters) of biodiesel;
- inject a measured amount of acetic acid (white vinegar) into the wash water;
- heat the wash water to a minimum of 120°F (50°C);
- supply the water at a pressure of 20 psi (≈140 kPa) with a flow rate of approximately 1 gallon (3.8 liters) per minute.

Figure 8.2.5-6 shows a close-up view of a spray nozzle of the type used for automatic plant watering. These nozzles and the associated fittings and pipe are available from most hardware stores including Home Depot. The nozzles are equipped with small control valves which allow the water spray to be adjusted to provide a slow, steady, and even pattern, ensuring that the four nozzles working together cover the entire biodiesel surface area with wash water droplets.

The nozzles are equipped with a barbed hose fitting and are connected to a "T" fitting, creating a continuous ring around the neck of the wash tank as shown in Figure 8.2.5-7.

Figure 8.2.5-7. These nozzles are equipped with a barbed hose fitting and are shown connected to a "T" fitting, creating a continuous ring around the neck of the wash tank.

Figure 8.2.5-8. An alternative to mist washing is a technique known as bubble washing. Softened water from the wash water heater is added to the raw biodiesel and then air is gently blown through the lower water layer, carrying a stream of air bubbles and a minute film of water through the biodiesel layer. An aquarium air compressor such as this model is required.

Figure 8.2.5-9. The air compressor is mounted above the flood level of the wash tank, creating a safety barrier by preventing biodiesel or water from being accidentally forced into the compressor.

An alternative to mist washing is a technique known as bubble washing. Softened water from the wash water heater is added to the raw biodiesel and then air is gently blown through the lower water layer, carrying a stream of air bubbles and a minute film of water through the biodiesel layer. When the air bubbles break at the surface of the biodiesel the water film drops to the bottom of the tank, carrying impurities from the biodiesel layer.

Bubble washing begins with a large aquarium air compressor such as the model shown in Figure 8.2.5-8. This model is equipped with a variable air flow control knob which allows adjustment of the air pressure. Referring back to Figure 8.2.5-2, the air compressor is mounted above the flood level of the wash tank, creating a safety barrier by preventing biodiesel or water from being accidentally forced into the compressor. The detail view in Figure 8.2.5-9 shows the compressor mounted next to an electrical switch which is used to control the power supply.

A large aquarium air stone such as the model shown in Figure 8.2.5-10 is connected by a length of air supply hose to the air compressor. After the wash tank is filled with biodiesel and 20 gallons (75 liters) of wash water, the air stone is lowered into the tank and the compressor is started, creating the "boiling appearance" on the surface of the biodiesel (Figure 8.2.5-4).

Figure 8.2.5-10. A large aquarium air stone such as the model shown is connected by a length of air supply hose to the air compressor. After the wash tank is filled with biodiesel and 20 gallons (75 liters) of wash water, the air stone is lowered into the tank and the compressor is started, creating the "boiling appearance" on the surface of the biodiesel (Figure 8.2.5-4).

Figure 8.2.5-11. Even the best micro-scale biodiesel producers will experience the formation of emulsion and soaps during the biodiesel washing stage. This is generally an avoidable situation, but when it occurs it can make separating the wash water from the biodiesel a tricky affair. An optional but simple solution is the addition of a top-mounted suction pump at the biodiesel washing station to assist when soap emulsion forms.

Even the best micro-scale biodiesel producers will experience the formation of emulsion and soaps during the biodiesel washing stage. Although this is generally an avoidable situation, it can make separating the wash water from the biodiesel a tricky affair.

An optional but simple solution is the addition of a top-mounted suction pump at the biodiesel washing station to assist when soap emulsion forms (Figures 8.2.5-11 and 8.2.5-12). This flexible impeller positive displacement pump is configured to draw biodiesel from the top of the wash tank, thereby avoiding the soap and emulsion layer, and then feed the washed fuel to the dryer stage. The pump is a 120-volt design which is available from most hardware and automotive stores.

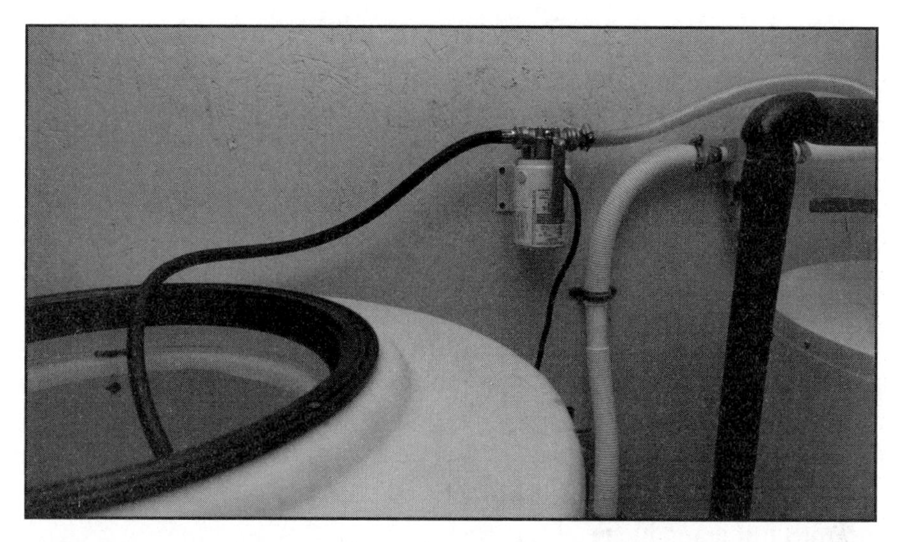

Figure 8.2.5-12. The top-mounted suction pump draws biodiesel from the top of the tank, feeding it into the dryer tank, leaving soap, emulsion, and water behind.

8.2.6
Fuel Dryer and Final Processing Unit

Once you have washed the biodiesel and drained the wash water the next stage is to pump the biodiesel from the washing unit into the fuel dryer. Although water washing removes impurities from the fuel, it leaves high levels of water suspended in the biodiesel. The fuel dryer will remove the water from the biodiesel as well as provide a final filtration before the fuel is sent for storage and testing. Removal of water is required to ensure fuel storage stability and prevent oxidation (rusting) of engine components.

Figure 8.2.6-1 is a schematic view of the fuel dryer unit, which uses the same 60-gallon water heater and circulation pump design used in the WVO treatment unit and biodiesel reactor.

Biodiesel from the fuel washing unit is directed to the inlet of circulation pump P5 and pressurized. It then flows to the junction of valves V3 and V4. During fuel drying mode, valve V3 is open and V4 is closed, causing biodiesel to spray into the top of the tank. The tank contains approximately 400 small pieces of stainless steel tubing that have been cut into 2" (50 mm) tubes and inserted through a tank inlet fitting. The biodiesel drains downward over the labyrinth of tubes and is drawn back into pump P5 for recirculation.

A ½" (12.7 mm) stainless steel tube (referred to as the "air inlet") is inserted into the tank so that it is approximately 6" (15 cm) above the bottom of the tank. (A machine shop or welder is required to make an adapter to allow the pipe to drop into the tank and be secured with a standard ¾" pipe thread fitting at the top of the tank.)

The air inlet is connected to a flexible hose which is in turn connected to an air filter and heated air blower unit of the type used in hot tubs, spas, and air blower baths. These air blowers contain a fan unit similar to those used in vacuum cleaners and are also equipped with a small electric heating element mounted in the output air stream. When connected to an industrial (1,000 watt) light dimmer switch, the fan and heater may be operated at variable speeds and temperatures. The heated air is pressurized by the blower and fed through the air filter, which captures any dust particles from the blower air intake and electrical commutator brushes.

The heated air is then blown down the fuel dryer air inlet and forced back up through the array of small tubes while biodiesel is flowing in the opposite, downward direction, passing over the large surface area of the

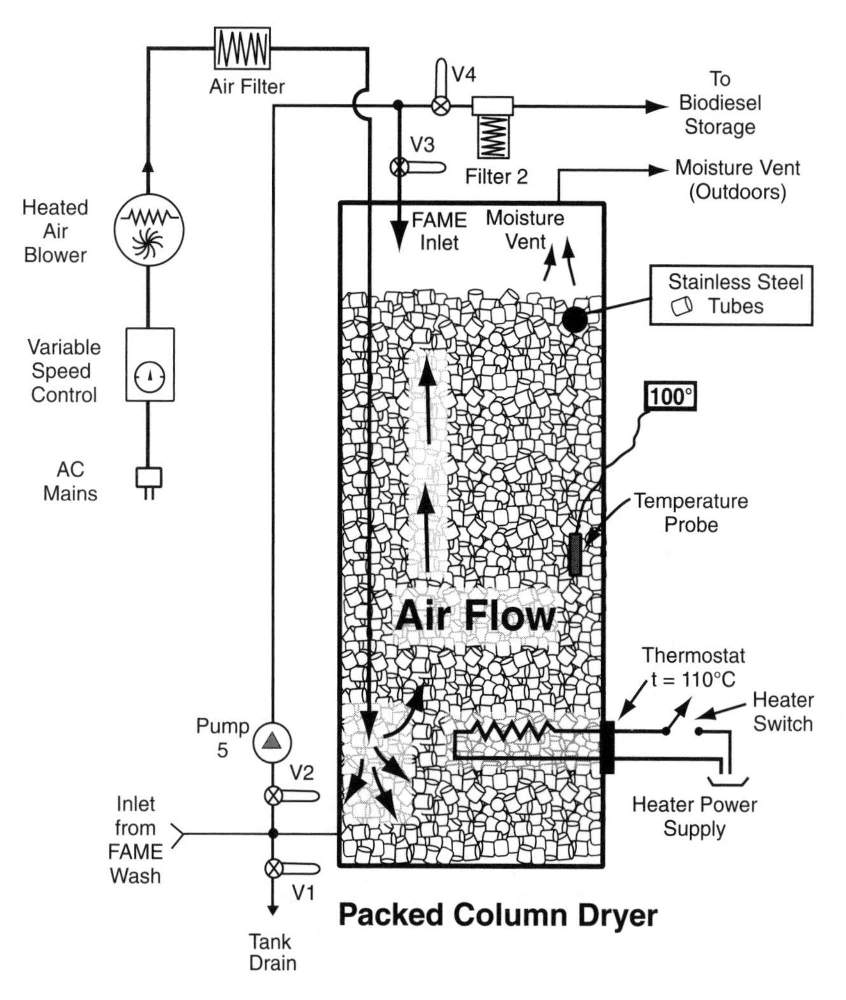

Figure 8.2.6-1. After you have washed the biodiesel and drained the wash water the next stage is to pump the biodiesel from the washing unit into the fuel dryer. Although water washing removes impurities from the fuel it leaves high levels of water suspended in the biodiesel. The fuel dryer will remove the water from the biodiesel as well as provide a final filtration before the fuel is sent for storage and testing.

stainless steel tubes. As the counter-current flow of heated air passes through the biodiesel, moisture is picked up and vented to the outdoors through the moisture vent.

To facilitate the operation of the fuel dryer, an electric heating element heats the biodiesel, preferably to above the boiling point of water (212°F/100°C).

Although this system may appear fairly complex, it is important to realize that many hobbyists attempt to dry their biodiesel by spraying or atomizing the fuel near a source of heat. Unfortunately, it is well known that any substance with a high ignition or flash point (such as biodiesel which is generally not flammable) can become explosive at fairly low temperatures if subjected to atomization. The process described here is known to be safe and is used as a dryer system in many chemical process technologies.

Once the fuel has been subjected to drying for a sufficient period of time, it is ready to be transferred to storage and tested for compliance with fuel quality standards. The transfer is started by switching off the tank heating element and air blower, opening valve V4, and closing valve V3, whereupon biodiesel will be pumped through Filter 2 and into the storage tanks.

The construction of the fuel dryer unit is shown in Figure 8.2.6-5. The unit uses a standard 60-gallon (227-liter) electric hot water tank and the same basic plumbing and circulation system used in the WVO treatment unit and biodiesel reactor tank.

Note that the dryer tank is elevated a few inches above the floor to facilitate draining of the tank in the event of contamination or other problems. During normal operation the bottom drain should not be required.

Figure 8.2.6-2. Dr. Marc Dube, P.Eng., Associate Professor at the University of Ottawa Chemical Engineering Department, is shown demonstrating a commercial counter-current flow packed column drying unit, on which the design of the fuel drying unit is based.

Figure 8.2.6-3. There are numerous designs of packed column dryers using different shapes and sizes of materials to increase the surface area of the fluid to be dried. Although stainless steel tubes have been suggested, it is possible to use other materials that do not react to biodiesel, such as glass marbles or steel mesh. Designs with no materials placed in the tank will also work, although drying time and energy consumption will rise accordingly.

Figure 8.2.6-4. This close-up view of a packed column dryer shows how the tubes are closely stacked and increase the surface area of the fuel as it passes from the top of the column to the bottom.

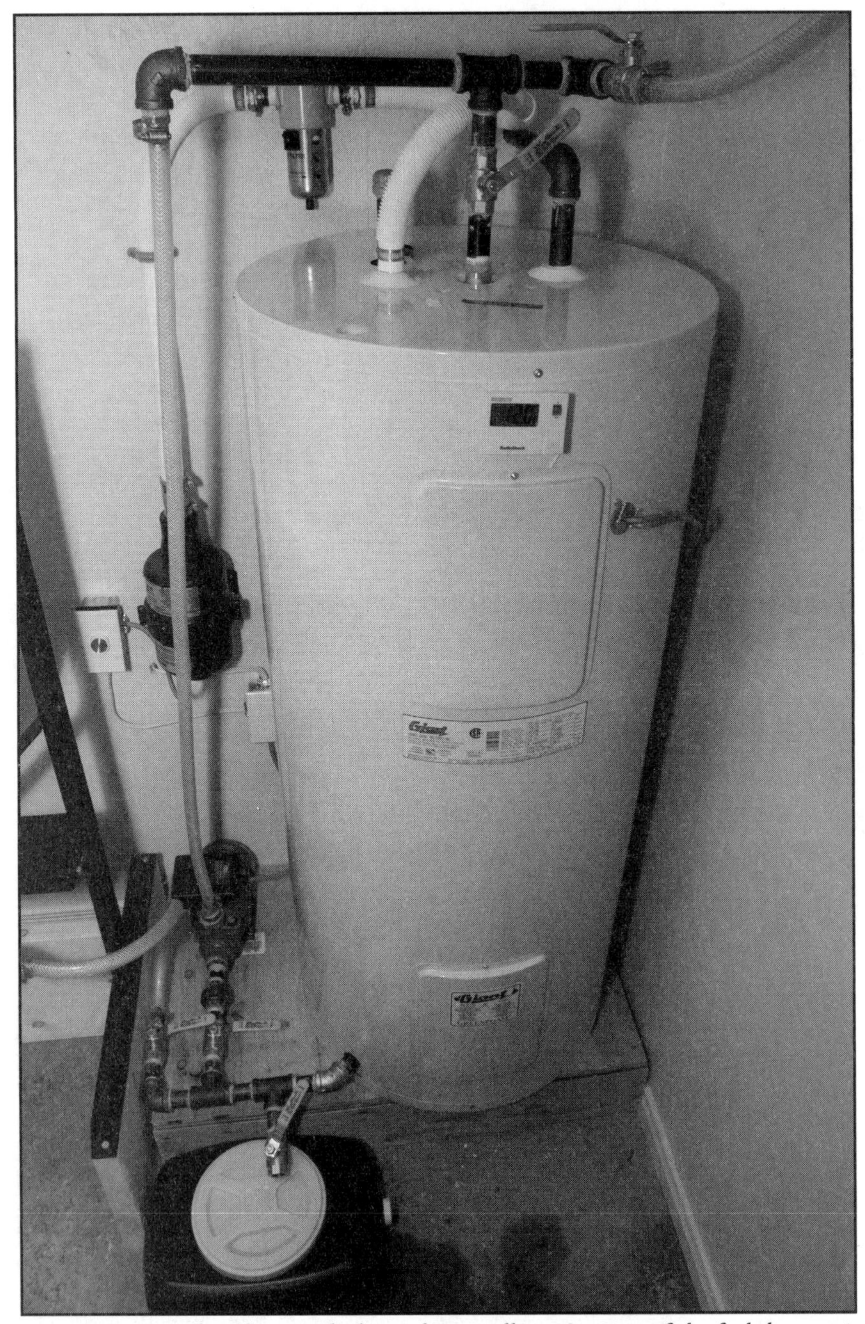

Figure 8.2.6-5. This photograph shows the overall construction of the fuel dryer unit, which uses a standard 60-gallon (227-liter) electric hot water tank. The unit is equipped with the same basic plumbing and circulation system used in the WVO treatment unit and biodiesel reactor tank.

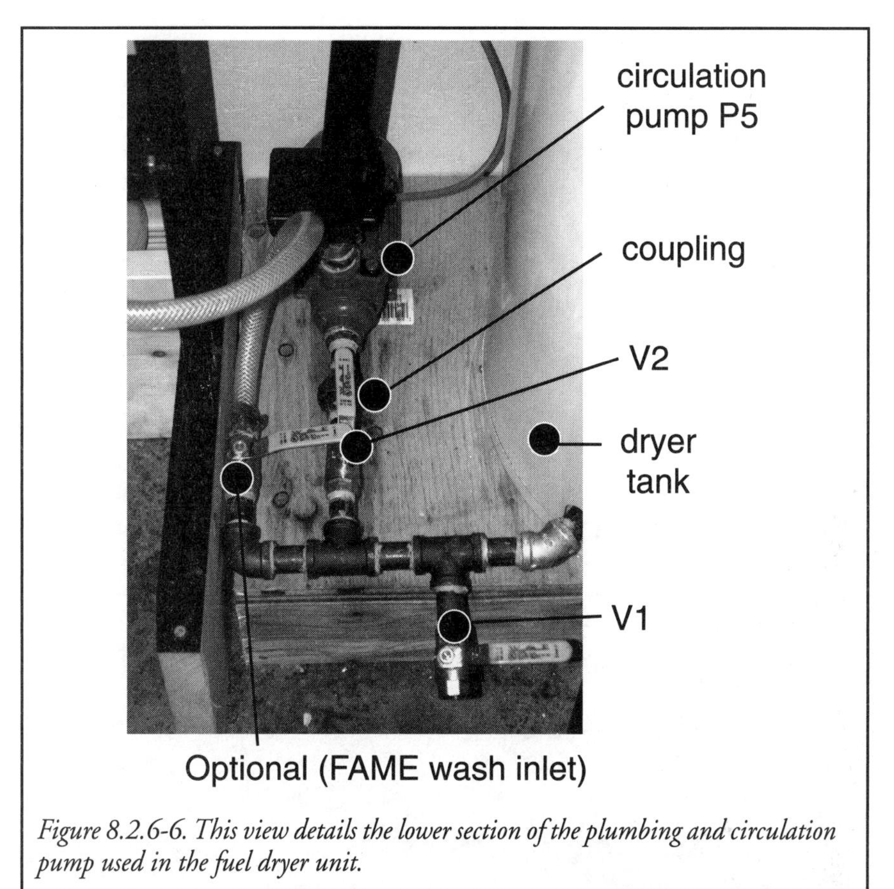

Figure 8.2.6-6. This view details the lower section of the plumbing and circulation pump used in the fuel dryer unit.

Figure 8.2.6-6 shows the lower section of the plumbing and circulation pump. The dryer tank outlet is connected to a plumbing tree which comprises a series of control valves.

Valve V1 is the tank drain valve. It should not be required under normal operating conditions but may be used if any debris or other contamination in the tank is suspected.

Valve V2 is a service valve which is only closed in the event of pump failure when the tank is full of biodiesel. Closing valve V2 allows the plumbing coupling to be opened and the pump to be removed while the dryer tank remains full of fuel.

An optional valve is shown at the inlet hose from the biodiesel wash unit. This valve is not required provided a control valve is placed at the wash unit outlet pipe as described in Chapter 8.2.5.

The circulation pump P5 is fitted with a "sight glass" braided pipe, although it is generally not required since a known amount of biodiesel will be pumped into the tank from the fuel washing unit. However, if the dryer tank remains partially filled (as a temporary storage tank), it is possible to overfill the unit, spilling biodiesel through the outdoor vent, in which case access to the sight glass is useful. In the interests of energy efficiency, I have covered the sight glass with pipe insulation, although the self-adhesive material has been removed, allowing the insulation to be pulled down for visual inspection during fuel transfer.

The upper section of the fuel dryer unit is shown in Figure 8.2.6-7. In this view, the biodiesel circulates through the insulated pipe and through throttling valve V3, which slows the flow of biodiesel through the packed column. This increases the moisture exchange between the fuel and the heated air, increasing energy efficiency and reducing drying time.

The tank is packed with the pieces of pipe which have been previously cut by a local hydraulic supply company that carries the stainless steel tubing.

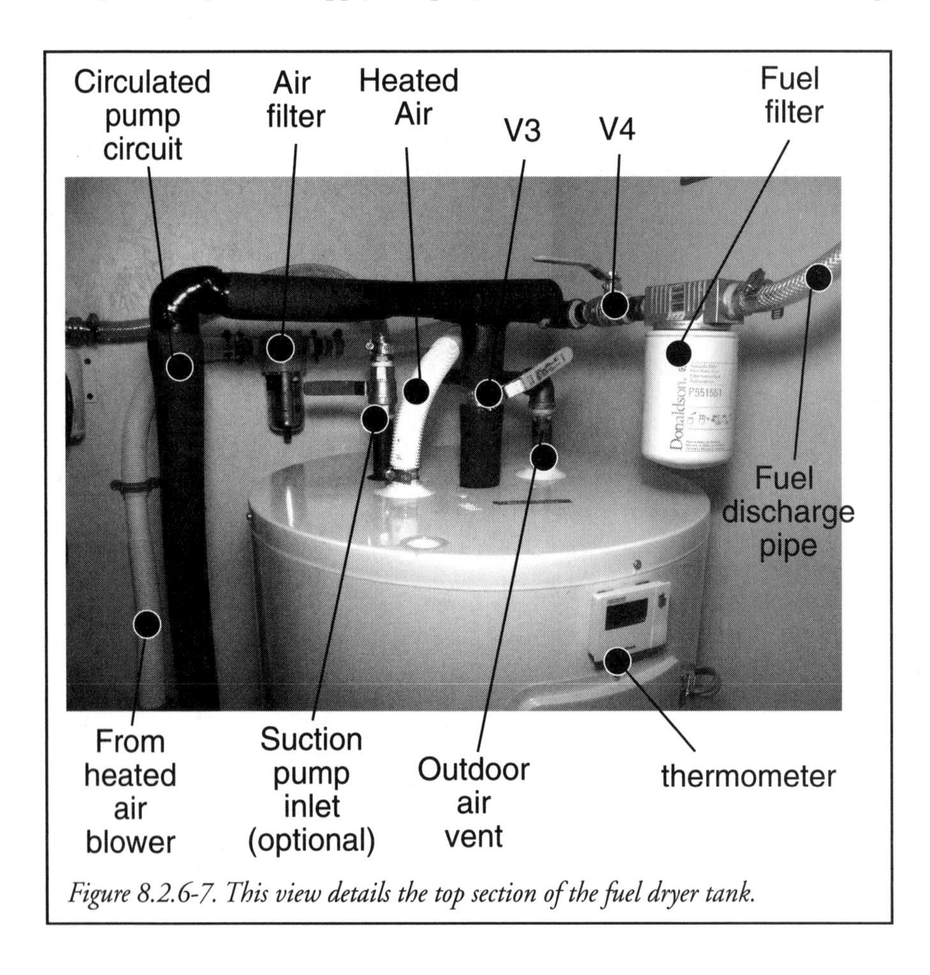

Figure 8.2.6-7. This view details the top section of the fuel dryer tank.

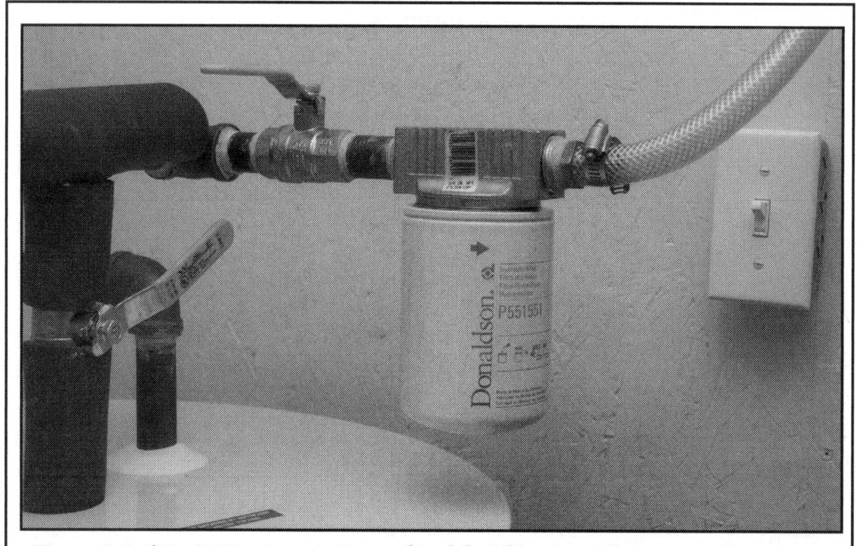

Figure 8.2.6-8. A 20-micron spin-on diesel fuel filter is used to remove impurities before the biodiesel is placed in storage.

Before packing the tank with the tubes it is necessary to install the heated air inlet tube which threads into one of the ¾" pipe fittings in the tank top and descends to approximately 6" (15 cm) above the bottom of the tank. After the air inlet tube is installed the packed column tubes may be dropped into the tank through the sacrificial anode fitting. It will take a bit of poking and stirring to get the maximum number of tubes into the tank. Be patient. The more tubes, the faster the dryer unit will operate.

The air inlet tube is connected to the outlet of the air filter while the inlet of the air filter is connected to the air blower unit shown in Figure 8.2.6-9. The air filter captures any dust or carbon brush material that is ejected by the air blower unit.

An outdoor air vent is connected to the tank in a similar manner as the WVO treatment unit. The outside vent portion should have a piece of stainless steel wool (pot scrubber) jammed into the air outlet to prevent insects from entering the dryer tank and contaminating the biodiesel.

Valve V4 is the biodiesel outlet control valve which directs "finished" fuel through the 20-micron diesel fuel filter before it enters the storage drums.

An optional valve called the "suction pump inlet" is used to draw biodiesel from the top of the wash tank in the event of emulsion and/or soap buildup during the washing stage. If waste vegetable oil has not been properly processed, allowing the free fatty acid level to remain high, a soap and/or emulsion

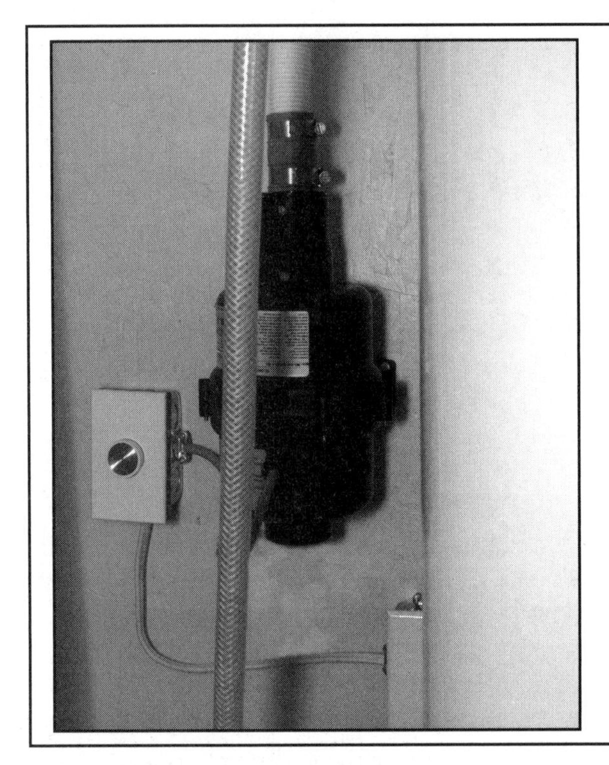

Figure 8.2.6-9. The heated air blower and variable speed control are shown above. This blower is commercially available from any swimming pool or spa manufacturer. It has the same blower assembly used in household vacuum cleaners and is equipped with a small air heater in the forward air outlet section of the unit.

layer will form which may make bottom draining difficult, as these layers may mix with the biodiesel phase. Using the top suction pump reduces the chance of phase mixing and allows the biodiesel to be drawn from the top. The wash tank can then be rinsed and readied for the next batch.

The heated air blower and variable speed control are shown in Figure 8.2.6-9. This blower is commercially available from any swimming pool or spa or jetted bath manufacturer or dealer. It has the same blower assembly used in household vacuum cleaners and is equipped with a small air heater in the forward air outlet section of the unit. (A vacuum cleaner or other similar unheated blower may be used in this application, although fuel drying time will be increased.) The blower is connected to a commercial 1000-watt dimmer switch so that the unit can be operated at a suitable speed and power level. Because of the various restrictions and back pressure applied to the blower it is important to check that air is flowing out of the vent pipe when the unit is put into operation. A few plastic streamers tied to the outlet vent will help determine when sufficient airflow has been reached.

If the fuel dryer is going to serve as a fuel storage tank as well, it is advisable to place a plug in the outdoor air vent or cover it with plastic to prevent water vapor from entering the tank and contaminating the biodiesel.

Figure 8.2.6-10. The author is shown getting ready to fill a 5-gallon fuel container, although larger 55-gallon tanks are more commonly used. Fuel quality issues do not end at the storage tank. From here, the biodiesel must be tested in the quality lab to ensure that it meets applicable standards. Additionally, fuel that is stored should be kept as cool, dark, and dry as possible. Use the fuel as quickly as you can to avoid oxidation and storage quality issues.

8.2.7
Fuel Dispensing Unit

It really doesn't matter whether the motivation to make biodiesel comes from trying to save the environment or save a few dollars, when the moment of truth arrives; pumping the first few gallons of quality fuel is an exhilarating experience. Perhaps the event is not as impressive as a rocket launch, but nevertheless, it really is time to celebrate disconnecting from the fuel pumps of society.

Figure 8.2.7-1. Our little biodiesel co-operative consists of five vehicles and a bunch of enthusiastic friends who are as much curious about the technology as they are being able to save some money.

Our little biodiesel co-operative consists of five vehicles and a bunch of enthusiastic friends who are as much curious about the technology as they are being able to save some money. With the biodiesel processor operating at full capacity, the system can process approximately 45 to 50 gallons (≈170 to 190 liters) of biodiesel per day, which is far in excess of our total requirements. However, processing, storing and handling that much fuel does take a fair amount of space, labor and care.

A fuel dispensing system can be as simple or complex as your needs dictate. Simple 5 gallon fuel totes are inexpensive and work well, but increase the amount off handling labor and increase the frequency of spills. An electric fuel dispensing system such as the model shown in Figure 8.2.7-3 and 8.2.7-4 keeps the dispensing method in familiar territory. This dispensing unit is

Figure 8.2.7-2. With the biodiesel processor operating at full capacity, the system can process approximately 45 to 50 gallons (≈170 to 190 liters) of biodiesel per day, which is far in excess of our total requirements. However, processing, storing and handling that much fuel does take a fair amount of space, labor and care.

Figure 8.2.7-3. A fuel dispensing system can be as simple or complex as your needs dicatate. Simple 5 gallon fuel totes are inexpensive and work well, but increase the amount off handling labor and increase the frequency of spills. An electric fuel dispensing system such as this model, keeps the dispensing method in familiar territory.

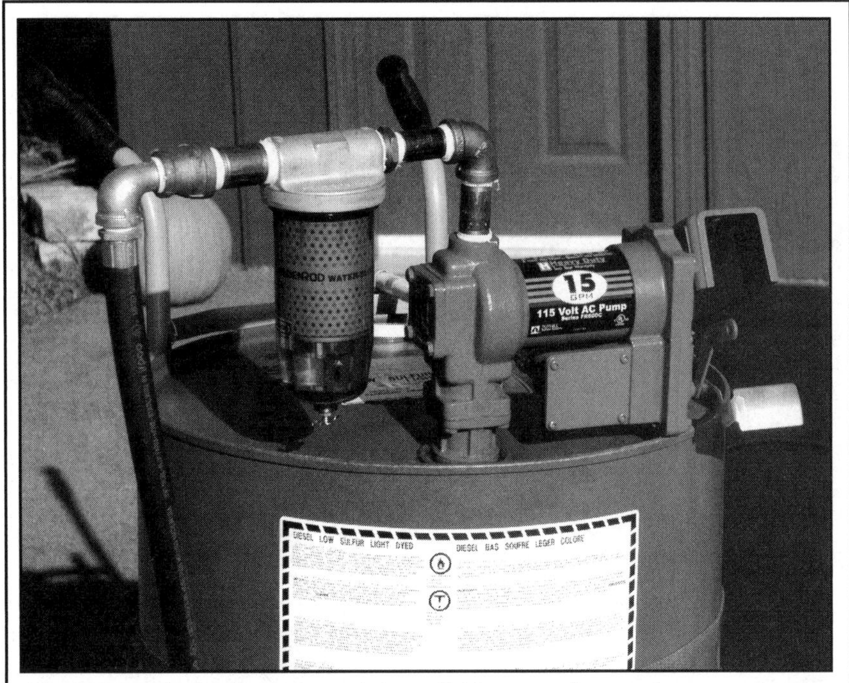

Figure 8.2.7-4. This dispensing unit is available at most farm and auto supply stores and comprises an explosion-proof fuel pump that is fitted to a standard 55 gallon (208 Liter) fuel drum.

available at most farm and auto supply stores and comprises an explosion-proof fuel pump that is fitted to a standard 55 gallon fuel drum. The pump may be driven by either 120 volt household power or through a 12 volt supply connection for in-vehicle use. A special water-absorbing filter known as an agglomerator is connected to the pump discharge, prior to feeding the fuel nozzle (Figure 8.2.7-5).

Incidentally, the 120 volt version can fill the Smart™ Car fuel tank in less than 20 seconds, making fill-ups a snap! Most fuel dispensing pumps are fitted with 20 foot (6 meter) hoses, allowing the storage tank to be semi-permanently installed, while being able to reach 2 cars in the garage without having to move tank.

Figure 8.2.7-6 shows a detail view of a typical fuel dispensing unit. A 55 gallon fuel drum must be purchased or "rented" by paying a drum deposit charge. It is strongly recommended that only a new or recently used diesel fuel drum be adopted for your fuel dispensing system. The drum can be mounted on a drum dolly or a drum cart such as the one shown in Figure

Figure 8.2.7-5. The 120 volt version pump shown here can fill the Smart™ Car fuel tank in less than 20 seconds, making fill-ups a snap! Most fuel dispensing pumps are fitted with 20 foot (6 meter) hoses, allowing the storage tank to be semi-permanently installed, while being able to reach 2 cars in the garage without having to move tank.

8.2.7-3. A fully loaded fuel drum is very heavy and is difficult to move using standard drum carts; ideally the drum can be placed in a convenient central location and not moved.

A commercial fuel dispensing pump is mounted to the drum using the threaded bung fitting. A suction pipe is lowered into the tank, which draws fuel into the pump intake as shown.

The agglomerator filter is fitted on the discharge side of the pump as shown. Standard black pipe fittings of the same type used in the biodiesel facility are required. The agglomerator filter is a disposable filter that is designed to agglomerate (group together or coalesce) water droplets suspended in the fuel and cause them to fall to the bottom of the sight glass bowl. A small drain valve is located at the bottom of the glass bowl which allows for drainage of any accumulated water. (Do not believe for one second that having an agglomerator will lessen the need for diligent fuel handling and storage. These filters will only remove relatively large water droplets and are therefore intended as precautionary devices only.

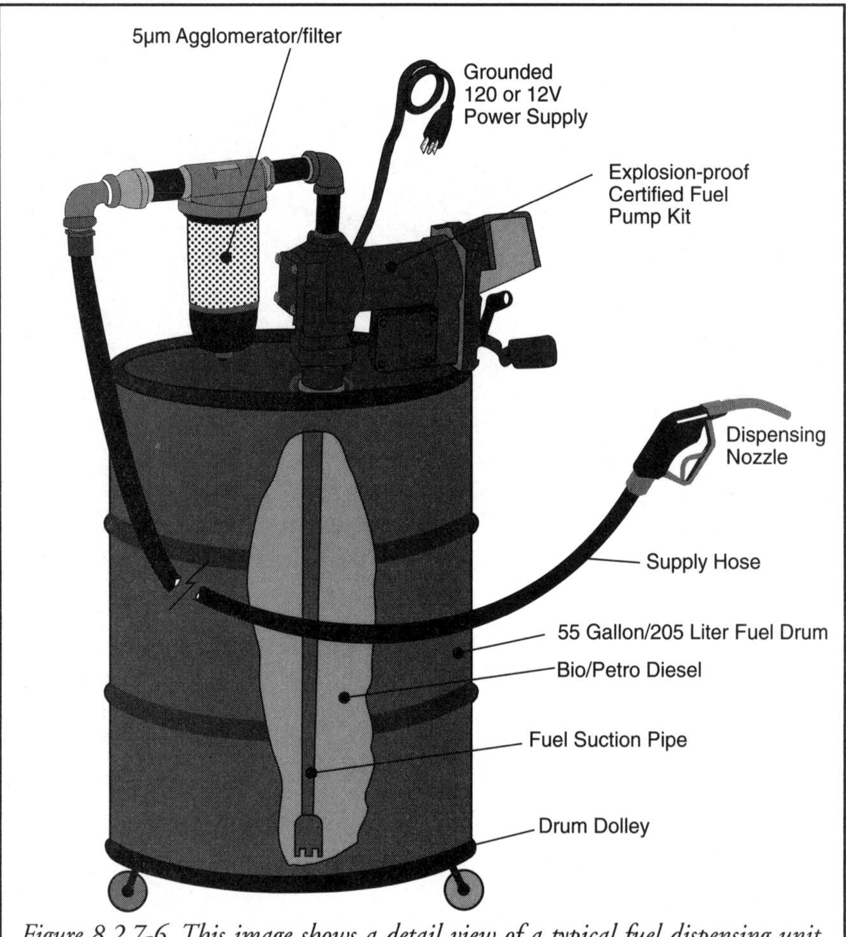

Figure 8.2.7-6. This image shows a detail view of a typical fuel dispensing unit. A 55 gallon fuel drum must be purchased or "rented" by paying a drum deposit charge. It is strongly recommended that only a new or recently used diesel fuel drum be adopted for your fuel dispensing system.

Cold Weather Issues

Biodiesel is subject to a gelling or freezing condition at its cold-temperature limits as discussed in Chapter 3.3. If you live in an area where cold weather is the norm, it will be necessary to adjust the concentration of biodiesel stored in the fuel dispensing system as a function of temperature. One very simple means of doing this is to create a "biodiesel blend thermometer" as shown in Figure 8.2.7-7. A series of 11 glass vials are filled with mixtures of petro and biodiesel, starting from left vial in the picture, No.2-D (straight diesel fuel), B10 (10% biodiesel/90% petrodiesel) through B100 (100% biodiesel). Each of the lids are marked with the appropriate concentration of fuel mixture.

Figure 8.2.7-7. Biodiesel is subject to a gelling or freezing condition at its cold-temperature limits as discussed in Chapter 3.3. If you live in an area where cold weather is the norm, it will be necessary to adjust the concentration of biodiesel stored in the fuel dispensing system as a function of temperature. One very simple means of doing this is to create a "biodiesel blend thermometer" as shown here.

The vials can be placed in a tin can and left in the same general location, out of direct sunlight, that you would normally store your diesel-powered vehicle. In the example shown in Figure 8.2.7-8, the No. 2-D sample (bottom vial) is free-flowing and gel free, while the B100 (top vial) is frozen solid at 5°F (-15 °C). At this temperature using refined, cold-pressed canola-based biodiesel, a concentration of B30 could be used as evidenced by the lack of crystal formation or any sign of fuel thickening.

It is important to make a biodiesel blend thermometer as the ambient local temperature and feedstock oil composition of the biodiesel will greatly affect the concentration of biodiesel that can be mixed into the fuel tank. It is very difficult to accurately calculate the blend levels due to these effects. For this reason, I keep a spare drum of No.2-D diesel fuel for splash blending on site.

An additional word of caution is to note where your vehicles will be stored during the day. For example, the biodiesel blend thermometer will give an accurate reading assuming your vehicle is stored in a garage at night. However, the car may be exposed to colder day-time temperatures if you are parked in a shaded outdoor parking lot all day long. It is always best to error on the side of caution and be a bit conservative when selecting the appropriate biodiesel blend level. There is nothing worse than having to call for a tow truck to tow a stalled, fuel-starved vehicle because you pushed the blending limits.

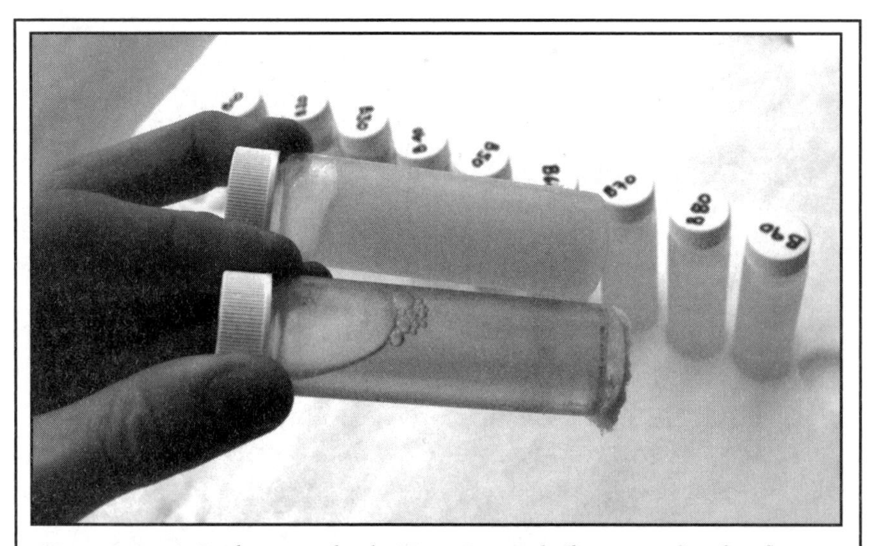

Figure 8.2.7-8. In this example, the No. 2-D sample (bottom vial) is free-flowing and gel free, while the B100 (top vial) is frozen solid at 5°F (-15 °C). At this temperature and using cold-pressed canola-based biodiesel a concentration of B30 could be used as evidenced by the lack of crystal formation or any sign of fuel thickening.

Blending Biodiesel with Petrodiesel

The National Biodiesel Board (NBB) commissioned a report on cold flow blending issues after the State of Minnesota established a requirement that all on-highway diesel fuels contain at least 2% biodiesel. In response to the need for proper blending and other cold temperature related issues, the NBB established a Biodiesel Cold Flow Consortium to study the blending properties of biodiesel and to report back their results. The study evaluated both "splash" and "proportional" blending techniques, of which only splash blending will be used by the micro and small-scale producer.

The results of the Consortium testing showed that biodiesel must be kept at least 10 °F (≈6 °C) above its cloud point temperature to ensure successful, homogenous blending with petrodiesel. For those who would like to learn more about the cold-flow blending study, a copy of the report can be downloaded at: http://www.biodiesel.org/resources/reportsdatabase/reports/gen/20050728_Gen-354.pdf.

For the micro and small-scale producer this report requires that a careful examination of the B100 biodiesel cloud point must be determined by carefully refrigerating and stirring a sample and noting the temperature at which the biodiesel begins to cloud. The pre-blending storage temperature of the

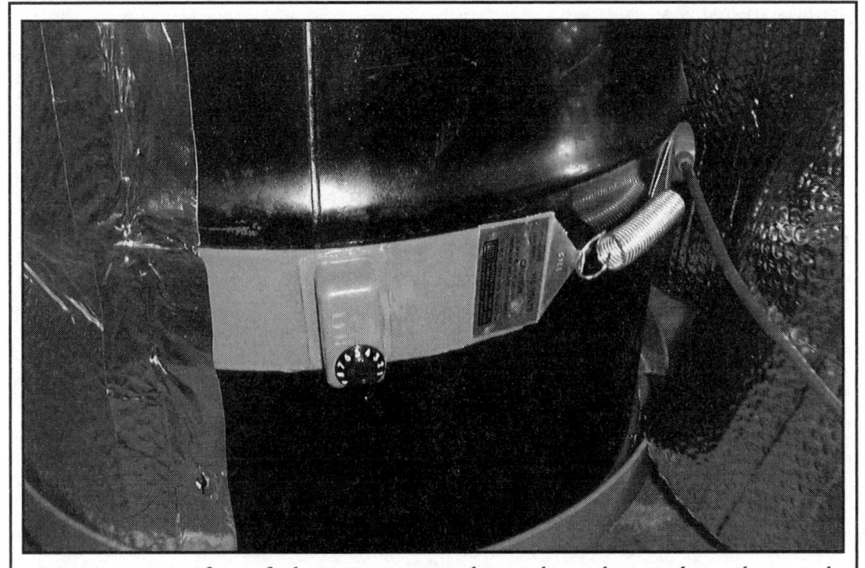

Figure 8.2.7-9. If your fuel requirements are larger than what can be easily carried into a warm house, an insulated band tank heater such as the model shown here can be added to the biodiesel storage tank. Band heaters are available from automotive and fuel suppliers and are designed to warm fuel for blending and prevention of cold weather gelling. (Courtesy Biodiesel Solutions)

biodiesel must be at least 10 °F (≈6 °C) above this temperature. Once the biodiesel has been blended, storage temperature is not as critical, provided the fuel remains cloud-free and fluid.

Depending on the volume of biodiesel that you produce, this may be as simple as bringing a few totes of biodiesel fuel into the house or warm garage. If your fuel requirements are larger than what can be easily carried, an insulated band tank heater such as the model shown in Figure 8.2.7-9 can be added to the biodiesel storage tank. Band heaters are available from automotive and fuel suppliers and are designed for exactly this application. A wrapping of insulation sleeving will reduce energy consumption and speed heating. (It is perfectly acceptable to allow the biodiesel to freeze during the winter storage period, provided the fuel is filtered prior to blending, to remove any fuel crystals that may have developed during the freezing state. It is always advisable to filter fuel when transferring from one tank to another.

Splash blending is really as simple as the name implies. Warmed biodiesel is poured into a storage or car fuel tank to the blend level desired. The effects of motion from pouring and road vibration will complete the blending, ensuring a homogenous mix between the two fuels.

Warm Weather Issues

People in warm and humid climates don't get off Scott-free either. Biodiesel that is subjected to poor, wet storage conditions will eventually cause the growth of bacteria at the biodiesel (or petrodiesel) water interface as was discussed in Chapter 3.3 and as shown in Figure 8.2.7-10.

Bio and petrodiesel also suffer from oxidation stability issues caused by contact with atmospheric oxygen and metallic contaminants, although biodiesel will degrade at a faster rate than petrodiesel.

Although biocides and fuel stabilizers can be added to the fuel, the best solution is to store the biodiesel or blended fuel in dry, full containers to limit the amount of contact with atmospheric oxygen, keeping them out of direct sunlight and using the fuel as quickly as possible.

Filter Plugging Issues

Fuel filter plugging is a problem normally related to poor fuel quality, almost certainly due to excessive free glycerin. In most cases, the cause can be directly co-related to inadequate transesterification and fuel washing (Figure 8.2.7-11 and 8.2.7-12).

In some cases, fuel filter plugging can be caused by the solvent capabilities of biodiesel, removing solid deposits within the fuel tank and flushing

Figure 8.2.7-10. Bacteria growth in the fuel will generate a "bio-slime" similar to the nasty layer that grows on sour cream or yoghurt that has been left in the fridge a bit longer than the expiration date. This growth can dislodge becoming trapped in the filter.

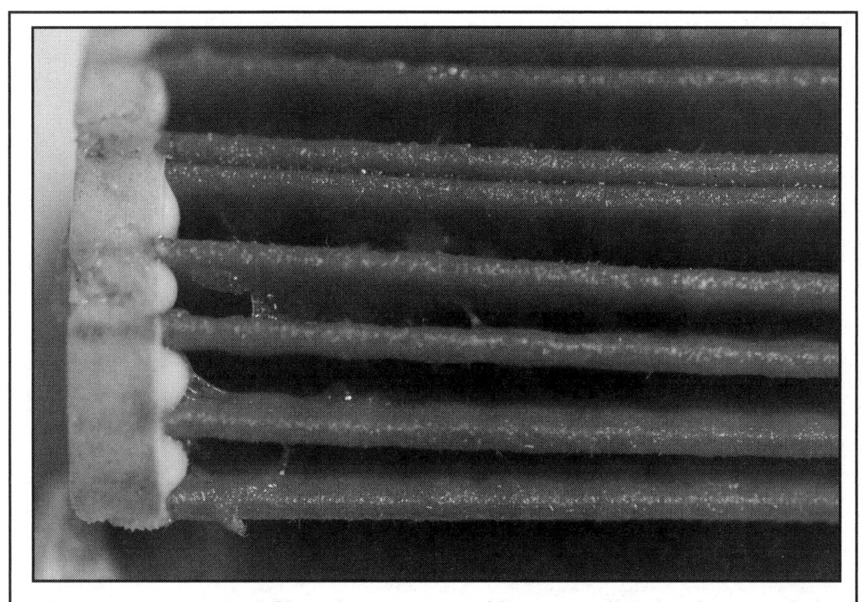

Figure 8.2.7-11. Fuel filter plugging is a problem normally related to poor fuel quality, almost certainly due to excessive free glycerin. In most cases, the cause can be directly co-related to inadequate transesterification and fuel washing. The top four causes of fuel filter plugging are:

- *poorly transesterified biodiesel or lack of washing*
- *solvent action of biodiesel dislodging sludge from vehicle fuel tank*
- *bacterial growth plugging filter with "bio-slime"*
- *high-concentration biodiesel use during cold weather exceeding fuel cloud point rating*

them into the filter.

Bacteria growth in the fuel will generate a "bio-slime" similar to the nasty layer that grows on sour cream or yoghurt that has been left in the fridge a bit longer than the expiration date. This growth can dislodge becoming trapped in the filter.

Cold weather filter plugging is caused by high-concentration biodiesel becoming a jelly-like substance when the fuel is operated below its cloud-point temperature.

Each of these conditions is detected by either a loss of vehicle power, rough operation or complete fuel starvation resulting in engine stalling. In some vehicles equipped with advanced electronic engine controls, the "check engine" light may also illuminate. A quick test for this condition is to allow the vehicle to sit for a minute or two, and then attempt to restart the vehicle.

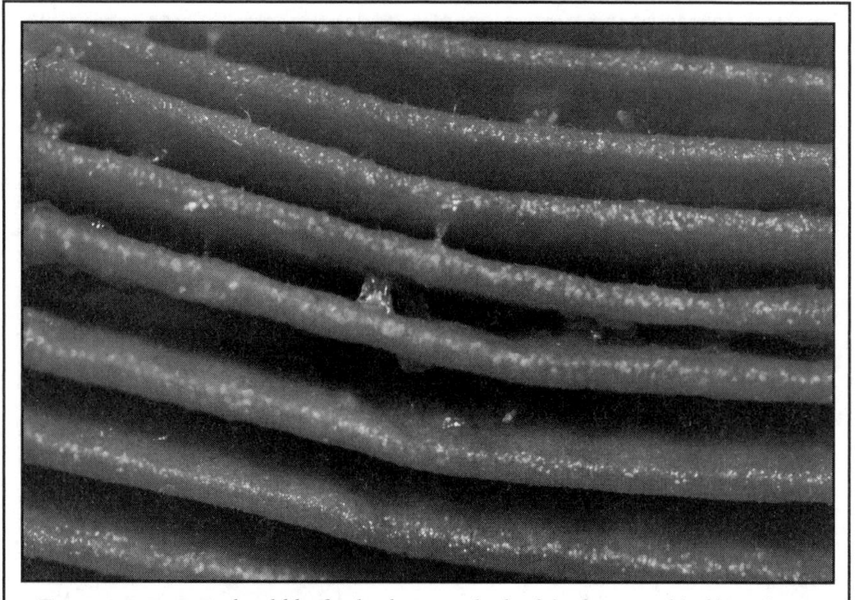

Figure 8.2.7-12. It should be fairly obvious which of the four possible filter plugging conditions have occurred. In any event, use caution when approaching a service garage or dealership, as they will be inclined to suggest some other, more expensive problem. If any of these conditions occur, change the fuel filter after following the remedial suggestions noted above.

If the engine starts and then quickly sputters or stalls, filter plugging is almost certainly the cause.

It should be fairly obvious which of the four possible filter plugging conditions have occurred. In the case of poor fuel quality the problem is obvious, stop using the fuel. Solvent extraction of fuel tank deposits can be solved by allowing the biodiesel to continue to remove deposits until the tank is clean, while living with the necessity of frequent filter changes. Alternatively, have the fuel tank steam cleaned at a local garage. Cold weather plugging is corrected by placing the disabled vehicle in a warm garage until the fuel has returned to a liquid state, at which time the biodiesel concentration should be lowered.

In any event, use caution when approaching a service garage or dealership, as they will be inclined to suggest some other, more expensive problem rather than simply changing the fuel filter. If any of these conditions occur, change the fuel filter after following the remedial suggestions noted above.

Happy motoring!

8.2.8 Glycerine Processing Unit

Figure 8.2.8-1 shows a schematic view of the glycerol processing unit. This device is very simple and allows the distillation and recovery of methanol along with the production of a partially refined glycerol that can be used in other applications.

Chapter 8.3 will describe the procedures necessary to convert raw, contaminated glycerol into a nontoxic product, while Chapter 11 will discuss some potential uses for this coproduct of the biodiesel production process.

In this design a cottage-sized 120-volt water heater is connected to an industrial light dimmer switch capable of handling 1,500 watts of electrical power. A digital "RMS" voltage meter, available from most automotive or hardware stores, is connected across the heater element terminals to record

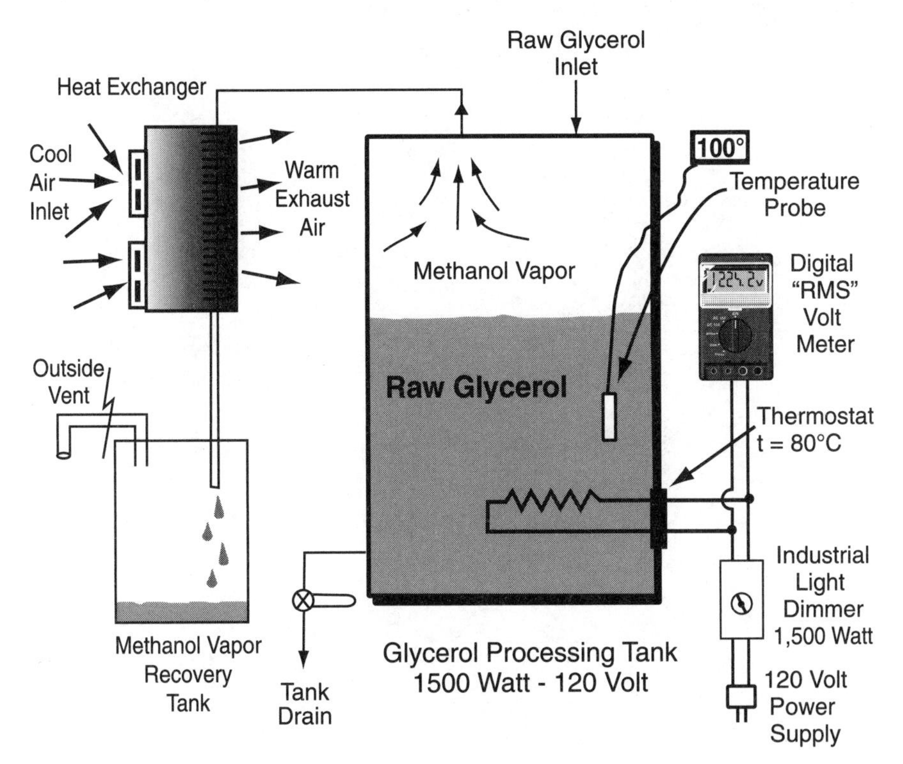

Figure 8.2.8-1. This drawing shows a schematic view of the glycerol processing unit. A very simple device, it allows the distillation and recovery of methanol along with the production of a partially refined glycerol that can be used in other applications.

the voltage applied to the heater element. (An RMS volt meter is capable of accurately measuring the "modulated" power that is output from the light dimmer switch. Although inexpensive non-RMS meters can be used, the measurements will only be accurate with that one brand of meter, making the sharing of data between a group of biodiesel producers impossible.) The combination of the light dimmer switch and the voltage meter allows the precise setting of power levels applied to the heater element and limits the chances of heater burnout due to excessive power settings. Glycerol does not transfer heat as quickly as water and therefore requires lower power densities and longer heating time.

Figure 8.2.8-2. The glycerol processing tank is fitted with an inlet pipe and cap as shown above. The cap is removed, a funnel is inserted into the pipe, and glycerol is poured into the tank.

The tank is fitted with an inlet pipe and cap as shown in Figure 8.2.8-2. The cap is removed, a funnel is inserted into the pipe, and glycerol is poured into the tank. Once the tank is filled the inlet pipe thread is checked to ensure that sufficient pipe tape is in place to ensure a vapor-proof seal, and the cap is then threaded back on.

The vapor recovery system used in the glycerol processor (Figure 8.2.8-3) is identical to that used in the biodiesel reaction tank described in Chapter 8.2.3. If sufficient space is available, it may be possible to plumb both the biodiesel reaction tank and the glycerol processing unit to the same vapor recovery system.

Lastly, complete the fabrication of the unit by adding a digital thermometer in the same manner as all other tanks used in the system. This will measure glycerol temperature during processing to ensure that methanol is being boiled-off for recovery.

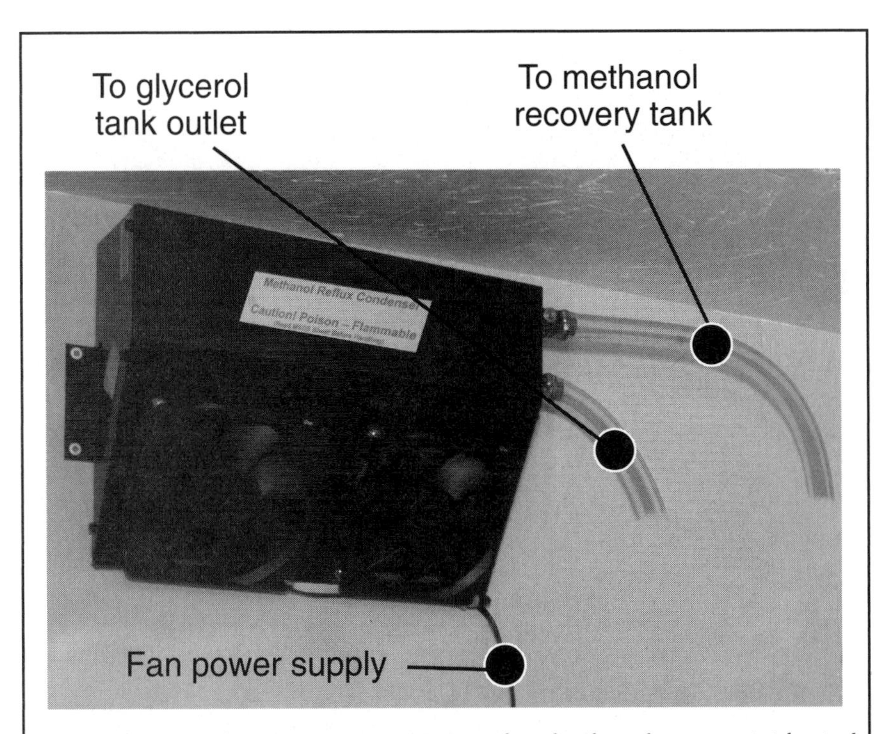

Figure 8.2.8-3. The vapor recovery system used in the glycerol processor is identical to that used in the biodiesel reaction tank described in Chapter 8.2.3. If sufficient space is available it may be possible to plumb both the biodiesel reaction tank and the glycerol processing unit to the same vapor recovery system.

8.2.9
Fuel Quality Laboratory

The production of biodiesel requires a number of chemical analysis steps that are conducted by the micro- and small-scale producer using a series of "wet chemistry" processes. Waste oils that are delivered to the facility must be tested for free fatty acid content prior to blending into the treatment unit; chemical titration is completed to calculate the correct amount of catalyst; and, finally, finished fuel must be tested against the applicable fuel quality standards.

Commercial biodiesel production facilities spend vast sums of money on advanced chemical analysis equipment such as gas chromatographs and spectrometers to reduce testing time and increase the accuracy of measurements. The small-scale producer has limited access to the necessary capital for this equipment but can spend more time and less money on analysis processing using wet chemistry techniques.

Chapter 8.3 will present a series of test "packages" that have been developed specifically for the micro- and small-scale biodiesel producer. Right now let's look at the suggested minimum work area requirements of the fuel quality laboratory in order to perform the package of chemical analysis tests.

Figure 8.2.9-1 shows the work area that has been set aside for the fuel quality laboratory. The space allocated is a mere 6' x 1.5' (1.8 m x 0.5 m) in area, with a common hallway shared with the biodiesel processing system (Figure 8.2.9-2).

The lab work area is a basic kitchen counter with a set of three cabinets used to hold large items such as paper towels, funnels, and a garbage receptacle. A drawer holds notepads, pens, a calculator, and other small items. A stainless steel bar sink is set into the counter and **is not** connected to a plumbing drain but feeds directly into a 5-gallon (≈20-liter) bucket. Waste water from the sink is added to the spent biodiesel wash water so that both can be processed at the same time. The sink is fed from a cold, potable water feed from the house supply but could easily be plumbed into an overhead water storage tank. Water use in the lab is kept to a minimum and is primarily for washing glassware with biodegradable soap. (It is inadvisable to wash dirty laboratory equipment with trace amounts of chemicals and reagents in the kitchen or where food preparation is done. If a washing sink is not practical in the laboratory area consider using a washing pan so that soiled water can

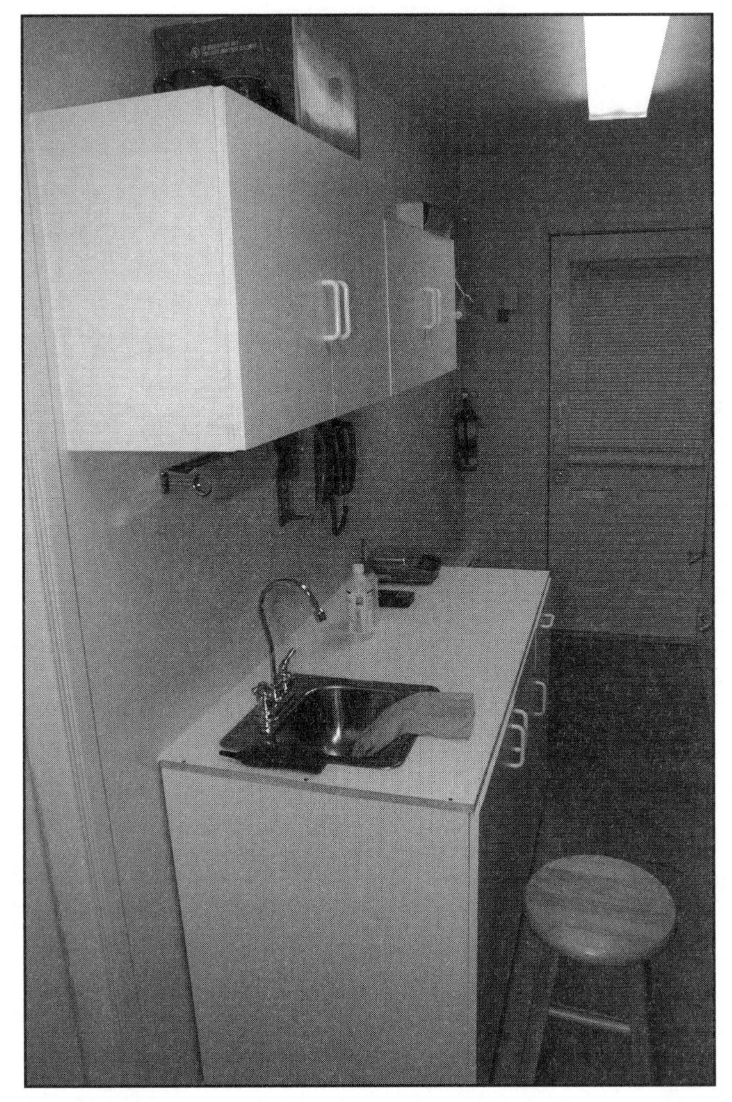

Figure 8.2.9-1. This photograph shows the work area that has been set aside for the fuel quality laboratory. The space allocated is a mere 6' x 1.5' (1.8 m x 0.5 m) in area, with a common hallway shared with the biodiesel processing system (Figure 8.2.9-2).

be retained and added to the spent biodiesel wash water.)

The overhead storage cabinet is used to hold a selection of laboratory glassware such as beakers, flasks, and wash bottles as well as the various reagents and test equipment required to perform the chemical analysis tests.

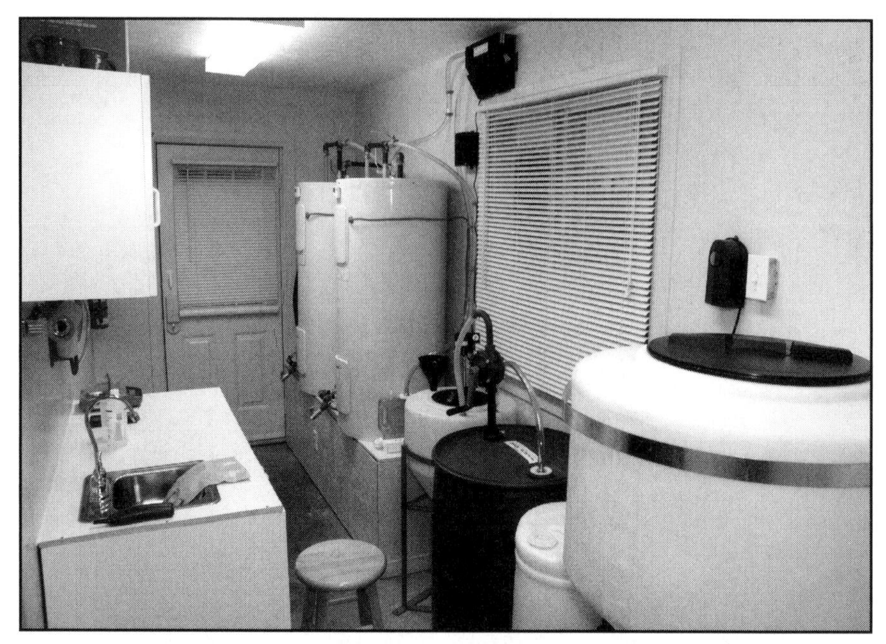

Figure 8.2.9-2. The fuel quality laboratory fits very compactly into the biodiesel production space. A bar stool is placed in the common hallway for working on chemical analysis.

Laboratory equipment is limited in an effort to keep capital costs manageable, with an electronic scale (Figure 8.2.9-3) possibly the most expensive item on the list.

Of course the laboratory would not be complete without the necessary safety equipment for handling the various chemicals. Safety is the theme of the next chapter (Figure 8.2.9-4).

Figure 8.2.9-3. Laboratory equipment is limited in an effort to keep capital costs manageable, with an electronic scale possibly the most expensive item on the list.

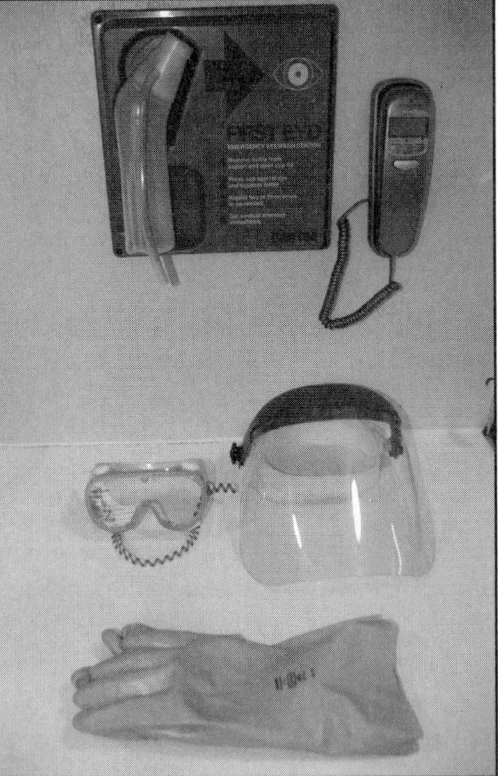

Figure 8.2.9-4. The laboratory would not be complete without the necessary safety equipment for handling the various chemicals. Safety is the theme of the next chapter.

8.2.10 Safety Equipment

Although properly manufactured biodiesel is environmentally friendly, non-toxic, and biodegradable, the same cannot be said for the materials used in its manufacture. Many hobbyist Websites and books make very little mention of the safety and environmental problems that can be caused by the chemicals and reagents used in the biodiesel production process.

Table 8.2.10-1 lists a number of the major chemicals that are used in the production of biodiesel. Each of these materials is known to have some degree of risk associated with it which should not be taken lightly.

All chemicals are registered under the CAS (Chemical Abstract Service) Chemical Registry System. The system is a universal registry for all known chemicals and it allows laboratories and other users to obtain information about the composition and appropriate safety and handling of specific materials. To use the CAS system to locate specific information on any chemical, first find the appropriate CAS number by typing "CAS (CHEMICAL NAME)" into Google search. Once the appropriate CAS number is found, type "MSDS CAS (CAS NUMBER)" into Google search to retrieve information on suppliers, safety information, and chemical composition, etc. Note that this search method will require a review of site header information to ensure you have the correct site and data located. An official option is to use the website www.msdsonline.com and pay for any data sheets required. A third option is to contact the chemical supplier or distributor and ask them for a copy of the MSDS sheets.

Safety is the theme of this section, and using the CAS registration system will allow you to locate the Material Safety Data Sheet (MSDS) for any chemical in the system registry. The MSDS is divided into several sections including the type of personal hazard expected from exposure to a given chemical, first aid measures, firefighting measures, handling and storage, ecological information, and disposal considerations.

It is strongly advised that before beginning any work in the analysis laboratory or biodiesel production workshop the MSDS data be obtained, printed, and placed in an emergency measures manual that is kept close at hand.

Having emergency measures in a booklet is fine, but in order to make sure that proper remedial action can be taken in the event of exposure, spill, or fire it is wise to develop a safety plan and make sure that everyone who works in

Item of Risk	CAS#	Type of Risk
Canola Oil [note 1]	120962-03-0	Flammable, Falls due to surface spills
Methanol	67-56-1	Poison, Flammable, Blindness due to splashing
Sodium Hydroxide	1310-73-2	Caustic
Potassium Hydroxide	1310-58-3	Caustic
Sodium Methoxide (liquid)	124-41-4	Caustic, Flammable
Potassium Methoxide (liquid)	865-33-8	Caustic, Flammable, Blindness due to splashing
Sodium Methoxide (crystalline)	124-41-4	Caustic, Poison, Highly Flammable, Reacts violently with water
Potassium Methoxide (crystalline)	865-33-8	Caustic, Poison, Flammable, Reacts violently with water
Phenolpthalein Reagent	201-004-7	Poison, Flammable
Isopropanol Alcohol	67-63-0	Poison, Flammable

Table 8.2.10-1. *Process chemicals used in biodiesel production and associated risks from accidental spills, exposure, or ignition.*

Note 1: *Waste oils are not given a CAS number because of the variability of their composition. For illustrative purposes, the CAS number for refined canola oil is included.*

the lab area is well informed about the safety manual and understands how to react and what equipment is required in the event of an accident. It is also strongly suggested that people working with the process chemicals take an accredited first aid course.

The work area of the fuel quality laboratory is shown in figure 8.2.10-1. It includes a wet sink and paper towel dispenser as well as an emergency

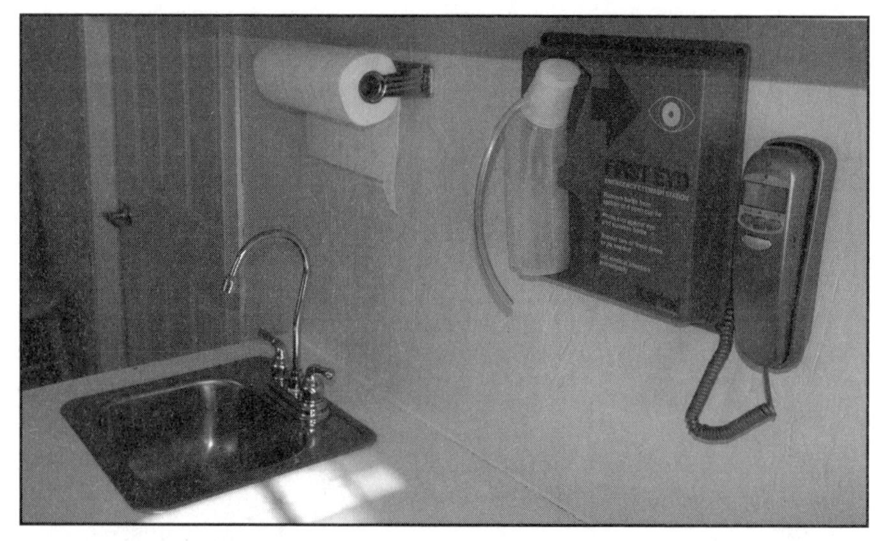

Figure 8.2.10-1. The work area of the fuel quality laboratory is shown in this view. It includes a wet sink and paper towel dispenser.

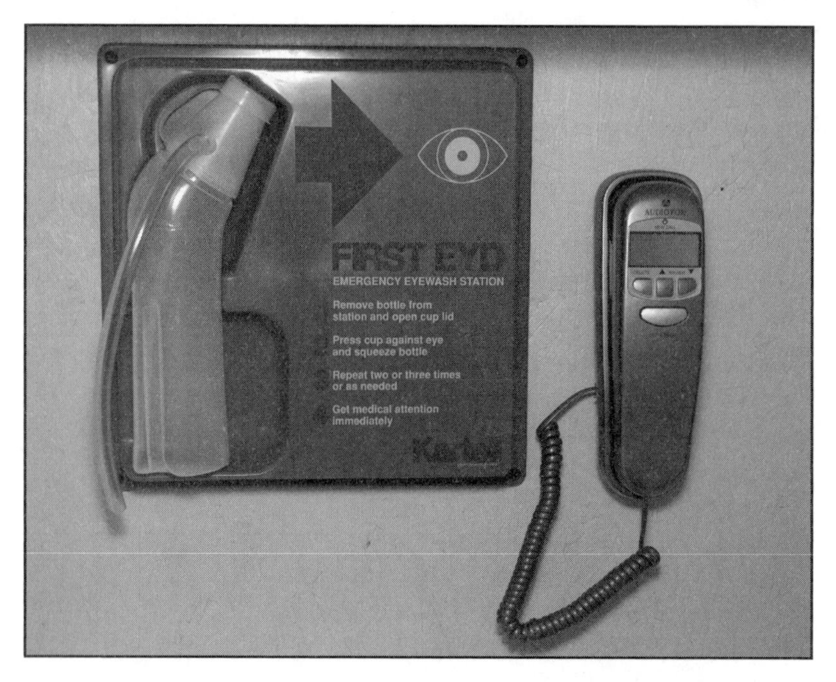

Figure 8.2.10-2. An easily located telephone with a list of emergency phone numbers and an emergency eyewash station are also an important part of the remedial safety equipment used in the fuel quality lab.

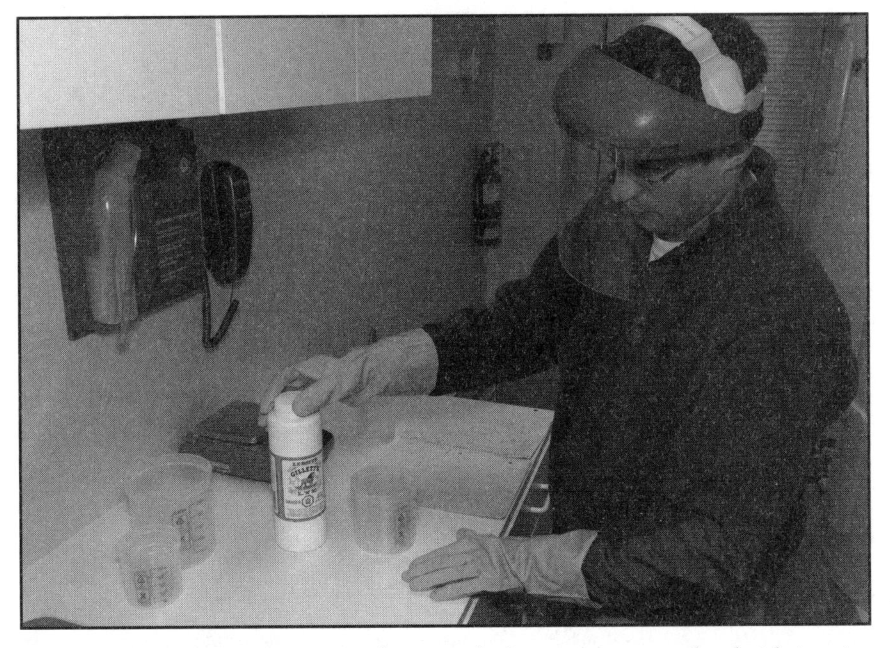

Figure 8.2.10-3. Safety requires a wall or ring of safety measures around each risk situation in the work area. Good safety measures start with a plan and address prevention first and remediation second.

eyewash station and telephone (Figure 8.2.10-2). Many people think of safety as a remedial measure, meaning that items such as the eyewash station are the key to a safe environment, but this isn't the case.

Safety requires a wall or ring of safety measures around each risk situation in the work area. Good safety measures start with a plan and address prevention first and remediation second:

1. Identify the risk and establish a preventive and remedial action plan.
2. Identify and acquire the appropriate preventive tools and equipment.
3. Identify and acquire the appropriate remedial tools and equipment.

For example, an identified risk is sodium methoxide getting into a person's eye. Once the action plan has identified the risk, the appropriate preventive tools can be determined, in this case full face shields (Figure 8.2.10-4) or eye protection glasses (Figure 8.2.10-5). Prevention is always the lowest risk and lowest cost approach in creating a safe environment, and identifying the risk and having the preventive tools available can prevent potential remedial costs.

Figure 8.2.10-4. Once your safety action plan has identified a given risk, the appropriate preventive tools such as full face shields or eye protection glasses can be determined.

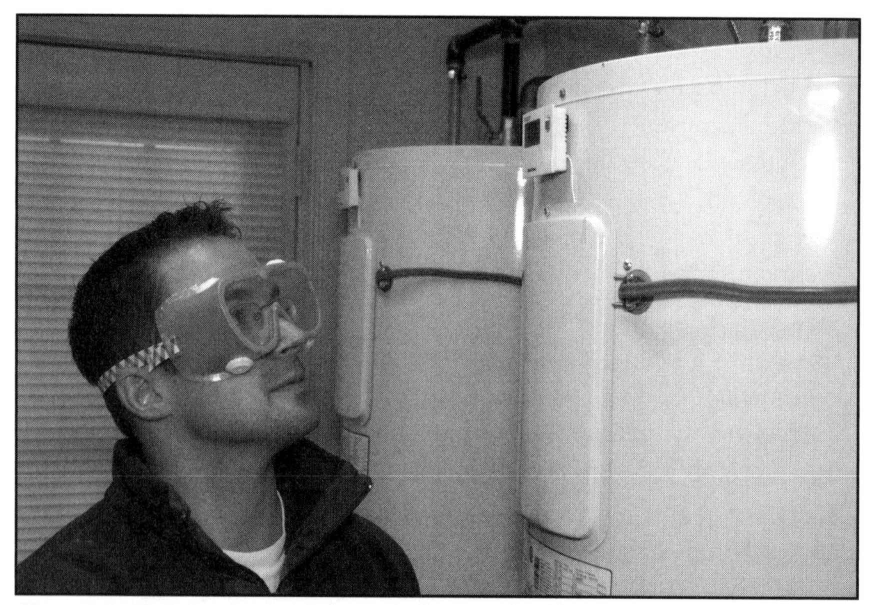

Figure 8.2.10-5. Prevention is always the lowest risk and lowest cost approach in creating a safe environment, and identifying the risk and having the preventive tools available is worth any cost.

However, even the best prevention methods fail and this is where the remedial side of the risk analysis is used. In the above example, we identified the risk (splashing sodium methoxide into eyes), determined the preventive measures (face/eye protection), and now we can develop the remedial plan to be used when the preventive plan fails.

The MSDS sheet for sodium methoxide states that the eye should be immediately flushed with water or cleansing solution from an eye wash station for a period of 15 minutes and that medical attention should be sought quickly. This remedial plan requires an eyewash station (Figure 8.2.10-2) and sufficient additional wash fluid to permit flushing of the eye for 15 minutes. In the event that an ambulance is required, there must be ready access to a telephone.

It does not matter whether the identified risk is fire, environmental, or human exposure in nature, the MSDS provides an excellent guide in planning your production workshop and developing a laboratory safety plan. However, in addition to the data listed in the MSDS there are several other considerations that must be taken into account during the construction of the laboratory and biodiesel processing area.

Room Ventilation

Biodiesel production involves exposure to varying concentrations of the chemicals listed above. Certain chemicals such as alcohol and phenolphthalein are flammable and will ignite if concentrations build above the lower explosive level (LEL); they are also highly toxic when inhaled. Face shields will protect against splashing and contact but do not prevent inhalation of toxic vapors. For this reason, the entire working area should be well sealed and built with an intake air vent at one end of the room. Controlling the air leakage into the building will ensure that air intake is from one location, providing maximum cross flow of fresh intake air.

At the other end of the room a high-capacity ventilation fan suitable for the room size (500 CFM/14 m^3/minute in my example) should be installed and equipped with automatic dampers (Figures 8.2.10-6 and 8.2.10-7). To ensure proper air flow the ventilation fan should be tested by injecting a small amount of smoke into the air intake so that it is driven through the room very rapidly. Any smoke pooling in the building envelope identifies "dead air" locations, which can be corrected by controlling air flow with louvers located at the intake and/or exhaust fan.

Figure 8.2.10-6. Certain chemicals such as alcohol and phenolphthalein are flammable and will ignite if concentrations build above the lower explosive level (LEL); they are also highly toxic when inhaled. A high-capacity ventilation fan will reduce exposure and concentration levels of airborne toxins.

Figure 8.2.10-7. The ventilation system should be equipped with automatic airflow-operated dampers to prevent rain or moisture ingress into the room.

Tank Ventilation

The WVO treatment unit, biodiesel reactor tank, methanol recovery tank, and fuel dryer all require positive ventilation to the outdoors. Ensure that vents point downward so that they do not become plugged with dirt and snow and ensure that coarse stainless steel wool mesh or screens are added to prevent insects from entering the plumbing lines (Figure 8.2.10-9).

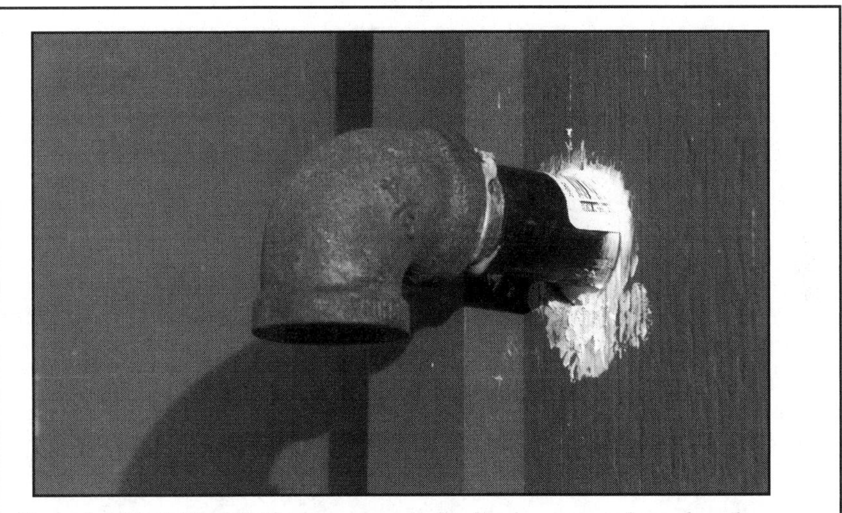

Figure 8.2.10-8. The WVO treatment unit, biodiesel reactor tank, methanol recovery tank, and fuel dryer all require positive ventilation to the outdoors.

Waste Disposal

Oily rags and paper towels may look innocent enough, but they will start to oxidize, releasing heat. If sufficient numbers are clumped together in a garbage can, the heat from oxidization will build up, causing more oxidization and additional heat. At some point the energy released from this slow, low-level process will cause the rags to self-ignite, starting a fire.

The best prevention is to discard or wash any oil rags as soon as possible. In the interim, store rags and other related garbage in approved fireproof garbage cans.

Fire Extinguishers

Two fire extinguishers are required in the biodiesel facility: a standard all-purpose dry Class A, B, C model for alcohol and other "typical" fires; and a foam extinguisher for fires that can't be extinguished by water such as those caused by the ignition of waste oil and biodiesel (Figure 8.2.10-9).

Figure 8.2.10-9. Two fire extinguishers are required in the biodiesel facility: a standard all-purpose dry Class A, B, C model for alcohol and other "typical" fires; and a foam extinguisher for fires that can't be extinguished by water such as those caused by the ignition of waste oil and biodiesel.

Antistatic Straps

A fire in early 2006 destroyed a Bakersfield, California biodiesel facility, thankfully without harming any of the personnel. The cause of the fire was investigated and is believed to have been a spark of static electricity as the methanol tanker truck was being connected to the building storage tank farm.

Static electricity can be caused by pulling a wool sweater over a shirt, rubbing shoes or boots on a concrete floor, or even brushing past someone. If the discharge of this electricity occurs anywhere near methanol or other flammable vapors, the result can be catastrophic.

A very simple solution is to use an antistatic strap such as the one shown in Figure 8.2.10-10a. These units are mandatory in the electronic industry and simply drain electric charge from your body to a nearby electrically grounded item such as the casing of the electric panel or a screw in a duplex outlet cover (Figure 8.2.10-10b). Do not attempt to make your own wrist strap unless you are able to solder a 100,000-ohm resistor in series with the ground wire. The resistor will prevent electrocution in the event that you come into contact with an electrically live device while wearing the wrist strap. The resistor limits the current flow in a shock condition, preventing a serious burn— or worse.

Grounding of Tanks and Other Exposed Metal

To further protect against the buildup of static electricity all exposed metal components in the laboratory must be properly grounded. Your electrician and electrical inspector will automatically require the grounding of the water heater tank chassis, circulation pump, submersible wash pump, ventilation fan, and other similar 120- or 240-volt appliances. While this wiring is being completed, have bare copper ground wires solidly attached to the sodium methoxide and biodiesel wash tank support legs, chemical mixer pump (assuming metallic construction), and other exposed metallic devices in proximity to flammable materials.

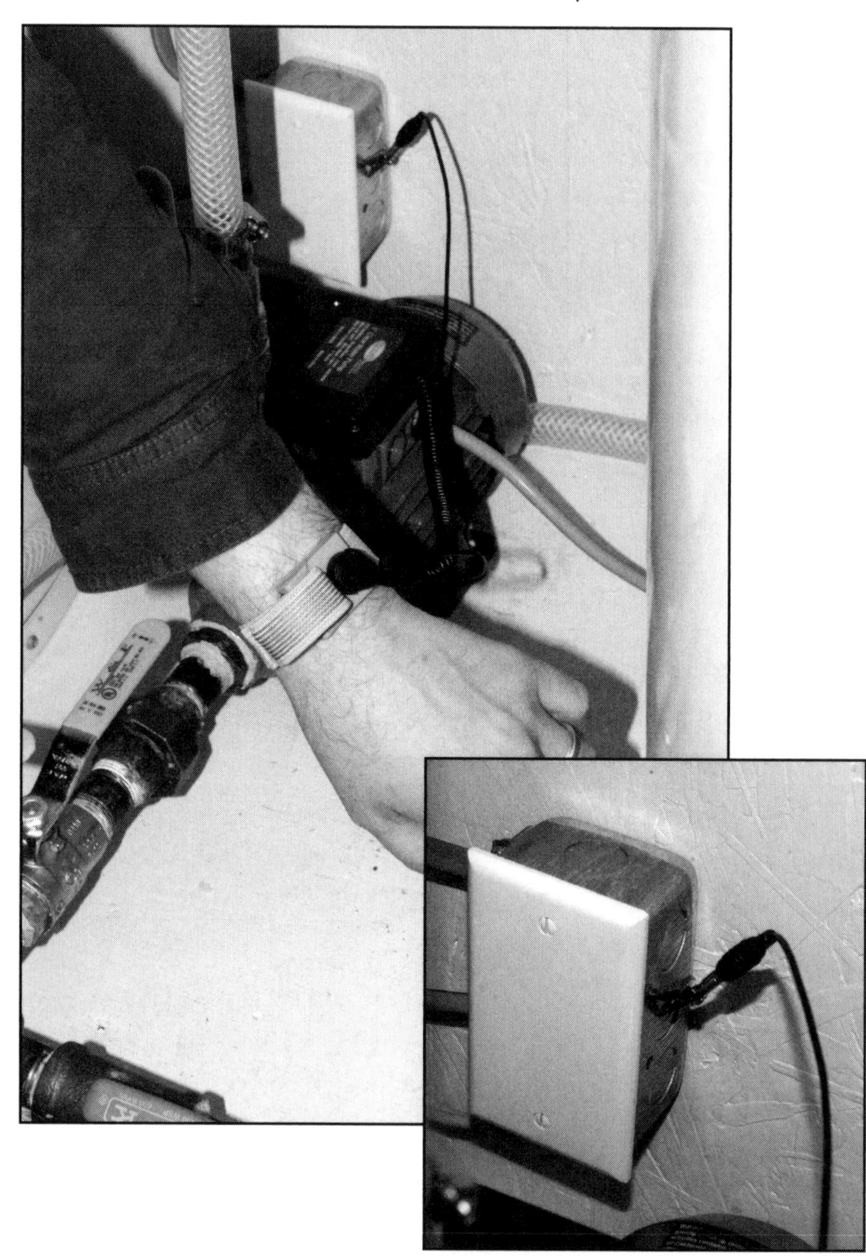

Figure 8.2.10a and b. Static electricity can be caused by pulling a wool sweater over a shirt, rubbing shoes or boots on a concrete floor, or even brushing past someone. If the discharge of this electricity occurs anywhere near methanol or other flammable vapors, the result can be catastrophic. A very simple solution is to use an antistatic strap such as the one shown above and to ensure that it is connected to an electrically grounded chassis such as a power outlet box.

First Aid Kit

A first aid kit such as the one shown in Figure 8.2.10-11 is a worthwhile investment and should remain in an accessible location near the exit door and close to fire extinguishers.

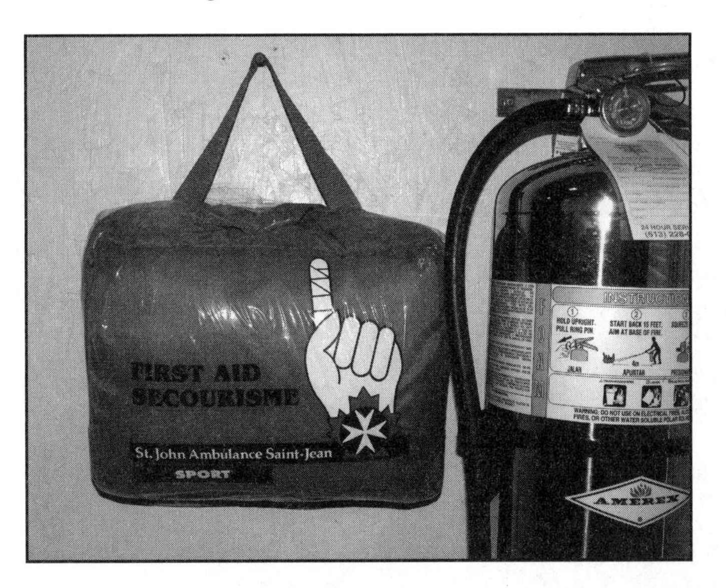

Figure 8.2.10-11. A first aid kit such as the one shown here is a worthwhile and inexpensive safety investment and should remain in an accessible location near the exit door and close to fire extinguishers.

Building Access

Ensure that the door to the building is lockable to prevent unauthorized access.

Personal Respirators (NIOSH Approved):

Check with your local safety supply store to ensure that you acquire a respirator that is suitable for the suite of chemicals that will be present in your facility. For sodium methoxide (powder only), the MSDS recommends a NIOSH type N95 filter or better (Figure 8.2.10-12).

Figure 8.2.10-12. Check with your local safety supply store to ensure that you acquire a respirator that is suitable for the suite of chemicals that will be present in your facility. For sodium methoxide (powder only), the MSDS recommends a NIOSH type N95 filter or better.

8.3 Production Procedures

8.3.1 Obtaining Waste Vegetable Oil

Waste Vegetable Oil (WVO) is a valuable commodity which can be converted into a product known as "yellow grease," used as a feedstock for numerous applications including pet and livestock feed production. The rendering and recycling industry has a firm grasp on the market, charging restaurants, French fry trucks, and commercial kitchens a fee to recycle their waste vegetable oil. The recycled grease is then processed or rendered and sold to the market.

Many fried food operations would be more than willing to have someone pick up their waste oil, saving the facility the cost of having a recycler haul the material away. Although the recycler is not concerned about the

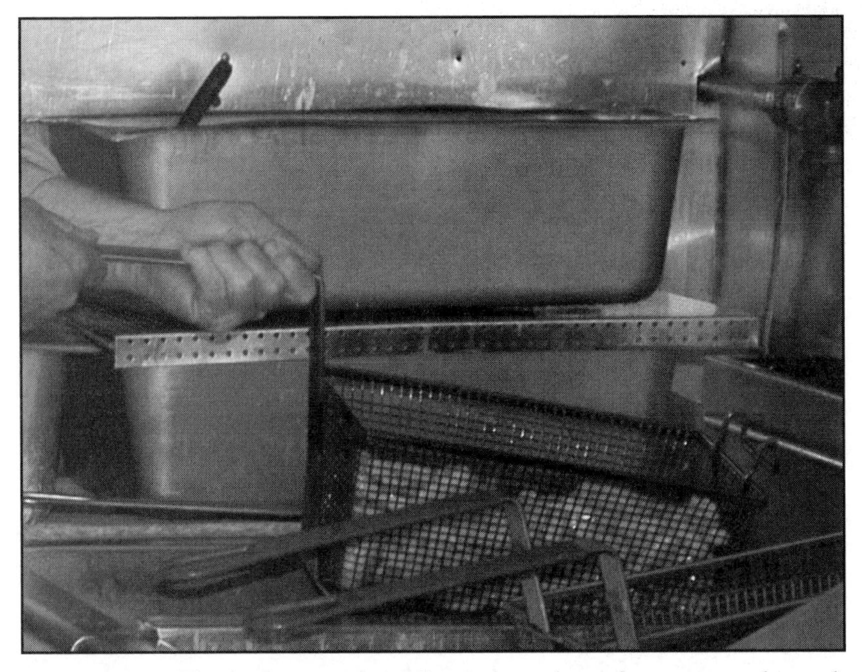

Figure 8.3.1-1. The rendering and recycling industry has a firm grasp on the market, charging restaurants, French fry trucks, and commercial kitchens a fee to recycle their waste vegetable oil.

Figure 8.3.1-2. Many fried food operations would be more than willing to have someone pick up their waste oil, saving the facility the cost of having a recycler haul the material away.

Figure 8.3.1-3. Although the recycler is not concerned about the type of oil or fat that is recycled, the micro-scale biodiesel producer is very concerned about feedstock quality. Paul Zammit, a budding biodiesel producer himself, has changed his restaurant's deep fryer to canola oil, recognizing the superior cold-weather operating characteristics of this feedstock.

type of oil or fat that is recycled, the micro-scale biodiesel producer is very concerned about feedstock quality. Canola, soybean, sunflower, and other non-hydrogenated or non-saturated oils make an excellent feedstock if cold weather performance issues are a concern. If you live in the southern part of the US, feedstock selection becomes less of an issue.

Of equal concern is ensuring that there is nothing other than oil or fat in the feedstock WVO. Since you are performing a service by collecting a "waste" product, inform your oil supplier of your concerns about feedstock quality and request that the following contaminants not be added to the oil:

- dish wash water or soaps;
- fryer or grill food scrapings;
- table scraps or waste products of any kind;
- fryer rinse water.

Figure 8.3.1-4. Canola, soybean, sunflower, and other non-hydrogenated or non-saturated oils make an excellent feedstock if cold weather performance issues are a concern. If you live in the southern part of the US, feedstock selection becomes less of an issue. Canola oil-based biodiesel (shown left) and peanut oil-based biodiesel (right) have both been stored at the same temperature of 45°F (7°C), yet the semi-saturated peanut oil has turned to a thick mass which can be used only sparingly during cold weather months. The canola-based biodiesel is able to withstand much colder operating temperatures.

Figure 8.3.1-5. This waste vegetable oil has been "contaminated" with fryer rinse water, which can be seen in the bottom half of the container. Explain to the oil supplier that you are concerned about the purity of the WVO and ask that foreign objects or other contaminants not be added to the oil.

Controlling Free Fatty Acid Levels

If your oil supplier is willing to work with you, it may be worthwhile explaining the benefits of using free fatty acid (FFA) or fryer oil test strips such as the 3M™ Shortening Monitor test kit (see Chapter 4.4). With use the fryer oil begins to deteriorate and its free fatty acid level increases. Overworked fryer oil results in a decline in food quality and also makes it more difficult for the micro-scale biodiesel producer to create quality fuel. The restaurant may be grateful for an unbiased test method for determining when the oil has reached the end of its life, and you will be grateful for easier processing.

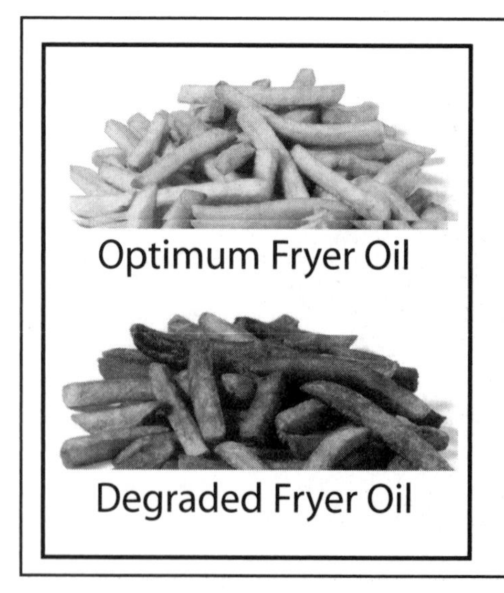

Figure 8.3.1-6. If your oil supplier is willing to work with you, it may be worthwhile explaining the benefits of using free fatty acid (FFA) or fryer oil test strips such as the 3M™ Shortening Monitor test kit (see Chapter 4.4).

Figure 8.3.1-7. With use the fryer oil it will begin to deteriorate and its free fatty acid level will increase. Overworked fryer oil will result in a decline in food quality and make it more difficult for the micro-scale biodiesel producer to create quality fuel.

When the FFA content of the WVO increases beyond 1%, ever-increasing amounts of catalyst are required to neutralize the acidity. When the FFA level increases beyond 2% soaps begin to form, causing excessive emulsion formation during the washing process. Although the FFA level can be reduced (see Chapter 8.3.5) it is better to obtain WVO that has the lowest possible acidity levels.

As a courtesy to your oil supplier, be punctual, taking care to pick up oil on a regular basis. This will ensure a good relationship and a consistent supply of feedstock.

Figure 8.3.1-8. As a courtesy to your oil supplier, be punctual, taking care to pick up oil on a regular basis. This will ensure a good relationship and a consistent supply of feedstock.

Once the oil has been delivered to your biodiesel processing facility, it is a good idea to test and record the FFA level of each pail of WVO received. Using this information will assist in determining why a given batch of biodiesel may have failed the transesterification process. Check the oil for water, foreign objects, and "bottom sludge," either leaving them behind when you pour the oil into another pail or allowing them to settle in a large settling tank, removing the "good" WVO after a period of time and composting or landfilling the remainder.

If there is a large variation in FFA level between oil samples, consider feeding a mixture of both high- and low-FFA oil into the waste oil treatment tank to "average out" the overall acidity reading.

Transferring WVO to the Treatment Unit

To proceed with transferring the WVO to the treatment tank follow these steps:

1. Open valve V1 to allow WVO to flow into the tank.
2. Close valve V2 to prevent draining the tank.
3. Open valve V3 to allow WVO to circulate within the treatment processor.
4. Open valve V4 to permit circulation of oil.
5. Close valve V5 to prevent transfer of WVO into the biodiesel reaction tank.
6. Place the WVO strainer and suction pipe into the waste oil pail.
7. Activate pump P1 to begin drawing WVO into the tank.

As the level of the WVO gets lower in the pail, air is sucked into the line, reducing the oil flow. This is not a concern. Simply turn off pump P1 and pour the remaining oil into the next full pail of WVO to be transferred into the treatment unit.

While filling the treatment tank, watch the sight glass tubing to ensure the tank is never more than about 80%-90% filled, leaving some air space at the top.

Sight Glass Volume Calibration

If this is the first time the waste oil treatment unit is being filled, it is necessary to perform a volume calibration to determine the relationship between the sight glass vertical level and the volume of fluid in the tank. Although this data is not required for the treatment unit, it is easiest to perform the procedure here and extrapolate the data to calibrate the biodiesel reactor.

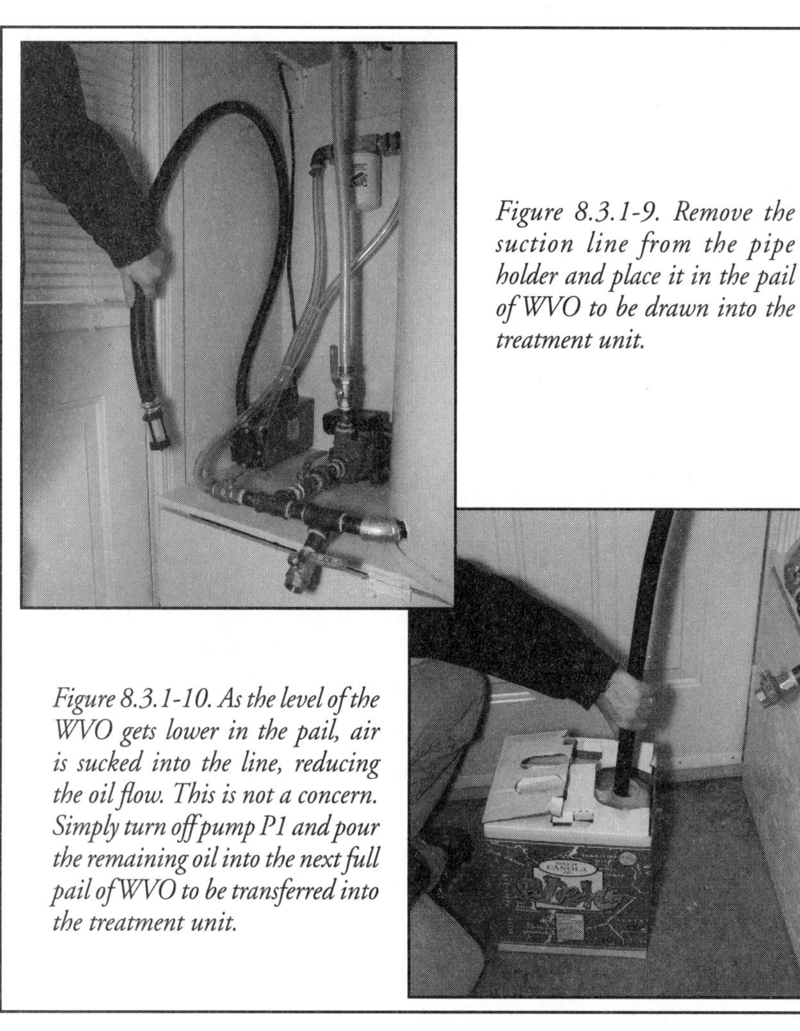

Figure 8.3.1-9. Remove the suction line from the pipe holder and place it in the pail of WVO to be drawn into the treatment unit.

Figure 8.3.1-10. As the level of the WVO gets lower in the pail, air is sucked into the line, reducing the oil flow. This is not a concern. Simply turn off pump P1 and pour the remaining oil into the next full pail of WVO to be transferred into the treatment unit.

The calibration process is very simple:

1. Start with a bucket that has volume levels embossed on the side. Metric units are recommended as they simplify the measurement process. (From this point forward in the book, I will use metric measurements as the primary units, with imperial measure as the secondary units.)
2. Fill the bucket with WVO to a level of 10 liters (2.6 gallons).
3. Draw the WVO into the treatment tank. Any leftover WVO should be measured and an identical amount added to the bucket so that the small amount remaining does not affect the measurement process.

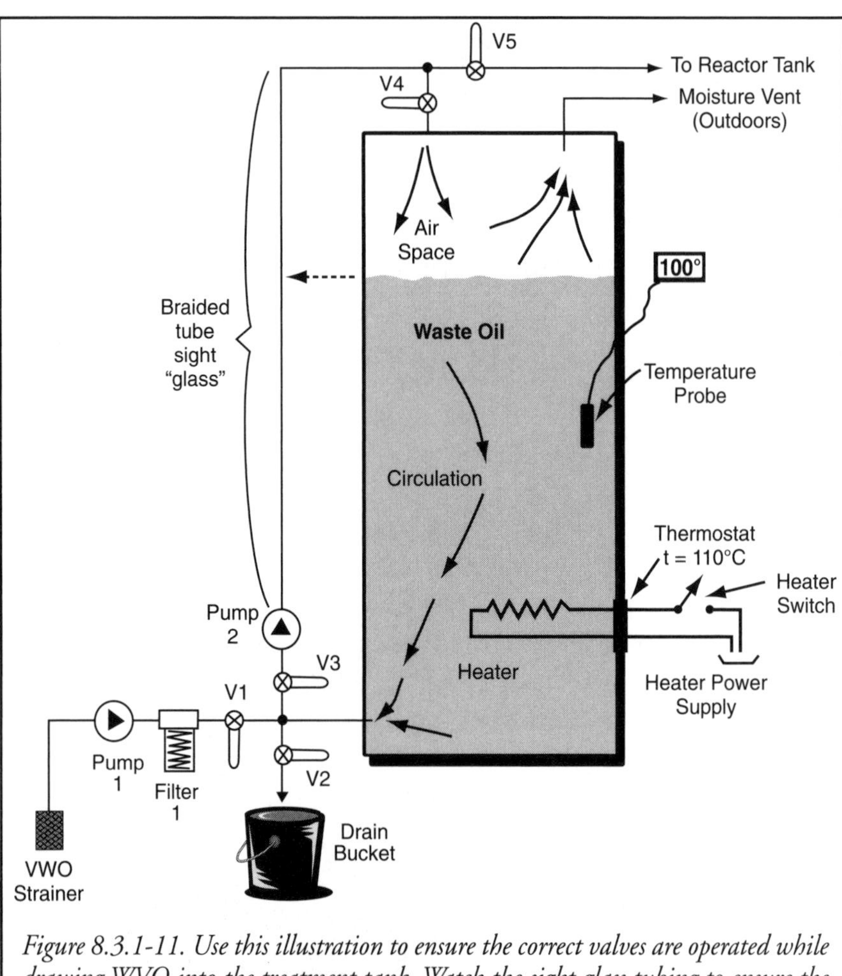

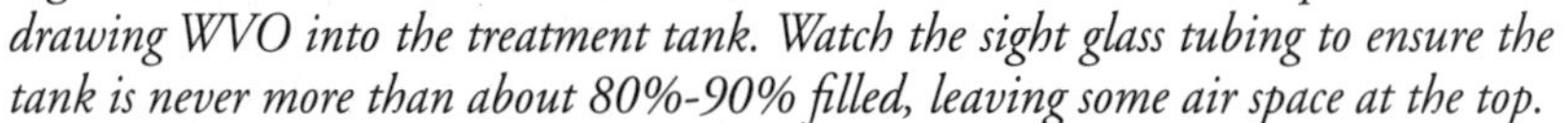
Figure 8.3.1-11. Use this illustration to ensure the correct valves are operated while drawing WVO into the treatment tank. Watch the sight glass tubing to ensure the tank is never more than about 80%-90% filled, leaving some air space at the top.

For example, if the suction draining of the first bucket resulted in 1 liter (0.26 gallon) of oil remaining, add another liter to the bucket and draw it into the tank. There should now be 10 liters of oil in the tank and 1 liter remaining in the bucket.

4. Mark the level of WVO in the sight glass to indicate the volume of WVO in the tank.
5. Continue filling the tank and measuring the vertical level until the tank is filled with 200 liters (53 gallons), assuming a 60-gallon tank is installed.

The vertical measurements can be marked on the sight glass with electrical tape or permanent marker, although the solvent effect of oil and grease will eventually cause the marks to fade. Be sure to keep the height-to-volume ratio data for future use.

The 227-liter (60-gallon) water tanks that are used in my system have a vertical-measurement-to-volume relationship as follows:

Vertical Measurement (Centimeters)	Vertical Measurement (Inches)	Tank Volume (Liters)	Tank Volume (Gallons)
39	15.4	50	13.2
70	27.6	100	26.4
83	32.7	120	31.7
101	39.8	150	39.6

The relationship between vertical height and volume is not linear because of the rounded shape of the tank as well as the volume of oil that resides inside the circulation pump.

It is most important to ensure accurate markings in order to identify the correct amount of WVO for each transesterification process (see Chapter 8.3-6).

Figure 8.3.1-12. When you have finished filling the WVO treatment unit, place the suction pipe on a pipe holder and place the strainer end of the pipe in a small bucket to avoid back dripping of WVO.

8.3.2
Cold Pressing Oilseed

There are numerous oilseeds that can be cold pressed directly into refined oil and protein meal. As discussed in Chapter 4, virtually any pressed oilseed can be used as a feedstock for biodiesel, while the meal may be used as animal feed supplement, fertilizer, aquaculture fish food, or direct combustion "biomass" fuel.

Figure 8.3.2-1. Biodiesel: Basics and Beyond photographer Lorraine found a new friend as we travelled across the prairies gathering research for this book. Some of you will know her as "Spa Girl" from her past exploits in my earlier book, The Renewable Energy Handbook. Lorraine's friend, the sunflower (left), is an excellent biodiesel feedstock.

The economics of using food-grade oilseed oil as a biodiesel feedstock may be difficult to justify unless there are mitigating circumstances, for example when the aggregate value of the extracted oil plus the commodity or heating value of the meal is greater than the current commodity price for the oilseed. However, the more likely event is either a soft commodity market for any oilseed variety or that a poor-quality seed crop which is unmarketable.

Such was the case in Ontario, Canada during the 2005 canola crop harvest, when the buyer/crusher rejected a very large percentage of the crop as subgrade. The reverberations through the growing community were felt

Figure 8.3.2-2. Dr. Martin Reaney contemplates a crop of canola in a Saskatchewan field. (The color image of this picture is magnificent with the yellow canola flowers and magnificent sky. It is no wonder Saskatchewan is referred to as "The Land of the Living Skies.")

Figure 8.3.2-3. The tiny canola seed is a staple of the Canadian prairies, with some 8+ million tonnes grown in Canada each year.

around the province, with calls on the government to intervene with significant crop insurance payments. Had all concerned created a special commodity price for "fuel-grade" canola (or any oilseed) more money would have ended up in farmers' pockets, alleviating the losses incurred in a horticultural roll of the dice with nature.

One enterprising dairy farm operator decided that perhaps all was not lost for the canola farmers and purchased 7,500 tonnes of the off-specification oil seed. "My plan is to try and develop a secondary use for the oilseed by using the oil for either straight fuel consumption [see Chapter 7.9] or to produce biodiesel," explains George Heinzle, who has a farm in Eastern Ontario. "Because my dairy cows require a protein supplement in their feed ration, I can use the meal to offset feed costs."

Figure 8.3.2-4. Every year approximately 5% or more of the total canola crop is declared "off-specification" and in effect wasted. Generating a special commodity price for "fuel grade" canola (or any oilseed) would put more money into farmers' pockets and help alleviate the losses incurred in a horticultural roll of the dice with nature.

Figure 8.3.2-5. Enterprising dairy farm operator George Heinzle of Ontario decided to purchase 7,500 tons of off-specification canola oilseed to produce biodiesel and provide protein supplements for his dairy herd.

Figure 8.3.2-6. George explains that he is only just starting to experiment with cold pressing, although he has some experience working with his sister-in-law, who produces cold-pressed oils for the specialty food industry.

George explains that he is only just starting to experiment with cold pressing, although he has some experience working with his sister-in-law, who produces cold-pressed oils for the specialty food industry. "I purchased a small, two-ton-per-day Chinese screw press (Figure 8.3.2-5) and a low-speed diesel engine (Figure 8.3.2-6) to begin my experiment," George continues. "After a fair bit of fiddling with the press I have finally got it to produce a reasonable amount of oil from both my canola (Figure 8.3.2-10) as well as some soybean that I have been testing. The screw press operates differently from an extrusion press but is less costly and easier to repair if there are problems. The Chinese unit I purchased is built like a tank. I have also fed the cows both 100% canola (Figure 8.2.3-12) and soy (Figure 8.3.2-13) meal and they like the flavor of both, although when I am ready to upgrade to full-scale operation they will only get a small percentage of the oilseed seed meal in their feed supplement."

George goes on to say that he is anxious to start working with the oil either in a straight oil system or, if he can get a small cooperative running, in a biodiesel production system. "I have a number of friends in Germany who have adapted their tractors and other farm equipment to straight oil systems. If I can help out local farmers by showing that there is a market for their waste crops, this will help everyone in the farming sector," he concludes.

Figure 8.3.2-7. George purchased a small, two-ton-per-day Chinese screw press and a low-speed diesel engine to begin his experiment. "After a fair bit of fiddling with the press I have finally got it to produce a reasonable amount of oil from both my canola and some soybean that I have been testing," George says.

Figure 8.3.2-8. These tiny canola seeds contain approximately 1/3 oil and a valuable protein meal that can be used for fertilizer, anaerobic digestion to make methane gas, feed supplement for animals and aquaculture. Sadly, even though these seeds are suitable for all of these products, the human food market has rejected them due to their color.

Figure 8.3.2-9. George is seen filling the small test hopper with canola seed during my visit to his facility.

Figure 8.3.2-10. Although the oil is expressed from the seeds very slowly, the process can continue 24 hours per day, yielding a large volume of oil in much the same manner as maple trees slowly drip their sweet nectar into collection buckets.

Figure 8.3.2-11. The Chinese screw press expresses flakes of protein meal which are collected in a receiver drum under the press frame.

Figure 8.3.2-12. Canola meal is dispensed from the screw press in a dry-to-the-touch flaky powder.

Figure 8.3.2-13. Soybean meal is expressed as large thin flakes which break into small pieces when handled. The oil content of either canola or soy meal is approximately 7%

Figure 8.3.2-14. George's "girls" are happily enjoying some feed ration that has been made with cold-pressed, off-specification oilseed that would otherwise have gone to waste.

Figure 8.3.2-15. The Peacock Industries facility in Hague, Saskatchewan uses a canola cousin, mustard, to produce an organic pesticide and nematocide from the protein meal and a coproduct of 150 liters of oil per day. As production increases, the company plans to produce biodiesel itself or sell the feedstock directly to an existing producer.

Figure 8.3.2-16. These twin extrusion presses can produce 150 liters (40 gallons) of oil and 200 kilograms (≈ 440 pounds) of meal per day.

Figure 8.3.2-17. The oil is expressed from the press and drops into a trough where it flows into a receiver tank that settles out any seed casing and meal that can foul the oil.

Figure 8.3.2-18. The meal "noodles" drop into a hopper and are fed by screw auger into a dryer bag and storage receiver.

Figure 8.3.2-19. The noodles travel through a screw auger before falling into the receiver bag the meal is allowed to dry.

Chapter 4.2 discusses Peacock Industries (Figure 8.3.2-15) and its production of an organic pesticide and nematocide from the meal portion of a canola seed cousin, mustard. The Peacock system uses an alternative to the screw press known as an extrusion press (Figure 8.3.2-16) which produces long spaghetti-like noodles of protein meal (Figures 8.3.2-18 and 8.3.2-19). Oil production from this 0.5-ton-per-day press provides approximately 150 liters per day of oil that can be used for any number of applications, including biodiesel. Assuming that the original oilseed is "off-specification" because of color, acid content, or frost damage, the resulting oil is not food grade and its value is negligible; fortunately biodiesel is not quite so fussy.

The recovered meal contains approximately 7% oil, making it an ideal combustion fuel that is compatible with many stoker/boiler units or pellet stoves. Because of their oil content, these pellets have a high energy density, providing greater heating value per unit of weight than traditional wood pellets.

Regardless of which seed type is used for the oilfeed stock or which press system is used, the oil only requires filtering to remove meal and seed particles before it is ready for transfer into the oil treatment unit as described earlier in Chapter 8.3.1.

Figure 8.3.2-20. Regardless of which seed type is used for the oil feedstock or which press system is used, the oil only requires filtering to remove meal and seed particles before it is ready for transfer into the oil treatment unit as described earlier in Chapter 8.3.1.

8.3.3 One-Liter Biodiesel Test Reaction

The one-liter biodiesel test reaction is a staple of classroom science experiments and summer ecofairs the world over. Unfortunately the simplicity of this test encourages many people to jump into biodiesel production. As I have stated before and continue to repeat, **it is exceptionally easy to make poor-quality biodiesel and leave a toxic mess** at the same time. This test program will demonstrate the problem clearly.

At the same time, this test can be useful in determining if the chemical analysis calculations required to make the correct concentration of sodium methoxide are accurate simply by scaling the production process run to blender size. This test is also required for one of the quality control tests and will be discussed later in Chapter 8.3.9.

Biodiesel is produced by the reaction of feedstock oils with an alcohol in the presence of a catalyst to produce fatty acid methyl esters (FAME) or biodiesel. The typical process, known as transesterification, is:

100 kg feedstock oil + 10 kg methanol ➡ 100 kg FAME + 10 kg glycerol

or in blender-size volume

1 liter feedstock oil + 200 ml methanol ➡ 1 liter FAME + 200 ml glycerol

Unfortunately, what this formula does not tell us is that the resulting FAME is contaminated with leftover reactants from the transesterification process, including methanol, catalyst, glycerin, unreacted or bonded glycerin resulting from incomplete reaction processes, and lastly soap produced by the presence of oils or fats in a caustic solution.

The one-liter test verifies the transesterification process in a kitchen blender, and as we discussed in Chapter 8.2.10 on safety issues, the test must not be done in your kitchen!

The Process

The following photomontage describes the one-liter biodiesel test reaction:

Figure 8.3.3-1. Step 1. Heat 1 liter (34 ounces) of cold-pressed or food-grade (not WVO) vegetable oil, heating it to a temperature of 50°C to 55°C (122°F to 131°F). Add the heated oil to the blender in the biodiesel processing area. Ensure that ventilation fans are turned on and all safety precautions are followed.

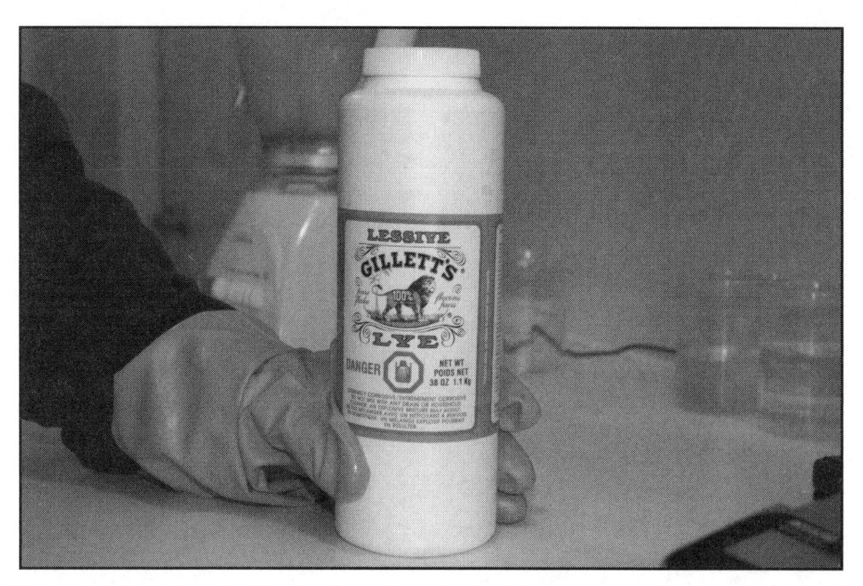

Figure 8.3.3-2. Step 2. Obtain the sodium hydroxide container. (Sodium hydroxide is commonly sold as a drain cleaner such as Gillett's Lye. Ensure that it is 100% concentration and that the container is dry. If the flakes are lumpy the product has been exposed to moisture, rendering it unusable.

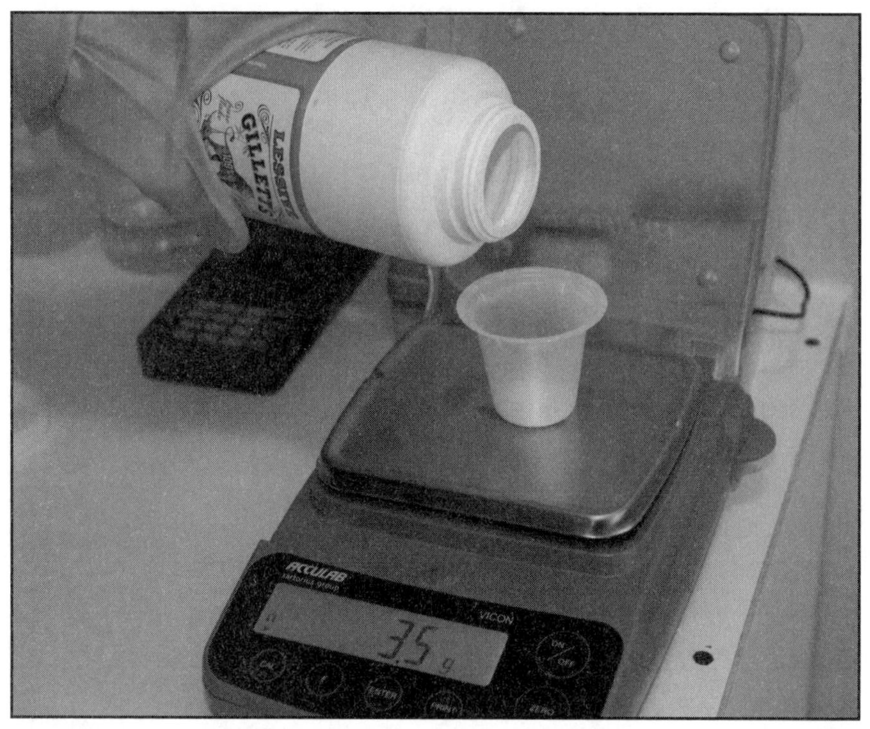

Figure 8.3.3-3. Step 3. Weigh 3.5 grams of sodium hydroxide, taking care to subtract or tare the weight of the measuring cup.

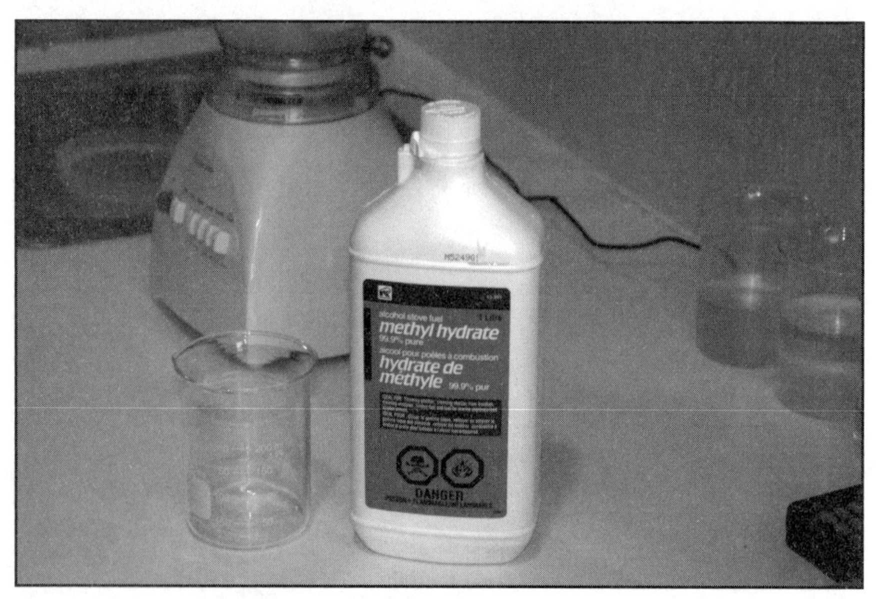

Figure 8.3.3-4. Step 4. Obtain the methanol container.

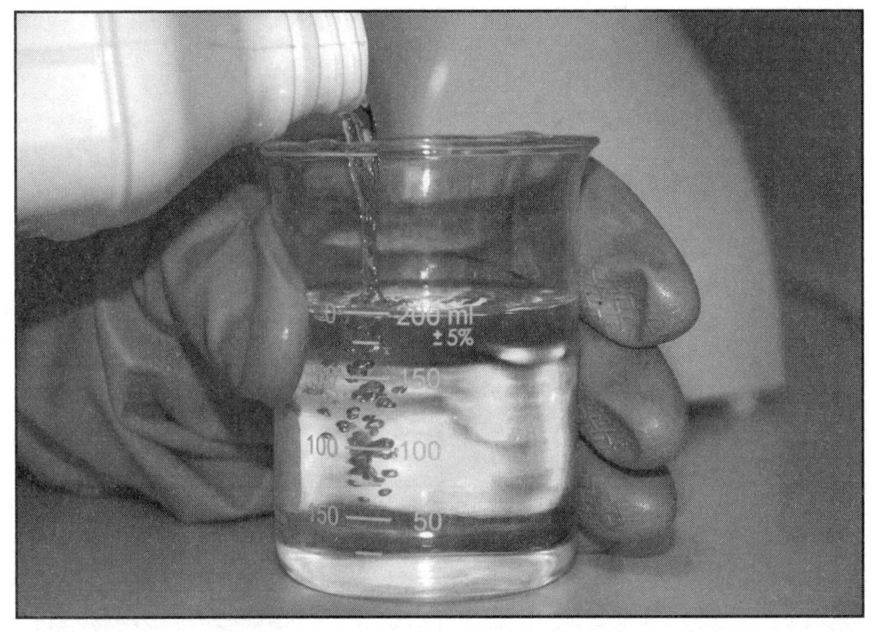

Figure 8.3.3-5. Step 5. Measure 200 milliliters of methanol into a beaker.

Figure 8.3.3-6. Step 6. Carefully add the sodium hydroxide to the methanol and stir using a plastic stirring stick to ensure that the beaker is not broken. Sodium hydroxide does not dissolve quickly, so be patient.

Figure 8.3.3-7. Step 7. Slowly add the sodium methoxide to the heated oil. Do not cause the oil to overflow the blender.

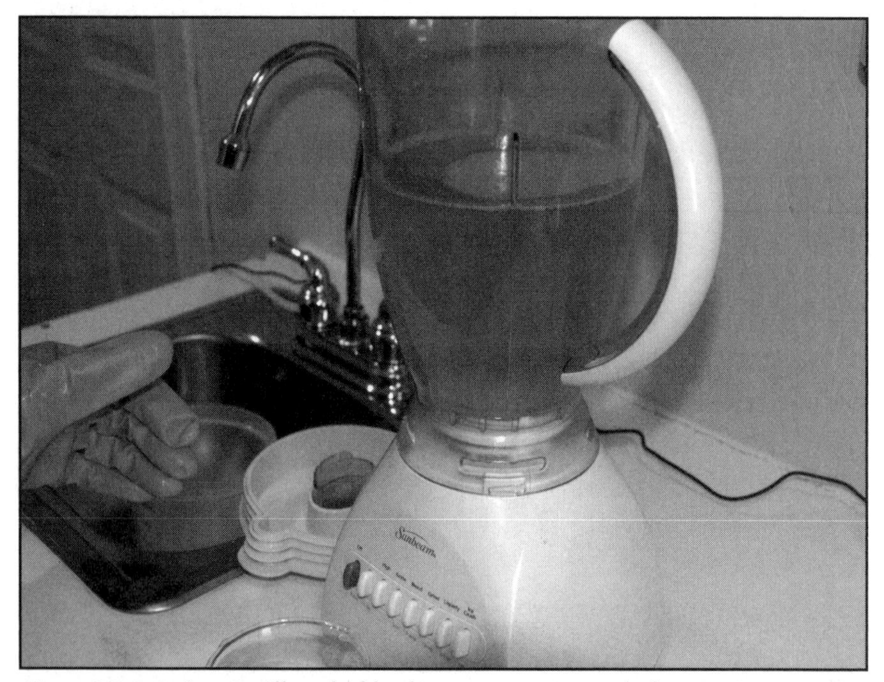

Figure 8.3.3-8. Step 8. Allow the blender to operate vigorously for 15 minutes, taking care not to cause the blender to overflow. Stop the blender at the end of this time.

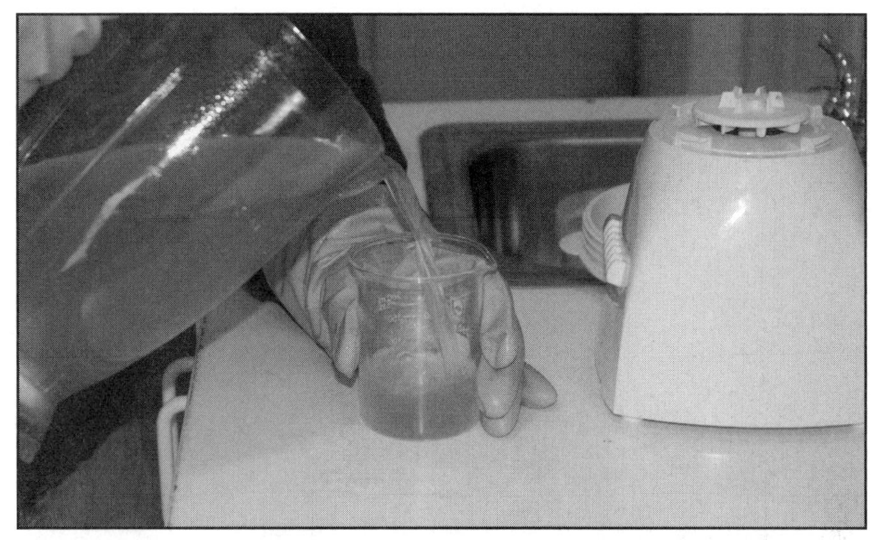

Figure 8.3.3-9. Step 9. Carefully pour the oil/sodium methoxide mixture into a beaker.

Figure 8.3.3-10. Step 10. Allow the beaker to settle for several hours

CAUTION: The beaker and remaining solution will continue to exude harmful fumes and remain flammable. Ensure the area remains well ventilated and off limits to unauthorized persons.

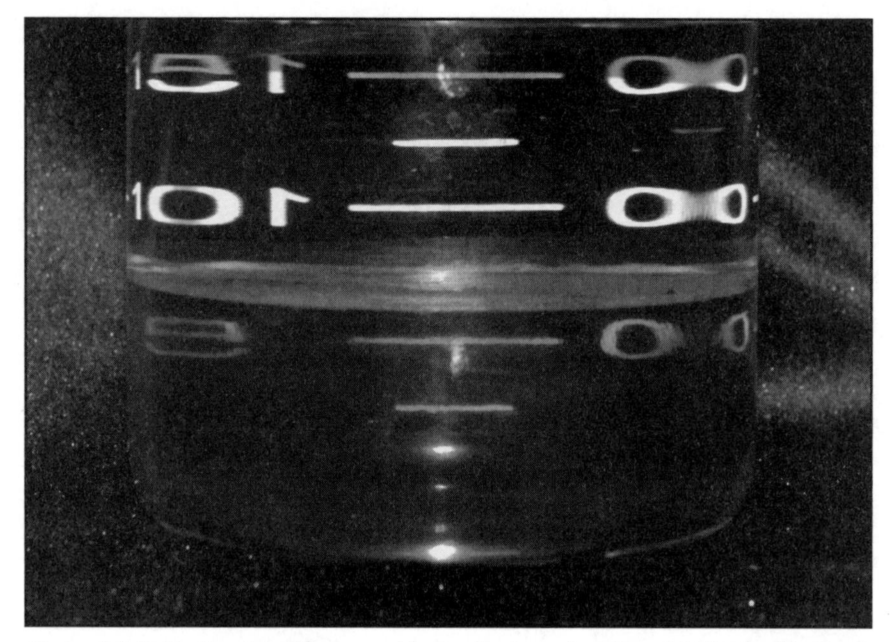

Figure 8.3.3-11. Step 11. After several hours the glycerol and raw biodiesel will separate into two distinct phases, with the heavier glycerol on the bottom and the raw biodiesel floating on top.

Test Summary

At this stage in the actual biodiesel process, the glycerin is removed by decanting it from the bottom of the reaction tank. Many home brewers then rid themselves of the glycerin by whatever means possible (check for a pile of dark, heavy jugs behind the shop to see the most common method of disposal) and then decant the raw biodiesel for use.

Here the biodiesel camp breaks into two distinct factions: those who wash their biodiesel and those who don't. In either case the process is wrong, as we shall learn.

The balance of Chapter 8.3 will tidy up the loose ends and show several procedures that greatly improve fuel quality and the treatment of the environment at the same time. Admittedly, my process does involve a bit more work and equipment, which is one reason why many end their story here.

Analysis	Method	Min	Max	Raw Un-Washed Biodiesel	Notes
Acid Number, mg KOH/gram	ASTM D664		0.80	0.11	
Ash, Sulfated, Mass %	ASTM D874		0.020	0.0	
Total Sulfur, Mass %	ASTM D5453		0.05	0.0006	
Cetane Number	ASTM D613	47		51.6	
Cloud Point, °C	ASTM D5773			-3	
Copper Corrosion	ASTM D130		No. 3	1a	
Vacuum Distillation End Point, ° C	ASTM D1160		360°C at 90%	525	1
Flash Point, °C	ASTM D93 Procedure A	130.0		25	2
Method for Determination of Free & Total Glycerin	ASTM D6584				
Free Glycerin, Mass %			0.020	0.98	3
Total Gylcerin, Mass %			0.240	1.407	4
Phosphorus (P), Mass %	ASTM D5185		0.001	0.1	5
Carbon Residue, Mass %	ASTM D4530		0.05	0.022	
Kinematic Viscosity, Centistokes at 40°C	ASTM D445	1.9	6.0	5.828	6
Water and Sediment	ASTM D1796		0.050	0	7
Total Methanol (%)	EN 14110		0.2	2	8

Table 8.3.3-1.

This table indicates the limits of various parameters according to ASTM D 6751 standards and compares them with a sample of raw, unwashed biodiesel. This raw biodiesel fails approximately 50% of the tests with several failing by a very wide margin.

Raw, Unwashed Biodiesel

Biodiesel that is used as is, or unwashed, is highly toxic, containing excessive levels of methanol, caustic catalyst reactants, glycerine, and other contaminants that exceed ASTM standards.

Instead of the "environmentally friendly" fuel they thought they were making, raw biodiesel users now have toxic fuel and unknown exhaust stream pollutants resulting from the fact that the vehicle's emission control system is not designed to process soap, sodium methoxide, glycerin, etc. In addition, the raw glycerol must be dealt with sooner or later. Methanol and the soup of other chemicals will eventually be dumped deliberately or leak out of the glycerin container over time. We have already debunked the myth that the methanol will magically evaporate if the container is placed outdoors; nor will this neutralize the caustic materials accumulated by the dissolved catalyst. And none of these issues even speaks to fuel quality and potential damage to a vehicle's engine. It is one thing to pour questionable fuel into a 1978 Volkswagen Rabbit that is worth less than the value of a real fill up; it is quite another to pour this same fuel into a shiny new BMW.

Dr. Andre Tremblay is Professor and Chair of the University of Ottawa Chemical Engineering Department. He and I performed a quick analysis of raw, unwashed biodiesel before sending it off to an accredited test laboratory for complete analysis and we quickly determined that this stuff was so far below any realistic quality standard that further consideration was of little value. Table 8.3.3-1 compares the sample of raw, unwashed biodiesel against the ASTM fuel standards and confirms our initial assessment.

Notes

1. High levels of di- and trigylcerides are the likely cause of the end point distillation temperature failure. Although unreacted gylceride measurement is not contained in the ASTM data tables it was found to be 2.56%, which indicates that the biodiesel reaction was not completed.
2. Biodiesel is not considered a flammable material and has handling and storage procedures that reflect this. Excess methanol levels have caused the flash point to drop precipitously.
3. Free glycerin levels are approximately 50 times the allowed levels, which indicates that the fuel has not been washed. Glycerin causes filter plugging and fuel injection system problems, particularly in newer engines with high pressure and mechanical tolerance.
4. Total glycerin is the sum of free glycerin and unreacted glycerin still "bound" to the biodiesel fuel molecules. High total glycerin is indicative of unwashed fuel and incomplete reaction.

5. Phosphorous levels are 100 times greater than the allowable limits. Phosphorous can be attributed to feedstock oil, waste materials from cooking or frying, or phosphoric acid used in the neutralization process of free fatty acids. It is unlikely that micro-scale producers can influence this value through process techniques.
6. I have indicated that viscosity has failed as the ASTM committee is currently evaluating lowering this level to 5 centistokes to more closely match diesel fuel parameters. High viscosity is known to damage fuel injection pumps, reduce atomization of fuel spray, and dilute engine lubricating oil.
7. Water and sediment is recorded as zero in this instance since the fuel has not been washed to remove contaminants. Washing fuel without subsequent drying will virtually guarantee a failure of this parameter.
8. Total methanol is not recorded in the ASTM standards but is inferred from the flashpoint level. I have provided the equivalent levels from the European standard, showing that total methanol has failed by a factor of 10 times.

Figure 8.3.3-12. Dr. Andre Tremblay is Professor and Chair of the University of Ottawa Chemical Engineering Department. He and I performed a quick analysis of raw, unwashed biodiesel before sending it off to an accredited test laboratory for complete analysis and we quickly determined that this stuff was so below any realistic quality standard that further consideration was of little value.

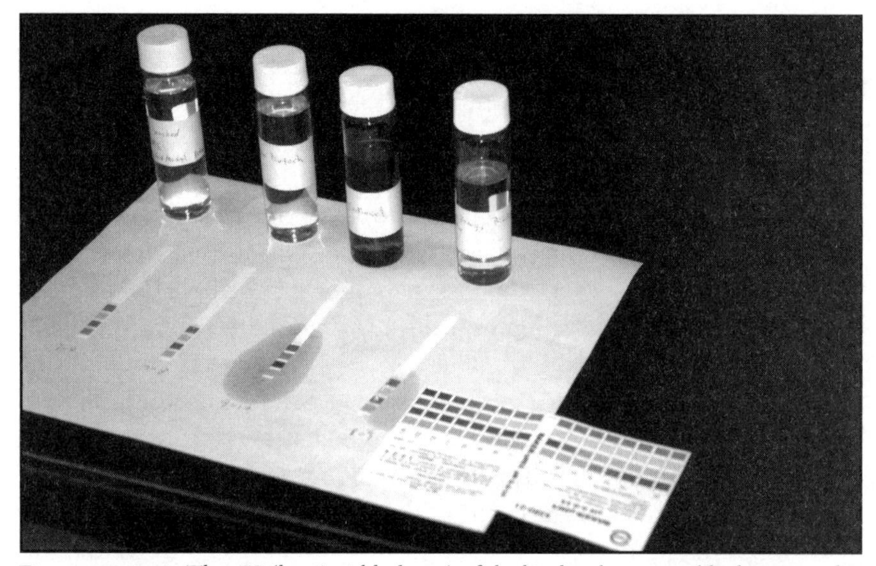

Figure 8.3.3-13. The pH (base/acid balance) of the biodiesel was quickly determined to be so high that engine damage would result as the highly basic fuel attacked metal and nonmetallic components. The raw, unwashed biodiesel was found to be exponentially more corrosive than the ASTM-certified samples tested alongside it.

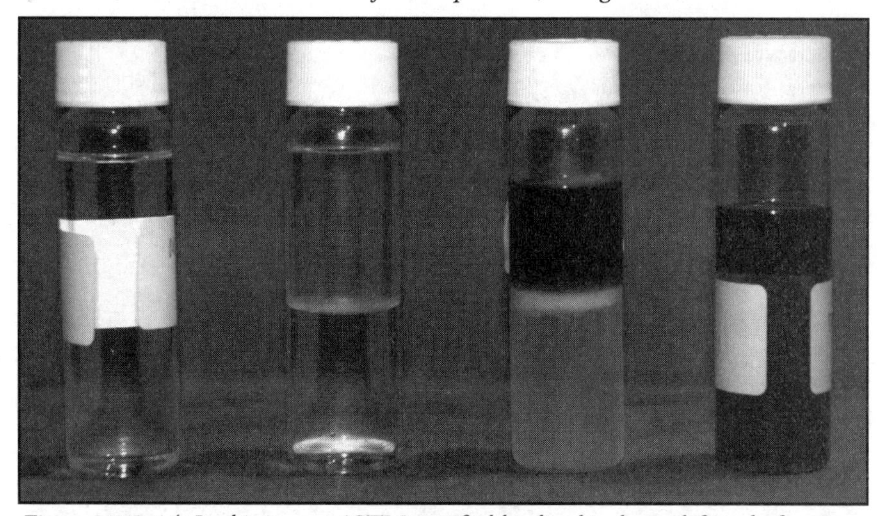

Figure 8.3.3-14. In this image, ASTM-certified biodiesel is shown left and after water washing (second from left). Note that the wash water is clear and that there are no soaps at the interface between the biodiesel and water. The tube shown far right is a sample of homemade biodiesel made from WVO. The sample shown second from right is the same homemade biodiesel after water washing. Notice the residual soaps in the wash water and at the interface layer between the water and "biodiesel." Additionally, the pH level of the biodiesel is a corrosive 10.5, indicating the presence of a considerable amount of unreacted sodium hydroxide.

Washed Biodiesel

The second group of biodiesel producers attempt to improve on the raw, unwashed biodiesel by washing their fuel. While there is no question that this action should improve fuel quality, there is little or no proof of the supposed improvement; nor do they provide a way to perform fuel quality tests. Assumptions are made; statements are made; proof is seldom given.

Things don't improve from the environmental side either. The mountain of glycerine remains stubbornly at the back of the shed while the little matter of toxic, contaminated wash water has now to be dealt with.

The process of washing raw biodiesel is designed to absorb anything that is not biodiesel. Fair enough. But what about the excess methanol and other contaminants that end up in the wash water? The government-accredited test centre that was used for the test shown in Table 8.3.3-1 indicated that one sample I sent for examination contained 2% methanol by volume. This translates into 3 liters (100 ounces, or approximately ¾ gallon) that ends up contaminating a large volume of wash water. And one biodiesel producer referred to in Chapter 7.1 thought that disposing of this waste down the sewer was no worse than flushing some household chemicals down the drain as you clean the sink! Do you remember the statement in Chapter 7 concerning methanol?

Methanol is registered as a hazardous waste product by the U.S. Environmental Protection Agency (40 CFR Ch. 1, Section 261.33). Methanol is also known as an "F-Listed chemical" that makes each waste it contaminates hazardous no matter what its concentration. Even one drop of an F-listed chemical on a shop rag, in adsorbent, or in used oil is enough to create a regulated hazardous waste that must be disposed of in accordance with federal, provincial and state laws.

The one-liter biodiesel test reaction demonstrates both the simplicity and the complexity of biodiesel production. It is very easy to make poor-quality fuel and wreck the environment at the same time. It takes more effort and time to improve the fuel quality, provide the necessary test programs to prove that the fuel is acceptable for its intended use, and protect the environment by detoxifying the waste streams.

8.3.4
Oil Pretreatment

As discussed in Chapter 4, vegetable oils contain varying amounts of free fatty acids (FFAs) which complicate the conversion of feedstock oil to biodiesel since they react with the catalyst to produce soaps that inhibit the transesterification process. Depending on the level of FFAs in the feedstock oil it may be necessary to undertake a pretreatment process. Regardless of whether the WVO feedstock is pretreated to reduce FFA levels or not, it is vitally important to ensure that it contains no water that would almost certainly stall the reaction process and create a thick viscous emulsion (Figure 8.3.4-1) rather than biodiesel.

Section of Chapter	Feedstock Oil/Fat	Free Fatty Acid Level	Relative Cost
4.2	Refined Vegetable Oil (Soybean, Canola)	0 - 1%	Highest
4.4	Waste Fryer Oil/Fat	2 - 7%	Low
4.4	Animal Fats (Beef Tallow, Lard)	5 - 30%	Low
4.5	Yellow Grease	7 - 30%	Moderate
4.6	Brown Grease	> 30%	Very Low

Table 8.3.4-1
Range of free fatty acids (FFAs) in various feedstock oils and fats (Source: 3M Food Services Business and Canada Challenge Inc.)

In general, the lower the feedstock cost the higher the free fatty acid level, which complicates the biodiesel production process and affects the resulting yield. The home-scale or micro-producer will almost certainly wish to focus on feedstock oils that are either free or of very low cost, virtually eliminating the best feedstock: refined vegetable oils.

The agricultural sector, and in particular oilseed growers, have access to off-specification or low-commodity-price crops which can provide the opportunity for cold pressing and recovery of low-FFA oil for use as a biodiesel feedstock.

When the FFA level is below 1% there is no requirement for pretreatment of the feedstock oil except for water removal, discussed below.

When the FFA level is above 1% and below 6% (as it generally is when recovered from restaurant deep fryers) it is necessary to neutralize these FFAs with the addition of excess amounts of alkali catalyst (discussed in Chapter 8.3.5) which will form soap. Although some of the catalyst is used up in the production of soap, there will be sufficient amounts remaining to ensure that the transesterification process continues to completion. The range of FFA level is not a hard and fast rule as other contaminants, including water, dictate the actual upper limit. If trace amounts of water are present in the feedstock oil the upper limit is severely reduced, often to a maximum level of less than 3%.

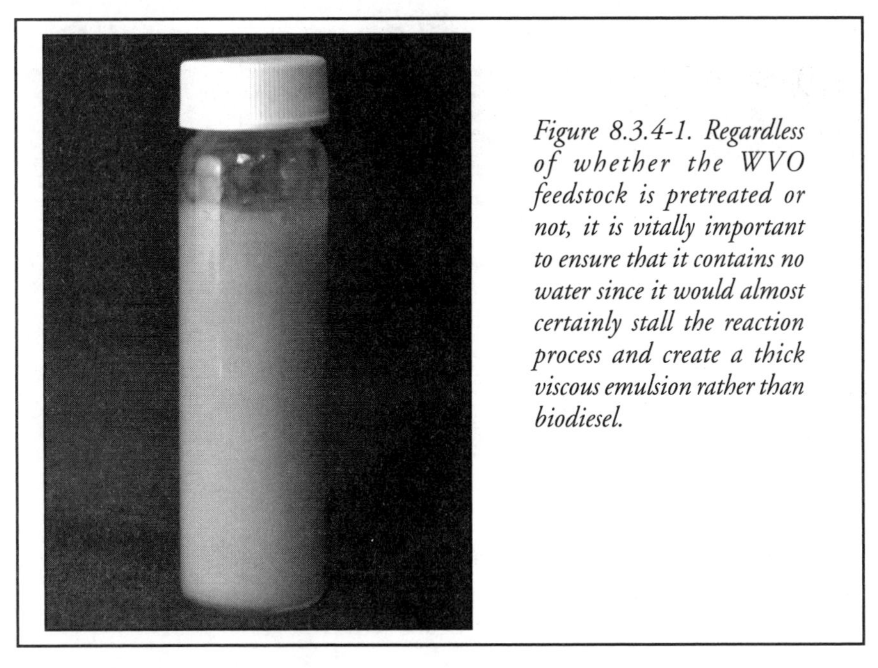

Figure 8.3.4-1. Regardless of whether the WVO feedstock is pretreated or not, it is vitally important to ensure that it contains no water since it would almost certainly stall the reaction process and create a thick viscous emulsion rather than biodiesel.

If the FFA level of the feedstock oil exceeds approximately 6%, adding more catalyst will lead to the formation of a jellylike substance rather than biodiesel because of excessive soap formation.

Although it is possible for the micro-producer to pretreat the WVO to reduce FFA levels below a given threshold level (ideally below 2%), few will take the time and trouble to complete the required processing, although one pretreatment method is discussed in Chapter 8.3.6. The simplest method of working with high FFA levels is simply to "dilute" the FFAs by adding additional feedstock oil with an average FFA reading that is lower than that

Figure 8.3.4-2. Although it is possible for the micro-producer to pretreat the WVO to reduce FFA levels below a given threshold level (ideally below 2%), few will take the time and trouble to complete the required processing. The simplest method of working with high FFA levels is simply to "dilute" the FFA level by adding additional feedstock oil with an average FFA reading that is lower than that of the original high-FFA feedstock. Use the 3M™ Shortening Monitor strips discussed in Chapter 4 to quickly determine FFA levels.

of the original high-FFA feedstock.

For example, assume that a mixture of 100 liters of WVO has an FFA reading of 7% and that this level should be lowered to ensure trouble-free transesterification. An additional 100 liters of WVO has been located with an FFA reading of 4%. Mixing the two feedstocks produces a blend with an FFA reading of 5.5%. If you have access to low-cost, refined vegetable oil it will offer the best dilution because it has an FFA level of effectively zero.

Dehydrating Oil Feedstock

To begin the WVO and cold-pressed oil dehydration process proceed with the following steps:

1. Ensure the WVO treatment unit is filled to a maximum of 90% of its capacity.
2. Close valve V1 to prevent backflow of WVO into suction pump1.
3. Ensure tank drain valve V2 is fully closed.
4. Open circulation valve V3.
5. Open tank circulation valve V4.
6. Close tank transfer valve V5.
7. Start circulation pump 2. Confirm that oil is flowing by watching fluid motion in sight glass.
8. Activate tank heater.

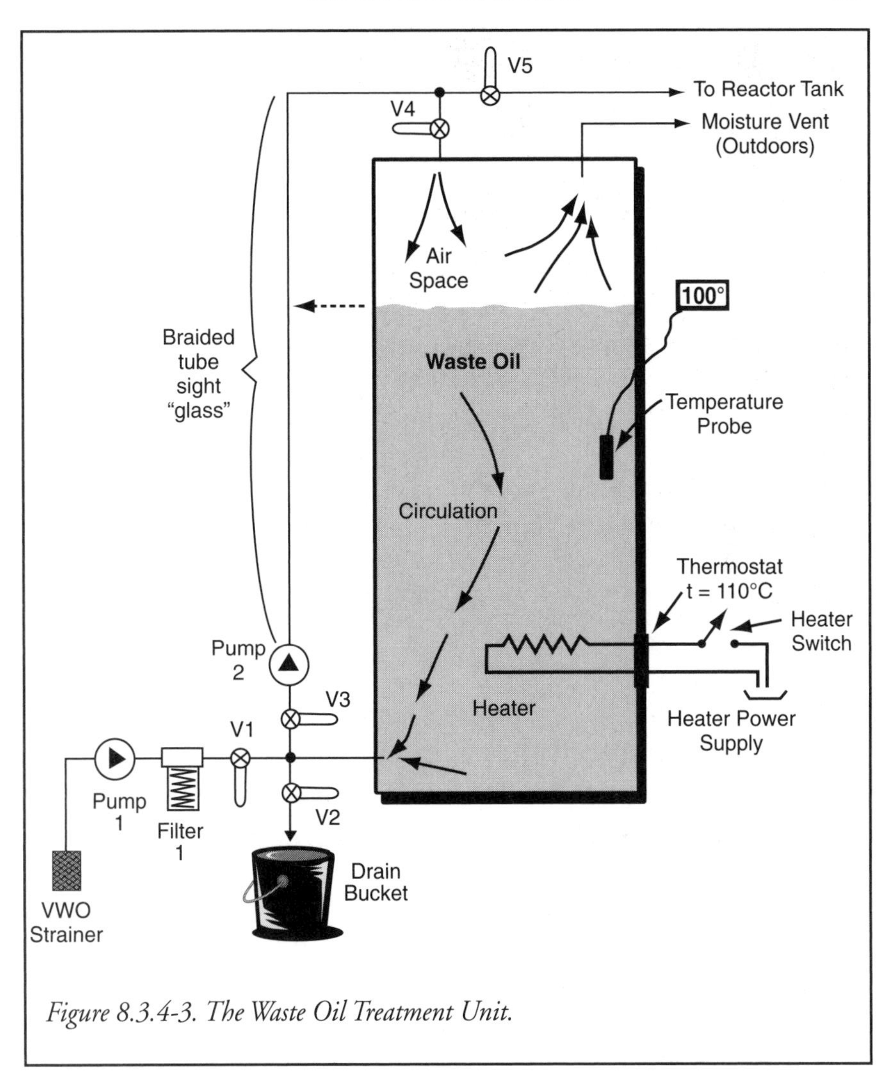

Figure 8.3.4-3. The Waste Oil Treatment Unit.

The oil begins heating and requires a number of hours to reach a minimum desired temperature of 70°C (158°F). When the oil reaches this temperature, turn off the heater and circulation pump and allow the contents to settle overnight.

Many home producers argue against the step of preheating WVO to dehydrate it since this uses energy and costs money and does not make any difference to the biodiesel reaction process. This statement is incorrect in several ways.

Firstly, the dehydration of the WVO prevents the possible loss of a batch of biodiesel, which wastes methanol and catalyst and also produces a jelly-like substance that is difficult to extract from the reaction tank and detoxify (methanol recovery and alkali neutralization) before waste disposal. And don't forget the time, trouble, and fuel required to collect the now-wasted feedstock oil.

Secondly, the energy required to heat the WVO to the dehydration level is minimal, in the order of 10 to 15 kilowatt-hours of energy with a retail value of under two dollars.

Thirdly, the WVO is being dehydrated in an insulated tank. Once the dehydration process is completed, the oil will still be at a fairly high temperature and ready for transfer to the biodiesel reaction tank, where the oil must be heated in any event. Assuming that the biodiesel reaction is completed in a short period of time after dehydration, this heating energy is recycled since the energy required in the reaction step is lowered. The real cost of energy to dehydrate the WVO is therefore about twenty-five cents at most. The simplest method to calculate the cost of energy (assuming electric resistance heating) is to multiply the heater electrical rating in kilowatts by the number of running hours to heat the oil and then multiply the total by the retail price of electricity. The retail price of electricity can be determined from your utility bill. For example:

3 kilowatt heater x 4.5 hours' heating time = 13.5 kilowatt-hours (kWh)

13.5 kWh x $0.10 = $1.35 cost of heating

It is possible to calculate the energy required to heat a given fluid by multiplying its weight by its specific heat by the change in temperature. The answer is in either calories or British Thermal Units (BTUs). The specific heat of vegetable oil in imperial measure is 0.4 BTU/lb-deg Fahrenheit. The result can then be converted to kilowatt-hours of electricity by dividing the total energy by the appropriate measure of energy per kilowatt hour, for example there are 3,413 BTU/kWh of electricity.

At this cost, dehydrating the WVO adds less than ½ cent per liter to the cost of biodiesel.

Fourthly, many researchers have provided conclusive test data proving that even small amounts of water in the oil feedstock cause problems during the reaction process and limit the amount of FFA that can be handled in the single-step reaction process[1].

After the oil has settled overnight, open valve V2 and drain a small amount (approximately 2 liters or 0.5 gallons) from the tank into a bucket (Figure 8.3.4-4) and save this WVO in the large oil recovery/settling drum discussed earlier. This oil will contain a small amount of water and other solids such as food particles that have settled to the bottom of the tank. The oil may be processed at a later time or composted if it proves to be of poor quality.

Figure 8.3.4-4. After the oil has settled overnight, drain a small amount (approximately 2 liters or 0.5 gallons) from the tank and save this WVO in the large oil recovery/settling drum discussed earlier. This oil will contain a small amount of water and other solids such as food particles that have settled to the bottom of the tank. The oil may be processed at a later time or composted if it proves to be of poor quality.

Transferring Treated WVO to the Biodiesel Reaction Unit

The oil is still hot and is now ready for transfer to the biodiesel reaction unit as follows:

1. Turn on circulation pump 2.
2. Ensure heater remains OFF. (If the heater is activated and exposed to air, it will quickly overheat and burn out.)
3. Open valve V5 allowing oil to flow into the biodiesel reaction unit.
4. Slowly close valve V4, forcing the circulating oil to transfer completely into the biodiesel reaction unit.
5. Fill the biodiesel reaction unit to the desired volume recorded on the sight glass. As the biodiesel tank continues to fill, "throttle" the flow of oil by opening valve V4 on the WVO treatment unit, causing oil to recirculate within it and slow oil transfer.
6. When the oil level in the biodiesel tank reaches the desired level, close valve V5 and open valve V4 on the WVO treatment unit.
7. Stop circulation pump 2 on the WVO treatment unit.

The WVO treatment unit can now be filled with the next charge of WVO for processing while the biodiesel reaction process is simultaneously started.

8.3.5
Sodium/Potassium Methoxide Preparation

Once the WVO or cold-pressed oil has been dehydrated in the waste oil treatment unit, it is transferred to the biodiesel reaction unit. A small sample of oil is taken from the WVO tank (WVO samples taken from the biodiesel reaction tank may be contaminated with reactant materials from previous processing) and is used to calculate the correct amount of catalyst required to complete the transesterification process. The calculation method is known as titration and is discussed below.

Figure 8.3.5-1. The sodium (or potassium) methoxide processor is shown above. In this unit, a catalyst of either sodium hydroxide or potassium hydroxide is reacted (mixed) with an alcohol (normally methanol) producing sodium or potassium methoxide which is mixed with WVO to produce FAME or biodiesel.

Safety Steps and Equipment Preparation

Many people incorrectly assume that respirators are not approved for this substance as methanol vapor cannot be filtered. This is false. NIOSH respirators equipped with type N95 or better filters are recommended where there is possible exposure to liquid sodium methoxide mist or dust[1].

1. Ensure that you are wearing all necessary safety equipment including a NIOSH-approved respirator for sodium or potassium methoxide. Powdered sodium methoxide is present after the solution has evaporated as can be seen in Figure 8.3.5-3.
2. Apply antistatic wrist strap.
3. Wear face shield or goggles.
4. Wear approved protective gloves.
5. Make sure that fire protection equipment is available and that all present know how to use it. Ensure that all present are familiar with safety procedures including emergency remedial actions.
6. NO SMOKING!
7. Activate the room ventilation fan.
8. Remove the lid from the mixing tank and set it aside. Be aware that the lid and tank surfaces may have accumulated sodium methoxide powder and must be considered dangerous (Figure 8.3.5-2).
9. Install the catalyst anticaking plug in the tank (Figure 8.3.5-3), taking care to ensure that the draw wire is taut and positioned so that it will not interfere with the mixing propeller.
10. Replace the mixing tank lid.

Making the Sodium Methoxide

The procedure for making the sodium methoxide is as follows:

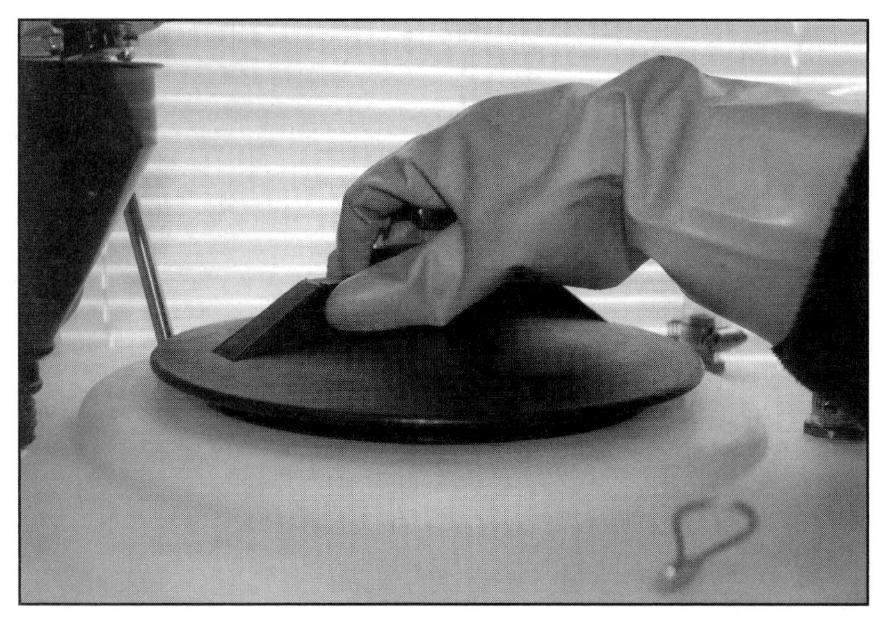

Figure 8.3.5-2. Remove lid from the mixing tank and set it aside. Be aware that the lid and tank surfaces may have accumulated sodium methoxide powder and must be considered dangerous.

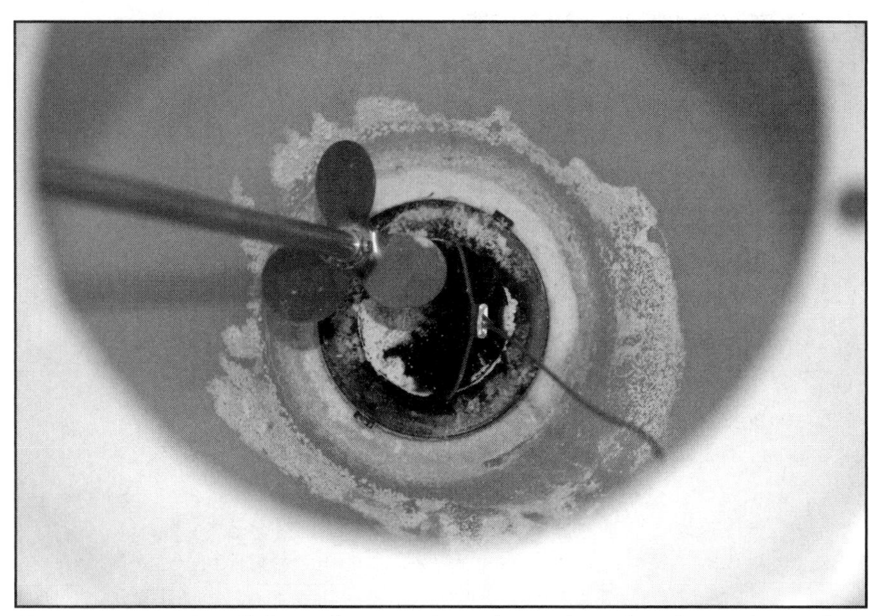

Figure 8.3.5-3. . Install the catalyst anticaking plug in the tank as shown above, taking care to ensure that the draw wire is taut and positioned so that it will not interfere with the mixing propeller.

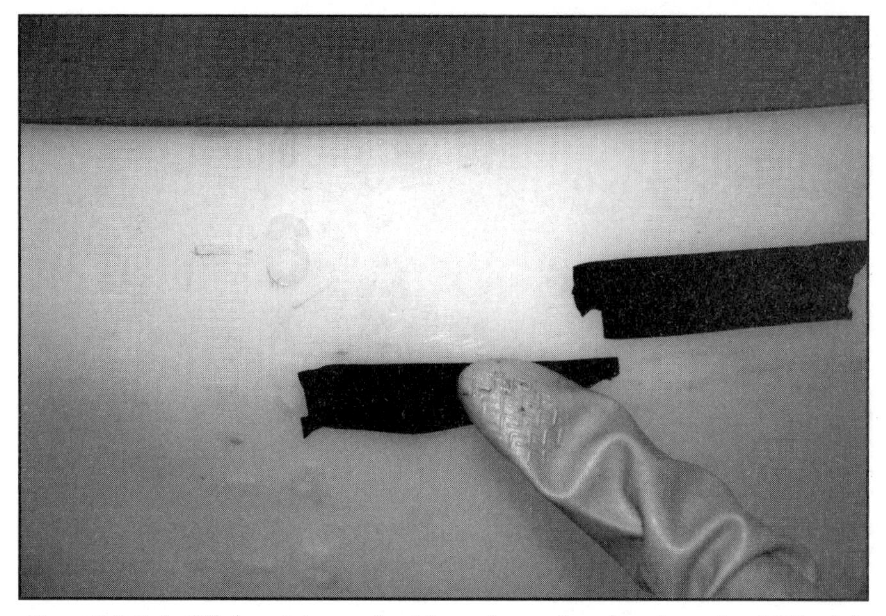

Figure 8.3.5-4. Fill the mixing tank with a volume of methanol equal to 20% of the volume of feedstock oil pumped into the biodiesel reaction tank.

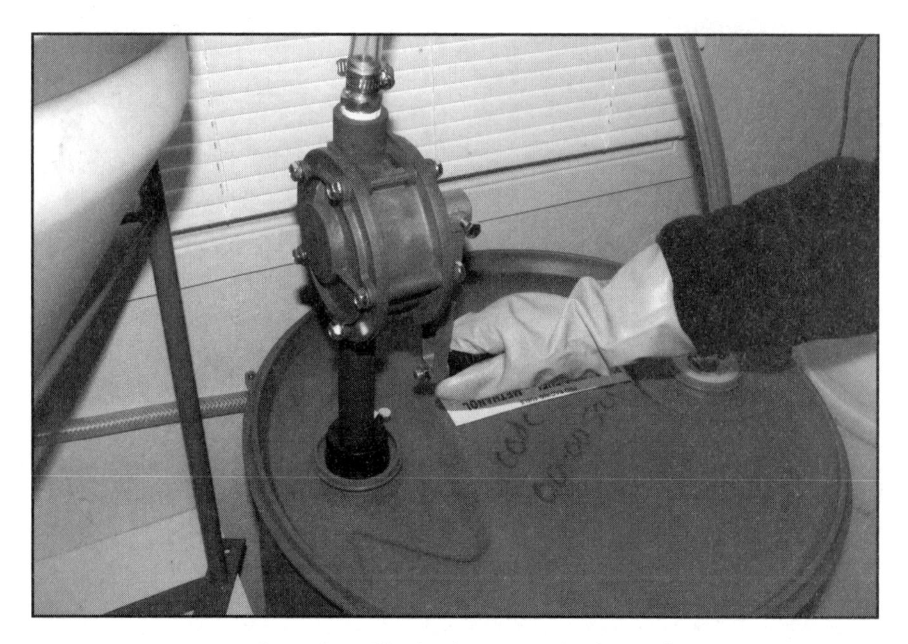

Figure 8.3.5-5. Pump the methanol by slowly turning the chemical mixing pump handle. Use care to ensure that the correct amount of alcohol is pumped into the tank.

Figure 8.3.5-6. Ensure that the sodium hydroxide catalyst is fresh and dry. It should not be clumpy and the crystals should be translucent and bright rather than white and opaque. Sodium hydroxide is very susceptible to degradation through the absorption of carbon dioxide and/or water vapor from the air.

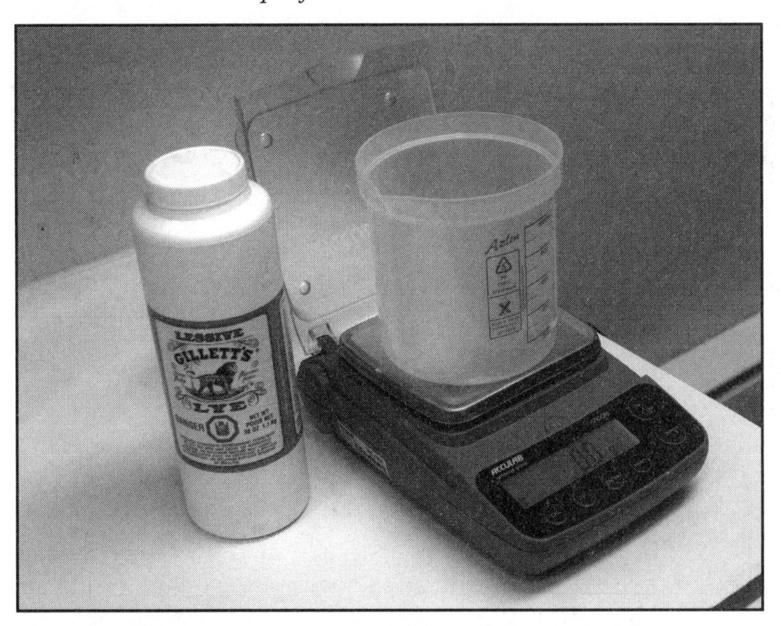

Figure 8.3.5-7. Use an electronic scale or triple beam balance which has an accuracy of at least +/-0.1 grams. Place a suitable plastic container on the scale/balance and zero or tare the container mass.

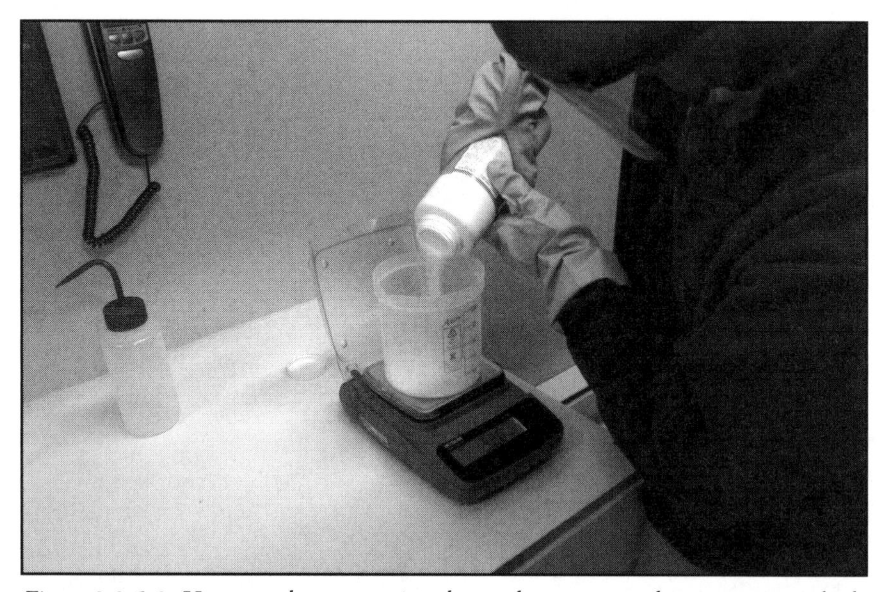

Figure 8.3.5-8. Use care when measuring the catalyst to ensure that you use exactly the correct amount according to the titration test. Too much catalyst may cause excessive soap formation; too little results in a reaction that is only partially completed, leaving high levels of triglycerides (oil) in the biodiesel.

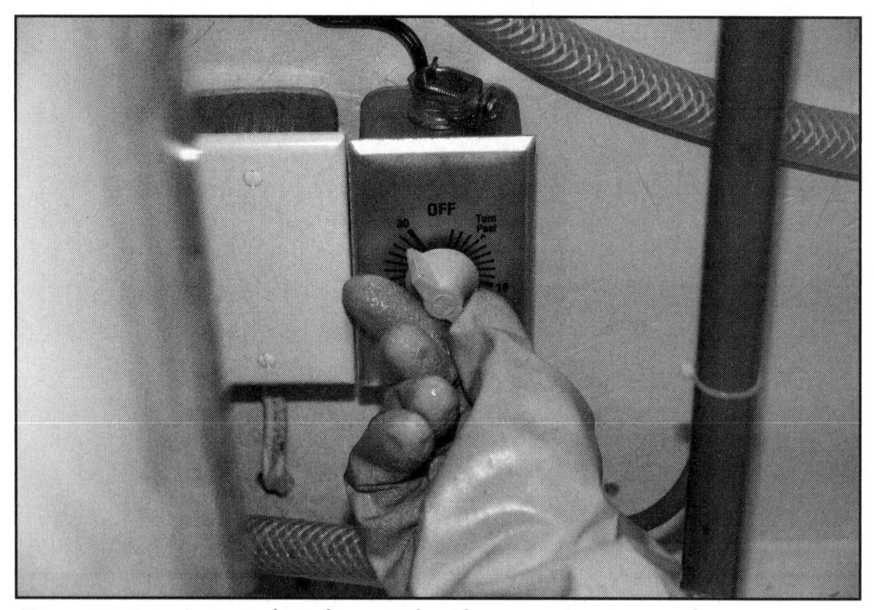

Figure 8.3.5-9. Activate the sodium methoxide mixing pump timer for 30 minutes

Figure 8.3.5-10. Remove the catalyst fill hopper plug, bearing in mind that it will contain catalyst residues and should be placed in a safe location where no one will accidentally touch it.

Figure 8.3.5-11. Add the catalyst to the fill hopper.

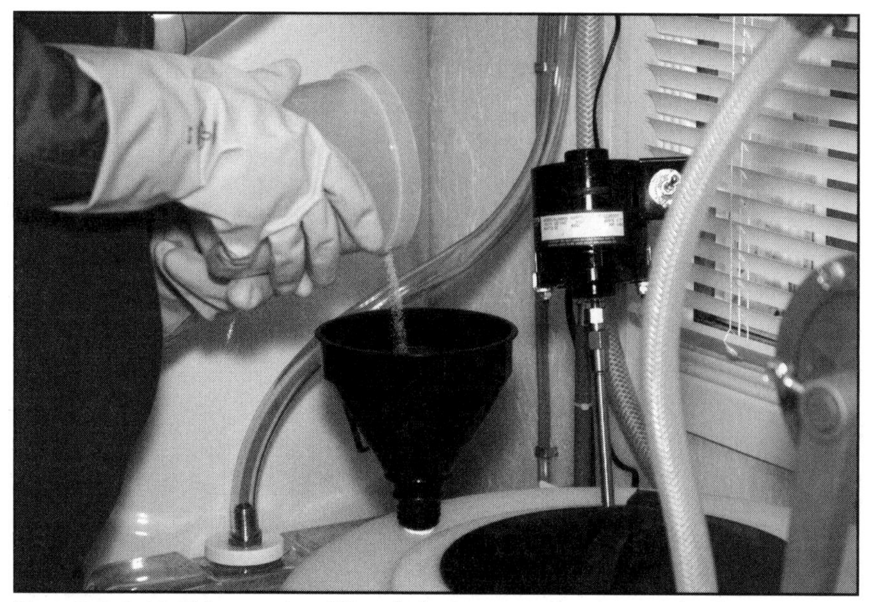

Figure 8.3.5-12. Take your time filling the hopper as catalyst does not dissolve in methanol quickly. If the catalyst clumps, use a small wooden dowel to dislodge any crystals.

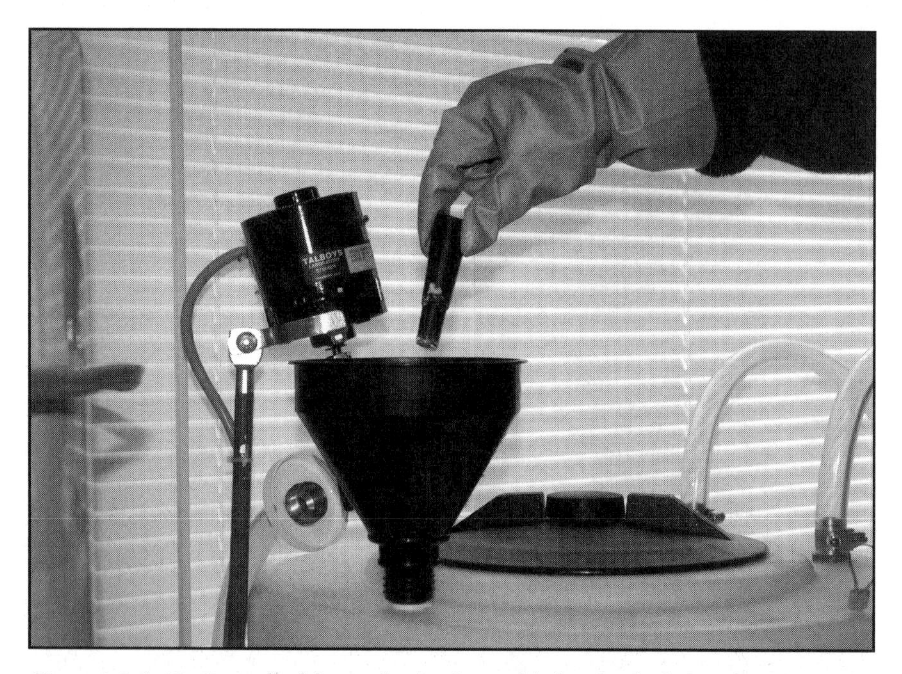

Figure 8.3.5-13. Once all of the catalyst has been added to the alcohol, replace the catalyst fill hopper plug and allow the solution to continue mixing until the timer stops.

1. Fill the mixing tank with a volume of methanol equal to 20% of the volume of feedstock oil pumped into the biodiesel reaction tank (Figure 8.3.5-4). (For example, if 150 liters of oil was pumped into the reaction tank, then 30 liters of methanol is pumped into the mixing tank.)
2. Pump the methanol by slowly turning the chemical mixing pump handle (Figure 8.3.5-5). Use care to ensure that the correct amount of alcohol is pumped into the tank.
3. Obtain the sodium hydroxide catalyst material (Figure 8.3.5-6) and ensure that it is fresh and dry. It should not be clumpy and the crystals should be translucent and bright rather than white and opaque. Sodium hydroxide is very susceptible to degradation through the absorption of carbon dioxide and/or water vapor from the air.
4. Use an electronic scale which has an accuracy of at least +/-0.1 grams (Figure 8.3.5-7). Place a suitable plastic container on the scale and zero or tare the container mass. A triple beam balance can also be used.
5. Carefully add sodium hydroxide crystals to the container until the mass calculated during the titration process noted below is reached (Figure 8.3.5-8).
6. Tightly cover the container of catalyst and return it to its storage cabinet.
7. Activate the sodium methoxide mixing pump timer for 30 minutes (Figure 8.3.5-9).
8. Remove the catalyst fill hopper plug, bearing in mind that it will contain catalyst residues (Figure 8.3.5-10).
9. Slowly add the catalyst to the fill hopper (Figure 8.3.5-11). Take your time. Catalyst does not dissolve in methanol well and taking extra time to fill the hopper is worthwhile. If the catalyst clumps, use a small wooden dowel to dislodge any crystals.
10. Once all of the catalyst has been added to the alcohol, replace the catalyst fill hopper plug (Figure 8.3.5-13) and allow the solution to continue mixing until the timer stops.

DO NOT REMOVE THE COVER OF THE TANK TO LOOK INSIDE AS CONSIDERABLE AMOUNTS OF TOXIC METHANOL VAPORS WILL HAVE ACCUMULATED.

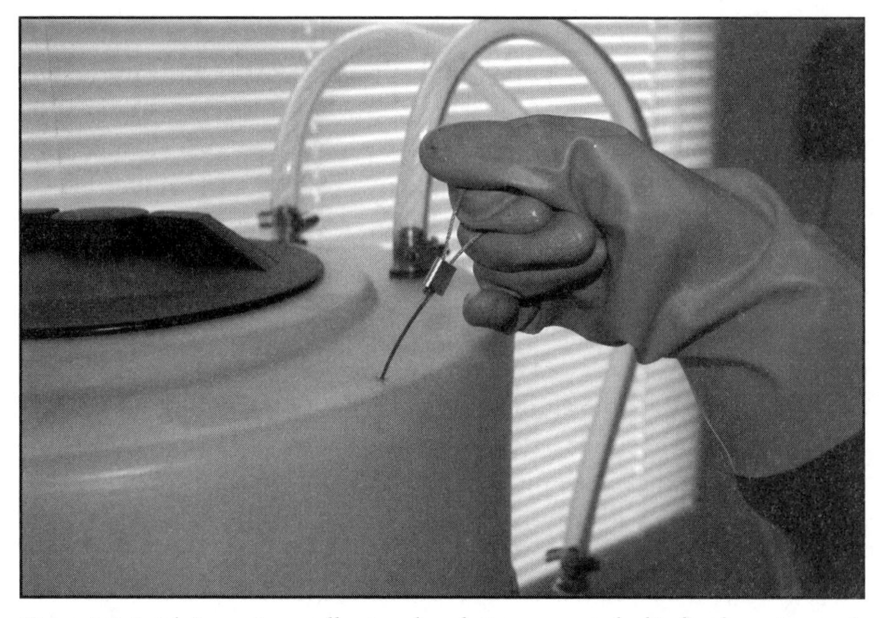

Figure 8.3.5-14. Just prior to allowing the solution to enter the biodiesel reaction tank, grip the anticaking release wire and give it a firm pull to remove the plug. You are now ready to proceed with the biodiesel reaction process described in Chapter 8.3.6.

The sodium/potassium methoxide may be left in the mixing tank until the feedstock oil in the biodiesel reaction tank has reached the operating temperature range of 55°C (131°F). Just prior to allowing the solution to enter the biodiesel reaction tank, grip the anticaking release wire (Figure 8.3.5-14) and give it a firm pull to remove the plug.

DO NOT ACTIVATE THE MIXER AFTER THE ANTICAKING DRAIN HAS BEEN REMOVED AS THE WIRE WILL BECOME TANGLED IN THE MIXING PROPELLER.

You are now ready to proceed with the biodiesel reaction process described in Chapter 8.3.6.

Choosing a Catalyst

Virtually all commercial biodiesel producers use either sodium hydroxide, potassium hydroxide, or a premixed solution of sodium methoxide as their catalyst of choice. Both sodium and potassium hydroxide are highly hygroscopic, absorbing water from the air during storage.

Potassium hydroxide is more expensive than sodium hydroxide and 2.6 times as much is required in the transesterification process; sodium hydroxide requirement is 3.5 grams per liter of oil feedstock while potassium hydroxide requires 9 grams per liter of oil. Potassium hydroxide may be more difficult to locate without having to purchase technical-grade material from a laboratory or specialty chemical supply store.

If you wish to continue with potassium hydroxide catalyst, use 2.6 times the amount specified for sodium hydroxide. For example, the reference solution discussed in the titration section requires 1 gram of sodium hydroxide, so use 2.6 grams of potassium hydroxide.

Although some people argue that potassium hydroxide catalyst can be neutralized with phosphoric acid to make fertilizer, the process is seldom undertaken and the value of any fertilizer produced is questionable. The quantities generated are small and there is the possibility that toxic methanol contamination will leach into the ground water. In addition, during the biodiesel washing process considerable amounts of phosphate are discharged with the neutralized water, which will not meet sewer effluent limits for phosphate[2].

Potassium hydroxide does have at least one advantage over sodium hydroxide. During the biodiesel phase separation process, raw glycerol is discharged from the reaction tank and is often placed in a carrier for later "processing." Recovered glycerol containing sodium hydroxide catalyst has a tendency to solidify after it cools below approximately 38°C (100°F). Glycerol made from potassium hydroxide will not. However, although this issue may seem important, it is irrelevant if the raw glycerol is poured directly into the glycerin processing unit during its liquid state.

Given the pros and cons of the two catalysts, I recommend the use of sodium methoxide.

Titration

If you can remember back to high school chemistry, you will recall the pH scale which is used to grade the strength of acids and their opposite cousins, "bases." The scale is numbered 0 through 14, with 7 being the halfway or "neutral" point on the scale. Pure water is neutral and therefore has a pH reading of 7. As the number drops below 7, the substance becomes an increasingly strong acid. Likewise, as the reading increases beyond 7, the substance becomes more basic. Virgin vegetable oil is by nature acidic and used fryer oils become increasingly stronger acids the longer and "harder" they are used.

The biodiesel transesterification process requires a catalyst such as sodium hydroxide to cause the "exchange" of glycerin for methanol molecules during the transesterification procedure. The titration process will determine the level of FFA contained in the oil as well as the correct amount of catalyst required to ensure that the transesterification process is "pushed" to completion, thereby producing biodiesel and not "glop."

An acidic substance can become neutralized by adding a basic substance of equal but opposite strength. Using this rule and the information derived from the titration process, it is possible to calculate the amount of sodium hydroxide necessary to neutralize the FFAs and ensure that the biodiesel conversion process proceeds correctly.

The titration process requires that a known volume of a solution of unknown acidic strength be placed in a jar with an indicating fluid known as phenolphthalein. Phenolphthalein is a commonly used indicator that is colorless when subjected to acids but turns bright red in the presence of bases. Adding an alkali solution of equal strength to the unknown acid will cause the solution to become neutral with a reading of pH 7. Adding just a bit more alkali to the mix will cause the solution to become slightly basic and the phenolphthalein will turn red.

Using this process, it is possible to slowly add a known volume of a known strength of alkali solution to the unknown acidic material and watch for the phenolphthalein to turn red. When this occurs, the amount of reference solution required to neutralize the acid is known.

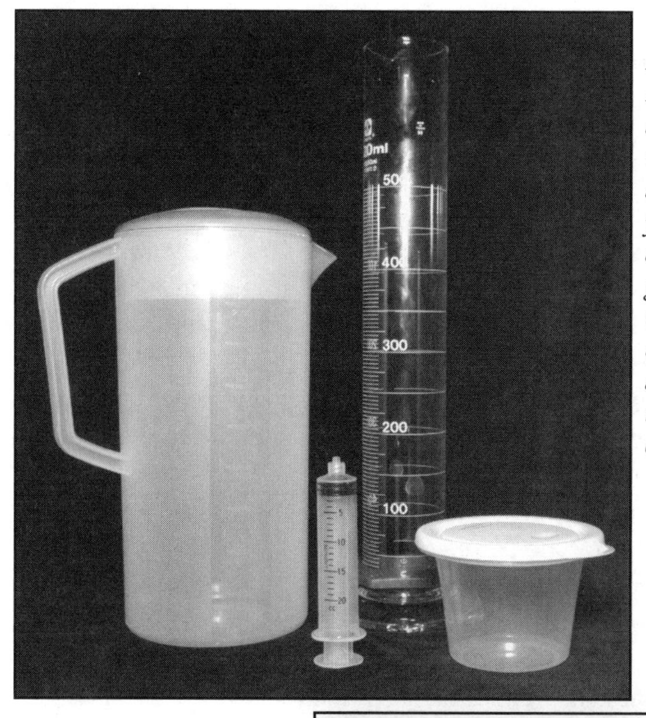

Figure 8.3.5-15. In order to perform a titration process to measure the free fatty acid level of WVO, you need a range of dedicated laboratory glassware such as beakers, graduated cylinders, syringes, and containers to store the various reference chemicals.

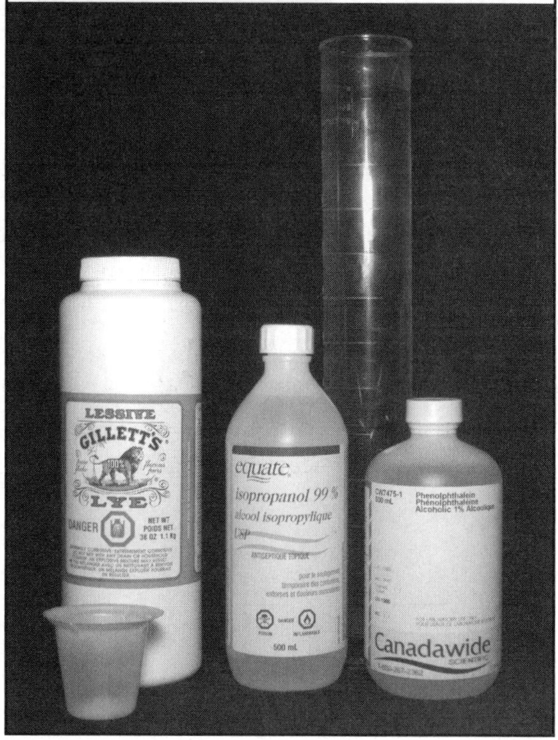

Figure 8.3.5-16. The essential chemicals required to perform a titration are (from left to right) sodium hydroxide, 99% isopropyl alcohol, laboratory-grade phenolphthalein, and distilled water (rear).

In order to perform a titration process you require the following chemicals, laboratory glassware, scale, and supplies:

- pH indictor known as phenolphthalein. Do not use the indicator from a swimming pool test kit known as "phenol red." Phenolphthalein is a laboratory-grade indicator and is more accurate than swimming pool phenol red.
- distilled water
- 100% pure sodium hydroxide. Use care to keep the sodium hydroxide dry, even preventing exposure to humid air by ensuring that the container remains tightly sealed.
- 99% isopropyl alcohol. (Use care when purchasing alcohol as there are numerous concentrations available. If in doubt, consult a pharmacist or chemist.)
- syringes graduated in 1 ml increments or preferably a graduated burette cylinder
- graduated cylinder or other accurate liquid measuring devices
- assorted small jars or beakers. At least one jar should have a sealable top to store 1 liter of fluid for a reference solution.
- a triple beam mass balance or electronic scale with a measurement accuracy of at least +/- 0.1 grams
- latex rubber gloves
- eye protection
- well-ventilated area

Procedure

Note: Ensure that all beakers, containers, and syringes are clean and used for only one chemical. Syringes and glassware may be reused, but only for one purpose or "step" outlined below:

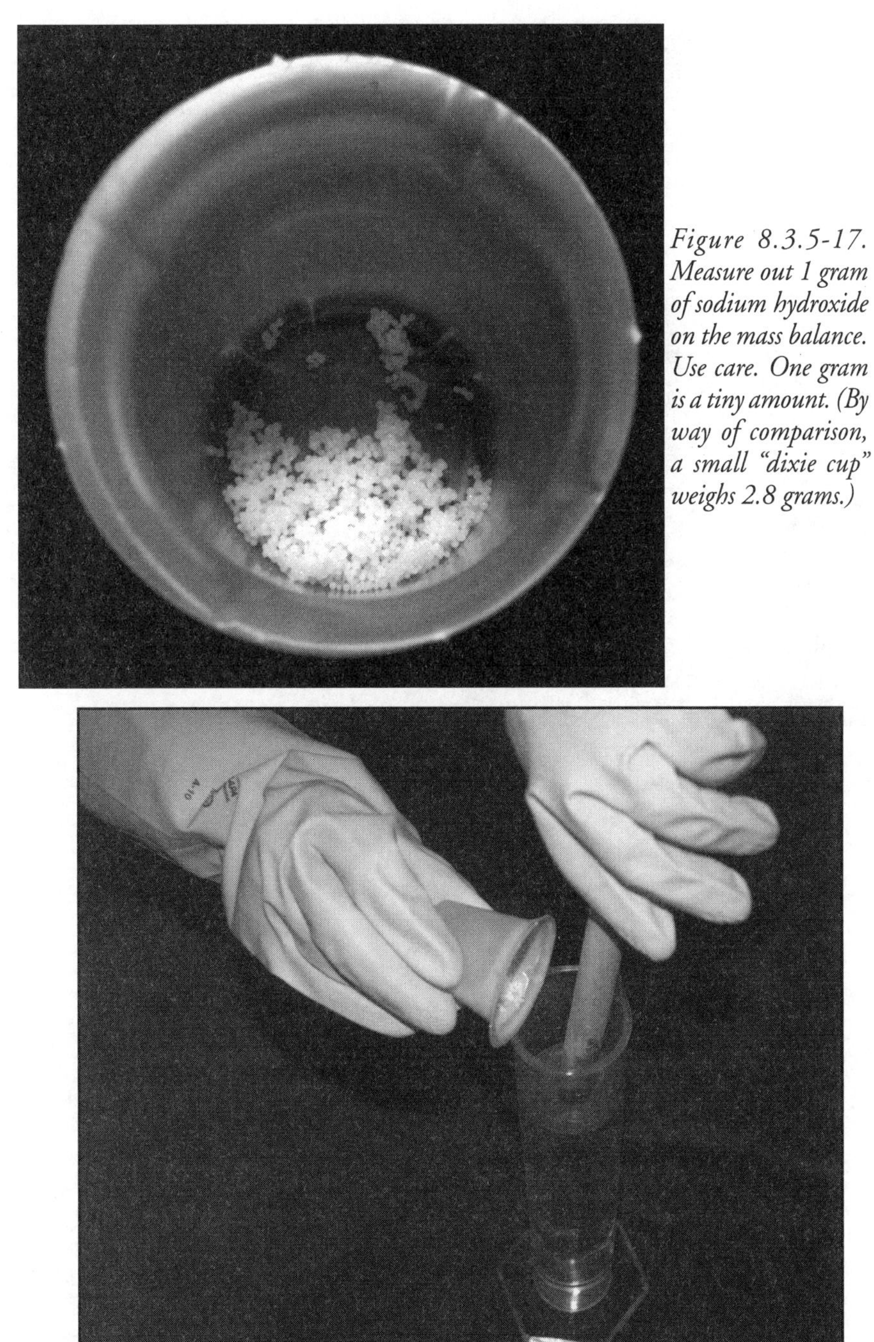

Figure 8.3.5-17. Measure out 1 gram of sodium hydroxide on the mass balance. Use care. One gram is a tiny amount. (By way of comparison, a small "dixie cup" weighs 2.8 grams.)

Figure 8.3.5-18. Measure 1 liter of the distilled water using a graduated cylinder and add to it 1 gram of sodium hydroxide.

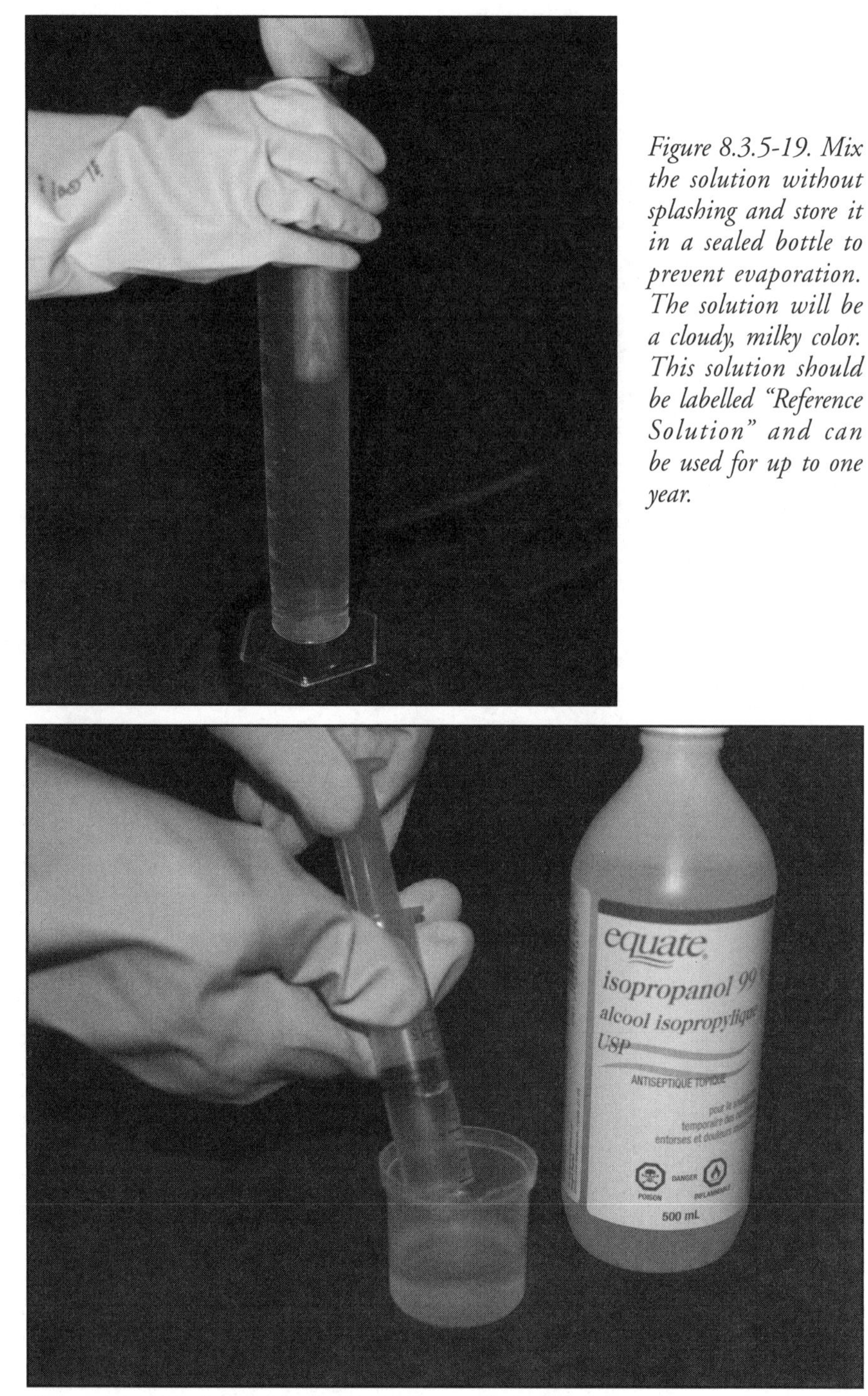

Figure 8.3.5-19. Mix the solution without splashing and store it in a sealed bottle to prevent evaporation. The solution will be a cloudy, milky color. This solution should be labelled "Reference Solution" and can be used for up to one year.

Figure 8.3.5-20. Draw 10 ml of 99% isopropyl alcohol.

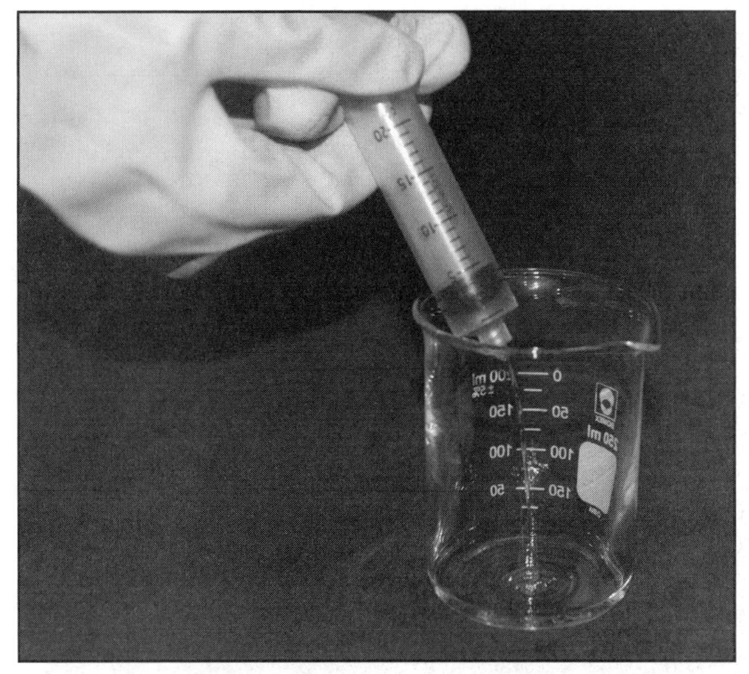

Figure 8.3.5-21. Place 10 ml of isopropyl alcohol in a small beaker.

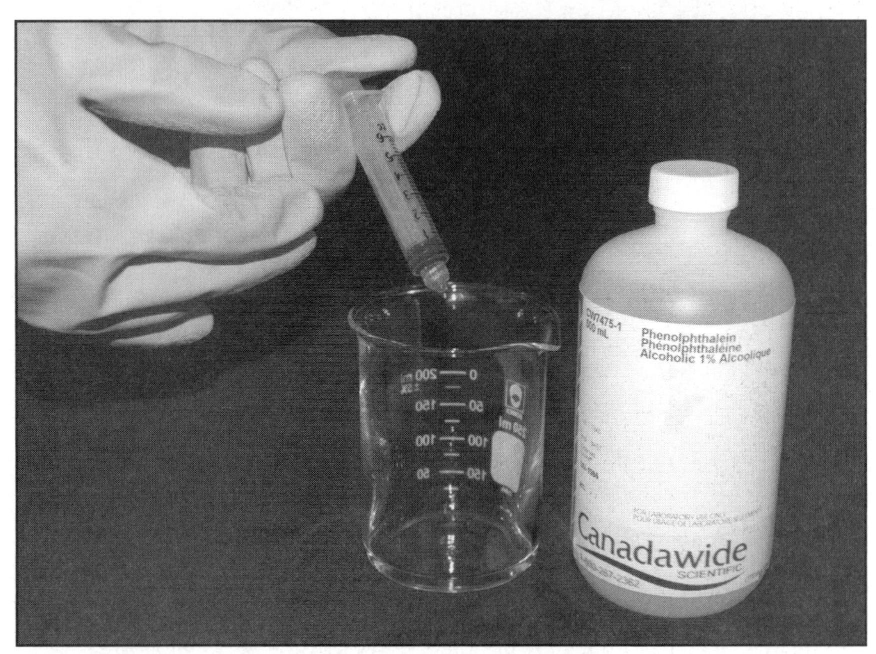

Figure 8.3.5-22. Add 2 drops of phenolphthalein indicator into the alcohol and swirl until well mixed.

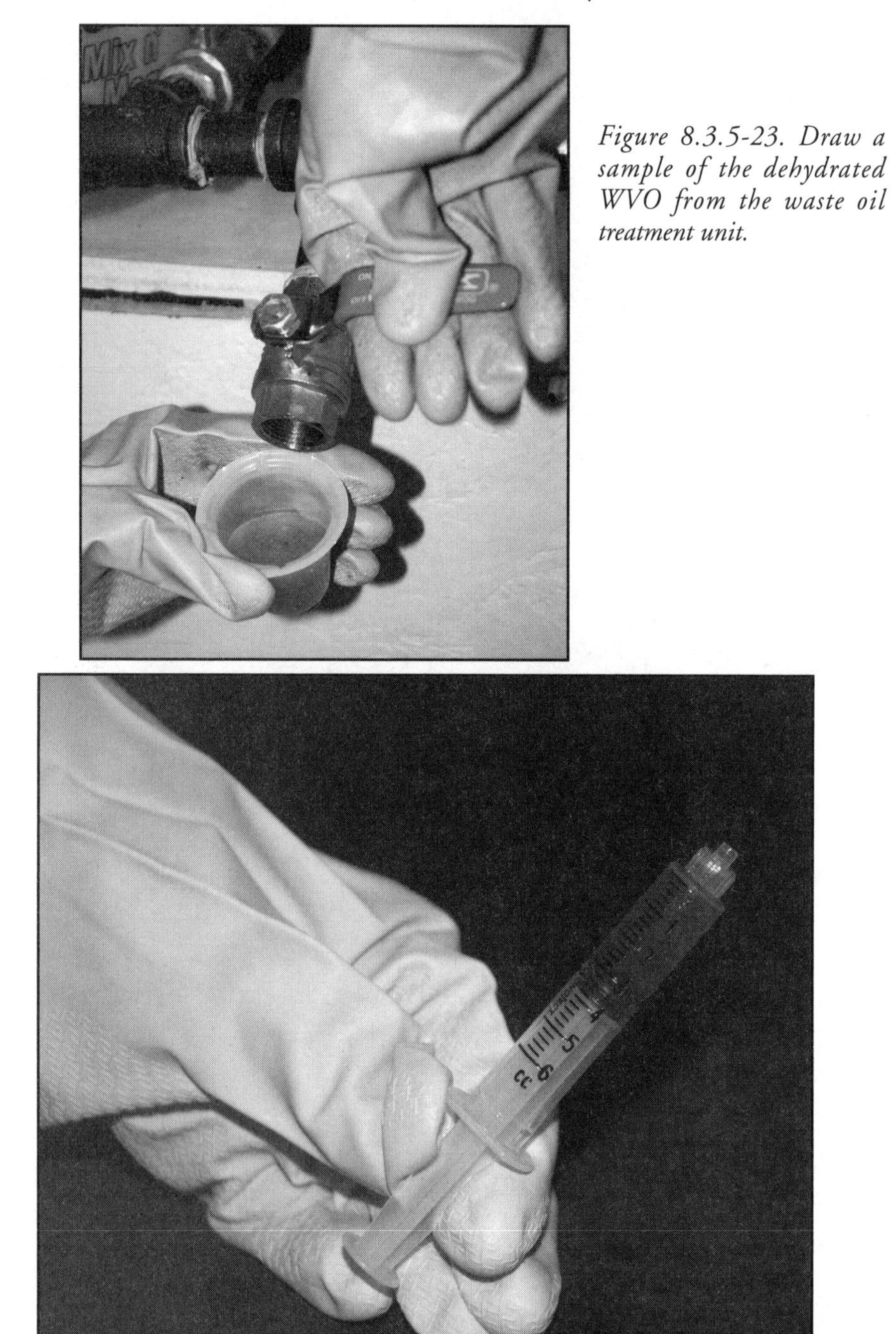

Figure 8.3.5-23. Draw a sample of the dehydrated WVO from the waste oil treatment unit.

Figure 8.3.5-24. Draw a few milliliters of WVO into a disposable syringe.

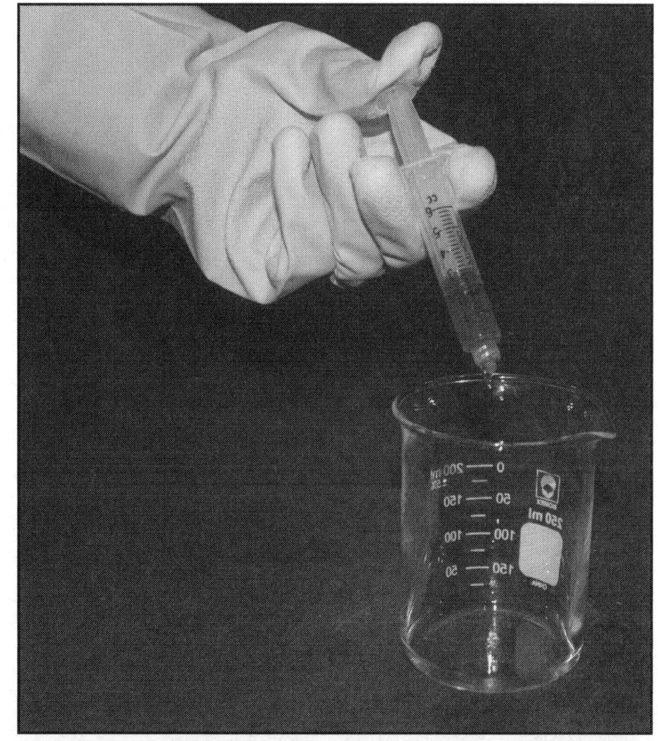

Figure 8.3.5-25. Add 1 ml of the dehydrated WVO to the phenolphthalein/alcohol solution.

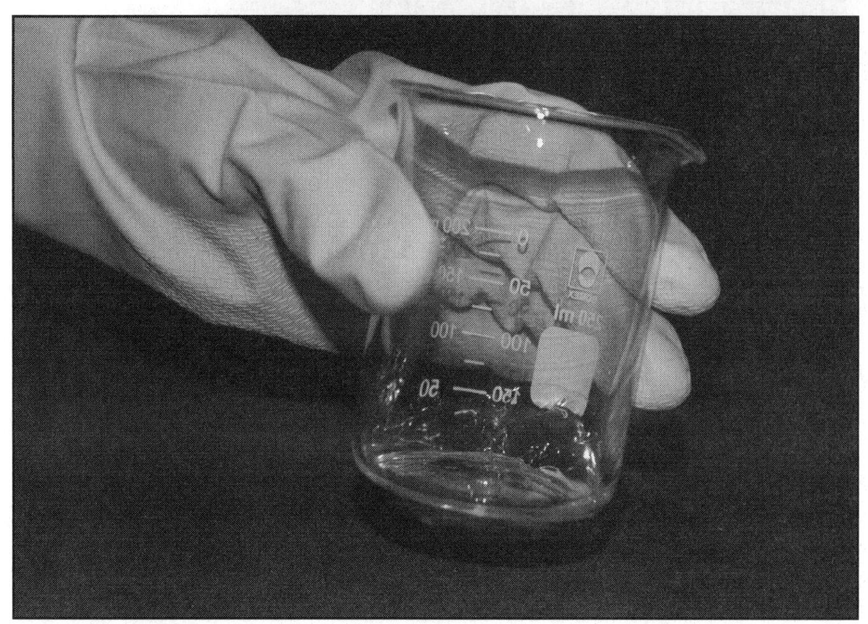

Figure 8.3.5-26. Vigorously swirl the beaker to ensure complete mixture of the phenolphthalein/alcohol/WVO solution.

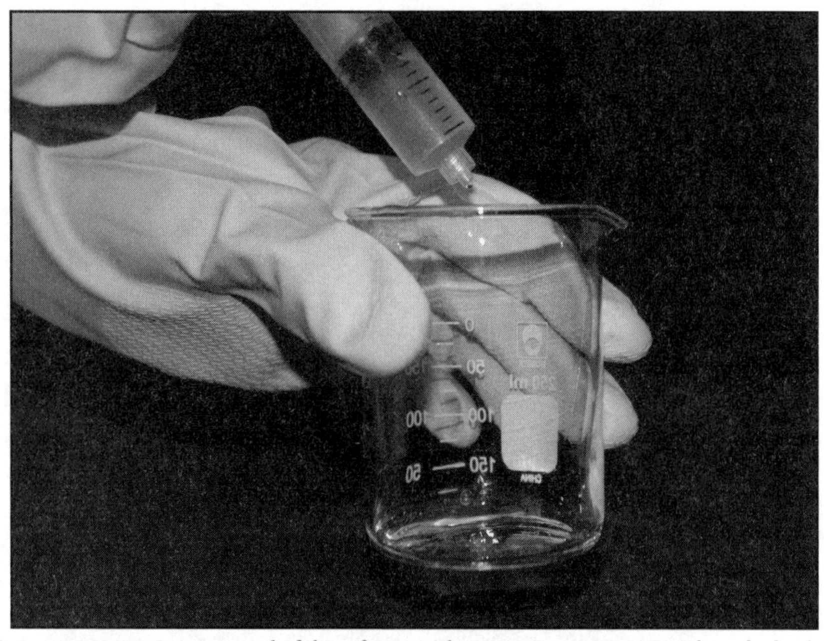

Figure 8.3.5-27. Draw 10 ml of the reference solution into a syringe. Swirling the beaker of phenolphthalein/alcohol/WVO in one hand, carefully drip 0.5 ml amounts of reference solution into the beaker at a time.

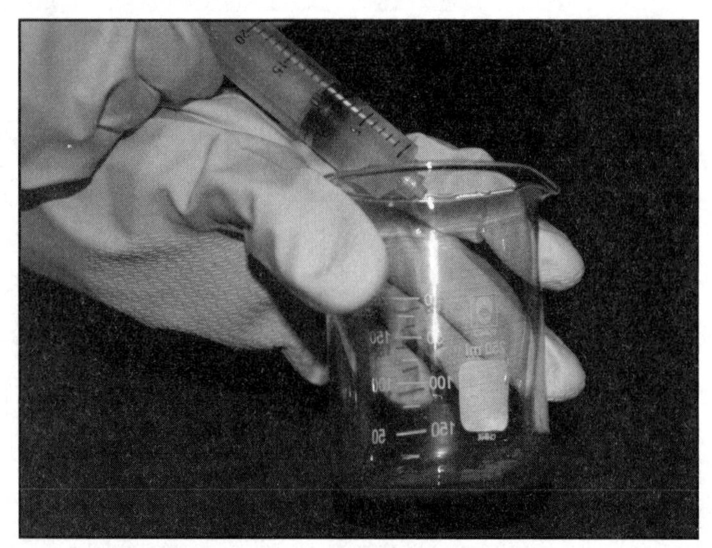

Figure 8.3.5-28. The solution will be a musty yellow at the start of this test and will turn a uniform bright pink when the titration is complete. The solution will turn pink when the reference solution is first injected but will return to yellow as you continue to swirl the beaker (Figure 8.3.5-27). The solution must be uniformly pink for at least 30 seconds before the titration is considered finished.

1. Measure out 1 gram of sodium hydroxide on the mass balance (Figure 8.3.5-17). One gram is a very small amount; be very careful with this measurement. Alternatively, measure out 4 grams of sodium hydroxide and make a larger (4 liter) batch of reference solution. Measuring a larger quantity will spread any measurement error over a larger sample, improving the accuracy of the solution.
2. Measure 1 liter of the distilled water using a graduated cylinder and add to it 1 gram of sodium hydroxide (Figure 8.3.5-18). Mix the solution without splashing (Figure 8.3.5-19) and store it in a sealed bottle to prevent evaporation. The solution will be a cloudy, milky color. This solution should be labelled "Reference Solution" and can be used for up to one year. (For disposal instructions see chapter 8.3.11.)
3. Draw 10 ml of 99% isopropyl alcohol (Figure 8.3.5-20) and place in a small beaker (Figure 8.3.5-21).
4. Add 2 drops of phenolphthalein indicator into the alcohol and swirl until well mixed (Figure 8.3.5-22).
5. Draw a sample of the dehydrated WVO from the WVO treatment unit (Figure 8.3.5-23) and use a disposable syringe to draw a sample of a few milliliters (Figure 8.3.5-24). Then add 1 ml to the phenolphthalein/alcohol solution (Figure 8.3.5-25). Use extreme care in ensuring that exactly 1 ml is dispensed. It is best to fill the syringe with 5 ml of oil and dispense down to 4 ml rather than trying to measure a single milliliter of oil.
6. Vigorously swirl the beaker to ensure complete mixture (Figure 8.3.5-26).
7. Draw 10 ml of the reference solution into a syringe. Swirling the beaker of phenolphthalein/alcohol/WVO in one hand, carefully drip 0.5 ml amounts of reference solution into the beaker at a time (Figure 8.3.5-27). The solution will be a musty yellow at the start of this test and will turn a uniform bright pink when the titration is complete. The solution will turn pink when the reference solution is first injected but will return to yellow as you continue to swirl the beaker (Figure 8.3.5-28). The solution must be uniformly pink for at least 30 seconds before the titration is considered finished.

Note that a disposable syringe can only provide an accurate measure of 0.5-ml amounts while the graduate burette shown in Figures 8.3.5-29 and 8.3.5-30 can increase the accuracy of the titration considerably because of the fine resolution of the unit, typically around 0.1 ml.

8. Record the amount of reference solution used to complete the titration process. The amount will be the amount of reference solution left in the syringe subtracted from the initial 10 ml. For example, if 7.5 ml remain in the syringe, then 2.5 ml were used (10-7.5 ml = 2.5 ml used).
9. Add 3.5 to the amount of reference solution recorded in Step 8 and then multiply this number times the number of liters of WVO used in the transesterification batch. This equals the mass in grams of sodium hydroxide required to complete the transesterification process.

   Using the example in Step 8 and assuming a WVO batch of 100 liters, 2.5 ml would be added to 3.5 yielding 6.0, multiplied by 100, resulting in 600 grams of sodium hydroxide required for every 100 litres of WVO. Use the balance beam or electronic scale to accurately measure the required amount of sodium hydroxide as described above.
10. It is recommended that you clean up the lab and repeat the test at least one more time to ensure accurate results. It is very easy to make a mistake and accuracy is what produces a high-quality fuel.
11. Store the measured sodium hydroxide in an airtight jar and set it aside in preparation for the next stage.

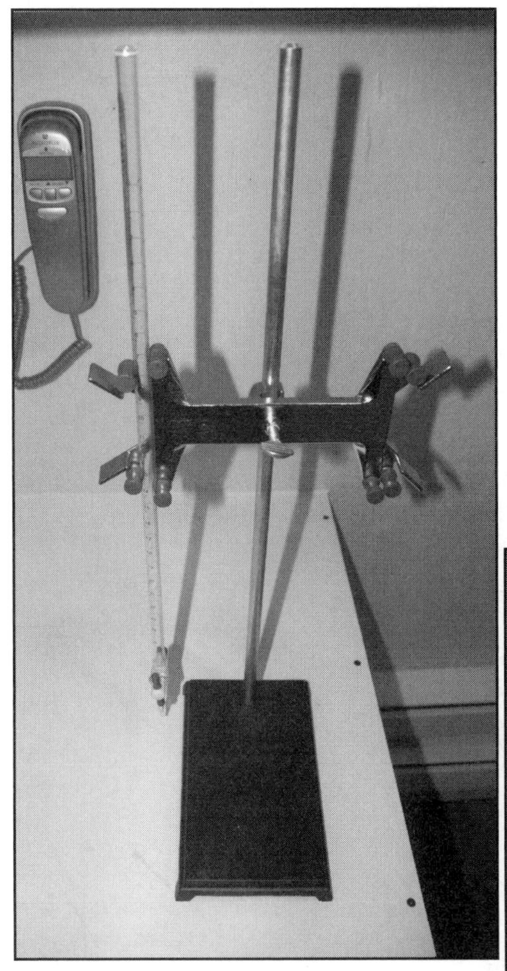

Figure 8.3.5-29. Note that a disposable syringe will can only provide an accurate measure of 0.5-ml amounts while the graduated burette seen here can increase the accuracy of the titration considerably.

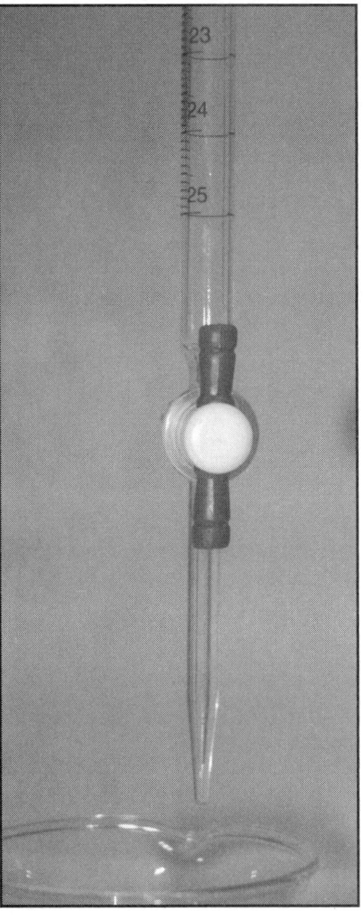

Figure 8.3.5-30. Because of the fine resolution of the graduations, typically around 0.1 ml, the burette is much more accurate than disposable syringes for the addition of reference solution during the titration process (see Step 7 above).

8.3.6
Reacting the Oil with Sodium Methoxide

At this point, the biodiesel reaction tank should be filled with the desired batch volume of either cold-pressed oilseed oil or dehydrated WVO (Figure 8.3.6-1). If the feedstock is oilseed oil, you can proceed directly to the transesterification process outlined below. If the feedstock is WVO, you must decide whether to proceed directly to transesterification or pretreat the WVO because of high FFA levels.

As discussed in Chapter 8.3.4, with WVO you may proceed directly to transesterification if the FFA level is below approximately 5% and the oil has had all residual water removed. If, on the other hand, the FFA level is above 5% or there is a risk of water contamination in the oil, pretreatment is required.

Figure 8.3.6-1. At this point, the biodiesel reaction tank should be filled with the desired batch volume of either cold-pressed oilseed oil or dehydrated WVO. The lower mark on the sight glass corresponds to 150 liters of feedstock oil. The second mark indicates the level after 30 liters of sodium methoxide (20% of the volume of oil) is added to the tank.

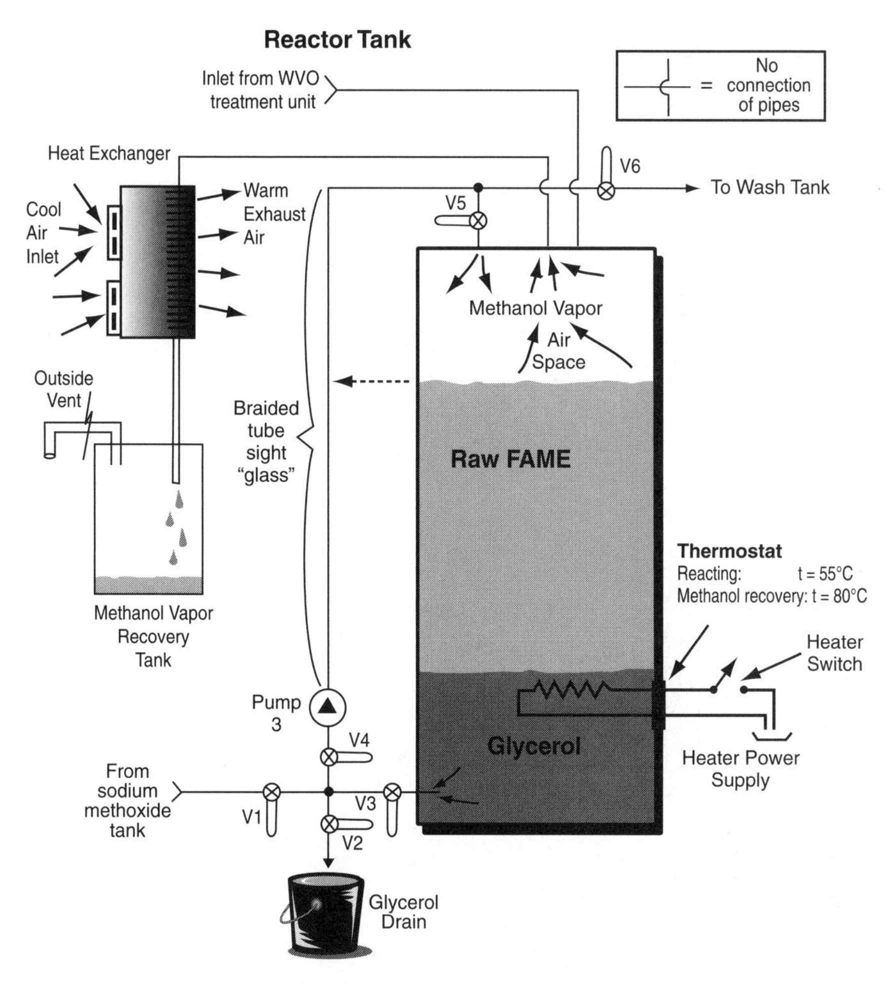

Figure 8.3.6-2. Ensure that valves V1, V2, and V6 are closed and valves V3, V4, and V5 are open and that the thermostat is set for approximately 55°C at the start of the biodiesel reaction process.

Pretreating High-FFA Feedstock with Acid Catalysis

As mentioned earlier, there are very few biodiesel producers who want to take the time or make the effort to pretreat high-FFA oils. The simplified dilution and dehydration process described in Chapter 8.3.4 yields satisfactory results in most cases, with perhaps a small amount of soap formation that can easily be dealt with in the washing stage discussed in Chapter 8.3.7.

An alternative to the dilution method of pretreating high-FFA feedstock is to perform a two-step process known as acid catalysis followed by alkali catalysis. This technique involves pretreatment of the WVO with an acid/alcohol mixture that converts the FFAs to biodiesel. When the FFA level has been lowered to approximately 1% or less, the standard alkali catalysis process is used to convert the remaining triglycerides into biodiesel.

For those who wish to consider this advanced technique, be warned that there are numerous pitfalls including the handling of highly concentrated sulfuric acid, greatly complicated waste stream processing, acid corrosion of pipes and fittings, and increased time and effort. The full procedure is described in Paper 016049 which is available from the American Society of Agricultural and Biological Engineers at http://asae.frymulti.com/default.asp and is not recommended for "home use."

The Transesterification Process

The biodiesel reaction process begins with the heating of the feedstock oil to approximately 55°C and the introduction of the sodium methoxide solution as follows:

1. Close valve V1, intake from sodium methoxide mixer.
2. Close valve V2, biodiesel reaction tank drain.
3. Open valve V3, tank circulation valve.
4. Open valve V4, optional valve, to allow oil circulation.
5. Open valve V5, tank top circulation valve.
6. Close valve V6, biodiesel transfer to wash tank valve.
7. Start circulation pump 3 (always start circulation pump prior to activating heater).
8. Activate tank heater and set thermostat to 55°C.
9. Start methanol recovery/heat exchanger fans.

Once the feedstock oil has reached the setpoint temperature of 55 °C, the sodium methoxide is admitted into the biodiesel reaction tank by opening valve V1. The sodium methoxide level should begin to drop as the suction from circulation pump 3 draws the solution into the reaction tank. It may be necessary to "throttle" valve V3 closed so that the suction strength builds to a level sufficient to draw the sodium methoxide into the reaction tank. (Depending on the type of circulation pump selected, it may be necessary to elevate the sodium methoxide tank exit above the inlet to the circulation pump in order to facilitate complete and rapid draining.)

Once the sodium methoxide has been drawn into the biodiesel reaction tank, valve V1 is closed and valve V3 remains fully open.

The circulation and mixing is allowed to continue for a period of one hour, after which time the heater is turned off followed by circulation pump 3.

Figure 8.3.6-3. The next three images were taken at Steve Anderson's facility when he was using a simple "open biodiesel processor" made from a 55-gallon drum. Although I do not accept this type of production technique, it does offer the ability to see the transesterification process in action. In this image, the sodium methoxide solution has just been added to the heated WVO.

Figure 8.3.6-4. The WVO is shown a couple of minutes into the transesterification process and has turned the color of chocolate milk.

Figure 8.3.6-5. The WVO has turned to a low-viscosity, dark, clear liquid after approximately 15 minutes of mixing, indicating that the transesterification process is nearing completion.

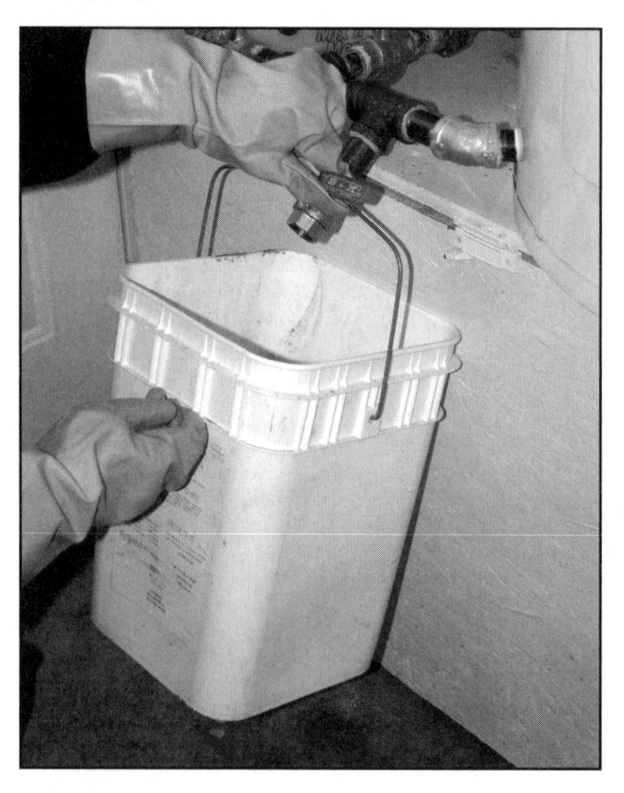

Figure 8.3.6-6. After the contents have settled overnight the reaction tank is drained of raw glycerol that has collected as the bottom phase.

Figure 8.3.6-7. The glycerol phase is instantly obvious as it is the first, dark, viscous liquid that drains from the reaction tank. One indication that the biodiesel reaction has worked correctly is the presence of two distinct phases with a "sharp" or obvious separation between them.

Figure 8.3.6-8. The FAME or raw biodiesel phase is also instantly recognizable since the fluid is lighter in color and of considerably lower viscosity, causing the flow rate from the tank drain to increase appreciably.

The contents of the reaction tank are allowed to settle overnight.

After the contents have settled overnight the reaction tank is drained of raw glycerol that has collected as the bottom phase (Figure 8.3.6-6). Provided the glycerol is handled quickly and not allowed to cool below 38°C it will remain liquid, allowing time for it to be transferred into the glycerin processing unit. Note that glycerol is toxic, containing large amounts of catalyst and methanol. Ensure that all necessary safety precautions are followed.

The glycerol phase is instantly obvious as it is the first, dark, viscous liquid that drains from the reaction tank (Figure 8.3.6-7).

As the glycerol continues to drain from the system you will notice the change from glycerol to the FAME or biodiesel phase, instantly recognizable since the fluid is lighter in color and of considerably lower viscosity, causing the flow rate from the tank drain to increase appreciably (Figure 8.3.6-8). Stop the flow immediately by closing tank drain valve V2. It is useful to take a moment and drain the glycerol into its treatment unit, allowing time for the contents of the reaction tank to settle. After a short wait, return to the reaction tank and slowly drain off any remaining glycerol by opening and

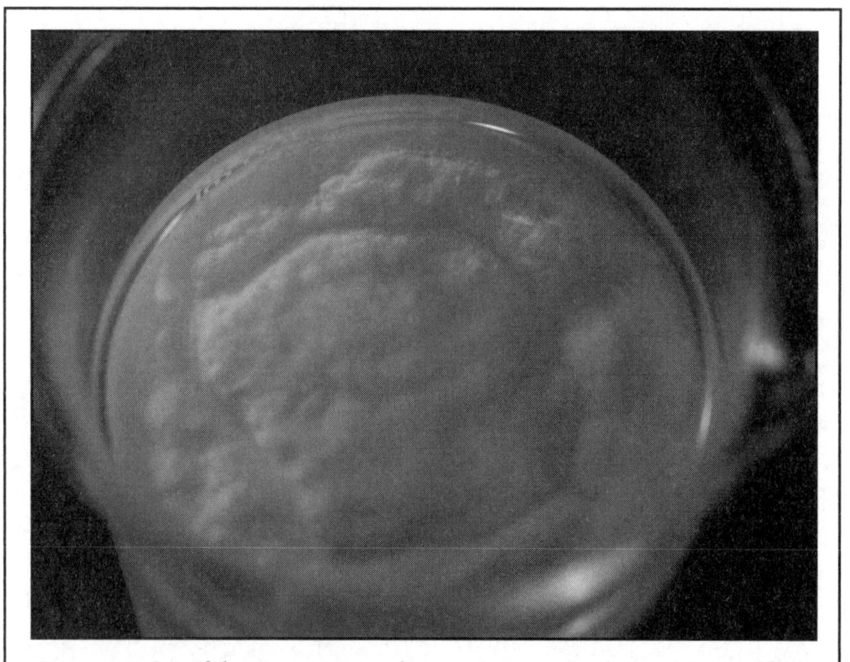

Figure 8.3.6-9. If the reaction process **has not** *run to completion, it is possible that a third layer of emulsion will form at the boundary between the glycerol and the FAME.*

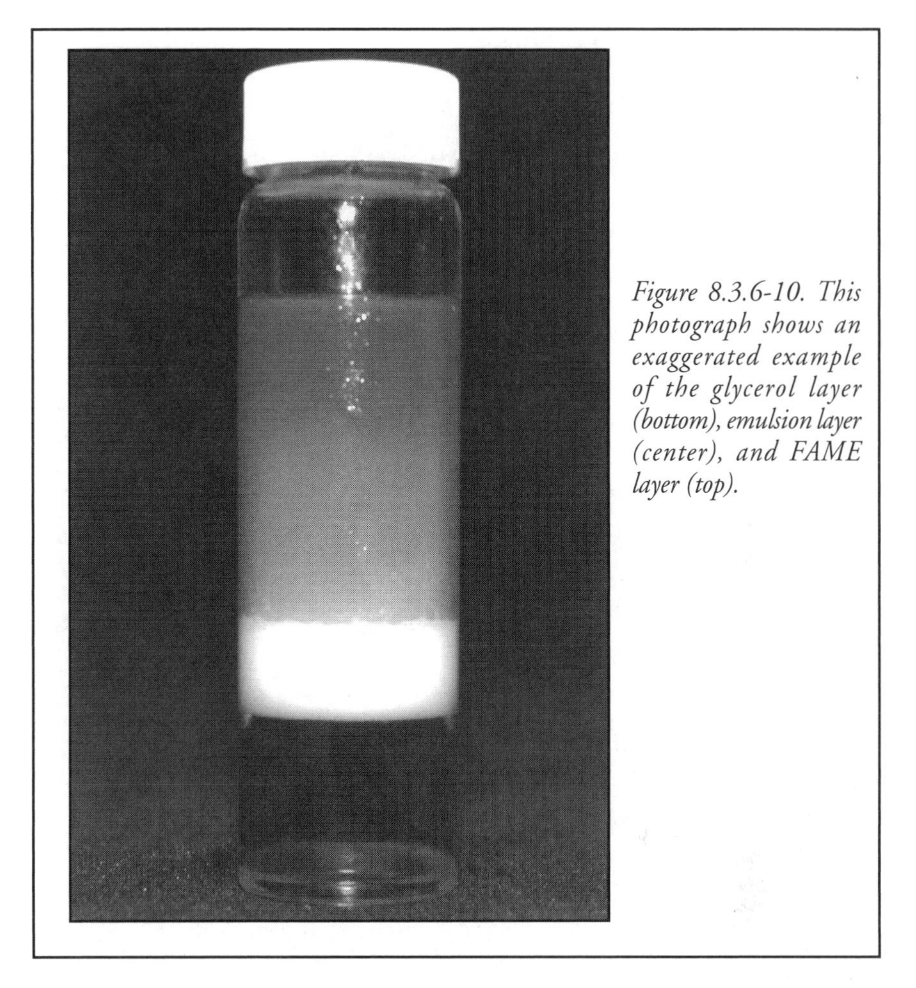

Figure 8.3.6-10. This photograph shows an exaggerated example of the glycerol layer (bottom), emulsion layer (center), and FAME layer (top).

closing valve V2. Deposit any remaining glycerol into its processing tank.

If the reaction process **has not** run to completion, it is possible that a third layer of emulsion will form at the boundary between the glycerol and the FAME (Figure 8.3.6-9). It is best to drain the emulsion layer with the glycerol for final processing. Figure 8.3.6-10 shows an exaggerated example of the glycerol layer (bottom), emulsion layer (center), and FAME layer (top).

The presence of an emulsion layer is indicative of an incomplete reaction and may be the result of an incorrect or inaccurate titration process, improper measurement of reactant chemicals, or insufficient mixing circulation time in the biodiesel reaction tank. The biodiesel produced from a batch with an emulsion layer requires a second-stage reaction as discussed below.

The FAME phase is now contained in the reaction tank.

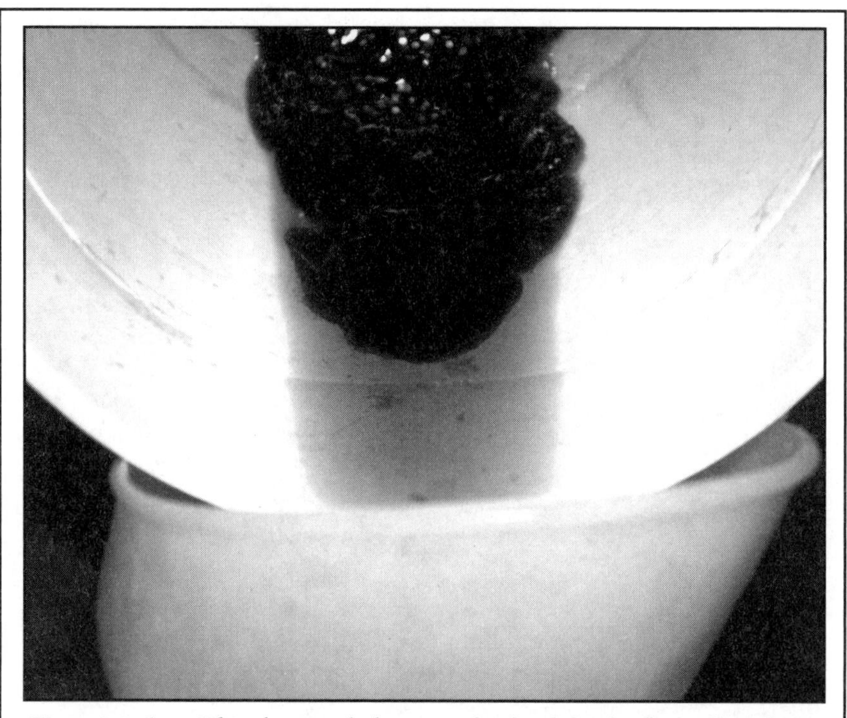

Figure 8.3.6-11. This photograph shows raw biodiesel that has been allowed to sit for a couple of days. The transesterification process has continued in this bucket (as it will in your fuel tank), causing glycerol to accumulate slowly but continuously. Unwashed biodiesel analyzed by an accredited laboratory was found to have glycerin levels 50 times in excess of ASTM standards.

This is the point where many home biodiesel producers drain the raw biodiesel and consider it acceptable for use. The biodiesel at this stage is very toxic and caustic and will certainly cause damage to modern automotive engines as discussed in earlier chapters. If you are still not convinced, refer to the end of Chapter 7.3 where unwashed biodiesel analyzed by an accredited laboratory was found to have glycerin levels 50 times in excess of ASTM standards (Figure 8.3.6-11) while methanol concentration was 10 times over the limit, making the fuel flammable and highly toxic. Even if you are not concerned about engine damage, **most** raw biodiesel users find themselves, sooner or later, sitting at the roadside in a vehicle disabled by a filter plugged with glycerin. This can require costly towing as well as being a pretty foolish waste of time considering that the problem can be overcome with just a bit more work.

Increasing the level of base catalyst above 3.5 grams per liter of oil does not increase the transesterification process yield and thereby lower total glycerin, and it may contribute to fuel washing problems as a result of soap formation. A batch of refined, food-grade canola oil was subjected to a reaction process using 4.5 grams of sodium hydroxide as the catalyst level mixed into the sodium methoxide solution. The resulting biodiesel was subjected to a two-hour circulation process (double the recommended time) and allowed to decant for 24 hours before methanol recovery and washing. The fuel was subsequently tested by an accredited laboratory and found to meet all of the ASTM requirements with the exception of total glycerin, which measured 0.672% mass compared to the maximum allowed under the standard of 0.24%. (Free glycerin measured 0.002% as expected as a result of the FAME washing process.)

The FAME was reprocessed in the reaction tank using 1 gram of sodium hydroxide and 5 litres of methanol in preparing the sodium methoxide solution. When reprocessed, the resulting total glycerin level was found to be within specification at 0.204% mass.

The Next Step – Second-Stage Reaction

If the goal of this process is to produce fuel that meets the ASTM fuel quality specification, it is necessary to reprocess the biodiesel to ensure that any remaining mono-, di-, and triglycerides are converted to biodiesel. The second-stage reaction ensures that gylceride-to-FAME conversion is completed and the fuel meets the total glycerin specification. The reprocessing can be performed by rerunning the transesterification process using sodium methoxide, produced without the requirement for a titration, at a ratio of 1 gram per liter of FAME with a volume of methanol equal to 5% of the FAME in the reactor vessel. For example, with 100 liters of FAME in the reaction tank use 100 grams of sodium methoxide and 5 liters of methanol to form the sodium methoxide solution. Proceed with the standard biodiesel reaction as noted above.

Methanol Recovery

After completing the biodiesel reaction it is necessary to recover excess methanol from the FAME phase in order to ensure that it does not contaminate the wash water. Removing methanol from the wash water requires a considerable amount of energy and tends to spread the toxic substance over additional material.

After the removal of glycerol the biodiesel is quite warm since the reaction tank retains a considerable amount of heat which can be reused to assist with the methanol recovery.

Popular texts claim that there are insignificant amounts of methanol in the biodiesel phase. Tests on unwashed biodiesel conducted by an accredited fuel laboratory showed conclusively that methanol concentration exceeded 2% in all cases. A typical 175-liter batch of biodiesel therefore has at least 2% by volume or 3.5 liters (≈1 gallon) of methanol concentration. Clearly this is not an insignificant amount of a toxic chemical and it can be easily and inexpensively recovered for future use. I have found it requires less than 3 kWh of electricity (say 30 cents) to recover the methanol, saving approximately $4.00 worth of methanol per batch or approximately 2.3 cents per liter (9 cents per gallon) of biodiesel produced.

To proceed with methanol recovery:

1. Ensure that valves V1, V2, and V6 are closed.
2. Ensure that valves V3, V4, and V5 are open.
3. Activate methanol heat recovery fans.
4. Start circulation pump 3.
5. Adjust tank heater thermostat to 80°C.

As the methanol reaches its boiling point of 64.5°C it turns into a vapor (methanol "steam") and flows through the upper vent of the tank into the heat exchanger. Fan pressure blows cool ambient air across the heat exchanger, cooling the methanol which condenses back to a liquid. As the heat exchanger is mounted above the tank and slightly angled, any accumulated liquid methanol will drip down the recovery tube and into the recovery tank. Excess methanol that does not condense will be vented through the outside vent pipe.

Once the methanol collection has stopped:

1. Turn off tank heater element.
2. Turn off circulation pump 3.

Allow heat exchanger fans to continue to operate for a several hours before turning them off, allowing the FAME to cool below the methanol boiling temperature. Carefully transfer the recovered methanol back into the methanol storage tank. The raw biodiesel is now ready for transfer into the fuel washing unit:

1. Ensure that the room ventilation fan is turned on to ensure that any residual methanol fumes are vented outdoors.
2. Open valve V6 to allow raw biodiesel to transfer to the fuel washing unit.
3. Activate circulation pump 3.
4. Slowly close valve V5 to ensure that all biodiesel is pumped into the fuel wash unit.

When fuel transfer is complete:

5. Turn off circulation pump 3.
6. Open valve V5.
7 Close valve V6.

The biodiesel reaction tank is now ready for processing the next batch of feedstock oil and the transferred batch is ready for washing.

8.3.7
Mist and Bubble Washing Steps

Biodiesel that has been transferred from the reaction tank into the fuel washing tank should be relatively free of methanol as a result of the distillation and recovery procedure conducted earlier. However, it will contain other contaminants including catalyst, soap, and free glycerin left over from the decanting process as well as emulsion and food particles, none of which is desirable in the final biodiesel fuel. The washing process removes the contaminants from the biodiesel and transfers them to the wash water.

Fuel washing is accomplished either by spraying the FAME with slightly softened, mineral-free water or by a gentle aeration process known as bubble washing. There is considerable debate about which method is better; however, I prefer mist washing because of minimal agitation of the biodiesel, resulting in less soap formation. From an efficacy standpoint either process yields excellent results, as will be discussed in Chapter 8.3.9.

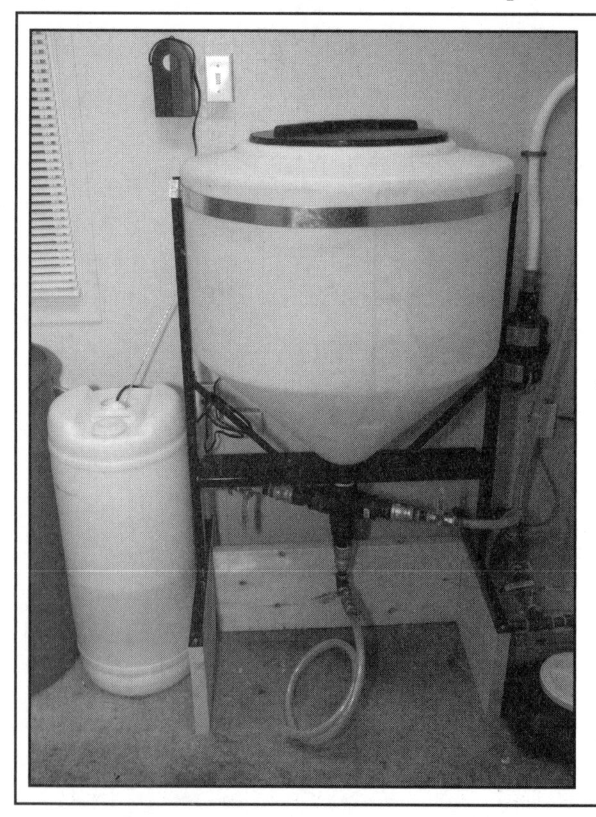

Figure 8.3.7-1. The fuel washing unit comprises a water heater and a fuel washing tank which is equipped for either bubble or mist washing systems, although some biodiesel producers prefer to use both technologies.

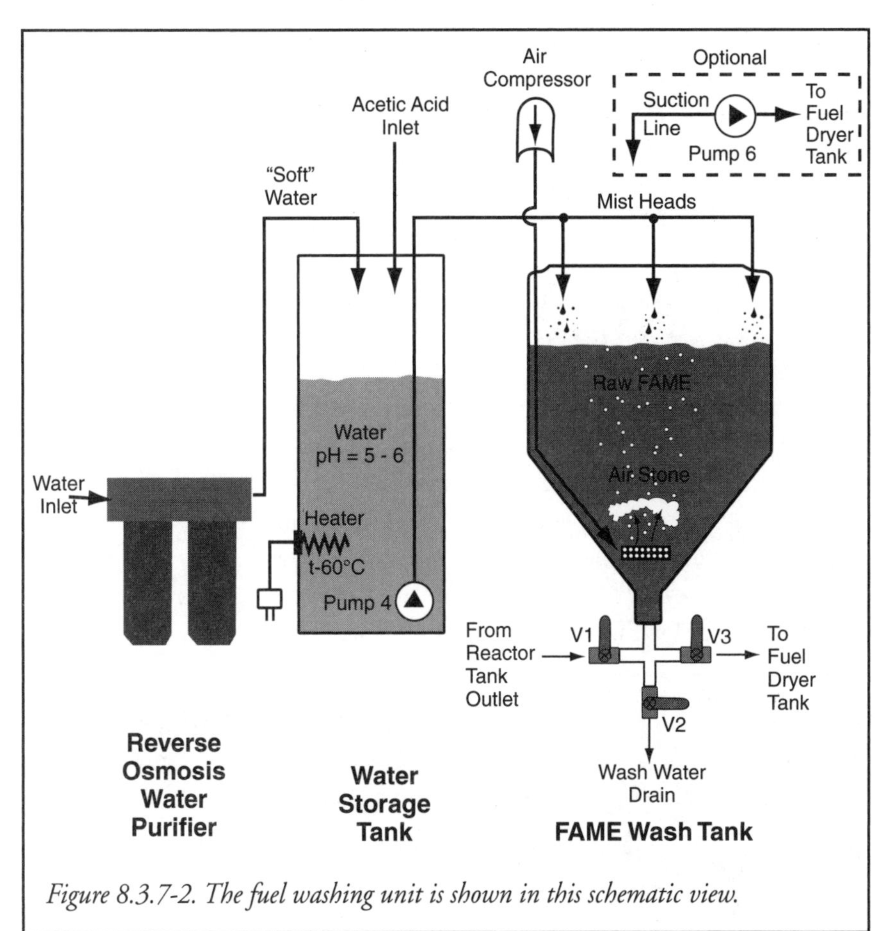

Figure 8.3.7-2. The fuel washing unit is shown in this schematic view.

The fuel washing unit is shown in Figure 8.3.7-1 and comprises a water tank filled with slightly softened (mildly acidic), demineralized, heated water. A pressure pump either transfers water to the wash tank directly (bubble washing) or sprays it through a series of "misting heads" which slowly mist water over the surface of the biodiesel (mist washing). Some biodiesel producers use both technologies.

Wash Water Supply

The fuel washing unit is shown in a schematic view in Figure 8.3.7-2. The system comprises a reverse osmosis water processing unit which provides demineralized water to the water tank. Other sources are softened, filtered rainwater or the potable water supply if it is known to be mineral-free.

The demineralized water is heated to a temperature of 60°C to prevent the precipitation of saturated fatty acid esters and to retard the formation of emulsions. The water is slightly softened by the addition of acid, which neutralizes the residual catalyst and removes salts formed during the production stages. The water softening process also eliminates calcium and magnesium contamination in the FAME[1]. I prefer acetic acid in a 5% concentration, also known as common white vinegar. In keeping with my environmental preferences it is known to be biodegradable and goes very nicely with your salad. Ten milliliters of vinegar for every one liter of wash water has been found to be an acceptable concentration.

Mist Washing Overview

If mist washing is the chosen system, a series of spray nozzles is configured around the circumference of the fuel washing tank as shown in Figure 8.3.7-3. The nozzles are supplied by either a submersible or an externally mounted water pressure pump connected to the wash water tank. I have experimented with both systems and have found that using a cottage-size, 120-volt water heater equipped with an external pressure pump requires less energy consumption than the polytank shown in Figure 8.3.7-1 which is equipped with a

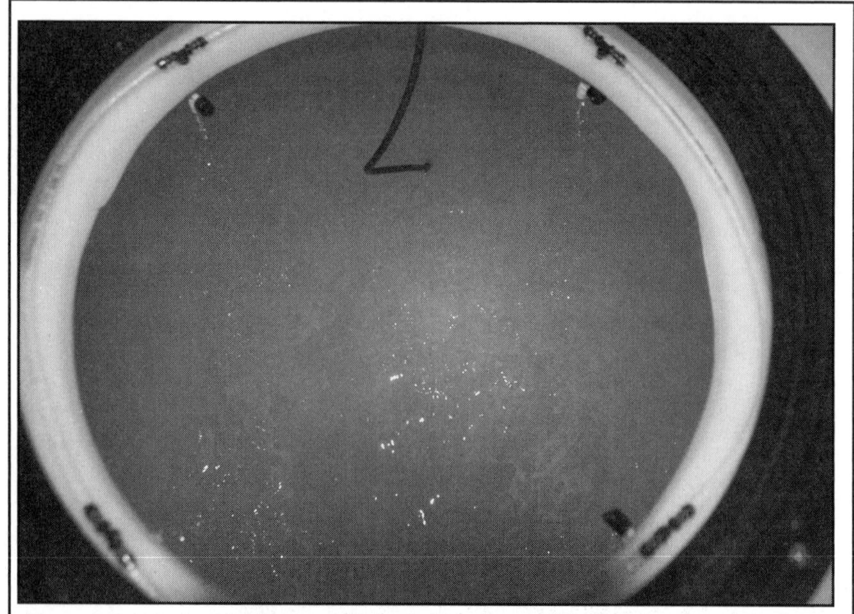

Figure 8.3.7-3. In this photograph, the fuel washing tank is operating with both mist and bubble washing technologies. A close examination of the picture shows that four nozzles are spraying wash water onto the biodiesel surface. The roiling surface of the fuel is caused by the bubble washing technique which uses compressed air and minute water bubbles to scrub the fuel of contaminants.

submersible water heater and pressure pump. In either case, it is necessary to develop a minimum of 10 psi (≈290 inches/24 feet of pump head) to ensure sufficient pressure to develop a light mist of water over the biodiesel surface (Figure 8.3.7-3). Pumps with high pressure and flow characteristics require a bypass valve to prevent excessively high water pressure damaging the spray nozzles or interconnecting hoses. To prevent high water pressure, connect a bypass valve between the outlet of the pressure pump and the wash water tank. A connection at the outlet of the pump supplies the nozzles. When the bypass valve is opened, the majority of the pressurized water will flow back to the tank and the spray nozzles will not function. As the bypass valve is closed, water pressure will build, directing water to the nozzles.

The wash water is then sprayed over the biodiesel fuel, absorbing contaminants as it falls to the bottom of the tank. I have found that using 50 liters of wash water for every 150 liters of fuel per wash and repeating the wash process 3 times has provided satisfactory results in accordance with ASTM standards. With a nominal 15 psi pressure applied to the spray nozzles, one wash cycle requires approximately 15 minutes to complete.

Bubble Washing Overview

The fuel being washed in Figure 8.3.7-3 is subjected to both mist and bubble washing. The surface of the fuel is roiling as a result of the application of air pressure through a submerged "air stone" of the type used in large aquariums (Figure 8.3.7-4). Prior to applying air pressure from a variable-flow aquarium compressor (Figure 8.3.7-5), 50 liters of wash water for every 150 liters of

Figure 8.3.7-4. The surface of the fuel is roiling during bubble washing as a result of the application of air pressure through a submerged "air stone" of the type used in large aquariums.

Figure 8.3.7-5. This type of large aquarium variable-pressure air compressor is used to supply the air stone submerged in the FAME. Although the bubble washing technique tends to cause more soap formation than mist washing, it is low cost and requires very little electrical energy or supervision.

fuel is added to the wash water tank. The compressor is started, allowing a **gentle stream** of air bubbles to rise from the bottom of the fuel washing tank, carrying along a minute amount of wash water. When the air bubbles reach the surface of the biodiesel they burst, allowing the water to fall back through the fuel, accumulating contaminants along the way. The bubble wash process can be operated for several hours or overnight for each wash cycle. The water is then changed and the process is repeated three or more times.

Pros and Cons of Each Technology

As with most things in life, there are pros and cons to each technology. The mist washing technique tends to be faster, allowing the wash water to be changed after each 60-minute cycle, thus completing the entire wash process in a few hours. Mist washing tends to be more gentle than bubble washing, preventing the buildup of soap within the fuel washing tank. On the other hand, mist washing requires a dedicated water heater tank, pressure pump, and other mechanical items that increase capital cost.

Bubble washing is very low cost, requiring only an aquarium air compressor and air stone. Wash water can be supplied by a bucket or hose from an existing hot water supply, with the acetic acid added directly to the fuel washing tank if a hose supply is used. Energy consumption is lower with bubble washing, requiring only a few watts of electrical energy to power an air compressor (approximately 5 cents per 24-hour operating period). In addition, the compressor can be left on for hours on end without concern for cycle time. On the downside, bubble washing tends to be harsher than mist washing, leading to increased soap formation, and the process requires considerably more time, often running for several hours per wash cycle.

Some biodiesel processors use both technologies, preferring to use the gentle spray of the mist washing system for the first wash and bubble washing for subsequent washing steps. In the end, as long as fuel meets the appropriate fuel quality standards, either technology can provide satisfactory results.

Fuel Washing Procedure

1. Fill the wash water heater tank with 1 liter of demineralized water for every 3 liters of biodiesel to be washed. Heat the water to approximately 60°C.
2. Add 10 milliliters of 5% white vinegar for every one liter of wash water.

3a. Mist Washing:
- Start the mist wash pump 4.
- Allow pump to operate until wash water has been drained from wash water supply tank.
- Go to Step 4.

3b. Bubble Washing:
- Gently add the wash water to the fuel washing tank.
- Lower the air stone (Figure 8.3.7-4) into the fuel washing tank.
- Activate compressor (Figure 8.3.7-5) at the lowest pressure setting to ensure adequate bubbling without causing excessive agitation of the fuel.
- Allow compressor to run a minimum of 4 to 6 hours or overnight.

4. Stop water circulation pump or air compressor and allow FAME to settle for approximately one hour.
5. Drain water from bottom of fuel washing tank by opening valve V2 as shown in Figures 8.3.7-6 and 8.3.7-7.
6. Measure the pH of the wash water using an electronic pH meter or pH strips. If the pH is greater than 7.9, proceed to Step 7. If it is equal to or less than 7.9, proceed to Step 8. For more information on the use of electronic pH meters, see section 8.3.10, Environmental Sustainability, page 516.
7. Process subsequent fuel washing stages in the same manner as noted above.
8. After the final wash step and water drain, the biodiesel is ready for transfer to the fuel dryer unit.

Figure 8.3.7-6. After each successive washing stage, wash water is drained from the bottom of the fuel washing tank by opening valve V2.

Figure 8.3.7-7. The wash water will be a milky color and may contain globules of soap that are formed when processing high-FFA WVO.

Figure 8.3.7-8. The first wash water is very milky and may have a soapy emulsion floating on top. Note the small layer of biodiesel that is floating on top of the water and soap layer. This is impossible to prevent and will be discussed in Chapter 8.3.10, Wash Water Management.

Figure 8.3.7-9. The wash water from the second washing step is normally soap-free and lighter in color owing to the reduced concentration of contaminants. Keep this wash water and use it for the first wash of the next batch of biodiesel that is to be processed.

Figure 8.3.7-10. The wash water from the third washing step is almost completely clear, indicating that the washing process is complete. Save this wash water for the second wash of the next batch of biodiesel and the first wash of the subsequent batch.

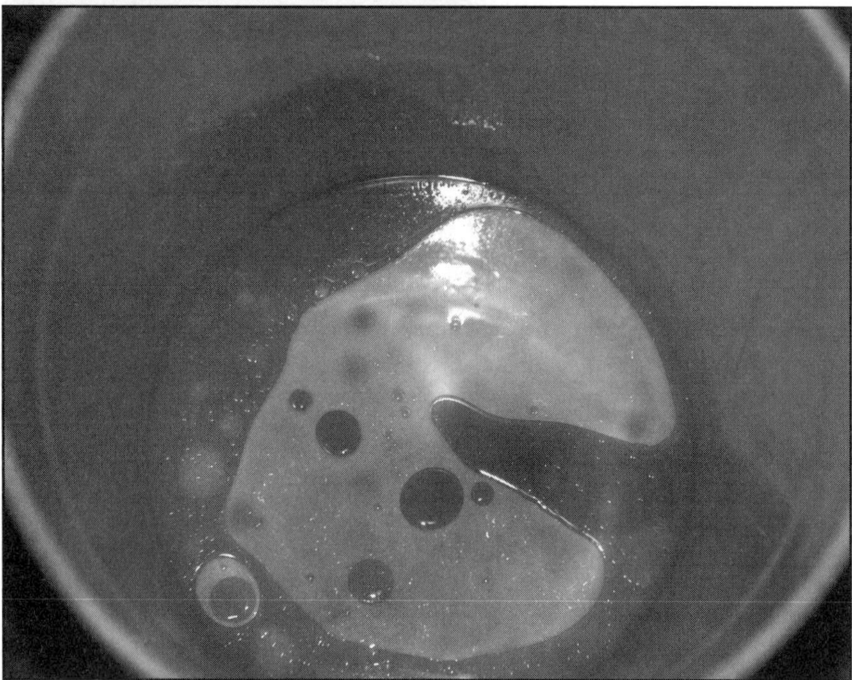

Figure 8.3.7-11. As wash water is drained from the fuel washing tank, biodiesel will start to run with the water. This is normal. When the water and biodiesel mix looks like this, with biodiesel just starting to flow, close valve V2 and continue with the next washing step or transfer biodiesel to the fuel dryer unit.

The first wash water will contain a considerable amount of catalyst and other materials, causing the water to be very milky in color. If soaps are formed during the washing stage they will float on the top of the wash water as shown in Figure 8.3.7-8. Note the small amount of biodiesel that has accumulated in the water during the wash water draining.

Subsequent wash steps will produce water that is progressively clearer as shown in Figure 8.3.7-9 (second wash water) and Figure 8.3.7-10 (third wash water). There is typically no soap formation in successive washing stages, which is one of the reasons that some biodiesel producers prefer to use the gentle mist washing process for first-stage washing and bubble washing for subsequent stages. I have personally found that since the equipment is very inexpensive bubble washing can be worthwhile if processing speed is of little concern. Remember that it can take several days to wash biodiesel using the bubble washing technique, which will slow production.

A small amount of biodiesel will always be present in the biodiesel wash water no matter how careful you are in the water decanting step (Figure 8.3.7-11). This and other waste water management issues will be dealt with in Chapter 8.3.10, Wash Water Management.

Because the wash water from successive washings is considerably less contaminated than the water used before it as noted in Figures 8.3.7-8, 8.3.7-9, and 8.3.7-10, it is possible to save the second wash water and use it for the first wash water of the next batch to be processed. Likewise, the third (as well as any subsequent) wash waters can be used for the second and first washing steps of subsequent batches. This process is known as counter-current washing; it permits recycling of water and reduces waste water treatment in later steps.

If the fuel washing process generates a considerable amount of soap, it may be difficult to decant the wash water without leaving a large amount of soap floating along the bottom of the biodiesel phase and coating the walls of the fuel washing tank (Figure 8.3.7-12). This situation can be controlled by careful titration and WVO drying before starting the transesterification process. The effect shown in Figure 8.3.7-12 was achieved by deliberately reacting WVO that had an FFA reading of 7% and was not dried in the waste oil treatment unit. The resulting biodiesel created a large amount of soapy emulsion that coated the fuel washing tank as shown. Subsequent efforts to drain the wash water from the tank resulted in large globules of soap being suspended in the water (Figure 8.3.7-13) and also floating along the bottom of the biodiesel phase (Figure 8.3.7-14).

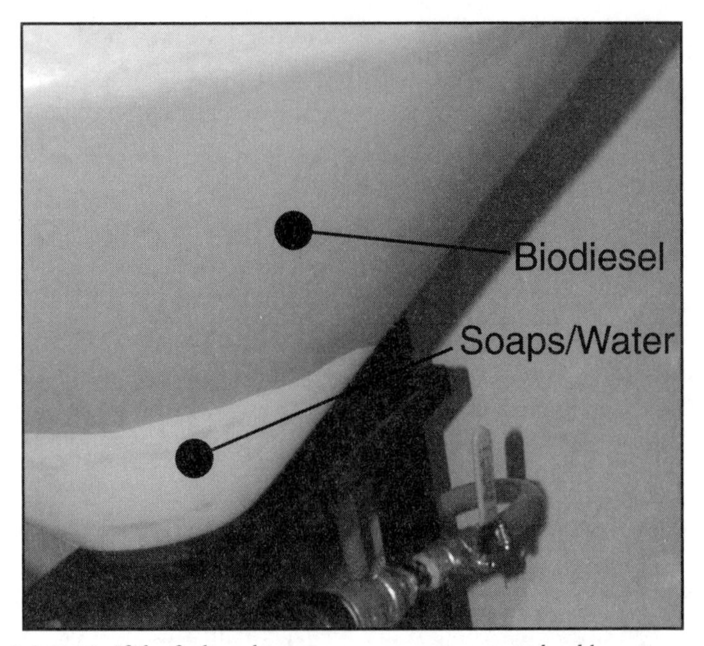

Figure 8.3.7-12. If the fuel washing process generates a considerable amount of soap, it may be difficult to decant the wash water without leaving a large amount of soap floating along the bottom of the biodiesel phase and coating the walls of the fuel washing tank.

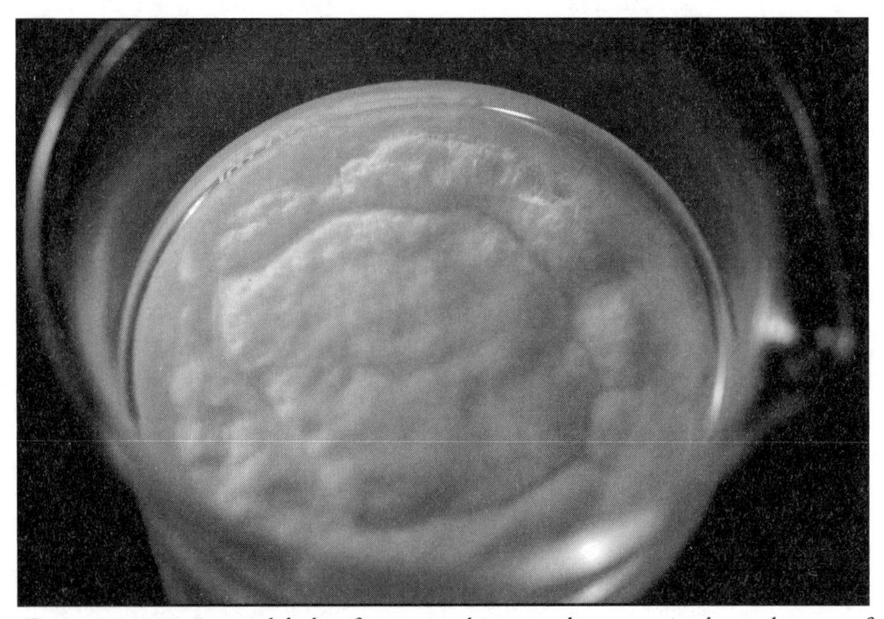

Figure 8.3.7-13. Large globules of soapy emulsion may be present in the wash water of biodiesel made with high-FFA or water-saturated WVO.

Figure 8.3.7-14. Soap floats between the wash water and biodiesel phases as shown in this image.

Transferring FAME to the Fuel Dryer Unit

If soap buildup occurs in the fuel washing system, the simplest means of removing the biodiesel without having to worry about messy filtration procedures is to use a top-mounted suction pump (Figures 8.3.7-15 and 8.3.7-16) to draw the biodiesel out of the top of the fuel washing tank and direct it into the fuel dryer unit. The pump shown in Figure 8.3.7-16 is a flexible impeller pump with a positive displacement design, providing considerable suction ability which can draw the biodiesel through the top access cover of the fuel washing tank. Direct the suction tube into the biodiesel phase, taking care not to draw any soap trapped along the walls of the tank and in the narrow "neck" of the tank drain area.

After removal of the biodiesel, the soap may be washed out with fresh water and placed in the waste water treatment container.

If the fuel washing process did not generate any significant soap buildup, the biodiesel can be transferred to the fuel dryer unit by opening valve V3 and using the bottom drain. Continue with fuel drying procedures as discussed in Chapter 8.3.8.

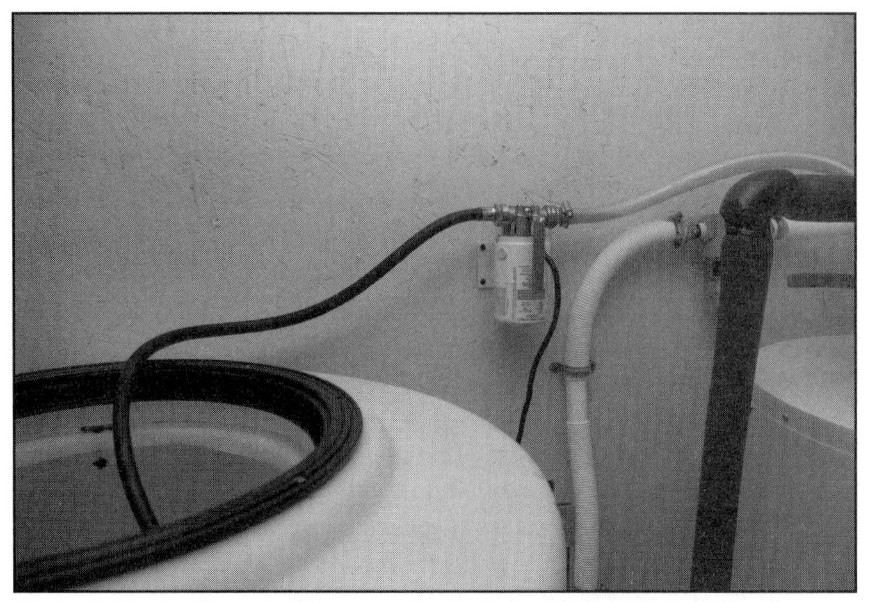

Figure 8.3.7-15. If soap buildup occurs in the fuel washing system, the simplest means of removing the biodiesel without having to worry about messy filtration procedures is to use a top-mounted suction pump as shown here.

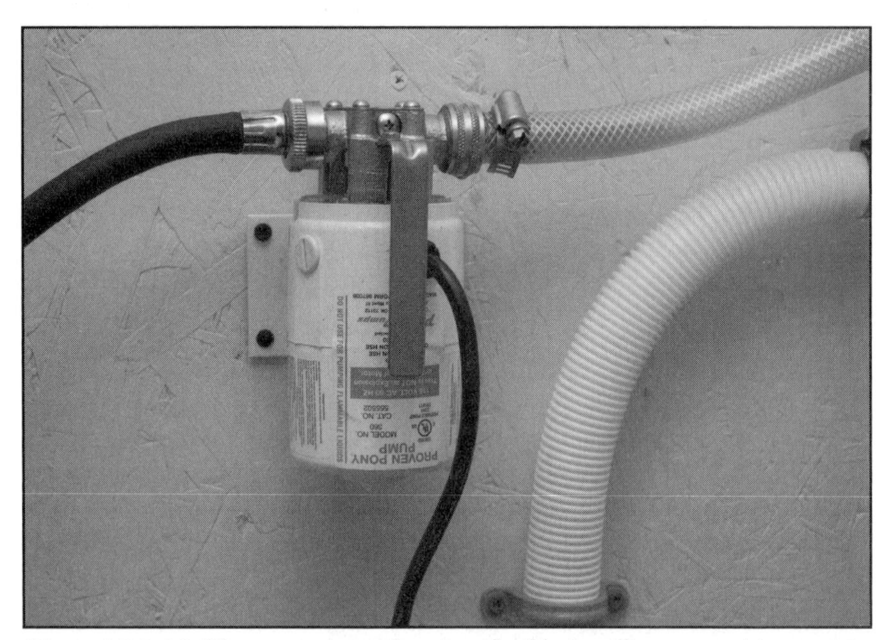

Figure 8.3.7-16. The pump shown here is a flexible impeller pump with a positive displacement design, providing considerable suction ability which can draw the biodiesel through the top access cover of the fuel washing tank.

8.3.8 Drying and Final Biodiesel Treatment

Biodiesel that has passed the fuel washing stage is saturated with water and is in excess of the ASTM limits. Water suspended in the biodiesel leads to storage degradation through bacterial growth and also promotes oxidation instability. It also contributes to oxidation (rusting) of the accurately machined parts in the modern diesel engine.

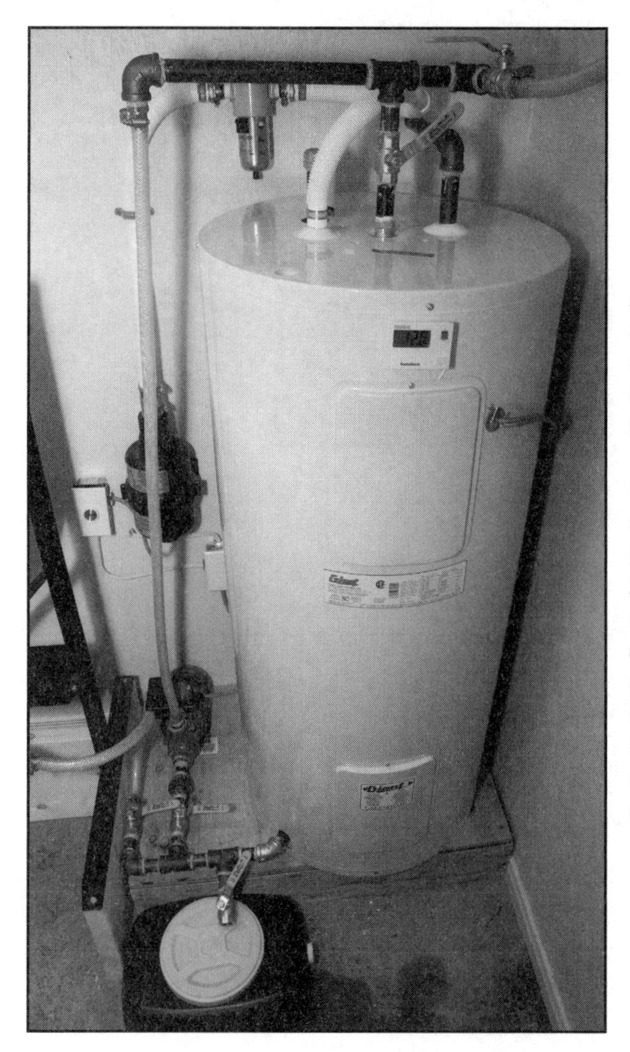

Figure 8.3.8-1. The fuel drying unit is an insulated water heater tank that is equipped in the same manner as a commercial "packed column" dryer (see Chapter 8.2.6). Biodiesel is heated and circulated in one direction while heated air is blown into the tank in the opposite direction, causing the evaporation of water molecules and drying the FAME. Moisture-laden air is then vented outdoors.

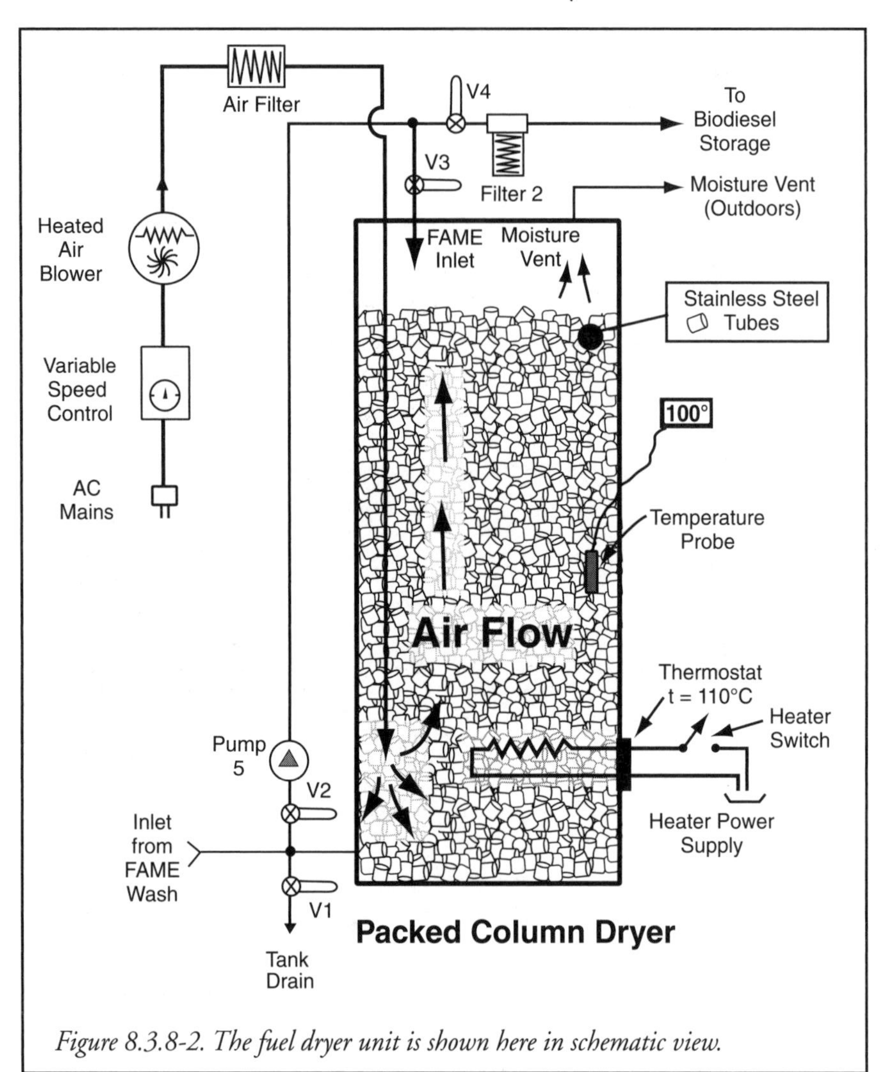

Figure 8.3.8-2. The fuel dryer unit is shown here in schematic view.

Fuel is transferred from the fuel washing unit to the fuel dryer and the drying process is started as follows:

1. Fuel dryer tank drain valve V1 is closed.
2. Fuel washing unit circulation valve V3 (if equipped) is opened.
3. Fuel dryer circulation valve V2 is opened.
4. Fuel dryer circulation valve V3 is opened.
5. Fuel dryer outlet control to storage valve V4 is closed.
6. Outdoor moisture vent plug is removed if one has been installed.

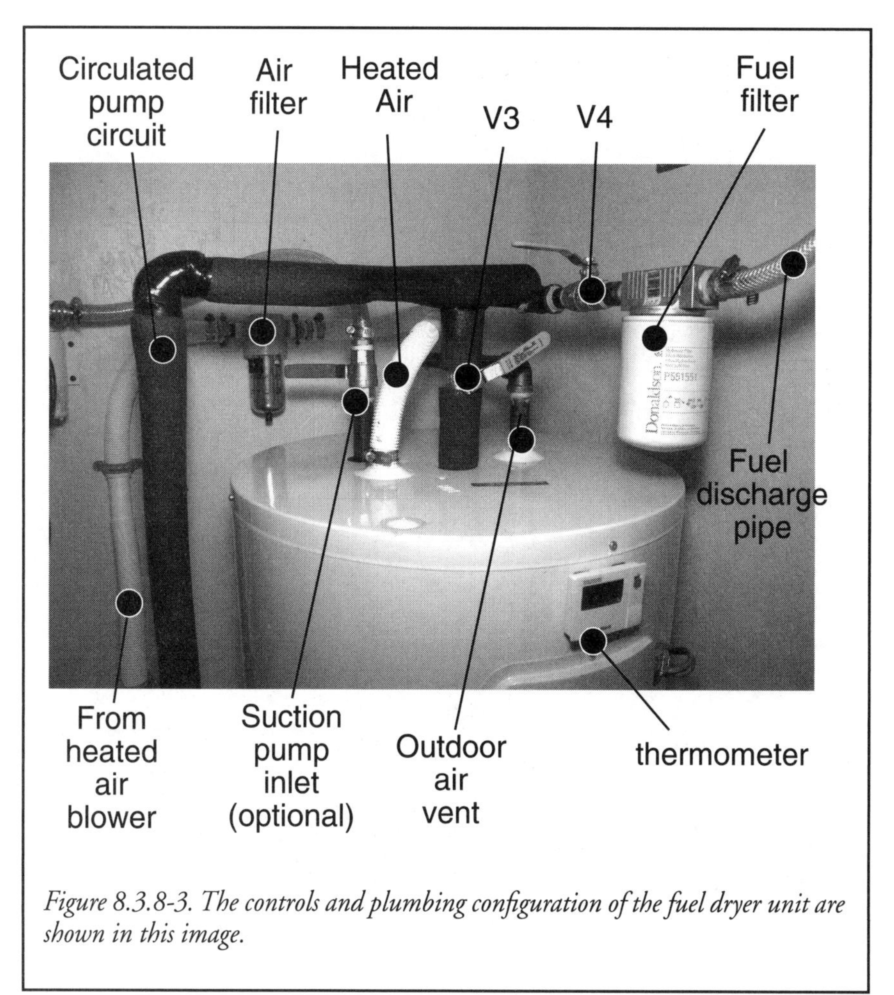

Figure 8.3.8-3. The controls and plumbing configuration of the fuel dryer unit are shown in this image.

7. Circulation pump 5 is activated.
8. Biodiesel from the fuel wash tank is drawn into the fuel dryer unit. Watch the vertical sight glass to determine when the fuel dryer is approximately 80% full. (Do not overfill the fuel dryer as there is a risk that biodiesel may be pumped out the moisture vent when air blower is operating.)
9. When fuel dryer is full, activate tank heater. It is not necessary to operate the tank heater above the boiling point of water (100°C) although the hotter the biodiesel the faster water will be driven from the fuel and the process completed. Any temperature above 60°C has been tested with satisfactory results.

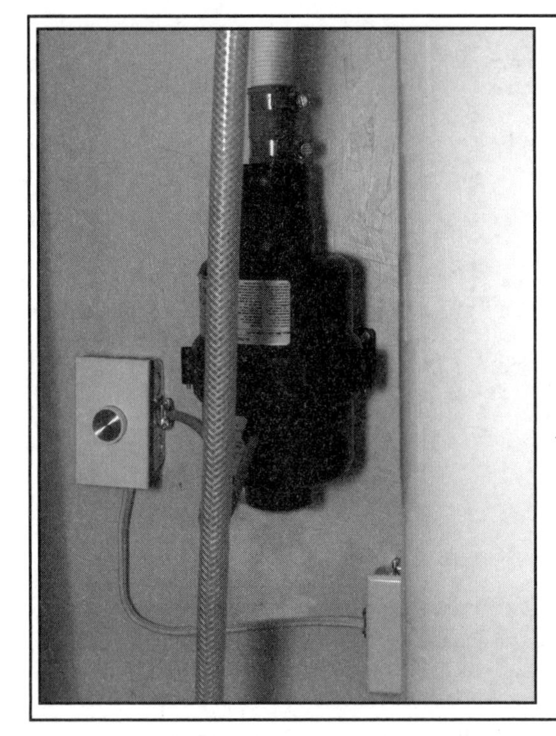

Figure 8.3.8-4. When biodiesel reaches approximately 60°C, activate the heated air blower by adjusting the variable speed control shown above. (The blower may be run below maximum operating speed provided that air flow can be detected coming through the outdoor moisture vent.)

FAME Temperature (°C)	Approximate Fuel Drying Time (minutes)
60	200
70	140
80	120
90	90
100	60

Table 8.3.8-1.
Relationship between fuel dryer operating temperature and time required to remove water from biodiesel.

10. When biodiesel reaches approximately 60°C, activate heated air blower by adjusting variable speed control (Figure 8.3.8-4). The blower may be run below maximum operating speed provided that air flow can be detected coming through the outdoor moisture vent. If air flow is restricted because of low blower operating speed, damage to the blower will result. Increase blower speed until satisfactory air flow has been achieved
11. Follow the operating time schedule shown in Table 8.3.8-1.

Completion of Drying Cycle

At the completion of the drying time specified in Table 8.3.8-1, follow the procedure noted below:

1. Turn off tank heater.
2. Turn off heated air blower.
3. Turn off circulation pump 5.

The biodiesel is now ready for storage and testing. If the fuel drying unit is being used as the final storage and dispensing unit, place a removable plug in the outdoor air vent to prevent moisture from entering the tank. Apply the ventilation plug only after the biodiesel has cooled to room temperature in order to prevent pressure from building up in the tank.

To dispense fuel or transfer it to another storage tank or tote:

1. Ensure valves V1 and V4 are closed.
2. Ensure that valves V2 and V3 are open.
3. Activate circulation pump 5.
4. Place fuel fill pipe (shown on output side of fuel filter in Figure 8.3.8-5) in fuel tote (Figure 8.3.8-6) or other storage drum.
5. Slowly open valve V4 to initiate fuel flow into storage tote.
6. Throttle valve V3 to its closed position to speed fuel flow.

The fuel is now ready for storage and quality control testing.

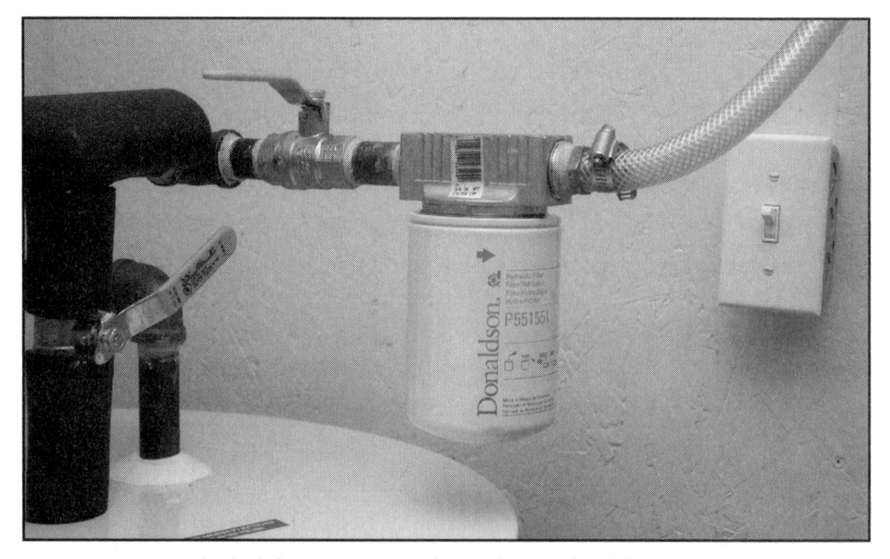

Figure 8.3.8-5. The fuel drying unit may be used as the final fuel storage tank if desired. When fuel is ready to be dispensed it is pumped through a spin-on fuel filter and dispensed through the supply pipe (right).

Figure 8.3.8-6. Completed biodiesel fuel may be stored in the fuel drying unit or in hand totes such as the model shown or in larger storage drums.

8.3.9
Quality Control Testing

As I have stated numerous times, it is deceptively easy to make poor-quality biodiesel fuel. Many a home-based biodiesel producer has questioned the need for high-quality fuel standards and has hidden behind a tale of "environmental sustainability" when the real reason for producing the fuel is to save money. If it were as easy as simply reacting vegetable oil and draining off the glycerol, wouldn't commercial biodiesel producers follow the same path?

Accredited fuel testing laboratories spend millions of dollars on the latest chemical analysis equipment to ensure that their fuel testing abilities meet the meticulous standards required by industry. Rigorous fuel testing and quality programs provide engine and vehicle manufacturers with a guarantee that biodiesel is fully compatible with their products and warranty claims will not result from its use. There are millions of diesel-powered vehicles in the world, and using poor-quality fuel could result in massive liability costs for the industry.

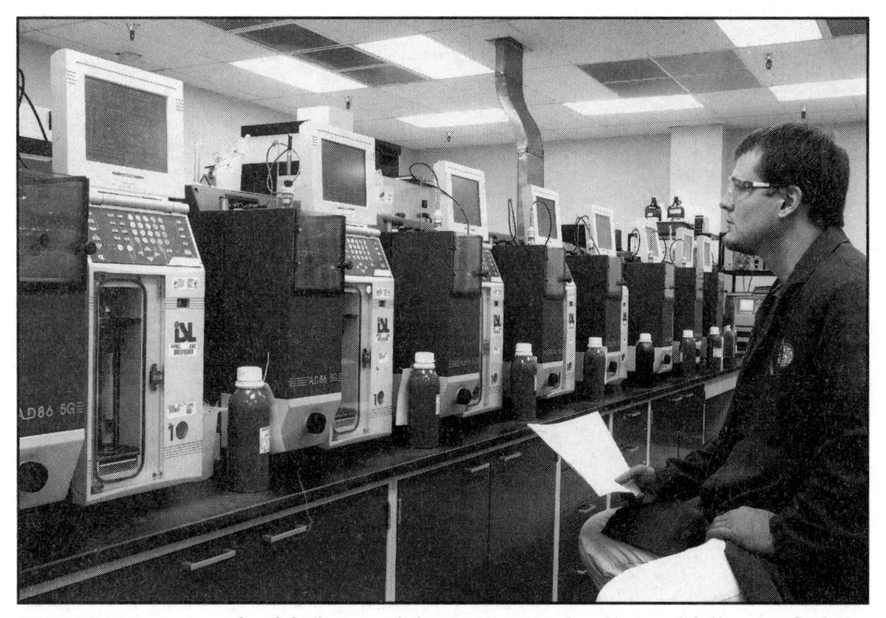

Figure 8.3.9-1. Accredited fuel testing laboratories spend millions of dollars on the latest chemical analysis equipment to ensure that their fuel testing abilities meet the meticulous standards developed by industry with the assistance of the ASTM or CGSB quality standards organizations.

Figure 8.3.9-2. . Rigorous fuel testing and quality programs provide engine and vehicle manufacturers with a guarantee that biodiesel is fully compatible with their products and warranty claims will not result from its use. "I would be out of business in a week if I sold off-specification fuel," says Dan Freeman of Dr. Dan's Alternative Fuel Werks. "The value of low-cost, off-specification fuel would be lost the first time a car stalled on the freeway or blew up a $1,000 fuel pump. It just isn't worth the risk to me or my customers."

Figure 8.3.9-3. Perhaps fuel quality doesn't matter to you if the vehicle is an older, high-mileage model such as this diesel truck that Steve Anderson found abandoned in the woods. "When I found it there were trees growing up through the floor boards," says Steve. "A bit of welding and it's as good as new." Perhaps, but a damaged engine or fuel system caused by poor-quality fuel isn't much of a risk when the vehicle is worth less than the cost of a fill-up.

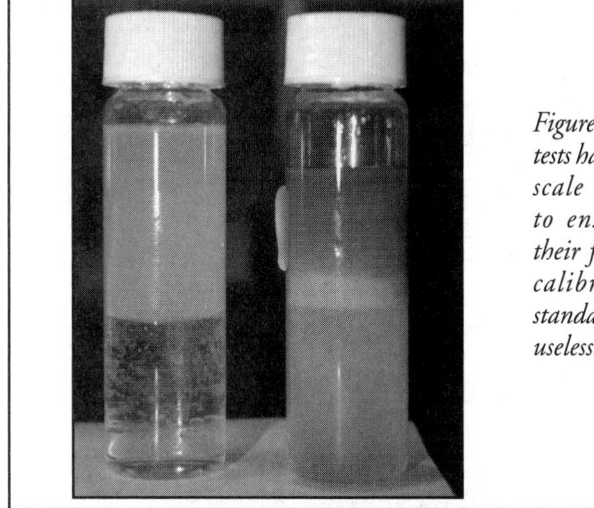

Figure 8.3.9-4. Several basic tests have been used by home-scale biodiesel producers to ensure the quality of their fuel, but without any calibration to published standards they are essentially useless.

Biodiesel producers wanting to meet the ASTM standard encounter a major obstacle: fuel test costs. Conducting the full suite of ASTM D 6751 tests may exceed $1,000 per test, regardless of whether the fuel batch is 100 or 10,000 liters in size. Obviously no one is going to conduct tests on homemade biodiesel at those rates. Well, almost no one.

A main objective of this book was to determine if fuel quality testing could be conducted to the same level as that of commercial test labs; and if not, to ascertain what level of test accuracy could be achieved by the micro- and small-scale producer. I have undertaken numerous fuel tests in order to calibrate the test procedures outlined below.

Certainly any testing is better than none. Several basic tests have been used by home-scale biodiesel producers to ensure the quality of their fuel, but without any calibration to published standards they are essentially useless. Consider the "water shake" test shown in Figure 8.3.9-4. This is a commonly touted test that determines the quality of fuel by the "sharpness" of separation and the number of phases that result when equal parts of water and biodiesel are shaken together. In this image, ASTM-grade biodiesel is shown left and raw, unwashed biodiesel (the type used by many micro-scale producers) is shown right. After a period of time the ASTM-grade fuel has settled into two distinct phases and the water has remained perfectly clear. (The biodiesel is slightly cloudy, however. This is expected and results from the absorption of approximately 2,000 parts per million of water. The fuel will become clear when the water is removed by fuel drying).

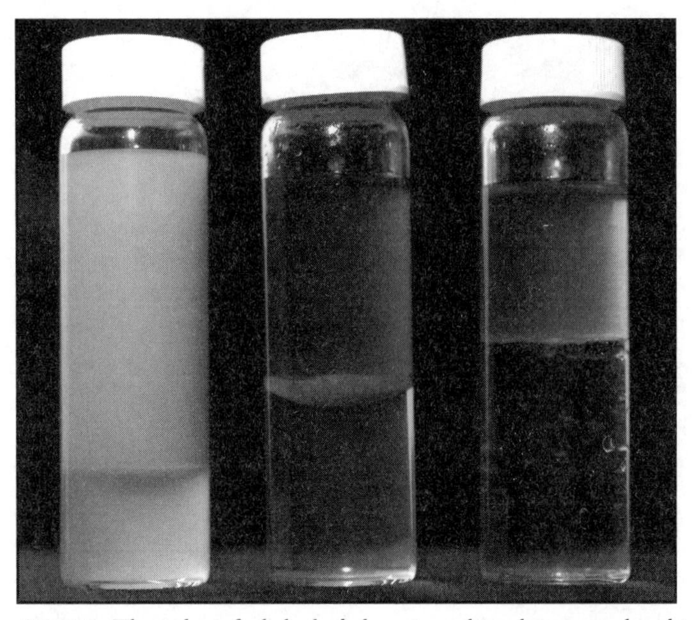

Figure 8.3.9-5. These three fuels looked the same when they were placed in the vials. In each case, the shake-the-fuel-with-water test proves nothing for two reasons: 1. in the first two vials it shows what we already knew because of incorrect production; 2. in the third vial the test is in fact false because although it looks to have passed the fuel is below specification.

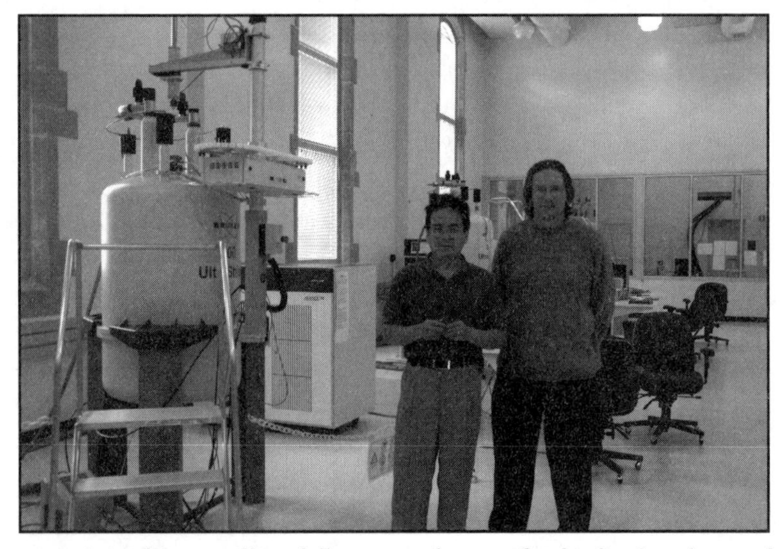

Figure 8.3.9-6. At about a million dollars a pop there are few biodiesel producers who have access to tools such as these used by Dr. Martin Reaney and his team at the University of Saskatchewan. Nevertheless, it is still possible to perform a reasonable level of fuel quality testing on a micro or small scale.

The homemade raw biodiesel has separated into three phases: water, soap/catalyst, and fuel, with the water turning milky and murky as a result of contamination from catalyst, soaps, and other materials. Even without the benefit of testing this fuel would have been known to be unacceptable because the fuel processing steps of washing and drying were not completed. In reality, there is no reason to run a test on any fuel that has not proceeded through all of the steps outlined in this book.

Now consider the sample biodiesel fuels shown in Figure 8.3.9-5. All three fuels looked the same when they were placed in the vials. After shaking with equal parts water, the first fuel sample turned into a mess of emulsion. Subsequent evaluation showed that the person attempting to produce biodiesel and performing the titration was using swimming pool pH indicator rather than technical-grade phenolphthalein; he was not following proper procedures to ensure an accurate measurement. In addition, the WVO was not being dried and was found to have significant levels of suspended water. When queried, this producer admitted to having a number of biodiesel batches fail in a similar manner.

The sample in the center of the photograph was successfully reacted but was neither washed nor dried, resulting in the presence of a soap layer and milky wash water. Subsequent analysis showed that total glycerin was 50 times in excess of ASTM standards, and several other readings were similarly "off-the-scale."

The sample to the right was washed and dried but was found to be well out of specification on several test parameters of the biodiesel fuel standards even though the visual indication from the "shake test" appeared to be acceptable. A review of the processing methods showed incomplete transesterification with mono-, di-, and triglyceride levels exceeding 5%. This "biodiesel" could almost be referred to as oil feedstock.

In each case, the shake-the-fuel-with-water test proves nothing other than to confirm that a shortcut in the production procedures has been taken; but then again, the fuel producers probably already knew that.

If biodiesel testing by third-party laboratories is too expensive and "shake-the-fuel" tests don't work, what can be done by the micro- and small-scale producer to ensure fuel quality?

One approach is for the commercial biodiesel industry, universities, and government to assist micro- and small-scale producers to improve their fuel production technology and procedures by offering subsidization of fuel test programs. The aim would be to help legitimize the small-scale industry by

Figure 8.3.9-7. Fuel quality matters. An engine that is starved of fuel because excess glycerin has plugged the filter and stalls on a freeway is more than an inconvenience and is totally unnecessary. "A shortcut is the longest distance between two points."

Figure 8.3.9-8. Fuel quality is not determined in the reaction tank alone. Fuel that is washed but not dried or is allowed to sit in a damp environment will quickly become saturated with water and develop bacterial growth. Although antibacterial agents can be added, it is much easier to produce quality fuel and store it in the proper manner.

improving production quality, safety, and environmental responsibility.

The Canadian federal government has developed its Biodiesel Initiative, which was announced as part of the Climate Change Plan for Canada. The intent of the program is to conduct fuel quality testing, establish an accessible database of test results, and develop industry protocols and standards for fuel analysis. Under this program, applicants are provided generous rebates on the cost of conducting early-stage fuel tests. Unfortunately, the program doesn't go far enough, as testing fees revert to standard cost after a few test iterations, depriving the micro-producer of an ongoing test program. (Those interested in the test program should contact the Alberta Research Council, Fuels and Lubricants Group, at http://www.biofuels.arc.ab.ca/Default.ksi).

After conducting numerous analyses using accredited fuel test facilities, I have found that several fuel samples produce similar and acceptable results for many of the required tests, regardless of who produced the biodiesel and from what feedstock it came from. For example, Acid Number was measured in two samples as 0.21 and 0.11 respectively, even through the first sample was WVO-based raw biodiesel and the second was cold-pressed canola, processed using the full procedures described herein. The maximum Acid Number allowed under ASTM D 6751 is 0.80, while the worst example of biodiesel I have measured is four times under this limit.

Comparing these results, it appears that there is a varying degree of risk related to passing or failing a given test. I have assumed that the "high-risk" tests should be completed first, while other tests can be deferred until accessible testing methods become available to the micro-scale producer.

It seems reasonable to assume that a condensed version of the ASTM D 6751 test could be created specifically for micro- and small-scale producers to drive them in the direction of some form of basic quality program, with staggered levels of sophistication en route to full ASTM compliance. Such a test program would be less costly and simpler to perform and would help corral the smaller producers and let them develop production processes to meet each step of the test program.

Of course this is heresy to the commercial fuel producers, but even they started without standards and testing programs when the rush for oil started at Spindletop decades ago. Why not help the small-scale industry develop as well, but with the support and guidance of their bigger brothers?

In the meantime, until every home shop is equipped with an imaginary $20 gas chromatograph (current models run into the several-thousands of dollars) I provide this simplified version of the ASTM D 6751 fuel test analysis that can be conducted by the micro-scale biodiesel producer.

The Simplified ASTM D 6751 Test Program

Introduction

Quality standards are required for the commercial use of any product, providing purchasers with an assurance that they are receiving proper value for their money. Obviously, no one would consider paying the full retail price for damaged goods, and biodiesel is no exception.

However, many stores will offer defective or marginal-quality products at fire sale prices and consumers will purchase them in an effort to save money. These discounted products are generally sold without warranty, the buyer accepting the risk in exchange for having paid a lower-than-retail price.

This is how most home biodiesel producers justify their relationship with poor-quality fuel; they willingly take the risk of engine damage in pursuit of lower fuel costs. With luck, the gamble pays off.

For those who are unwilling to damage a late-model or higher-value vehicle but are reluctant to spend the money on accredited fuel testing of **each batch** of biodiesel produced, the following simplified tests provide a high degree of quality assurance. (As every batch of biodiesel is produced from different feedstock and under new production conditions, it is necessary to test individual batches to assure compliance with either ASTM or your own fuel quality standards.)

Many of the tests required by ASTM standard D 6751 are beyond the scope of basic chemistry and are difficult if not impossible for the micro-producer to replicate. One example is the measure of cetane or the ability of the fuel to ignite smoothly and quickly, helping to ensure good cold-starting capabilities and smooth overall operation of the engine. The test for cetane is conducted either in a specially designed engine that is able to measure the specific combustion pattern of the fuel or by analysis in a gas chromatograph. Obviously this equipment is beyond the reach of the micro- or small-scale producer.

On the other hand, virtually all biodiesel fuels (including fuels I have tested that are produced from varying feedstocks) have cetane ratings above the ASTM minimum requirement of 47, indicating that this test is of marginal value in any event.

Taking into account financial or technical limits, I have removed several tests from this quality assurance program, especially if there is an alternate means of verifying that fuel-related problems with diesel-powered vehicles are limited.

The desire to produce fuel that is certified to ASTM standards is tempered by the cost of obtaining certification. If fuel is to be sold on the market, certification is a must; for those who consume their own fuel or produce it in a local cooperative for local consumption, perhaps ASTM certification is not as important. The following tests provide a measure of confidence in the fuel that is produced and for a consumer who is aware of the limitations they are perfectly adequate.

Excluded Tests

The specific tests that have been eliminated and the reasons for their elimination are:

- Acid Number
- Sulfated Ash
- Cetane Number
- Distillation Temperature
- Flash Point
- Carbon Residue
- Total Sulfur
- Water and Sediment Volume

Acid Number

Acid number is a value that relates to the neutralization of free fatty acids and mineral acids present in the biodiesel. Research has shown that mineral acids used in the fuel production process (such as sulfuric acid to pretreat WVO) are a much higher concern than FFAs[1]. The ASTM limit is 0.8 units and the maximum reading I have measured on various fuel samples has not exceeded 0.21 units. Accordingly, this test has been removed. The acid number can be correlated with the copper strip corrosion test.

Sulfated Ash

Sulfated ash in the fuel can result in engine deposits and abrasive wear as a result of contamination from suspended solids and catalyst residue[2]. Proper fuel washing should result in a near zero level as recorded in my analysis.

Cetane Number

As discussed above, cetane number is a measure of the ignition capability of the fuel. Most feedstocks produce biodiesel with a cetane number that is greater than the minimum 47 called for in the ASTM standard[3].

Distillation Temperature

Distillation of fossil fuels shows that the fuel is made up of hundreds of compounds, each with its own boiling point. This analysis is required in determining the composition of the fuel source. Biodiesel, on the other hand, will exhibit a nearly flat distillation temperature "curve" with possible spikes at the low-temperature end due to excess methanol (which should be recovered and/or washed out) and at the high end of the temperature scale which are probably the result of incomplete reaction (excess mono-, di-, and triglycerides) of the oil feedstock. The total glycerin test noted below can test for these products, eliminating the need for distillation testing[4].

Flash Point

Flash point is a measure of the flammability of biodiesel. A low-temperature flash point is due to excess methanol in the fuel. Thermal recovery of methanol and water washing of the biodiesel will virtually eliminate methanol from the fuel and drive the flash point reading well below the minimum ASTM requirement of 130.0°C. (Although the ASTM standard does not record methanol content, the European biodiesel standard EN 14214 requires total methanol not in excess of 0.2%. After thermal recovery of methanol and water washing of the FAME, maximum methanol content has never exceeded 0.02% in my experience.)

Carbon Residue

Carbon residue is not considered an important characteristic in fossil diesel fuel but is one of the more important issues when assessing biodiesel because excessive carbon residue is indicative of deposit buildup on fuel injectors and in the combustion chamber[5]. Although there is no simplified method for determining carbon residue, all samples I have measured of properly reacted and washed FAME using WVO have given results of under 0.022% mass compared to the ASTM limit of 0.05%.

Total Sulfur

Although cold-pressed vegetable oils that are not pretreated with sulfuric acid produce biodiesel that is essentially sulfur-free[6], WVO and pretreated oils may have sulfur levels that vary from batch to batch. However, I have never observed levels higher than 0.0008% mass compared to the ASTM limit of 0.05%.

Water and Sediment Volume

Biodiesel must not contain water in concentrations above 0.05% by volume to prevent bacteriological growth which forms slime and results in fuel filter blockage. In all tests conducted on biodiesel processed by the fuel dryer unit described in Chapter 8.2.6, water and sediment volumes were reported as 0.0%.

Some popular texts and Web sites offer a method of measuring water content by weighing a sample of biodiesel before and after heating to 100°C. The intent is to "boil off" water and weigh the difference in mass resulting from water evaporation. This process will not work because measurement accuracy to 0.05% or less requires a scale of exceptional resolution, well beyond the limits of the micro-scale laboratory.

Included Tests

Tests that can be included in the simplified quality program are:

- Cloud Point
- Copper Corrosion
- Density
- Glycerine (Free)
- Glycerin (Bonded and Total)
- Kinematic Viscosity

Cloud Point

As the name implies, this is the temperature at which wax crystals first appear in the biodiesel when it is cooled in accordance with the ASTM standard. The effect of cloud point is a major concern for those in northerly climates where winter temperatures can cause the fuel to gel, causing temporary fuel filter and line blocking. Although the test can be conducted based on the ASTM program, a more practical test is to subject samples of varying concentrations of biodiesel and No. 2 Diesel fuel to the expected cold weather temperature or place them in a freezer to determine the highest concentration of biodiesel that can be safely used in your geographical area.

Biodiesel made from cold-pressed oilseed will exhibit the best cold-weather performance while saturated fats such as lard will be the worst feedstock choice for winter operation.

Figure 8.3.9-9. As the name implies, this is the temperature at which wax crystals first appear in the biodiesel when it is cooled in accordance with the ASTM standard. The effect of cloud point is a major concern for those in northerly climates where winter temperatures can cause the fuel to gel, causing temporary fuel filter and line blocking.

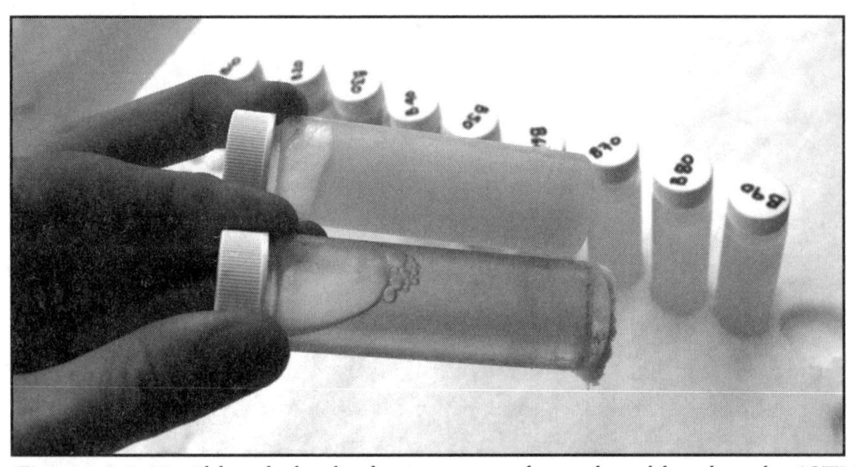

Figure 8.3.9-10. Although the cloud point test can be conducted based on the ASTM program, a more practical test is to subject samples of varying concentrations of biodiesel and No. 2 Diesel fuel to the expected cold weather temperature or place them in a freezer in order to determine the highest safe concentration of biodiesel that can be used in your geographical area.

Copper Corrosion

The copper strip corrosion test replicates the tendency of biodiesel to cause corrosion to nonferrous parts made from copper, zinc, brass, and bronze. Copper strips are polished, placed in a sample of the biodiesel, and heated to 50°C for a period of three hours. The strips are then washed and compared to test standards which can be purchased from ASTM. The test is designated as Standard D 130-04.

The copper strip corrosion test can correlate with acid number, with readings greater than number 1a or 1b being indicative of high levels of sulfur and acids[7]. In practice, the acid number is unlikely to exceed number 1 if the fuel is properly reacted and washed[8].

Figure 8.3.9-11. To perform the copper strip corrosion test, contact any electronic supply store or hobby shop and purchase a small sheet of blank printed circuit board material. This $5.00 sheet measures 4" x 6" (10cm x 15 cm) and is large enough to make at least 24 test strips measuring 2" x 0.5" (5cm x 1.2 cm) in size.

Figure 8.3.9-12. On a hot plate, heat a sample of the biodiesel to a temperature of 50°C. Use Pyrex™ or other heat-resistant glass to ensure the vessel does not influence the test results.

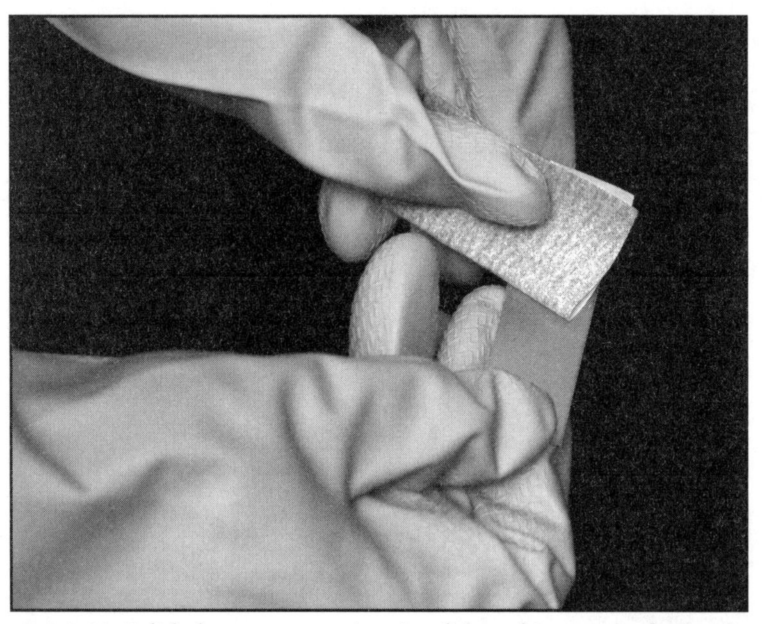

Figure 8.3.9-13. Polish the copper test strip using dish washing soap and either fine steel wool or 400 grit wet or dry polishing sandpaper. Rinse the copper strip in water and place it in the heated biodiesel.

Figure 8.3.9-14. Remove the copper strip from the biodiesel after three hours. Rinse in water.

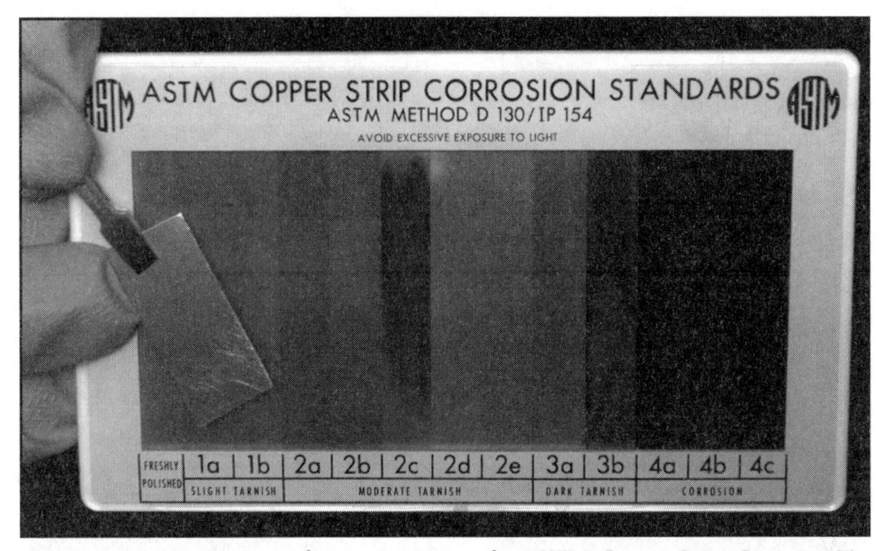

Figure 8.3.9-15. Compare the copper strip to the ASTM Copper Strip Corrosion Test Standard and locate the standard strip that best illustrates the level of corrosion on the test strip.

Density

Density is the mass per unit volume of a given substance. One pound of feathers has a much lower density than a pound of lead owing to the greater volume of feathers required to equal the same mass.

Biodiesel generally has higher densities than petroleum diesel fuels, with levels in the range of 860 to 900 kg/m^3 when measured at 15°C. The density of biodiesel is affected by the parent oil or fat used as the feedstock and is greatly impacted by the presence of contaminants. Methanol will have a tendency to drive the density downward, while excessive amounts of mono-, di-, and triglycerides will drive the density upwards. For this reason, measuring the density of biodiesel that has not been subjected to methanol recovery as well as washing and drying is of no value.

Density is measured by taking a sample of biodiesel which has been buffered to 15°C and placing it in a graduated cylinder. A calibrated glass hydrometer is then inserted into the biodiesel and a reading is taken from the scale inside the stem.

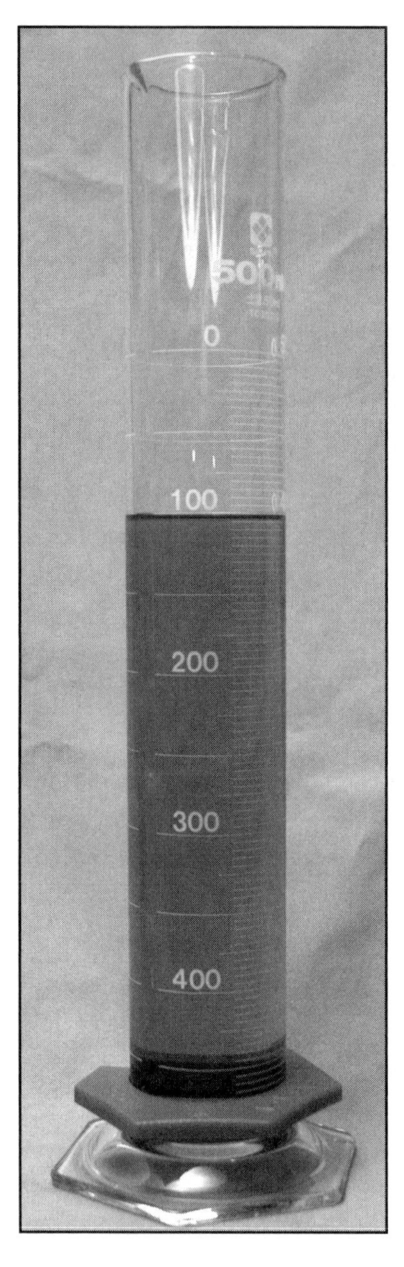

Figure 8.3.9-16. (Left) Density is measured by taking a sample of biodiesel which has been buffered to 15°C and placing it in a graduated cylinder.

Figure 8.3.9-17. (Right) A calibrated glass hydrometer with an operating range of between 800 and 1,000 kg/m³ is then inserted into the biodiesel and a reading is taken from the scale inside the stem.

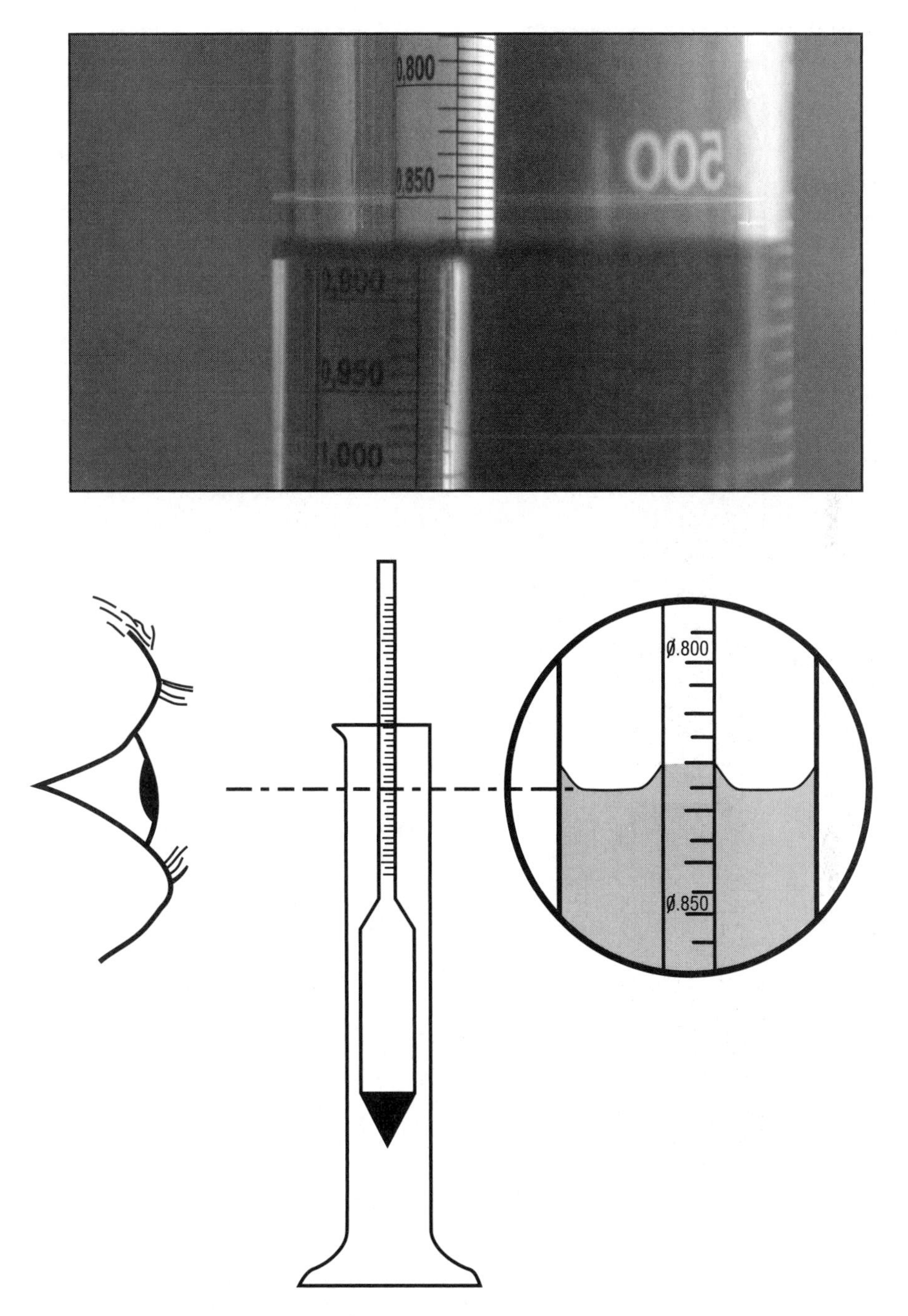

Figure 8.3.9-18. Use care when reading the scale of the hydrometer. Make sure you view it at eye level, ignoring the meniscus (fluid that creeps up the wall of the graduated cylinder) when taking the reading.

Glycerin (Free)

High levels of free glycerin above the ASTM limit of 0.02% mass are indicative of insufficient washing of the biodiesel. As discussed earlier, free glycerin separates from the biodiesel fuel and collects on the bottom of the fuel tank where it attracts water, forming sludge and bacterial slime which cause filter plugging and fuel injection system damage.

I have recorded glycerin levels of 0.098% in unwashed fuel while properly washed fuels typically record levels in the order of 0.002%, an improvement of nearly 50 times and approximately 10 times better than the standard requires.

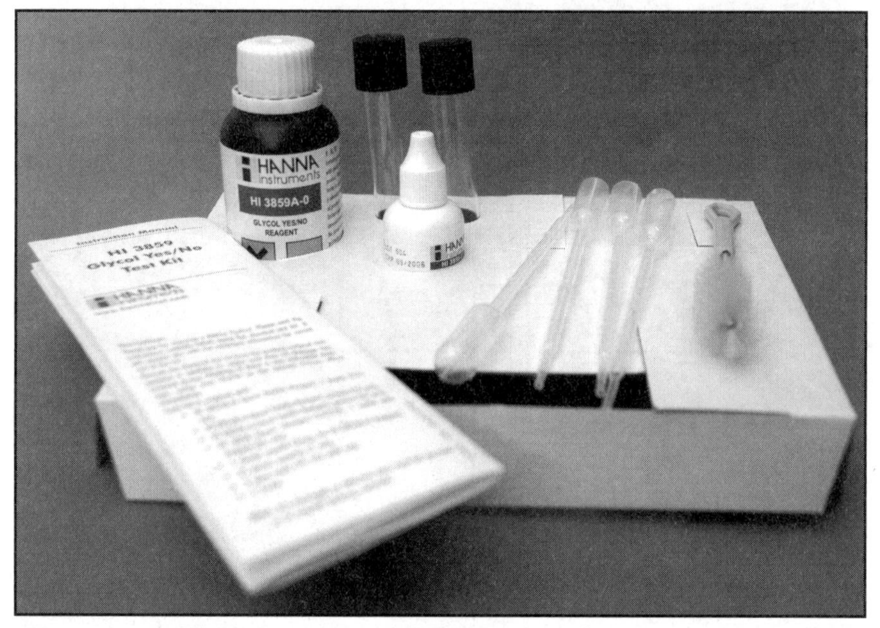

Figure 8.3.9-19. The Hanna Instruments Company (www.hannainst.com) manufactures its HI 3859 glycol antifreeze detection kit for the automotive industry. Leakage in the engine's cooling system may introduce ethylene glycol into the lubricating oil, which can lead to engine failure. By slightly modifying this test kit procedure, it is possible to detect the presence of free glycerin in biodiesel. The test kit will not work for bonded glycerin.

Figure 8.3.9-20. A 0.5 ml sample of biodiesel is mixed with 2 ml of fresh crankcase engine oil. The engine oil does not interfere with the test and dilutes the biodiesel concentration by a factor of 4 times, allowing the test kit to operate in the sensitivity range of 0 to 0.03 % (0 to 300 ppm) necessary to perform the ASTM test rather than in its standard calibration range of 0 to 75 ppm (0 to 0.0075%). The limit for free glycerin is 0.02% (200 ppm).

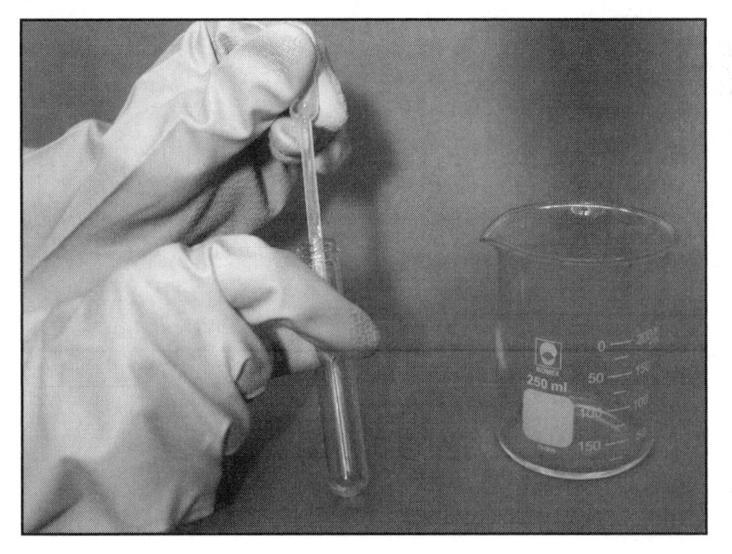

Figure 8.3.9-21. A 0.5 ml sample of the biodiesel/engine oil mixture is added to the test tube included with the kit. The test instructions are then followed to completion, requiring the addition of several reagent chemicals, mixing, and waiting.

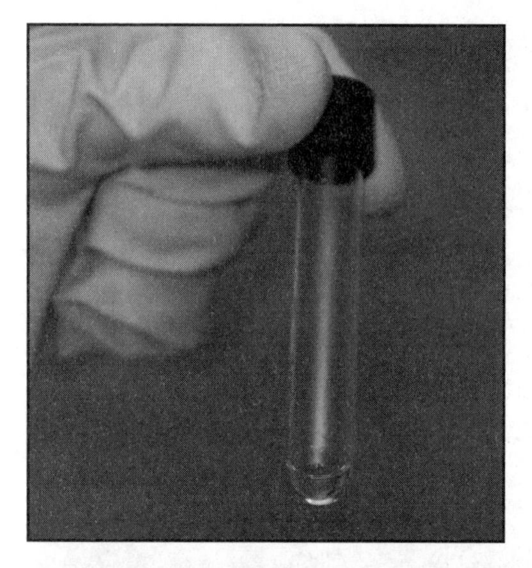

Figure 8.3.9-22. After a period of 20 minutes the test tube contents change color in relation to the amount of glycerin present in the biodiesel fuel, with a color of yellow-brown to faint light purple being acceptable as noted in Table 8.3.9-1.

Glycerin (ppm)	Glycerin (% volume)	Color
0 - 120	0 - 0.012	yellow-brown (within specification)
120 - 300	0.012 - 0.030	light purple (within specification)
> 300	> 0.030	dark purple (outside specification)

Table 8.3.9-1.
Relationship of color to free glycerin content in biodiesel when using the modified instructions described above with the Hanna Instruments Test Kit Model HI 3859.

Glycerin (Bonded and Total)

The bonded and total glycerin level of biodiesel is the sum of free glycerin in fuel and glycerol that remains bonded in the form of mono-, di-, and triglycerides. The presence of high levels of total glycerin can lead to the formation of deposits or coking on the fuel injection nozzles, pistons, and valves[9]. High kinematic viscosity levels are indicative of high gylceride content in the biodiesel.

Keeping bonded glycerin levels at a low level requires careful attention to the biodiesel reaction process. I have discovered that it has not been possible to produce biodiesel with a total glycerin level below 0.672% mass with a single pass through the biodiesel reaction unit. Having adjusted numerous process parameters, including base catalyst levels above 3.5 g/liter of feedstock oil, I believe it is not possible for the reaction to proceed to completion without decanting the first reaction stage glycerol and then proceeding with a second pass through the reactor using a reduced sodium methoxide concentration as discussed in Chapter 8.3.6.

Although it appears to be a lot of additional work to lower total glycerin by a seemingly trivial amount, a reading of 0.672% mass is 2.8 times over the ASTM limit of 0.24% mass.

Testing for bonded glycerin content is a time-consuming process that involves running a sample of biodiesel through a one-liter transesterification using the same process described in Chapter 8.3.3 and observing the level of glycerin formation after a 24-hour waiting period. If the one-liter test is conducted with biodiesel that has a reading above the bonded glycerin level called for in the ASTM standard, and if all of the residual mono-, di-, and triglycerides are reacted into biodiesel during the test, a small layer of glycerin will remain at the bottom of the sample. The same test performed using biodiesel that passes the ASTM total glycerin test will leave no trace of glycerin.

Kinematic Viscosity

Kinematic viscosity is a measure of the resistance to flow of a given fluid. You will recall that vegetable oils cannot be used as fuel for diesel engines unless their viscosity has been lowered either by heating the oil or by transesterification. Even after the transesterification process, biodiesel tends to be slightly more viscous than petrodiesel fuel.

High measures of viscosity may be indicative of high total glycerin levels, which can have an impact on fuel injection pumps by increasing resistance to fuel flow, leading to higher fuel pressure and injection volumes.

Kinematic viscosity is measured using a Zahn cup-type viscosimeter as shown in Figure 8.3.9-23. A sample of biodiesel heated to 40°C is drawn into the cup of the viscosimeter, which is quickly lifted while a stopwatch is started. The watch is stopped when the steady flow of liquid "breaks" at the outlet of the cup. This process is repeated several times until the readings are consistent. The reading is converted from Zahn seconds to centistokes according to the viscosimeter manufacturer's manual and the results are compared to the ASTM standard range of 1.9 to 6.0 centistokes.

Figure 8.3.9-23. Kinematic viscosity is measured using a Zahn cup-type viscosimeter as shown in this picture.

Figure 8.3.9-24. A sample of biodiesel is heated to 40°C and the cup of the viscosimeter is filled and quickly lifted out. A stop watch is started as soon as the top of the viscosimeter cup breaks the surface of the biodiesel.

Figure 8.3.9-25. The watch is stopped when the steady flow of liquid "breaks" at the outlet of the cup. This process is repeated several times until the readings are consistent. The reading is converted from Zahn seconds to centistokes according to the viscosimeter manufacturer's manual and the results are compared to the ASTM standard range of 1.9 to 6.0 centistokes.

Quality Test Summary:

There is no guarantee that any biodiesel fuel tested using this simplified procedure will meet the ASTM D 6751 quality standard. However, after performing a number of tests on biodiesel that has been produced following the procedures outlined in this book and comparing the results to accredited fuel test laboratory results, I am quite certain the two will be very close and the amount of fuel-based risk will be minimal.

The sample fuel analysis shown on the next page provides evidence that meeting ASTM D 6751 is possible for the micro and small-scale biodiesel producer.

Analysis	Method	Min	Max	Biodiesel: Basics and Beyond
Acid Number, mg KOH/gram	ASTM D664		0.80	0.21
Ash, Sulfated, Mass %	ASTM D874		0.020	0.001
Total Sulfur, Mass %	ASTM D5453		0.05	0.0008
Cetane Number	ASTM D613	47		53.6
Cloud Point, °C	ASTM D5773			-2.4
Copper Corrosion	ASTM D130		No. 3	1a
Vacuum Distillation End Point, ° C	ASTM D1160		360°C at 90%	359.2
Flash Point, °C	ASTM D93 Procedure A	130.0		179.0
Method for Determination of Free & Total Glycerin	ASTM D6584			
Free Glycerin, Mass %			0.020	0.002
Total Gylcerin, Mass %			0.240	0.204
Phosphorus (P), Mass %	ASTM D5185		0.001	0.0002
Carbon Residue, Mass %	ASTM D4530		0.05	0.0006
Kinematic Viscosity, Centistokes at 40°C	ASTM D445	1.9	6.0	5.007
Water and Sediment	ASTM D1796		0.050	0
Total Methanol (%)	EN 14110		0.2	0.003

Table 8.3.9-2. *This table indicates the limits of various parameters according to ASTM D 6751 standards and compares them with biodiesel produced using the procedures recommended in Biodiesel: Basics and Beyond. This fuel is incompliance with the requirements of the standard.*

There is no question that there is room for improvement in my test procedure, and I challenge an enterprising chemist to produce a complete wet-chemistry test kit that can be sold to the micro- and small-scale biodiesel industry. As everyone is aware, this is a growing sector all around the world and could become a profitable niche business.

8.3.10 Environmental Sustainability

Glycerin (or more correctly "unrefined glycerin" or glycerol) as well as toxic, contaminated wash water are the micro-scale biodiesel producer's equivalent to the nuclear or industrial waste that pours out of the back of power plants and industrial facilities while the folks in the front office smile and offer self-congratulatory pats on the back for their outstanding environmental stewardship. It just doesn't add up.

Figure 8.3.10-1. Glycerin (or more correctly "unrefined glycerin" or glycerol) as well as toxic, contaminated wash water are the micro-scale biodiesel producer's equivalent to the nuclear or industrial waste that pours out the back of power plants and industrial facilities.

If the production of biodiesel (or any product for that matter) is going to be considered "environmentally friendly" the entire life cycle of the product, its raw material and energy inputs as well as waste outputs, must be considered as part of the sustainability equation.

Fortunately for the micro- and small-scale producer of biodiesel, feedstock materials, which make up approximately 75% of the total input, can be considered a waste product and their being recycled into a fuel can be a good thing provided all of the steps in the production process are properly adhered to.

Methanol, which makes up approximately 20% of the total raw material input, requires natural gas and considerable energy inputs in its manufacture. For this reason biodiesel does not completely eliminate life cycle greenhouse gas emissions, but compared with petroleum-based diesel fuel it does offer a reduction of approximately 75% reduction. The alternative to methanol is ethanol, produced from grains or wood cellulose materials, which requires considerably fewer petroleum fuel inputs in its manufacture. Many Web sites and books refer to the use of "green methanol" and ethanol alcohols in the production of biodiesel, but virtually no one does for technical reasons and because of cost.

There is considerable discussion about the total amount of electrical energy required in the production of biodiesel and many use this "exorbitant" cost as the basis for opting (perhaps weasling is a better choice of words) out of WVO drying, methanol recovery, and fuel drying. As discussed in earlier chapters, this energy input is insignificant because the use of insulated tanks and thermal energy recycling (transferring warm materials from one process to the next) virtually eliminates the need for supplementary energy beyond what is required for the basic transesterification process.

In addition to the main raw materials used in the production of the fuel itself, there are several other chemicals and products used during processing that require storage and disposal, including the chemicals required in the titration process, pH meter buffers, cleaners, floor cleaners, dirty rags and the like.

Glycerin Detoxification

The next time you are chatting with any micro-scale biodiesel producers, ask what they do with the waste glycerol. Chances are they will fall into one of several camps:

- There is a huge mountain of waste glycerol pails behind the garage or shed.
- They check their nails and avoid answering.
- They mutter something about sending the stuff to the landfill because it will biodegrade.
- They plan to make soap from it.
- They suggest that dumping it in the "back forty" works for them.

Sadly, the processing of waste streams is one area where most biodiesel producers tend to fall short, propagating the myth that glycerol (and wash water) are safe and biodegradable and that there is no problem simply dump-

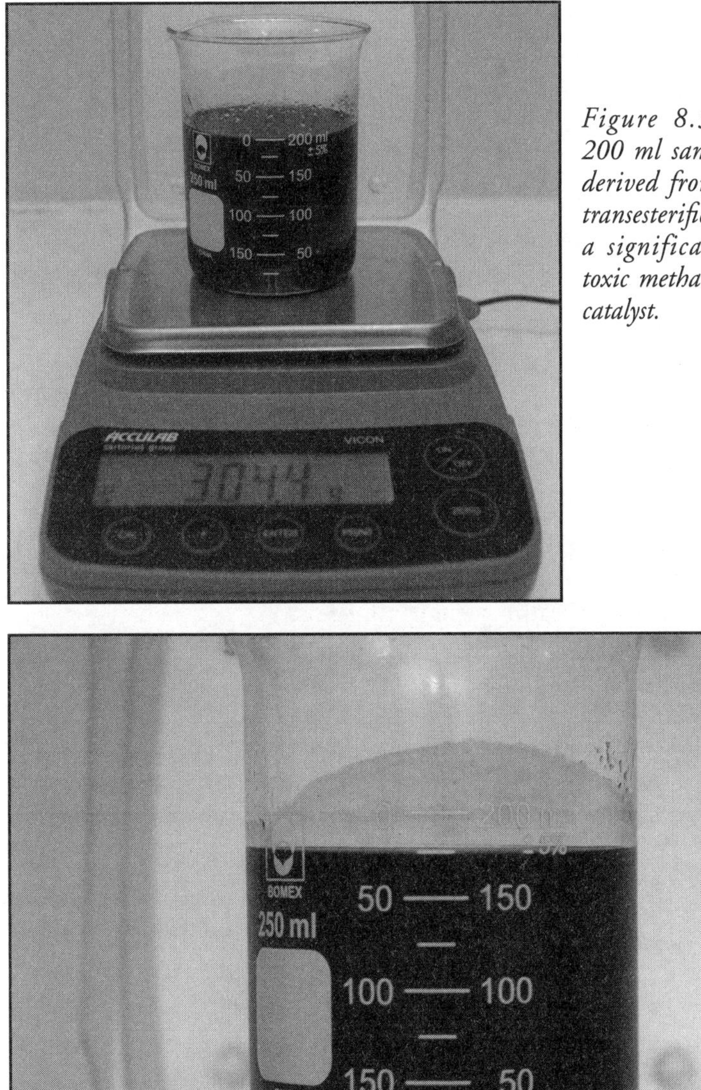

Figure 8.3.10-2. This 200 ml sample of glycerol derived from WVO-based transesterification contains a significant amount of toxic methanol and caustic catalyst.

Figure 8.3.10-3. Methanol **does not** *evaporate from glycerol that has been placed in the sun. Heating the glycerol to 70-80°C will boil off the methanol as witnessed by the drop in volume.*

ing the stuff hither and yon.

Glycerol removed from the biodiesel reaction contains a very significant amount of toxic and flammable methanol as well as catalyst and soap that must be processed before it can be claimed to be biodegradable, let alone useful.

The glycerol phase that is removed from the biodiesel reaction tank will remain a viscous liquid above a temperature of approximately 38°C (100°F) when the catalyst used in the reaction is sodium hydroxide. Below this temperature, the glycerol will start to gel. Biodiesel produced with potassium hydroxide catalyst produces glycerol that remains fluid at much lower temperatures. Regardless of which catalyst is used, the glycerol generally remains fairly warm and liquid during the phase separation process described in Chapter 8.3.6. As the glycerol is drawn off from the reaction tank at a fairly warm temperature, there is also a possibility of breathing methanol vapors during the handling and transfer process. It is therefore important to get the glycerol into the processing unit as quickly as possible in order to prevent solidification and avoid breathing methanol fumes.

The warm glycerol is transferred to the processing unit the through an inlet pipe and cap as shown in Figure 8.2.8-2. The cap is removed, a funnel

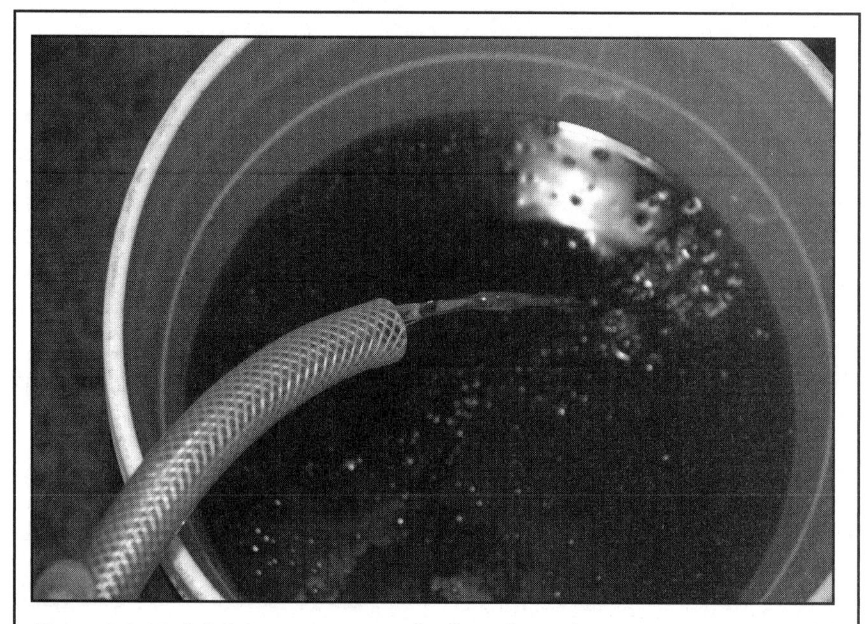

Figure 8.3.10-4. It is important to get the glycerol into the processing unit as quickly as possible after phase separation in order to prevent solidification and avoid breathing methanol fumes.

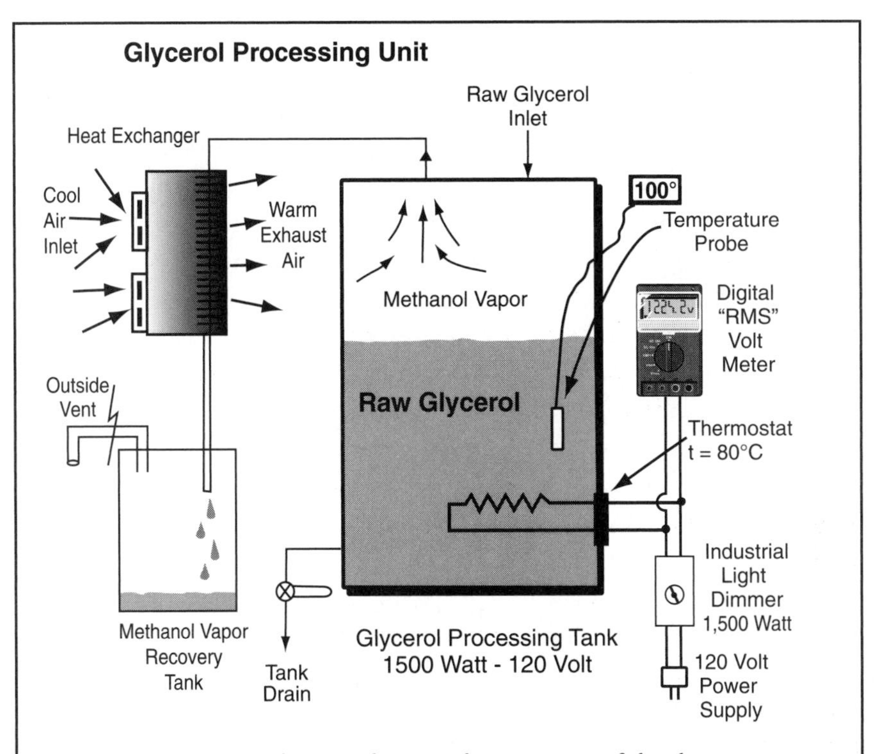

Figure 8.3.10-5. This drawing shows a schematic view of the glycerin processing unit. It is a simple device that allows the distillation and recovery of methanol as well as the production of a partially refined glycerol that can be used in other applications.

is inserted into the pipe, and glycerol is carefully poured into the tank. Once the tank is filled, the inlet pipe thread is checked for sufficient pipe tape to ensure a vapor-proof seal, and the cap is then threaded back on.

The vapor recovery system used in the glycerol processor is identical to that used in the biodiesel reaction tank described in Chapter 8.2.3. If sufficient space is available, it may be possible to plumb both the biodiesel reaction tank and the glycerol processing unit to the same vapor recovery system.

Once the tank is filled, **methanol recovery** can begin:

1. Ensure glycerin tank drain valve is closed.
2. Activate heat exchanger cooling fans.
3. Turn on RMS volt meter, making sure it is set to measure "AC Volts" in the range of 120 volts.
4. Slowly activate light dimmer switch and adjust to 60 volts.
5. Allow tank to warm up while monitoring glycerol temperature. When the glycerol temperature stabilizes, slowly increase voltage in

10-volt increments, allowing the tank temperature to stabilize again. Stop adjusting the heater input voltage when the tank temperature reaches between 70 and 80°C. It is very important to increase glycerol temperature slowly. Attempting to heat the glycerol too quickly may cause the electric heating element to burn out.

6. When the glycerol temperature reaches approximately 60°C, methanol will begin boiling and vapor will condense in the recovery tank.
7. Allow methanol recovery to complete and then turn off the tank heater.
8. Allow the methanol to cool to approximately 50°C and drain it into a bucket, taking care to leave room for water and mixing.

The glycerol has now been stripped of methanol and contains food particles, dirt, and other contaminants as well as caustic catalyst. The removal of catalyst is achieved through a neutralization process that returns the glycerol pH (acid/base ratio) back to 7 or neutral.

In order to measure pH, you can purchase an electronic pH meter such as the model shown in Figure 8.3.10-6 or pH test strips that change color in the presence of acids and bases. The color change is compared to a reference chart in order to determine pH levels.

In either case, it is not possible to measure the pH level of glycerol directly as the electronic recording device or measurement strips will become blackened in the tarlike glycerol substance. It is better to take a sample of glycerol and mix it with a small amount of distilled water first and then use the meter/test strip to take the reading from the water phase (Figure 8.3.10-10).

After methanol recovery, **caustic (catalyst) neutralization** can begin:

1. Making sure that the bucket is not filled to capacity, add approximately 500 ml of 5% acetic acid (white vinegar) to the glycerol and stir thoroughly. (I have chosen to use acetic acid rather than other more traditional mineral acids in an effort to develop the most environmentally friendly processes and reduce chemical usage.)
2. Take a small sample of the glycerol/acetic acid mix, add a small amount of distilled water, stir, and then measure the pH. Glycerol that has high levels of unneutralized catalyst will have a reading greater than 7.
3. If the pH measures higher than 7, continue with step 1 and repeat until pH equals 7.
4. The glycerol is now neutralized and non-toxic and may be used in various applications as discussed in Chapter 10.

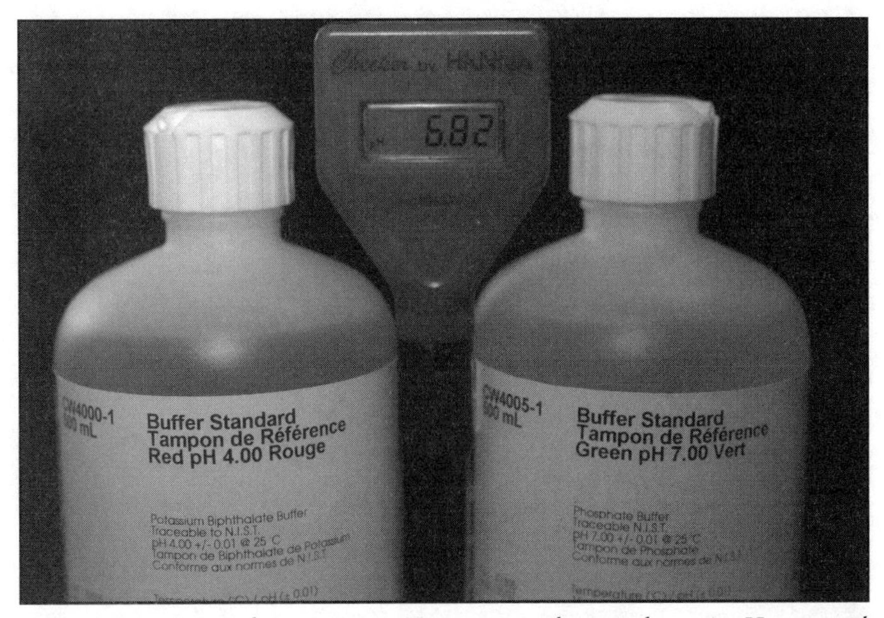

Figure 8.3.10-6. In order to measure pH you can purchase an electronic pH meter such as this inexpensive model or pH test strips that change color in the presence of acids and bases. It is a good idea to purchase calibration fluids along with the meter to ensure its long-term measurement accuracy.

Figure 8.3.10-7. Never dip an electronic pH meter or tests strips directly into the tarlike glycerol.

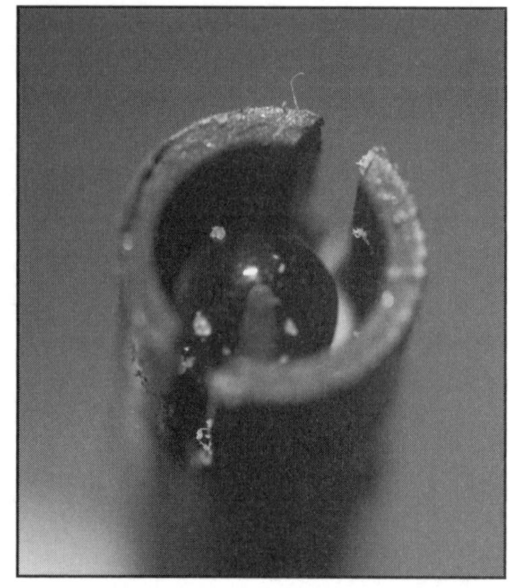

Figure 8.3.10-8. (Right) Electronic pH meters have a special sensing unit encased in glass which must be kept clean and moist.

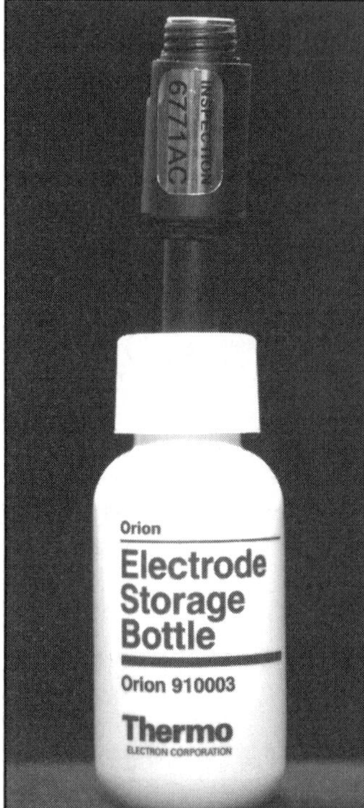

Figure 8.3.10-9. (Left) When the measurement is complete, remove the pH electrode, rinse it in probe cleaner, and then place it in a bottle filled with electrode storage fluid.

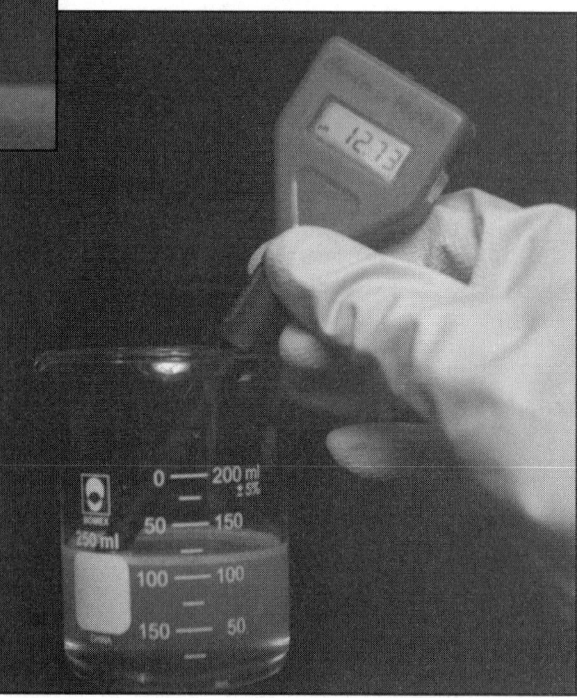

Figure 8.3.10-10. (Right) When measuring glycerol pH, place a sample in a beaker and add a small amount of distilled water. Stir the mixture and measure the pH in the water phase. Note that this raw glycerol measures a skin-burning 12.73.

Wash Water Management

Wash water from the FAME washing process should be completely free of methanol before you proceed to the step of caustic (catalyst) neutralization. While it is possible to recover methanol from the wash water, it is much simpler and requires considerably less energy to recover it at the biodiesel reaction stage as discussed in Chapter 8.3.6.

Many home-based biodiesel producers have told me that there is an insignificant amount of methanol in the wash water and for this reason disposal is perfectly safe without having to subject the FAME to distillation for methanol recovery. Figures 8.3.10-12 through 8.3.10-14 refute these claims.

Figure 8.3.10-11. Wash water recovered from the FAME washing process described in Chapter 8.3.7 should be free of methanol prior to neutralization of caustic catalyst materials.

Figure 8.3.10-12. To test the residual methanol content in wash water, a 200 ml sample weighing 185.1 grams was measured and placed in a beaker and weighed, giving a total mass of 307.2 grams (beaker mass = 122.1 grams).

Figure 8.3.10-13. The sample wash water was heated to 75°C; approximately 10 degrees above the boiling point of methanol, allowing it evaporate.

Figure 8.3.10-14. The final sample was weighed and found to have a mass of 294.2 grams, indicating that 13 grams or 7.02% of the mass of wash water was composed of methanol. At a density of 791.8 kg/m^3 this gives a methanol volume content of approximately 8.9%. Since the first test wash requires 50 liters of water, this equates to a total methanol volume of 4.45 litres or just over 1 gallon. Clearly, this is not an insignificant amount. Methanol is an environmental toxin, and failing to recover it is a waste of money. At current electricity rates, it would cost considerably less in energy to recover over four dollars' worth of methanol.

Figure 8.3.10-15. In an attempt to inexpensively measure methanol content in wash water, Dr. Tremblay of the University of Ottawa Chemical Engineering Department is shown filtering a sample of wash water for analysis.

Figure 8.3.10-16. It was hoped that wash water could be filtered sufficiently well using a coalescing filter to produce a sample for measurement that did not contain dissolved particles of soap.

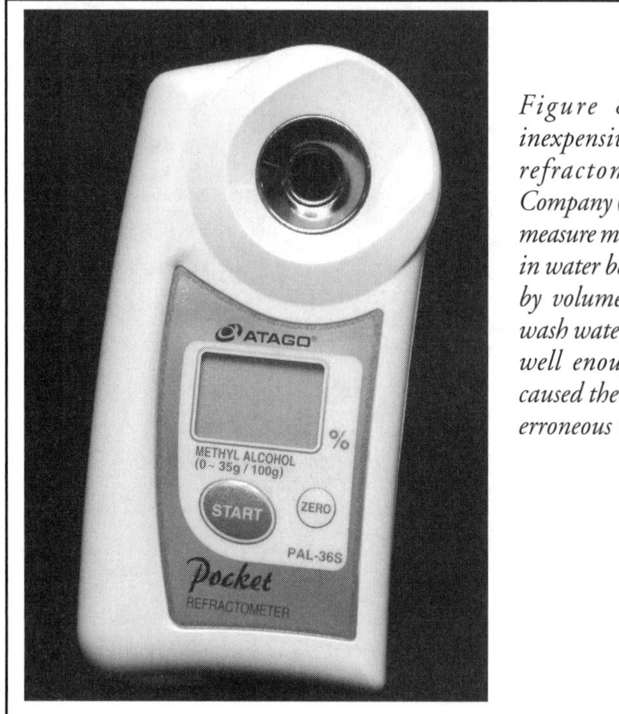

Figure 8.3.10-17. This inexpensive "pocket" infrared refractometer from Atago Company (www.atago.net) can measure methanol concentration in water between 0% and 35% by volume. Unfortunately the wash water could not be filtered well enough and the sample caused the refractometer to give erroneous readings.

To begin wash water catalyst neutralization proceed as follows:

1. Add approximately 1 liter of 5% acetic acid (white vinegar) to each 50 liters of wash water. Stir the contents.
2. Measure the pH of the wash water. If the pH reading is above 7, repeat step 1.
3. Once the wash water has been neutralized (pH =7), it can be disposed of in any municipal sanitary sewer.

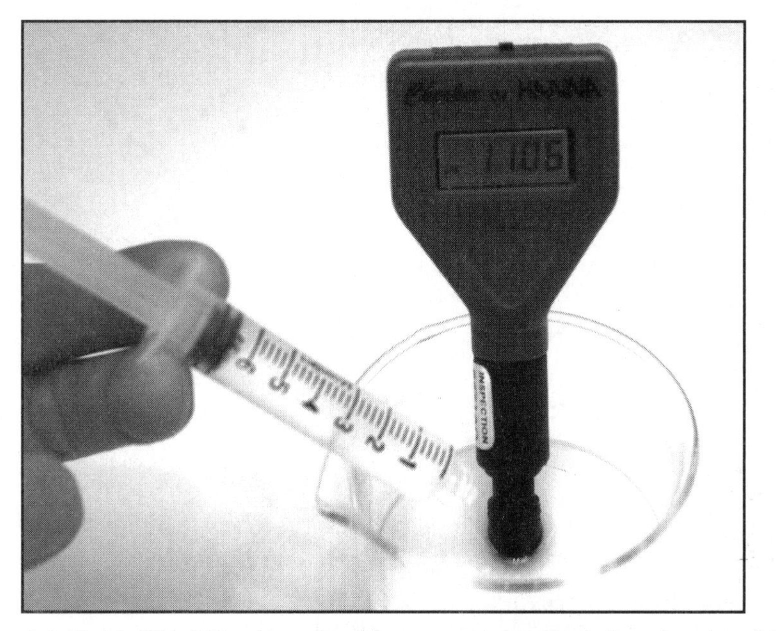

Figure 8.3.10-18. This 200 ml sample of first-stage wash water is free of methanol but is caustic, with a pH reading of 11.06. A dosage of 5 milliliters of 5% acetic acid (white vinegar) is about to be injected into the wash water. The dosage rate is approximately 2.5% acetic acid to wash water by volume.

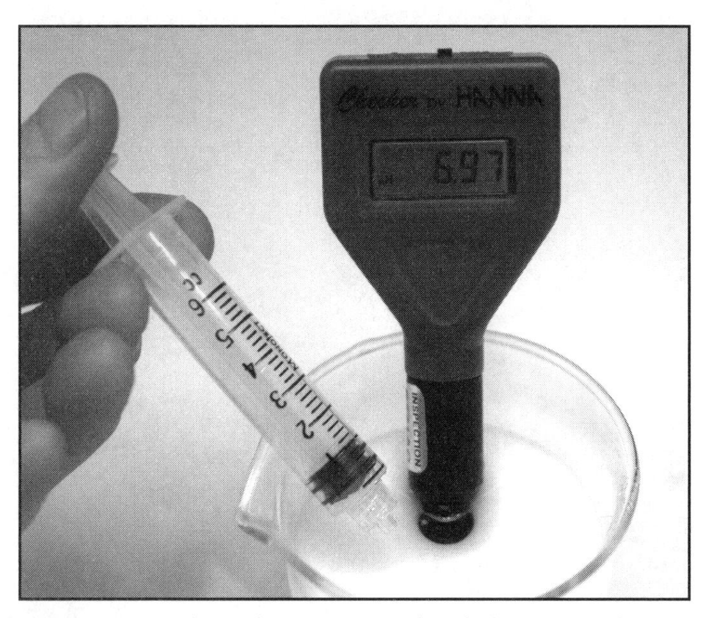

Figure 8.3.10-19. Once the wash water is stirred with the acetic acid it is neutralized (pH reading 6.97) and ready for disposal.

Used Process Chemical Disposal

Many of the chemicals used in the processing or measurement of biodiesel have a shelf life which is printed on the container label. It is also possible that chemicals may become contaminated or may spill, requiring cleanup and segregation of the potentially hazardous waste rags.

The MSDS provides information on storage and disposal of these waste chemicals, but regulations change between state, provincial, and federal agencies and between chemical types, making generalized statements difficult to make. For example, the disposal information for phenolphthalein (a known cancer-causing chemical) is as follows:

"Whatever cannot be saved for recovery or recycling should be managed in an appropriate and approved waste disposal facility. Processing, use or contamination of this product may change the waste management options. State and local disposal regulations may differ from federal disposal regulations. Dispose of container and unused contents in accordance with federal, state and local requirements."

Although this may sound obvious, the question remains: what exactly do you do with these materials? When this situation arises, it is best to contact a local waste management or disposal company which is aware of local regulations and can accept the material for recycling or safe disposal.

9 Home Heating with Biodiesel

9.1 Overview

Road taxes and hype to the contrary, there is virtually no difference between No. 2 diesel fuel and home heating oil. As a result many people wonder, "Why not use low-emission biodiesel in place of petroleum heating oils?" The desire to make the switch is also driven by the knowledge that micro-scale producers make biodiesel for less than half the cost of its petroleum counterpart.

According to the US Energy Information Agency, approximately 7.7 million homes consumed No. 2 heating oil in the year 2000. Consumption was shown to be 6.7 billion gallons (25.4 billion liters), making this a desirable market for commercial biodiesel producers. Even if they offered their product at a blend level of B20, this new biodiesel market would amount to a staggering 1.3 billion gallons (4.9 billion liters).

The commercial industry is only just beginning to service the market and most biodiesel fuel is supplied by smaller firms such as Frontier Energy of China Village, Maine that are developing a niche market by selling biodiesel blended with No. 2 heating oil. This fuel is a "bio-heat" alternative to fossil fuel heating oils and is marketed to customers willing to a pay a small premium for fuel that is domestically produced and more environmentally friendly. The major heating oil distributors are not yet fully engaged in this market as the price of commercially-produced biodiesel is still a bit higher than petroleum-based heating oil. Once the commodity price for petro-heating oil matches biodiesel pricing, this situation will quickly change. For the micro or small-scale biodiesel producer, it already has.

Biodiesel is commonly used in Italy for home and commercial building heating but is a new concept on this side of the ocean. North Americans have ignored environmental issues and focused on price, being able to purchase fossil fuels at below true cost as a result of flawed government policy[1].

Studies conducted by the Brookhaven National Laboratory in December 2001 and field trials around the US northeast found that when biodiesel was blended with petroleum diesel at levels of 20% or less, few operational difficulties arose. At high biodiesel blend levels the expected problems of seal incompatibility and cold weather gelling became evident. At the same time, testing showed that soy-based biodiesel and fuel oil were very close in thermal performance and that sulfur emissions were virtually eliminated when using biodiesel heating oil.

Figure 9.1-1. The commercial market for biodiesel heating oil is just developing and most of this fuel is supplied by smaller firms such as Frontier Energy of China Village, Maine that are developing a niche market by selling biodiesel blended with No. 2 heating oil. This fuel is a "bio-heat" alternative to fossil fuel heating oils and is sold to customers willing to a pay a small premium for fuel that is domestically produced and more environmentally friendly. (Courtesy Frontier Energy)

Figure 9.1-2. When the commercial industry hesitates, community-based cooperative fuel suppliers fill the void. Many homeowners will happily pay to have their home heating needs supplied by renewable fuels. This small-town cooperative purchases biodiesel in small quantities and blends it with home heating oil for the benefit of its members.

Figure 9.1-3. The desire to make the switch from fossil to bio-heating fuel is driven by the knowledge that micro-scale producers make biodiesel for less than half the cost of its petroleum counterpart. Here Steve Anderson loads a pail of home-brewed fuel into the "tank" of an oil furnace. "I have a great deal here," Steve says. "I supply the heat and I get half of the shop rent-free. I can't beat the price, especially when I make the fuel for about 30 cents a liter [$1.14 per gallon]."

9.2
Biodiesel Heat without Electricity

For those who are considering building a home or cottage off the electrical grid (off-grid) the question of how to heat the building always arises. In the past, off-grid homeowners had to settle for heating their houses with either wood or propane, with most people choosing a combination of both fuels. Wood tends to be used as the primary renewable fuel source when everyone is home, while propane fills the gaps in the energy supply, providing backup space heat, cooking and hot water fuel, and the energy to power a gas clothes dryer. It is a necessary but often uncomfortable situation for people who are concerned about their carbon footprint on the environment.

One option that many people are unaware of is the old-fashioned oil heater. These units were designed to provide space heating where access to electricity was either impossible or impractical. Heating oil is gravity-fed into an oil pot burner, which is normally lit once per heating season. A damper/

Figure 9.2.1. For those who are considering building a home or cottage off the electrical grid (off-grid), the question of how to heat the building always arises. In the past, off-grid homeowners had to settle for heating their houses with either wood or propane, with most people choosing a combination of both fuels. One option that many people are unaware of is the old-fashioned oil heater fueled with biodiesel. (Courtsey Enterprise-Fawcett)

Figure 9.2-2. Wood tends to be used as the primary heat source in off-grid homes while propane fills the gaps in the energy supply, providing backup space heat, cooking and hot water fuel, and the energy to power a gas clothes dryer. An alternative energy source for space heating is an oil-stove heater powered by renewable biodiesel fuel. (Courtesy Enterprise-Fawcett)

thermostat control regulates the fire pot temperature, providing a radiant heating system similar to that of a woodstove.

Stove Builder International Inc. has one of the most complete lines of oil heaters. They can be seen on the Drolet Web site at http://www.drolet.ca/index-en.aspx. I contacted company president Marc-Antoine Cantin about the product line and found that his company is doing a healthy business selling these models around the world. "We export our oil stoves around the world and do a very good business in New Zealand, where they have very strict emission standards for heating appliances," Marc-Antoine explained. "Many people think of oil stoves as ugly, dirty units, but they can achieve heating efficiencies to match midrange-efficiency oil furnaces."

Looking over the Stove Builder International Inc. catalog of oil heaters, I have to agree that these models are not something that were designed 50 years ago and then forgotten. Without a doubt Marc-Antoine has done a good job of matching engineering and product aesthetics. Naturally he was interested the minute I suggested testing biodiesel fuel in the stoves. "To be truthful, I had never heard of the term "biodiesel" until you called," he explains later. "However, the concept of using a biodegradable and low-emission fuel in our stoves was very interesting." Samples of biodiesel were shipped to the Stove Builder International Inc. factory in Quebec City, Canada and tests were conducted to determine how the biodiesel would react in a gravity-fed burner assembly.

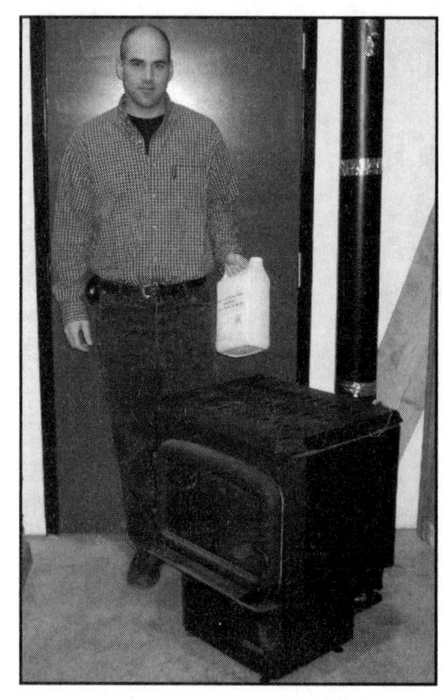

Figure 9.2-3. (left) Stove Builder International Inc. has one of the most complete lines of oil heaters. They can be seen on their Drolet distribution web site at http://www.drolet.ca/index-en.aspx. I contacted company president Marc-Antoine Cantin, shown standing beside a test unit, about the product line and found that his company is doing a healthy business selling these models around the world.

Figure 9.2.4. (above) Looking over the Stove Builder International Inc. catalog of oil heaters, I have to agree that these models are not something that were designed 50 years ago and then forgotten. Without a doubt Marc-Antoine has done a good job of matching engineering and product aesthetics.

Figure 9.2.5. (left) "To be truthful, I had never heard of the term "biodiesel" until you called," explains Marc-Antoine. "However, the concept of using a biodegradable and low-emission fuel in our stoves was very interesting."

Marc-Antoine explains: "Our engineering manager, Pierre Pleau, had just finished testing one of our oil stoves with a regulatory agency audit, so we knew the stove was 100% ready to go. We measured the viscosity of the biodiesel and found that it was just slightly higher than that of No. 2 heating oil. Although it did not cause any problems with operation or emission output, the higher viscosity limited the maximum heat output. We adjusted the internal carburetor settings to match the fuel viscosity and everything worked as well as with normal heating oil."

Those wanting to lower their overall carbon fuel emissions, whether on- or off-grid, might consider one of the bio-heat options marketed by Stove Builder International Inc.

Figure 9.2-6. The flame front of a biodiesel-fired heating stove manufactured by Stove Builder International shows a remarkably clean-burning fire.

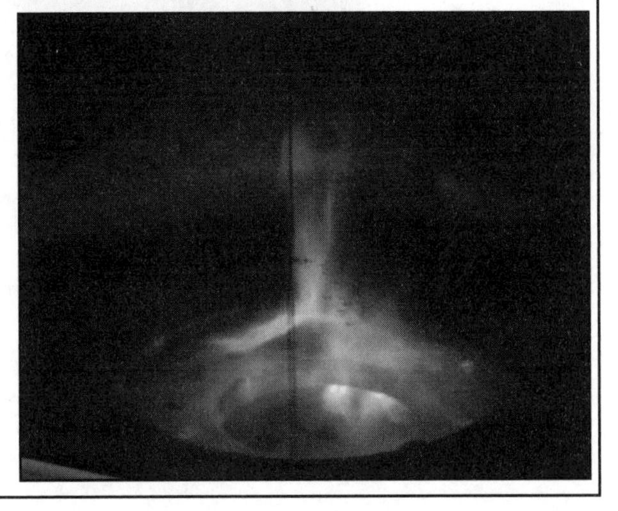

Figure 9.2-7. The thermal and emission profile of the biodiesel-fired stove is very similar that of the same stove burning No. 2 heating oil. The efficiency of these units can easily compete with mid-efficiency oil-burning furnaces while requiring no electricity to operate.

9.3
Handling/Storage/Blending

This will no doubt be the shortest section of the book. Handling biodiesel for home heating is no different than for vehicular use. If your oil tank is located indoors in a reasonably warm environment, fuel gelling due to low-temperature operation should not be a major problem. If the indoor tank storage area is unheated and you are using high-concentration blends of biodiesel produced from yellow grease, saturated oils, or animal fats, gelling may be a problem.

For those with unheated indoor or outdoor fuel oil storage tanks, I strongly recommend performing a fuel blend thermal analysis using the fuel thermometer described in Chapter 8.3.9. It is far better to be conservative when it comes to cold weather performance than to wake up one morning to find the house is cold and the oil tank is filled with frozen B100 fuel.

Splash blending of biodiesel works well with furnace oil, but be sure that the biodiesel has been warmed before mixing it with the cold heating oil.

Ensure that the tank caps and seals are properly fitted and that water cannot drip directly onto seals and gaskets. Because the furnace/stove will be turned off for several months of the year, it is wise to fill the tank to capacity in order to prevent moisture from condensing in the air space immediately above the fuel level. Bacterial growth can cause slime and sludge buildup in a heating oil tank just as in an automotive fuel tank, leading to filter plugging.

9.4
Not Quite Heating Ideas

There is always room at the end of chapters like this for a few ideas that are a bit tangential but still part of the picture. Biodiesel is a pretty universal fuel and there are those who might wonder just how far you can go in using it around the home.

Some applications are not so obvious. Biodiesel makes a wonderful antiseizing agent for nuts and bolts that are too stubborn to loosen with just elbow grease. It also makes a wonderful solvent for removing sticky labels from appliances and containers. Just remember that it can damage paintwork, so be sure to rinse the area with soap and water after biodiesel is applied.

The following photomontage shows a few other applications in the home, some that work and some that don't quite cut it.

Figure 9.4-1. The kerosene (No. 1 heating oil) space heater is often used in workshops, cottages, or other places where a little supplementary heat is required.

Figure 9.4-2. When biodiesel is used in place of kerosene, the burner of this heater is only partially operational because of the high viscosity of the fuel. Do not attempt to adjust the flame carburetor, as doing so may cause misadjustment, releasing excessive products of combustion including poisonous carbon monoxide into the air. This is one device where biodiesel does not belong.

Figure 9.4-3. The tabletop oil lamp is a sentimental reminder of days gone by and the reason why everyone has switched to the electric compact fluorescent lamp. These oil lamps are more for ambience than for lighting and work wonderfully well with biodiesel that has been made following the procedures in this book. Never burn raw biodiesel in a closed space as there will be numerous products burned, including soap, catalyst, and other fuel contaminants, that should not be there. Who knows whether or not the emission profile will be safe?

Figure 9.4-4. The Tikki lamp is a favorite of summer barbeques. Now you can use low-emission biodiesel as part of the party fun.

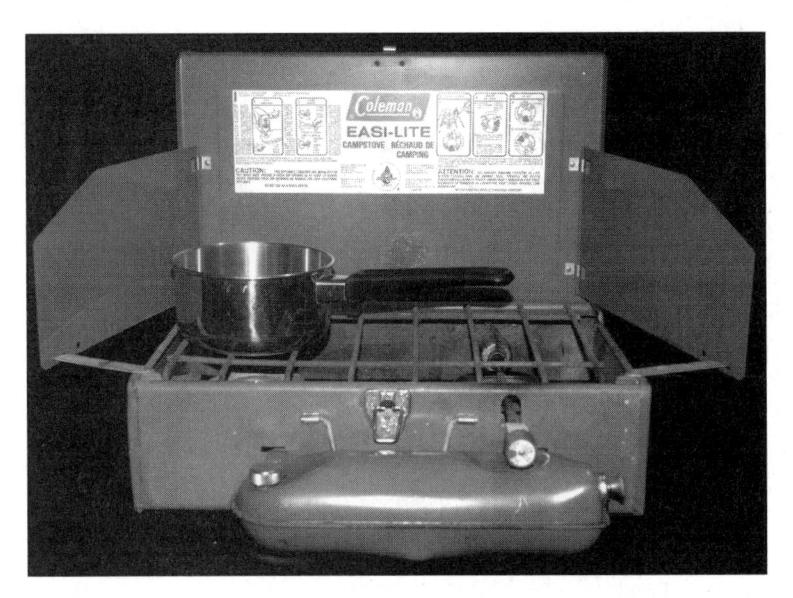

Figure 9.4-5. Sorry, the old standby Coleman™ stove is not a candidate for biodiesel because of the high viscosity of the fuel.

10
Glycerin

Every 150-liter (40-gallon) batch of WVO that is converted into biodiesel generates approximately 30 litres (8 gallons) of raw glycerol. Of this amount, some 8% is methanol and other contaminants. After first-stage methanol recovery the yield of semirefined glycerine is approximately 27 litres (7.1 gallons). Assuming an average density of 1,250 kg/m^3 this is approximately equal to 34 kilograms (75 pounds) of glycerine, an enormous amount when one considers that the average bar of soap weighs 75 grams. These figures translate into a yield of 453 bars of soap for each batch of biodiesel produced!

Clearly, disposing of this much glycerin, no matter how dirty you and your family gets, is a problem, and it is one that few people have bothered to tackle head-on. Well, let me roll up my sleeves and provide my completely unbiased opinion about how this coproduct of the biodiesel production process can be put to use.

10.1 Soap Production

I personally believe that this entire section should be labelled a "Myth Buster." While I am sure it is possible to make soap from WVO-based glycerin, there are numerous technical, aesthetic, and financial reasons why this is not a recommended procedure; but I digress.

In order to clarify why I am not excited about WVO-based glycerin as a soap stock, I would like to take a few moments to show you how typical glycerin soaps are produced, using the following photomontage.

Figure 10.1-1. Glycerin soap has been a staple around the world for a long, long time with companies such as Pears™ producing a high-quality and reasonably priced product.

Figure 10.1-2. In the last decade or two, numerous specialty soap companies have sprung up around North America, offering all manner of hygiene, aromatherapy, and massage products.

Figure 10.1-3. These are cunning business people who create the illusion of quality products and charge outrageous prices. This bar of soap is made almost completely with chemical products, the only natural ingredients listed being the finishing materials and scents. At over $7.00 a bar, I would expect something much better.

Figure 10.1-4. To the rescue: I have asked Danielle Shewfelt (left) and a slightly mischievous Lorraine Kemp (right) to help create soap from all-natural products (if possible) and to provide a costed bill of materials. They will produce the first batch of soap using a commercial melt-and-pour soap base, and then they will tackle the WVO-based glycerin.

Figure 10.1-5. The first thing that any soapmaker requires is tools, starting with this two-stage soap kettle. It operates at a high temperature to quickly melt the soap base and then switches to a cooler temperature which maintains the soap in a liquid state while allowing the soap to set quickly when it is poured into the mould.

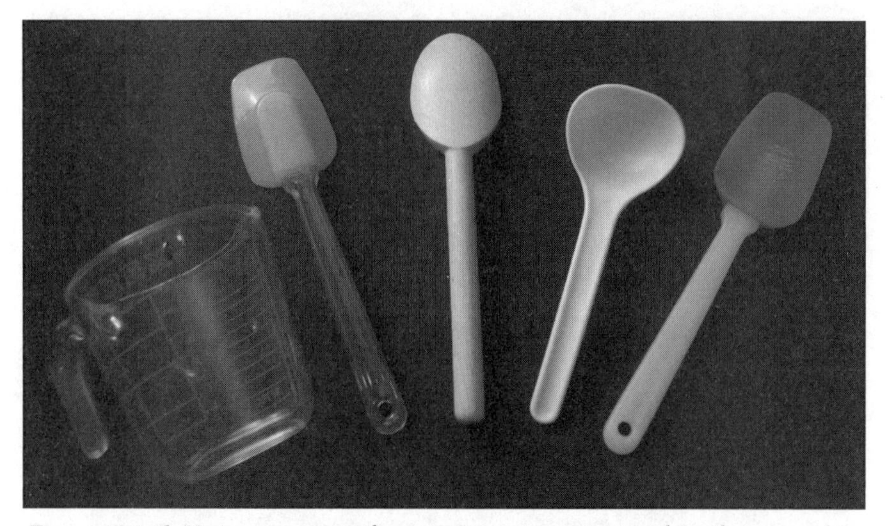

Figure 10.1-6. Various spoons and measuring cups are required in the soapmaking business.

Figure 10.1-7. Soap moulds can be store-bought such as the ones shown or you can use muffin tins or discarded yogurt containers.

Figure 10.1-8. This is a craft store brand of glycerin-based clear soap stock (glycerin). It can also be purchased in bulk from wholesale suppliers at a significantly reduced cost.

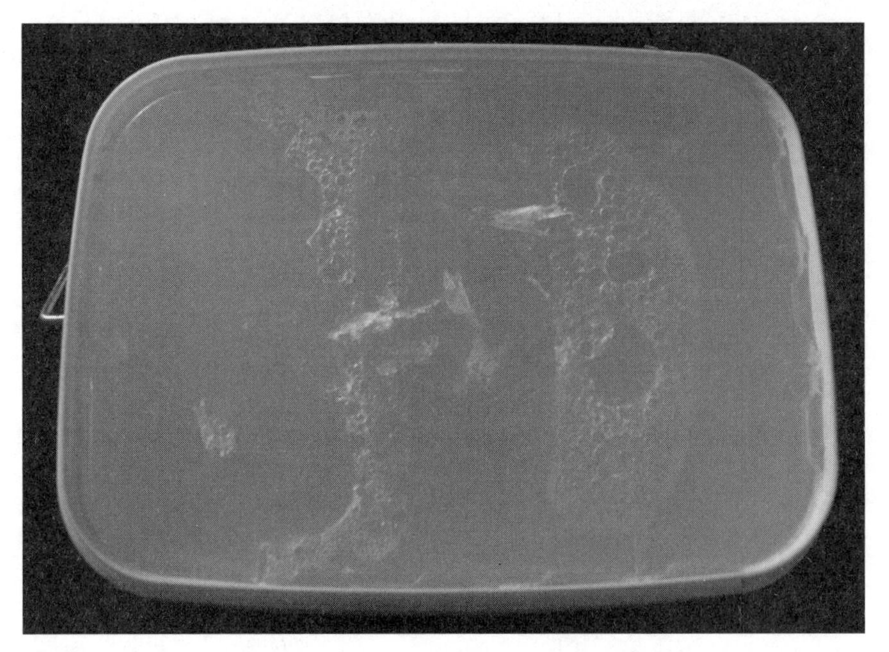

Figure 10.1-9. The refined glycerin is translucent and has a very mild, subtle odour.

Figure 10.1-10. Soap dyes can be used to create thematic colors to match the essential oil fragrances. For example, the color orange works well with a citrus fragrance.

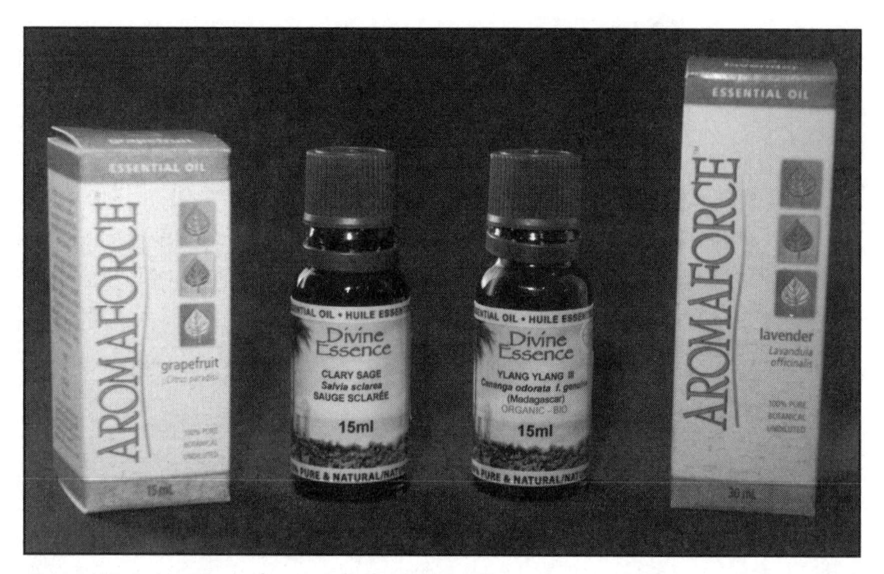

Figure 10.1-11. Natural essential oils are the only way to provide fragrance in homemade soaps. After all, why go to all the work and then use artificial fragrances? Some of these oils, such as peppermint, are very reasonable in price while the more exotic Ylang Ylang is quite pricey.

Figure 10.1-12. To get things started, the glycerin is placed on a cutting board and cut into slices.

Figure 10.1-13. The slices of glycerin are then cut into cubes.

Figure 10.1-14. The glycerin cubes are placed in the soap kettle and the unit is switched on to a high-heat setting.

Figure 10.1-15. After a few minutes the glycerin begins to melt.

Figure 10.1-16. Once the glycerin has melted, the kettle will switch to the low-heat setting and the liquid glycerin can be ladled into a measuring cup.

Figure 10.1-17. Color dyes are added to the liquid glycerin.

Figure 10.1-18. The mixture is thoroughly stirred.

Figure 10.1-19. Aroma oils are added to create the desired fragrance.

Figure 10.1-20. The final mixture is continuously stirred, forming a homogenous soap base while you wait for the materials to reach the correct viscosity to begin pouring.

Figure 10.1-21. The soap is poured into soap moulds and allowed to cool, after which it can be popped out of the moulds and used.

Figure 10.1-22. More advanced soaps include exotic items, for example seaweed or coarse salt with aqua colors and undertone fragrances used in a marine soap bar. You are only limited by your imagination when creating wonderful soap products for gifts and personal use as these Danielle and Lorraine do at a cost of about $0.60 per bar when purchasing materials from a local craft store and about half this amount when purchasing in bulk.

WVO-Based Glycerin

I am assuming that you have recovered the methanol from the glycerin as discussed in Chapter 8.3.10. If you have not, the raw glycerol will melt in the soap kettle (Figure 10.1-23) and the methanol will begin to boil off when the temperature reaches approximately 65°C. If the methanol has not been previously recovered, the raw glycerol will look similar to that shown in Figure 10.1-24. Should glycerol boiling occur take great care, as the methanol fumes that are exuded are very poisonous and flammable.

Figure 10.1-23. Raw glycerol can be melted in a soap kettle and additional sodium hydroxide added to create a hard soap stock.

Figure 10.1-24. If the methanol has not been previously recovered, the raw glycerol will begin to boil at approximately 65°C. Should this occur take great care, as the methanol fumes that are exuded are very poisonous and flammable.

Figure 10.1-25. Once the raw glycerol has been processed by melting & methanol recovery, you see why I am concerned about WVO-based glycerin being used as a soap stock. Firstly, the semirefined glycerin has the smell of dirty socks and French fries all rolled into one. The second problem becomes evident as soon as you touch the stuff. The texture of the glycerin is very gritty, feeling almost dirty, certainly not the same polished material we worked with in Figure 10.1-9.

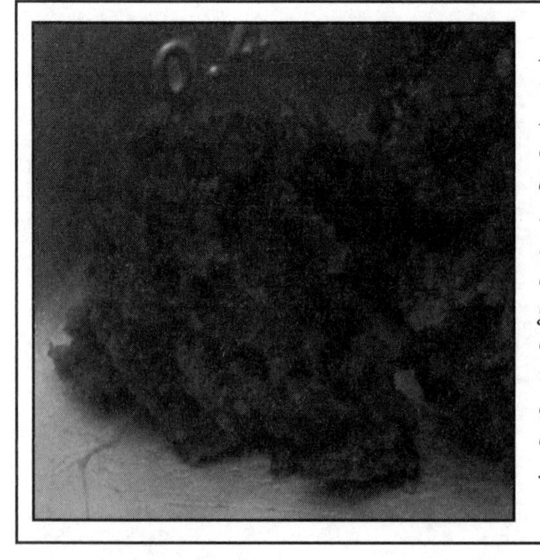

Figure 10.1-26. Experimenting with the WVO-based glycerin, Lorraine found that no amount of soap dye would color the dark brown mass and that it took approximately three to four times the amount of essential oil to mask the smell, greatly increasing the cost. A couple of small batches of soap were produced, but no one was either pleased with the outcome or happy about washing with someone else's lunch scraps.

Once the glycerol has been processed by melting and methanol recovery, you see why I am concerned about WVO-based glycerin being used as a soap stock. Firstly, the semirefined glycerin has the smell of dirty socks and French fries all rolled into one. The second problem becomes evident as soon as you touch the stuff. The texture of the glycerin is very gritty, feeling almost dirty (Figures 10.1-25 and 10.1-26), and certainly not the same polished material we worked with in Figure 10.1-9.

Experimenting with the WVO-based glycerin, Lorraine found that no amount of soap dye would color the dark brown mass and that it took approximately three to four times the amount of essential oil to mask the smell, greatly increasing the cost. A couple of small batches of soap were produced, but no one was either pleased with the outcome or happy about washing with someone else's lunch scraps.

The larger-scale biodiesel producer might have access to glycerin refining and bleaching processes, but that technology is not within the scope of this book.

An average micro-scale producer may use four or five batches of WVO per year to produce biodiesel, which would yield enough glycerin to produce 1,500 or more bars of soap. I would suggest that even if you were to refine the raw glycerol into usable glycerin for soapmaking, the volume you would produce would be too large for single-home production/consumption and too small for a commercial soapmaking venture, even a simple one at the local farmers' market. To my way of thinking, this is a dead-end application for this very coarse, low-grade product.

10.2
Dust Suppression

If soap making doesn't quite make the grade as a good use for raw glycerol, perhaps using its humectant properties could make it suitable for use as a dust suppressant. Everyone has at one time or another driven down a gravel country road only to see a plume of dust billow from the rear of the vehicle. Besides making the car dusty, this annoying attribute of otherwise quaint, peaceful roads reduces vehicle life because of abrasion to numerous rotating parts and requires people living near the roads to keep their windows closed and to frequently clean both inside and outside the house.

Townships have been dealing with the problem for years. Up until the late 1960s, the solution was to dump waste motor oil on the gravel. Once this was determined not to be such a good idea, a search was started to find a suitable replacement. The salt called calcium chloride was discovered to have humectant capabilities, retaining moisture and coalescing dust particles into larger bits of grit that do not become airborne when a car passes.

Figure 10.2-1. Everyone has at one time or another driven down a gravel country road only to see a plume of dust billow from the rear of the vehicle. Besides making the car dusty, this annoying attribute of otherwise quaint, peaceful roads reduces vehicle life because of abrasion to numerous rotating parts and requires people living near the roads to keep their windows closed.

Although this material works well, it is not the ideal solution either. Calcium chloride will wash away over time and requires yearly applications; nor is it an environmentally benign substance.

In the equestrian world, both indoor arenas and outdoor rings generate tremendous amounts of dust. Spraying with water will work, but the requirement for frequent applications makes this a nuisance. Calcium chloride, on the other hand, is known to have the side effect of drying the frog of horses' hooves, as well as being expensive to apply.

Glycerin that has had the methanol recovered and the caustic catalyst neutralized is an ideal humectant, offering the features of calcium chloride without its drying side effects. This affords an excellent opportunity for the biodiesel producer who is burdened with a surplus of the product; it may even be possible to sell your time and materials, applying the "special dust-suppressing product" at equestrian parks.

Figure 10.2-2. In the equestrian world, both indoor arenas and outdoor rings generate tremendous amounts of dust. Spraying with water will work, but the requirement for frequent applications makes this a nuisance.

Figure 10.2-3. Glycerin that has had the methanol recovered and the caustic catalyst neutralized is an ideal humectant, offering the dust-suppressing features of calcium chloride without its drying side effects. This affords an excellent opportunity for the biodiesel producer who is burdened with a surplus of the product.

I have found that after neutralizing the catalyst in the glycerin and adding an equal amount of water the diluted glycerin can be applied to dirt roads and riding arenas as an excellent dust suppressant. A chemical or weed sprayer is all you need, filled with a mixture of water, and glycerin diluted to the extent necessary to work your particular model. This allows you to rid yourself of an unlimited amount of glycerin, as a local riding arena can consume more than you will ever produce. Likewise a cottage road or other dirt track will absorb as much material as you can spray.

10.3
Composting

Glycerin can be difficult to compost in large blocks or amounts. My preferred method of composting the material is to proceed as if I were going to spray a dirt road, simply spraying the material over an area that is exposed to the sun and allows the glycerin to break down and dissipate. The film of material that is sprayed on the ground degrades in a brief time and no unsightly blocks of glycerin "mush" are left in the composting area.

10.4
Landfilling

Even if the Glycerin has had methanol recovered and catalyst neutralized it cannot be disposed of in subsurface septic systems or in sanitary sewers. As a general rule, if the glycerin does not contain any free liquids, including methanol and/or water, it may be possible to dispose of it at regulated landfill sites. For guidance on its disposal, contact your local municipal, state, or provincial waste management program officer.

10.5 Combusting

It may be possible to combust glycerin in a co-fired biomass furnace or boiler; however, there is very little credible data regarding the emission profile when glycerine is used in standard wood stoves, outdoor wood burners, or other similar combustion appliances. Current research shows that the poisonous gas acrolein is given off during the low-temperature combustion found in typical wood or other biomass-fired heating appliances when fired with glycerin.

The following information is taken from the MSDS data for acrolein and clearly demonstrates that exposure to this product is dangerous and therefore, glycerin should not be combusted, as inhalation and exposure to smoke cannot be avoided:

Acrolein (CAS# 107-02-8) - Potential Health Effects

Inhalation: Toxic. Highly irritating. It can injure the lungs and bronchial airways. Symptoms include severe irritation of the mucous membranes, burning of the throat, cough, difficulty breathing, tightness in the chest, nausea, vomiting and diarrhea, pulmonary edema, high blood pressure, and unconsciousness. A weak sensitizer; may cause asthmatic reaction. Inhalation of high concentrations can cause permanent lung damage. Fatalities have occurred from exposure to levels as low as 10 ppm; 1 ppm produces marked eye, respiratory, and mucous membrane irritation in less than 5 minutes.

Ingestion: May produce severe irritation of the mouth and gastrointestinal tract.

Skin Contact: Causes severe irritation with redness, pain, and possibly skin burns. Contact may cause sensitization dermatitis.

Eye Contact: Causes severe irritation, extensive tearing, pus-like discharge, corneal damage and damage to area around eyelids. May cause corneal burns.

Chronic Exposure: The violent irritating effects of acrolein generally prevent any chronic toxicity. Repeated inhalation can sensitize some individuals, resulting in an asthmatic response.

Aggravation of Pre-existing Conditions: Persons with pre-existing respiratory disorders may be more susceptible to the effects of this substance.

Conclusion

Having read this entire book, you will be well aware of my bias against people who take shortcuts in an effort to get a job done quickly. Biodiesel production is not simply mixing a couple of chemicals together and driving off into the sunset in the hope of saving the environment. This notion is not only flawed, it is outright deceitful and misleads honest people into thinking that biodiesel is the holy grail of environmental sustainability when in fact it is not.

Biodiesel using waste feedstocks and off-specification oilseeds has its place, but the benefits of all biofuels could be eclipsed if governments mandated massive public transit infrastructure construction, improved vehicle fuel efficiencies, and increased fuel taxes. Resource consumption, ego, and our endless demands and entitlements are all part of society's problem. Those who think that biofuels are going to solve our energy, transportation, and environmental troubles are truly fooling themselves.

There are legitimate reasons for developing micro- and small-scale levels of biodiesel production. People are grappling with ways of producing fuel and are yearning for process information and ways to improve fuel quality. The farming sector would certainly produce their own fuels if they could be sure of quality and safety issues. Developing a link with the national biodiesel associations will prove to be a step in the right direction.

Some readers of **Biodiesel: Basics and Beyond** may be surprised at the complexity and amount of work involved in producing a quality fuel while leaving the environment in the same condition it was in beforehand.

The amount of work will lead some to decide not to produce their own biodiesel. To those people I offer my congratulations for having made an informed decision that is right for you.

For those who do go on to produce their own fuel, I suggest that this book is not The Answer to biodiesel production and related process technology but rather a stepping stone on the journey. There are loose ends that I have not been able to address and I hope that people who do continue on this path provide me with the benefit of their knowledge and discoveries so that I can verify and publish the data for all to explore.

I wish you bon voyage on your biodiesel odyssey!

To provide feedback to William Kemp, email him at:
biodieselbasics@aztext.com

And be sure to visit the Biodiesel: Basics and Beyond website at:
www.thebiodieselbook.com

Glossary

Agglomerator – A special filter designed to agglomerate or combine smaller water molecules into larger ones causing the accumulated water to remain in a drain bowl and preventing the water from entering the fuel system.

Ash - Ash is the residue left once the fuel has been completely combusted. Excessive ash can lead to engine damage due to scoring and abrasion of metal parts.

Atomization – The ability of fuel to spray into a fine mist, which ensures complete combustion, leading to improved performance and fuel economy.

Atomize – To finely spray

Biodiesel - A biodegradable transportation fuel for use in diesel engines that is produced through transesterification of organically derived oils or fats. Biodiesel is used as a component of diesel fuel. In the future it may be used as a replacement for diesel.

Bonded Glycerin – The level of glycerin that remains in the final product biodiesel as mono-, di- and triglycerides. High-levels of bonded glycerin in the biodiesel are indicative of an incomplete biodiesel reaction.

Capacity factor – The percentage time that an engine or generator is available for use, in a given period of time.

Cetane Number - The measure of the fuel's ability to ignite quickly. A higher cetane number is better, although most feedstock oils will produce biodiesel with a cetane rating that is better than petroleum diesel fuel.

Cloud point - The temperature at which wax crystals inherent in the fuel begin to form, which may lead to filter plugging, engine fuel starvation and stalling.

Coking – The buildup up of carbon or coke on fuel injection nozzles and piston, generally attributed to excessive free glycerin in the biodiesel from improper fuel washing or high, mono-, di- and triglyceride levels due to incomplete oil to biodiesel reaction.

Compression ignition – An engine that uses the heat of compression to ignite fuel rather than the spark plug used in the Otto-cycle engine. Also known as a Diesel engine.

Copper Strip Corrosion - Copper is particularly susceptible to chemical corrosion, thus providing a rapid means of indicating the corrosive nature of diesel fuel to non-ferrous metals such as copper, brass, bronze and zinc.

CRFA – Canadian Renewable Fuels Association (Canada)

Distillation Temperature - A sample of the diesel fuel is subjected to progressive heating, and the initial boiling point (IBP) is recorded as well as the temperature at each 10% distillation of the sample, concluding with the final boiling point (FBP). Although the distillation curve is not as important in biodiesel fuel, the test can indicate excessive levels of methanol indicated by abnormally low IBP levels. Likewise, the indication of very high FBP levels may be indicative of excessive levels of mono-, di- and triglycerides.

FAME – Fatty Acid Methyl Esters, also known as Biodiesel

Fatty acids – Chains of hydrocarbons that vary in carbon length depending on the oil feedstock. Fatty Acids can be either bonded to the parent oil or they may be Free Fatty Acids.

Final boiling point (FBP) – The point at which the last fraction or compound in a sample begins to boil. In biodiesel, a high FBP temperature is generally indicative of a high concentration of gylcerides, indicating incomplete reaction of the feedstock to biodiesel.

Flashpoint - The minimum temperature at which a liquid gives off a vapor in sufficient concentration to ignite if tested in accordance with the definition of "flashpoint" in ASTM Standard D93 – Procedure A. A low flashpoint is indicative of excessive levels of methanol present in the fuel.

Flooded suction – Means that the intake of a pump must be constantly supplied with fluid in order for the pump to operate. A pump that requires a flooded suction has no ability to draw a liquid up a vertical distance.

Free Glycerin – The amount of glycerin that remains suspended in the final biodiesel phase. Excessive levels of free glycerin are indicative of incomplete washing of the biodiesel.

Free Fatty Acid (FFA) – The fatty acid level or concentration found in Waste Vegetable Oil that occurs as a result of cooking and thermal degradation of deep fryer oils.

Freezing point – The point at which a liquid turns into a solid.

Glycerol - An alcohol that can combine with up to three fatty acids to form mono-, di-, and triglycerides.

HDPE plastic – High Density Polyethylene plastic.

Hydrocarbons – Substances containing only hydrogen and carbon. Fossil fuels are made up of hydrocarbons. Some hydrocarbon compounds are major air pollutants.

Hydrophilic- Means having a strong attraction to water. Hydrophilic molecules are soluble in water and travel with the groundwater.

Ignition delay - The delay time between the opening of the fuel injector (signifying the initial fuel spray into the cylinder) and ignition of the fuel.

Initial Boiling Point (IBP) – The point at which the first fraction or compound in a sample begins to boil. In biodiesel, a low IBP temperature generally signifies the presence of alcohol (typically methanol).

Kinematic Viscosity - The measure of a fluid's resistance to flow measured over time

Lower Explosive Limit (LEL) – The lowest concentration of a flammable substance mixed with air will not combust or explode. At concentrations above this level, combustion will occur. The LEL is expressed as a percentage of the ratio of flammable substance to air.

Lubricity - The capacity to reduce engine wear from friction

Methanol - A colorless alcohol that is very toxic and extremely flammable. It is used to produce a number of products including MTBE, fuels, and solvents. Methanol is acutely toxic and can cause blindness.

Micro-producer – A producer of biodiesel who is capable of a maximum production volume of (0-10,000 gallons/year)(0-38,000 liters). In the context used in this book, the micro-producer is considered a home-based, non-commercial producer.

NBB – National Biodiesel Board (USA)

Polyunsaturated - A polyunsaturated organic compound is one in which more than one double bond exists within the representative molecule. They are often found in liquid oils of vegetable origin. Common sources of polyunsaturated fatty acids are safflower, sesame and sunflower seeds, corn and soybeans, many nuts and seeds and their oils. Canola, olive and peanut oils and avocados are sources of monounsaturated fatty acids.

Pour point – The lowest temperature at which biodiesel will continue to flow.

Raw Biodiesel – See Raw FAME.

Raw Biodiesel - Raw biodiesel is the term used to distinguish non-ASTM D 6751 certified biodiesel from certified fuels. As a general rule, raw biodiesel is taken to mean the biodiesel layer that is removed immediately after separation from the glycerol layer. This fuel contains large amounts of contaminants, methanol, soaps, catalyst residue and other materials that make the fuel completely incompatible with modern diesel engines.

Raw FAME - FAME that is directly taken from the reaction process and which is known to contain high levels of contaminants such as methanol, catalyst, free glycerin, bonded glycerin, soap, water. Diesel engines and related equipment are not compatible with Raw FAME, although many home-scale producers erroneously state otherwise. Raw Biodiesel is an equivalent term.

Saturated - If each carbon atom has 2 associated hydrogen atoms, the fatty acid is known to be saturated. They are generally solid at room temperature (eg, butter, lard) and are commonly found in meats and foods of animal origin.

Small Scale Producer - A producer of biodiesel who is capable of a maximum production volume of (10,000-500,000 gallons/year (38,000-1.9 million liters). In the context used in this book, the small scale producer is considered an agricultural or community-based producer of biodiesel that is not selling their product in a traditional retail arrangement. The retail sale of on-road fuel is governed by regulations that are outside the scope of this book.

Soap - A natural cleaning agent produced by the reaction of a fat or oil and an alkali.

Sodium Hydroxide – Is a catalyst used to assist (catalyze) the Transesterification of oils or fats and alcohol into Biodiesel. It is a strong alkali (base) that is also called caustic soda or lye. It is produced by the electrolysis of brine (seawater).

Sodium Methoxide – A solution made from the combination of Sodium Hydroxide and Methanol. Used in the Transesterification of oils and fats to create Biodiesel.

Suction head – As defined herein means the ability of a pump to draw a liquid against the force of gravity. The standard technical definition is the head on the suction side of the pump. This is subtracted from the discharge head to determine the head being produced by the pump. It is a sum of the static, pressure and friction heads.

Sulfur - A pale yellow chemical element. Sulfur compounds have a strong rotten egg-like smell. .A major contributor to atmospheric smog and its concentration in all petroleum fuels has been steadily regulated downwards

Titration - A method, or process, of using a standard solution of known pH (the titer) to determine the strength of a second solution of unknown pH level. The function of Titration as it applies to Biodiesel production is to determine the amount of alkali Catalyst required to neutralize the Free Fatty Acid level of Waste Vegetable Oil (WVO).

Total Glycerin – The sum of the bonded and free glycerin in the final biodiesel product.

Transesterification – (simplified) Transesterification is the process of exchanging a gylceride of an oil compound with an alcohol in the presence of an acidic or basic catalyst. The process of "reacting sodium or potassium methoxide with an oil or fat to create FAME or biodiesel.

Triglycerides - Chemical compounds formed from one molecule of glycerol and three fatty acids

Unsaturated – A fat that is liquid at room temperature. Vegetable oils are unsaturated fats. Unsaturated fats include polyunsaturated fats, and monounsaturated fats. They include canola, sunflower and soybean oils.

Virgin Vegetable Oil – Vegetable oil that has not been used in the preparation of food or other products. (See also WVO).

Viscosity – The thickness of a fluid. Viscosity generally increases as temperature decreases; it is inversely proportional to temperature.

Water and Sediment – Water that is solublized (dissolved) in biodiesel fuel will contribute to oxidation (rusting) of metal components and can contribute to microbial growth while the fuel is in storage.

WVO – Waste Vegetable Oil. Although the term implies oil from a plant source it is also understood to include used animal fats from restaurant deep fryers and commercial cooking facilities. Used as a low or no-cost feedstock for micro-scale biodiesel production.

Notes

Chapter 2.1

1. Jose Etcheverry, Keith Stewart, and Steven Hall, "Bright Future," <http://www.davidsuzuki.org/files/Climate/Ontario/brightfuture.pdf>, September 2003.
2. Ibid.
3. William H. Kemp, *$mart Power: An Urban Guide to Renewable Energy and Efficiency* (Tamworth, Ontario: Aztext Press, 2004).
4. The Energuide for Houses program is part of the Canadian Government climate change policy under the Kyoto Protocol. Homeowners who take part in the program undergo a home energy audit and receive rebate funds based on energy efficiency upgrade performance.
5. Etcheverry, Stewart, and Hall, "Bright Future."
6. Electricity Supply and Conservation Task Force, "Tough Choices: Assessing Ontario's Power Needs," <http://www.energy.gov.on.ca/english/pdf/electricity/TaskForceReport.pdf>, 14 January 2004.
7. Ibid.

Chapter 2.2

1. Ontario Medical Association, "The Illness Costs of Air Pollution," <http://www.oma.org/phealth/smogmain.htm>, June 2005.
2. *New Scientist,* June 2003. Electrical power generation using coal is rising not only in the developing world but also in North America. While clean coal technologies are being touted as a saviour for this sector of the power generation industry, the moniker is fraught with error as the technology currently has no way of dealing with the massive generation of carbon dioxide and scrubbing technology does not completely rid stack emissions of airborne pollutants.
3. Although electrical energy demand is increasing and many countries are attempting to develop their renewable and clean-power generation technologies, high demand is driving traditional power generating companies to seek new markets for their products.
4. Intergovernmental Panel on Climate Change, "Climate Change 2001: Synthesis Report," <http://www.grida.no/climate/ipcc_tar/vol4/english/index.htm>, 2001.
5. "Changing Science: Climatology," *The Economist,* 10 December 2005, 91.

Chapter 2.3

1. Energy Information Administration, "Country Analysis Briefs: United States," <http://www.eia.doe.gov/emeu/cabs/Usa/Profile.html>, November 2005.
2. Ibid.
3. Ibid.
4. Energy Information Administration, "Country Analysis Briefs: Canada," <http://www.eia.doe.gov/emeu/cabs/Canada/Profile.html>, April 2006.
5. Ibid; and Natural Resources Canada, "Canadian Natural Gas Winter 2005-06 Outlook," <http://www2.nrcan.gc.ca/es/erb/CMFiles/WINTER_MARKET_OUTLOOK_2005_ENGLISH-206KCY-25112005-2389.pdf>, November 2005.
6. Energy Information Administration, "Country Analysis Briefs: Mexico," <http://www.eia.doe.gov/emeu/cabs/Mexico/Profile.html>, December 2005.
7. Mathew R. Simmons, *Twilight in the Desert: The Coming Saudi Oil Shock and the World Economy* (New Jersey: John Wiley & Sons, 2005), 290.
8. Energy Information Administration, "Country Analysis Briefs: Mexico."
9. Energy Information Administration, "Persian Gulf Oil and Gas Exports Fact Sheet," <http://www.eia.doe.gov/emeu/cabs/pgulf.html>, November 2004.
10. Ibid.
11. Simmons, *Twilight in the Desert,* 283.

12. Patrick Clawson and Simon Henderson, "Reducing Vulnerability to Middle East Energy Shocks: A Key Element in Strengthening U.S. Energy Security," <http://www.washingtoninstitute.org/pubPDFs/PolicyFocus49.pdf>, November 2005.
13. Ibid.

Chapter 2.4
1. Patrick Brethour, "Only conservation efforts will keep a lid on energy costs: IEA," The Globe and Mail, 3 December 2005, B6.
2. Energy Information Administration, "Country Analysis Briefs: United States," <http://www.eia.doe.gov/emeu/cabs/Usa/Oil.html>, November 2005.
3. Even though the SPR is essentially a large underground "tank" located in the salt caverns in Louisiana, oil pumping is limited by extraction rates according to EIA data: "The SPR has a maximum drawdown capability of 4.3 million bbl/d for 90 days, with oil beginning to arrive in the marketplace 15 days after a presidential decision to initiate a drawdown. The SPR drawdown rate declines to 3.2 million bbl/d from days 91-120, to 2.2 million bbl/d for days 121-150, and to 1.3 million bbl/d for days 151-180" (ibid.).
4. Ibid.
5. Mathew R. Simmons, *Twilight in the Desert: The Coming Saudi Oil Shock and the World Economy* (New Jersey: John Wiley & Sons, 2005), 69.
6. Tom Bergin, "Shell Cuts Oil Reserves Again," *Energy Bulletin*, <http://www.energybulletin.net/4220.html>, 3 February 2005.

Chapter 2.5
1. Wintershall AG, "Great variety of crude oil grades," <http://www.wintershall.biz/34.html?&L=1>, 2006.
2. Ibid.
3. Energy Information Administration, "Country Analysis Briefs: United States," <http://www.eia.doe.gov/emeu/cabs/Usa/Oil.html>, November 2005.
4. Canadian Renewable Fuels Association, "Fuel Change: Presentation to the Industry Committee hearings on Gasoline Prices in Canada," <http://www.greenfuels.org/fuelchange/index.htm>, 22 September 2005.
5. Husky Energy, "Better for your car. Better for the environment.," <www.huskyenergy.ca/products/downloads/Ethanol.pdf>, 2006.
6. Canadian Renewable Fuels Association, "Ethanol: Reducing Greenhouse Emissions by the Equivalent of 200,000 Cars," <http://www.greenfuels.org/ethanol/index.htm>, 28 March 2006.
7. Husky Energy, "Better for your car."
8. Canadian Renewable Fuels Association, "Ethanol."
9. Refer to Murtagh & Associates, *The Online Distillery Network for Distilleries and Fuel-Ethanol Plants Worldwide*, <http://www.distill.com>, 2005.
10. National Biodiesel Board, "Specifications for Biodiesel (B100)," <http://www.biodiesel.org/pdf_files/fuelfactsheets/BDSpec.PDF>, December 2001.
11. Mercedes-Benz, "Mercedes-Benz Debuts High-Tech Diesel Car For Canadian Market," <http://www.mercedes-benz.ca/index.cfm?NewsID=144&id=3246>, 8 April 2004.
12. Refer to the "Factsheets" page on the National Biodiesel Board's website (http://www.biodiesel.org/resources/fuelfactsheets).
13. Refer to the National Biodiesel Board's website (http://www.biodiesel.org).
14. Ibid.

Chapter 3.2

1. Refer to Automotive Industry Data website (http://www.eagleaid.com/index.htm).
2. Ibid.
3. Diesel Technology Forum, "Demand for Diesels: The European Experience," <http://www.diesel-forum.org/fileadmin/templates/whitepapers/EuropeanExperience.PDF>, July 2001.
4. Refer to Automotive Industry Data website (http://www.eagleaid.com/index.htm).
5. Diesel Technology Forum, "Demand for Diesels: The European Experience."
6. United States Environmental Protection Agency, "A Comprehensive Analysis of Diesel Exhaust Emissions: Draft Technical Report," <http://www.epa.gov/otaq/models/analysis/biodsl/p02001.pdf>, October 2002.
7. Diesel Technology Forum, "Demand for Diesels: The European Experience."

Chapter 3.3

1. Martin Mittelbach and Claudia Remschmidt, Biodiesel: The Comprehensive Handbook Martin Mittlebach. (Austria: Boersedruck Ges.m.b.h, Vienna 2004), 1.
2. Dow AgroSciences, "Glossary," <www.dowagro.com/natreon/resource/glossary.htm>, 2006.
3. Mittelbach and Remschmidt, *Biodiesel.*
4. Ibid.
5. Martin Mittlebach, in discussion with the author, Ottawa, 2005.
6. While rebuilding of distribution is a major obstacle to overcome, it is important to remember that hydrogen is not a fuel, but rather an energy carrier similar to a battery. Energy must come from somewhere to convert either natural gas or other hydrocarbons to hydrogen, providing no net energy benefit and producing further emissions. Hydrogen developed from the breakdown of water by electrolysis may be better for the environment, provided the input energy source is in itself clean or renewable.
7. Canadian Renewable Fuels Association, "Everything you wanted to know about biodiesel, but were afraid to ask..." <http://www.greenfuels.org/biodiesel/everything.htm#d>, 30 March 2005.
8. Note that the ASTM D 6751 standard for biodiesel makes reference to the fact that there is little practical knowledge about using high concentrations of biodiesel.
9. Mittelbach and Remschmidt, *Biodiesel,* 6.
10. Union for the Promotion of Oil and Protein Plants, "Status Report Biodiesel: Production and Marketing in Germany 2005," <http://www.biodiesel.org/resources/reportsdatabase/viewall.asp>, June 2005.
11. National Biodiesel Board, "National Biodiesel Board, DOE, USDA Officials Dispute Biofuels Study," <http://www.biodiesel.org/resources/pressreleases/gen/20050721_pimentel_response.pdf>, 21 July 2005.
12. Union for the Promotion of Oil and Protein Plants, "Status Report Biodiesel."
13. Although cloud point temperature is a useful parameter, it is far simpler to subject known concentrations of biodiesel/petroleum diesel blends to the ambient temperatures expected and observe which concentration is still above the cloud point rating.
14. Ethanol also presents problems with biodiesel production because it forms an azeotrope with water, making recovery and recycling very difficult.
15. Blending biodiesel with petroleum diesel is the preferred and simplest means of increasing the cold weather performance of biodiesel.
16. National Biodiesel Board, "Biodiesel cold weather blending study," Report No. 20050728.
17. Every manufacturer has a fuel warranty policy concerning not only biodiesel but any off-specification fuel. Using biodiesel that causes damage to a fuel system will not be covered under warranty. Likewise sand, dirt, or other contaminants that cause damage to a fuel system when operating with petroleum diesel would not be covered under warranty either.
18. Refer to the National Biodiesel Board's website (www.biodiesel.org).

Chapter 4

1. Jon Van Gerpen et al., *Building a Successful Biodiesel Business* (University of Idaho, Biodiesel Workshop Handbook, 2005).

Chapter 5

1. Jon Van Gerpen et al., *Building a Successful Biodiesel Business* (University of Idaho, Biodiesel Workshop Handbook, 2005).
2. Ibid, 52.
3. Ibid.

Chapter 6.1

1. Refer to the Engine Manufacturers Association website (http://www.enginemanufacturers.org).
2. Engine Manufacturers Association, "Technical Statement on the Use of Biodiesel Fuel in Compression Ignition Engines," <http://www.enginemanufacturers.org/admin/library/upload/297.pdf>, February 2005.
3. Engine Manufacturers Association, "Technical Statement on the Use of Biodiesel."
4. National Biodiesel Conference, Fort Lauderdale, Florida. January 30 to February 2, 2005, as reported by Mr. Steve Howell of MARC IV Consulting and Chair of the ASTM Task Force on biodiesel Standards.
5. Engine Manufacturers Association, «Technical Statement on the Use of Biodiesel.»

Chapter 6.2

1. Jeffrey Gail Rothermel, «Investigation of Transesterification Reaction Rates and Engine Exhaust Emissions of Biodiesel Fuels,» (master's thesis, Iowa State University, 2003), <http://72.14.207.104/search?q=cache:mYdp9m4laBoJ:www.me.iastate.edu/biodiesel/Technical%2520Papers/Roth%2520T%2520of%2520C.pdf+exhaust+gas+temperature+biodiesel&hl=en&gl=ca&ct=clnk&cd=1>.
2. United States Environmental Protection Agency, «A Comprehensive Analysis of Diesel Exhaust Emissions: Draft Technical Report,» <http://www.epa.gov/otaq/models/analysis/biodsl/p02001.pdf>, October 2002.

Chapter 7.1

1. Joshua Tickell and Kaia Tickell, *From the Fryer to the Fuel Tank: The Complete Guide to Using Vegetable Oil as an Alternative Fuel* (Tickell Energy Consulting, 1999), 69.
2. Jon Van Gerpen et al., *Building a Successful Biodiesel Business* (2005), 71
3. United States Environmental Protection Agency, «Pollution Prevention (P2) for auto repair and fleet maintenance,» <http://www.epa.gov/region9/p2/autofleet/>, 27 February 2006.
4. Tickell and Tickell, *From the Fryer to the Fuel Tank*, 69.
5. Maria Alovert, «Bubblewashing Biodiesel & Dealing With Wash Problems,» <http://www.biodieselcommunity.org/bubblewashingbiodiesel/>, 2005.
6. Journey to Forever, «Make Your Own Biodiesel,» <http://journeytoforever.org/biodiesel_make.html>, 2 April 2006.
7. Tickell and Tickell, *From the Fryer to the Fuel Tank*, 119.

Chapter 7.3

1. Test of Steve Anderson's raw biodiesel fuel conducted under report number GO-2005-4803 by Alberta Research Council. December, 2005.

Chapter 7.9

1. Iowa State University Department of Mechanical Engineering, «Biodiesel Education: Raw Oils,»

<http://www.me.iastate.edu/biodiesel/Pages/A.html>, 2003.
2. Ibid.
3. Refer to Wartsila's website (http://www.wartsila.com).
4.http://www.wartsila.com/en,powerplants,0,product,801640516810256,4009656286697540,,.htm
5. Renewable Energy Report. Issue 81. May 30, 2005

Chapter 8.3.4
1. Jon Van Gerpen et al., *Building a Successful Biodiesel Business* (2005), 87.

Chapter 8.3.5
1. See MSDS for Sodium Methoxide.
2. Jon Van Gerpen et al., *Building a Successful Biodiesel Business* (2005), 53.

Chapter 8.3.7
1. Jon Van Gerpen et al., *Building a Successful Biodiesel Business* (2005), 66.

Chapter 8.3.9
1. Martin Mittelbach and Claudia Remschmidt, *Biodiesel: The Comprehensive Handbook* (Martin Mittlebach, 2004), 131.
2. Ibid, 144.
3. Ibid, 141.
4. Ibid, 160.
5. Ibid, 140.
6. Ibid, 139.
7. Ibid, 147.
8. Ibid.
9. Ibid, 122.

Chapter 9
1. National Biodiesel Board, «Bioheat Offers Cost-Competitive, Environmentally Friendly Alternative to Regular Home Heating Oil,» <http://www.biodiesel.org/resources/pressreleases/hom/20041123_bioheat.pdf>, 23 November 2004.

Appendix 1
ASTM D 6751

Table 1 Detailed Requirements for Biodiesel (B100)[A]

Property	Test Method[B]	Limits	Units
Flash point (closed cup)	D 93	130.0 min	°C
Water and sediment	D 2709	0.050 max	% volume
Kinematic viscosity, 40°C	D 445	1.9-6.0[C]	mm^2/s
Sulfated ash	D 874	0.020 max	% mass
Sulfur[D]	D 5453	0.05 max	% mass
Copper strip corrosion	D 130	No. 3 max	
Cetane number	D 613	47 min.	
Cloud point	D 2500	Report[E]	°C
Carbon residue[F]	D 4530	0.050 max	% mass
Acid number	D 664	0.80 max	mg KOH/g
Free glycerin	D 6584	0.020	% mass
Total glycerin	D 6584	0.240	% mass
Phosphorus content	D 4951	0.001 max	% mass
Distillation temperature	D 1160	360 max	°C
Atmospheric equivalent temperature. 90% recovered			

[A] To meet special operating conditions, modifications of individual limiting requirements may be agreed upon between purchaser, seller and manufacturer.

[B] The test methods indicated are approved referee methods. Other acceptable methods are indicated in 5.1.

[C] See X1.3.1. The 6.0 mm^2/s upper viscosity limit is higher than petrodiesel and should be taken into consideration when blending.

[D] Other sulfur limits can apply in selected areas in the United States and in other countries.

[E] The cloud point of biodiesel is generally higher than petrodiesel and should be taken into consideration when blending.

[F] Carbon residue shall be run on the 100% sample (see 5.1.10).

Appendix 2
Resource Guide

Micro-Scale Producer Supply Information:

Air Blower with Heater

Plumbing Mart
www.plumbingmart.com
1-800-325-2960

Order a replacement "air bath blower" for the fuel dryer system.

Chemical Pump

Chemical pumps (for methanol) are available at most hardware stores. Ensure that pump model is compatible with methanol alcohol.

Cone Bottom Dispensing Tanks

Diverse Plastics Group Inc.
8485 Parkhill Drive
Milton, Ontario, Canada
L9T 5E9
Phone: (800)-685-3174
or (905)-864-1746
Fax: (905)-864-4895
www.diverseplastics.com
support@diverseplastics.com

15 Gallon Sodium Methoxide Tank part number: CB 15245
65 Gallon Wash Tank part number: CB 15645

Order tank stand for delivery with specified tank.

Laboratory – Supplies

Midland Scientific Inc.
1202 South 11th Street
Omaha, NE 68108
Phone (800) 642-5263
Fax (402) 346-7694
Contact: Sterling Greni,
BioFuels Mktg. Mgr.
info@midlandssci.com
www.midlandsci.com

Canadawide Scientific
2300 Walkley Road,
Ottawa, Ontario
K1G 6B1
Phone (800)-267-2362
Fax (800)-814-5162
Contact: Christina Moretto, CSR x 224
cmoretto@canadawide.ca
www.canadawide.ca

Suppliers of:

- *chemical reagents for titration testing*
- *chemical catalyst mixer (Talboys Model 105 Laboratory Stirrer)*
- *safety equipment, eyewash station, gloves, face shields*
- *beakers, graduated cylinders, assorted lab glassware*
- *viscosimeter, density float, pH meter*
- *electronic or balance beam scale*

Laboratory – Testing Services for Fuel Quality

National Tribology Services
1711 Orbit Way
Minden, NV 89423
Phone (775) 783-4688
Fax (775) 783-4651
Contact: Alan Coombs,
Acct. Mgr. – Sales
Alan.combs@net
www.biodieseltesting.com

Alberta Research Council
Fuels and Lubricants Group
250 Karl Clark Road,
Edmonton, Alberta
T6N 1E4
Phone (780)-450-5108
Fax (780)-988-9053
Contact: Deni Sarnelli, Manager
sarnelli@arc.ab.ca
www.exchange@arc.ab.ca

Methanol:

Methanol is available from any commercial fuel supply company or distributor in 205 litre drums. Ensure that it is 99.9% pure methanol and that the container has not been opened prior to delivery, by ensuring the drum seal is intact. Methanol will absorb atmospheric water, which can cause the transesterification reaction to fail.

Plumbing and Fittings:

Ace Hardware
Home Depot
Harbor Freight Tools
Lowes Hardware
True Value
Suppliers of:

- *Water heater tanks (recommended water heater tank should have 4 x ¾ inch fittings on top of tank. Many models will have 3 fittings on top and one on the side wall, which can be used if recommended model is not available). Giant Industries produces one such model.*
- *¾ inch threaded black pipe and nipples*
- *ball valves*
- *pipe unions*
- *¾ inch brass thread to hose barbs (hose size ½ inch i.d.)*
- *hose clamps*
- *braided, reinforced, ¾ inch hose*
- *yellow black pipe tape and/or pipe dope sealant*

Pumps:

Circulation Pumps
Northern Tool and Equipment Company
P.O. Box 1499 Burnsville, MN
553377-0499
Item number 109955

Any circulation pump with a die-cast aluminum chassis may be used in this application.

Feedstock Oil Suction Pump

Fill-Rite Industries
8825 Aviation Drive
Fort Wayne, Indiana
46809
Telephone: 219-747-7524

Model FR1604
This pump is also available from farm and automotive distribution companies

Fuel Delivery Pump and Supply Nozzle

Fill-Rite Industries
8825 Aviation Drive
Fort Wayne, Indiana
46809
Telephone: 219-747-7524

Model FR600C – 15 GPM, 120 Volts
This pump is also available from farm and automotive distribution companies

Top Suction Pump – Wash Tank

Proven Pony Pump
Model 360

Available from hardware stores and pump distributors

Sodium Hydroxide:

Sodium hydroxide (lye) is available from most general hardware stores under the name of Gillettes® or Red Devil® brand lye. Look for it in the plumbing department which is used as a drain cleaner or with household cleaning supplies. Sodium hydroxide will absorb water and carbon dioxide from the atmosphere and will become "clumpy" and solid white in color when it is past its best before date. Try to encourage your supplier to order fresh stock for your requirements. Keep lid tightly sealed and in a cool, dry location.

Tank Thermometer:

Any electronic indoor/outdoor thermometer with a maximum temperature reading of at least 80 °C (176 °F).

Small-Scale and Larger Producer Supply Information:

Additives:

Fuel Quality Services Inc.
P.O. Box 1380
Flowery Branch, GA 30542
Phone (770) 967-9790
Fax (770) 967-9982
Contact: Howard Chesneau, President
sales@fqsinc.com
www.fqsinc.com

Primrose Oil Company Inc.
11444 Denton Drive
Dallas, TX 75229
Phone (800) 275-2772
Fax (972) 241-4188
Contact: Scott Crawford, V.P. of Sales
info@primrose.com
www.primrose.com

Adsorbent:

Dallas Specialty Adsorbents
1402 Fabricon Boulevard
Jeffersonville, IN 47130
Phone (812) 283-6675 x 204
Fax (812) 285-7560
Contact: Bryan Bertram, Director Industrial Sales
bbertram@dallasgrp.com
www.dallasgrp.com

Alcohol:

Ashland's Methanol Business Group
6332 Mayville Drive
St. Louis, MO 63129-5021
Phone (314) 846-6183
Fax (314) 846-5806
Contact: Bradley Bauman, Sr. Acct. Mgr.
bbauman@ashland.com
www.ashland.com/adc

Water Treatment:

Vitusa Products Inc.
Phone (908) 665-2900
Contact: Bill Lewis, VP/Market Mgr.
blewis@vitusaproducts.com
www.ritusaproducts.com

Dryers – Flash:

Barr-Rosin Inc.
92 Prevost
Rosemere, QC J7G 2S2
Phone (450) 437-5252
Fax (450) 437-6740
Contact: Francis Chartrand, Engineer
Francis_chartrand@barr-rosin.ca
www.barr-rosin.com

Gasification – Heat Exchangers:

Hoffmann, Inc.
6001 49th St. S.
Muscatine, IA 52761
Phone (563) 263-4733
Fax (563) 263-0919
Contact: Paul Reed, Sales Mgr.
Hoffmann@hoffmanninc.com
www.hoffmanninc.com

Laboratory – Supplies:

Midland Scientific Inc.
1202 South 11th Street
Omaha, NE 68108
Phone (800) 642-5263
Fax (402) 346-7694
Contact: Sterling Greni, BioFuels Mktg. Mgr.
info@midlandssci.com
www.midlandsci.com

Canadawide Scientific
2300 Walkley Road,
Ottawa, Ontario
K1G 6B1
Phone (800)-267-2362
Fax (800)-814-5162
Contact: Christina Moretto, CSR x 224
cmoretto@canadawide.ca
www.canadawide.ca

Laboratory – Testing Services:

National Tribology Services
1711 Orbit Way
Minden, NV 89423
Phone (775) 783-4688
Fax (775) 783-4651
Contact: Alan Coombs, Acct. Mgr. – Sales
Alan.combs@net
www.biodieseltesting.com

Alberta Research Council
Fuels and Lubricants Group
250 Karl Clark Road,
Edmonton, Alberta
T6N 1E4
Phone (780)-450-5108
Fax (780)-988-9053
Contact: Deni Sarnelli, Manager
sarnelli@arc.ab.ca
www.exchange@arc.ab.ca

Tanks:

Apache Stainless Equipment Corporation
200 West Industrial Drive
Beaver Dam, WI 53916
Phone (920) 887-3721 x337
Fax (920) 887-0206
Contact: Michael Peterson, Tank Division Sales Mgr.
mpeterson@apachestainless.com
www.apachestainless.com

Valves:

Alfa Laval Inc.
5400 International Trade Drive
Richmond, VA 23231
Phone (618) 281-5512
Fax (618) 545-2196
Contact: John Piazza,
Regional Sales Mgr.
John.piazza@alfalaval.com
www.alfalaval.us

TankSafe Inc.
Phone: (403) 291-3937
www.tanksafe.com

Animal Fats (First Use):

New Land Trading Company
Phone (800) 760-8573 x13
Contact: Jacob C. Davis, Sales Mgr.
jcdavis@newlandtradingcompany.com
www.newlandtradingcompany.com

Oils – Seed:

ADM Oils & Fats
4666 Faries Parkway
Decatur, IL 62526
Phone (217) 451-2251
Fax (217) 424-5835
Contact: Kris Kappenman, Technical Oils Marketing Mgr.
kappenman@admworld.com
www.admworld.com

Recycled Fats & Oils:

Grease Mart
87 Avondale Pl.
Buffalo, NY 14210
Phone (716) 823-3755
Contact: Edward Mack, Founder
info@greasemart.com

Publications:

Biodiesel Plant Development Handbook
308 2nd Avenue North, Suite 304
Grand Forks, ND 58203
Phone (866) 746-8385
Fax (701) 746-5367
Contact: Jessica Beaudry, Admin Svcs.
jbeaudry@bbibiofuels.com
www.bbibiofuels.com

Websites:

Biodiesel Magazine
308 2nd Avenue North, Suite 304
Grand Forks, ND 58203
Phone (866) 746-8385
Fax (701) 746-5367
Contact: Matthew Spoor, Sales Director
mspoor@bbibiofulels.com
www.bbibiofuels.com

National Biodiesel Board
P.O. Box 104898
Jefferson City, MO 65110-4898
Phone (573) 635-3893
Fax (573) 635-3893
info@biodiesel.org
www.biodiesel.org

Appendix 3

Production of Common Oil-Producing Crops Based on International Harvest Averages.

Plant Name	Pounds Oil per Acre	Kilograms Oil per Hectare
Oil Palm	4,585	5,000
Coconut	2,070	2,260
Jatropha	1,460	1,590
Rapeseed/Canola	915	1,000
Peanut	815	890
Sunflower	720	800
Safflower	605	655
Soybean	345	375
Hemp	280	305
Corn	125	145

Index

D

E

F

G

M

N

O

P

R

S

About the Author

William (Bill) Kemp is V.P. Engineering of Powerbase Automation Systems Inc. where he leads the development of low environmental impact hydroelectric and agricultural biogas systems. Bill is a leading expert in small- and mid-scale renewable energy technologies. He is the author of the best selling books ***The Renewable Energy Handbook*** and ***$mart Power: An Urban Guide to Renewable Energy and Efficiency***. Mr. Kemp is a coauthor of the David Suzuki Foundation report ***Smart Generation: Powering Ontario with Renewable Energy***. In addition he has published numerous articles on small-scale private power and is chairman of electrical safety standards committees with the Canadian Standards Association. He and his wife Lorraine live off the electrical grid on their hobby/horse farm in eastern Ontario.

Environmental Stewardship

In our private lives, the principals in Aztext Press try to minimize our impact on the planet, living off the grid and producing our electricity through renewable energy sources and looking at all facets of our lives, including our personal transportation, heating, cooling, food choices, etc.

We continue to evaluate our books to try minimize their impact, and pick the right mix of recycled stock that still allow photographs to be clear and crisp and offer a quality product to motivate our readers.

In the last 10 years we have planted over 2,200 trees, and each year make a commitment to plant more and nurture existing trees so they thrive and maximize their potential to act like the lungs of the planet and clean the air.